Contents

Climate Change and Variability: A Global Outlook

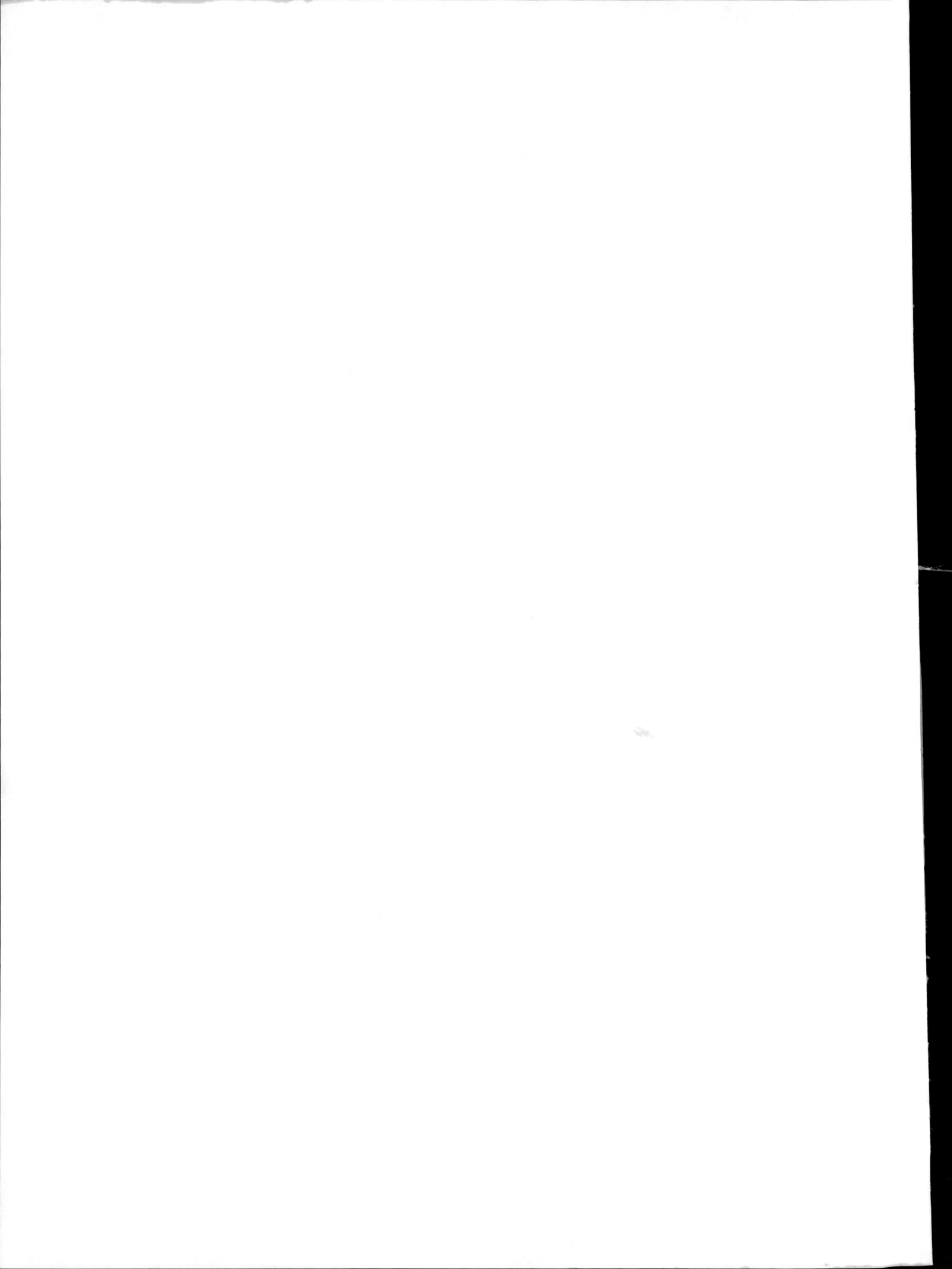

Climate Change and Variability: A Global Outlook

Edited by **Andrew Hyman**

SYRAWOOD
PUBLISHING HOUSE

New York

Published by Syrawood Publishing House,
750 Third Avenue, 9th Floor,
New York, NY 10017, USA
www.syrawoodpublishinghouse.com

Climate Change and Variability: A Global Outlook
Edited by Andrew Hyman

International Standard Book Number: 978-1-68286-040-3 (Hardback)

Printed in the United States of America.

Preface

The study of climate change has advanced rapidly in the past few years. The changes witnessed in weather patterns over the years have had a considerable impact on the environment. This book provides comprehensive insights into the study of climate change. It explores all the important aspects of climate variability in the present day scenario through lucid explanations of various topics, such as ocean, atmosphere and ice dynamics, carbon cycle, greenhouse gases, climate modeling, etc. This book will serve as a valuable source of reference for graduate and post graduate students. Those in search of information to further their knowledge will be greatly assisted by this book.

Various studies have approached the subject by analyzing it with a single perspective, but the present book provides diverse methodologies and techniques to address this field. This book contains theories and applications needed for understanding the subject from different perspectives. The aim is to keep the readers informed about the progress in the field; therefore, the contributions were carefully examined to compile novel researches by specialists from across the globe.

Indeed, the job of the editor is the most crucial and challenging in compiling all chapters into a single book. In the end, I would extend my sincere thanks to the chapter authors for their profound work. I am also thankful for the support provided by my family and colleagues during the compilation of this book.

Editor

Hindcasting the continuum of Dansgaard–Oeschger variability: mechanisms, patterns and timing

L. Menviel[1,2]**, A. Timmermann**[3]**, T. Friedrich**[3]**, and M. H. England**[1,2]

[1]Climate Change Research Centre, University of New South Wales, Sydney, Australia
[2]ARC Centre of Excellence in Climate System Science, Australia
[3]International Pacific Research Center, University of Hawaii, Honolulu, USA

Correspondence to: L. Menviel (l.menviel@unsw.edu.au)

Abstract. Millennial-scale variability associated with Dansgaard–Oeschger events is arguably one of the most puzzling climate phenomena ever discovered in paleoclimate archives. Here, we set out to elucidate the underlying dynamics by conducting a transient global hindcast simulation with a 3-D intermediate complexity earth system model covering the period 50 to 30 ka BP. The model is forced by time-varying external boundary conditions (greenhouse gases, orbital forcing, and ice-sheet orography and albedo) and anomalous North Atlantic freshwater fluxes, which mimic the effects of changing northern hemispheric ice volume on millennial timescales. Together these forcings generate a realistic global climate trajectory, as demonstrated by an extensive model/paleo data comparison. Our results are consistent with the idea that variations in ice-sheet calving and subsequent changes of the Atlantic Meridional Overturning Circulation were the main drivers for the continuum of glacial millennial-scale variability seen in paleorecords across the globe.

1 Introduction

The glacial climate system during Marine Isotope Stage 3 (MIS3, 59.4–27.8 ka BP) experienced massive variability on timescales of centuries to millennia (Masson-Delmotte et al., 2013), referred to as Dansgaard–Oeschger (DO) variability. DO events are characterized by rapid northern hemispheric transitions from cold (stadial) to warm (interstadial) conditions (Fig. 1, black line), subsequent gradual cooling and final rapid transitions to cold conditions (Dansgaard et al., 1993). The origin of this prominent variability still remains elusive with proposed mechanisms invoking internal ocean–sea-ice climate instabilities (Timmermann et al., 2003; Dokken et al., 2013), coupled synchronized ocean–ice-sheet variability (Schulz et al., 2002), North Atlantic sea ice (Li et al., 2005, 2010) and sea-ice–ice-shelf fluctuations (Petersen et al., 2013) as well as externally solar driven reorganizations of the ocean circulation (Braun et al., 2008).

An important element of DO events is the corresponding variability of ice-rafted debris (an indicator for iceberg surges) (Bond and Lotti, 1995; Sarnthein et al., 2001), illustrated here in a high-resolution ice-rafted debris (IRD) composite record from the Irminger and Iceland seas cores SO82-5 (van Kreveld et al., 2000) and PS2644 (Voelker et al., 2000) (Fig. 1, upper panel, orange line). We observe that within the age uncertainties, all DO stadials between 30–50 ka BP were accompanied by iceberg surges, which originated from the adjacent northern hemispheric ice sheets. According to these marine records and other data sets from the northern North Atlantic (e.g., Elliot et al., 1998; Grousset et al., 2001; Elliot et al., 2002), the iceberg calving increased during interstadial periods and peaked at the end of the stadials, after which it decayed rapidly. This sequence of events is consistent with the notion of a freshwater-driven throttling of oceanic convection (Sarnthein et al., 2001), meridional mass and heat transport and subsequent sea-ice expansion, which caused the gradual cooling during interstadials. Moreover, possible feedbacks between the Atlantic meridional overturning circulation (AMOC) and ice sheets may have further modulated the evolution of iceberg and freshwater discharges into the North Atlantic as previously suggested (Schulz et al., 2002;

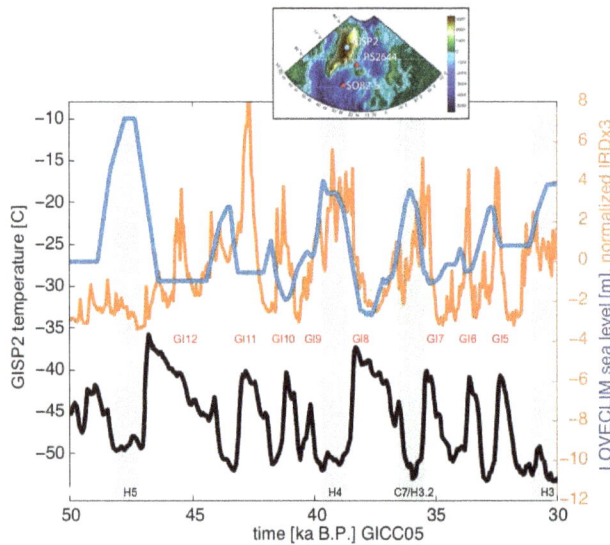

Fig. 1. Upper panel: composite IRD record (orange) obtained from the average of the normalized IRD records of SO82-5 (van Kreveld et al., 2000) and PS2644 (Voelker et al., 2000) from the northern North Atlantic (see map inlay). Integral of freshwater flux forcing used in the LOVECLIM MIS3 hindcast simulation (blue) (see Fig. 2, upper panel). Lower panel: GISP2 reconstructed central Greenland temperatures (Alley, 2000). Greenland interstadials (GI) are highlighted by red labels. The main Heinrich events are represented by gray bars and the light green bar marks are the Greenland stadial C7 just prior to GI7. The age shift between the GISP2 record (Alley, 2000) and the NGRIP record on the GICC05 timescale (Andersen et al., 2006; Svensson et al., 2006) is determined. To project the paleorecords onto the common GICC05 age model, we apply this shift to the GISP2 record and to IRD records of cores SO82-5 and PS2644, whose original age models were partly based on correlations with GISP2. Subsequently the IRD composite (orange) was calculated.

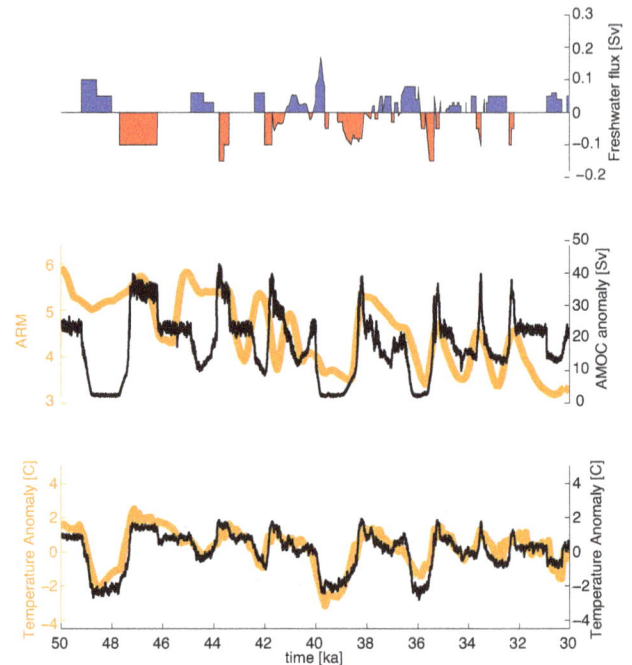

Fig. 2. From top to bottom: time series of North Atlantic freshwater forcing (Sv) applied to LOVECLIM in the region 55–10° W, 50–65° N; simulated maximum meridional overturning circulation in the North Atlantic (Sv) compared to North Atlantic marine sediment cores ARM data (Kissel et al., 2008); and simulated SST anomalies off the Iberian margin (15–8° W, 37–43° N) compared to alkenone-based SST anomalies from marine sediment core MD01-2444 (Martrat et al., 2007). Model results are in black and paleoproxy records in orange.

Timmermann et al., 2003; Shaffer et al., 2004; Alvarez-Solas et al., 2010; Marcott et al., 2011; Alvarez-Solas et al., 2013).

In addition to DO-related IRD variability, there is widespread sedimentary evidence (e.g Heinrich, 1988; Zahn et al., 1997; van Kreveld et al., 2000; Schönfeld et al., 2003; Hemming, 2004; Hodell et al., 2010) for massive iceberg surges that originated mainly from the Laurentide ice sheet and extended far into the eastern North Atlantic (e.g., Grousset et al., 1993). During these so-called Heinrich events a large amount of freshwater was released into the North Atlantic, causing a weakening of the AMOC, as suggested by paleoproxy data (Kissel et al., 2008; Sarnthein et al., 1995, 2001; Vidal et al., 1997; Zahn et al., 1997) and numerous climate modeling experiments (Stouffer et al., 2007; Krebs and Timmermann, 2007; Kageyama et al., 2013). Climate models further document that the corresponding changes in meridional oceanic heat transport, SST and atmospheric circulation are consistent with paleodata evidence of an interhemispheric temperature seesaw (e.g., Stenni et al., 2011),

large-scale drying of Eurasia (e.g., Harrison and Sánchez-Goñi, 2010) and the northern tropics (e.g., Wang et al., 2001; Deplazes et al., 2013) and increased precipitation in parts of the southern hemispheric tropics (Garcin et al., 2006; Wang et al., 2007; Kanner et al., 2013).

Based on their very distinctive sedimentological characteristics, Heinrich stadials (HS) have often been regarded as dynamically different from other DO stadial–interstadial transitions. However, given the fact that both Heinrich and DO stadials were accompanied by (i) large-scale oceanic changes (Fig. 2), (ii) IRD layers (Fig. 1) and (iii) similar global teleconnections, we hypothesize here that Heinrich and DO stadials are part of a continuum of variability that is generated through ice-sheet-driven AMOC changes.

In this paper we set out to simulate the time evolution of DO and Heinrich variability for the period 50–30 ka BP using an intermediate complexity global climate model. We will compare the simulated variability with high-resolution paleoclimate reconstructions. The model simulation is based on the underlying assumption that the continuum of MIS3 climate variability on centennial to millennial timescales can be generated by a suitable North Atlantic freshwater forcing and the associated AMOC response.

The paper is organized as follows: in Sect. 2 the model and experimental setup are described. In Sect. 3 we discuss the patterns of variability associated with DO cycles, the abruptness of stadial–interstadial transitions as well as the timing of Heinrich stadials. We also derive a common age scale that allows for a better comparison between paleoproxy records and model simulations. The paper concludes with a synthesis and discussion of the main results.

2   Model and experimental setup

One of the key goals of our study is to simulate the sequence of millennial-scale events during the period 50–30 ka BP and to determine the corresponding global teleconnections. For this task we have chosen the intermediate complexity earth system model LOVECLIM (Timm and Timmermann, 2007; Menviel et al., 2008; Timmermann et al., 2009b; Goosse et al., 2010). The ocean component of LOVE-CLIM (CLIO) consists of a free-surface primitive equation model with a horizontal resolution of 3° longitude, 3° latitude, and 20 depth layers. The 3-D atmospheric component (ECBilt) is a spectral T21, three-level model based on quasi-geostrophic equations of motion and ageostrophic corrections. LOVECLIM also includes a dynamic–thermodynamic sea-ice model, a land surface scheme, a dynamic global vegetation model (VECODE, Brovkin et al., 1997) and a marine carbon cycle model (LOCH, Menviel et al., 2008; Mouchet, 2011).

Initial conditions for the transient run were obtained by conducting an equilibrium spin-up simulation using an atmospheric CO_2 content of 207.5 ppmv, orbital forcing for the time 50 ka BP and an estimate of the 50 ka BP ice-sheet orography and albedo which were obtained from a 130 ka off-line ice-sheet model simulation (Abe-Ouchi et al., 2007). In the subsequent transient run greenhouse gases, orbital and ice-sheet forcing were updated continuously following the methodology of Timm et al. (2008). Note that our coupled model does not include an interactive ice sheet. Therefore, freshwater withholding from the ocean during phases of ice-sheet growth and freshwater release into the ocean as a result of ice-sheet calving and ablation are not explicitly captured. To mimic the time evolution of these terms and their effect on the ocean circulation, we apply an anomalous North Atlantic freshwater forcing $F(t)$ to the North Atlantic region 55–10° W, 50–65° N. Negative forcing anomalies can be interpreted as periods of ice-sheet growth and excess evaporation over precipitation, whereas positive freshwater anomalies represent times of negative net mass balance of the northern hemispheric ice sheet, associated for instance with massive iceberg calving events or surface ablation.

The freshwater forcing time series $F(t)$ is obtained through an iterative procedure, that optimizes the anomalous freshwater flux such that the simulated temperature anomalies $T_s(t)$ in the eastern subtropical North Atlantic best match

the target alkenone-based SST anomalies $T_r(t)$ reconstructed from the Iberian margin core MD01-2443 (Martrat et al., 2007) (Fig. 2, lower panel, orange line). Starting from an initial guess of the freshwater forcing $F_i(t) = -\alpha T_r(t)$, a series of about $j < 5$ experiments was conducted every 1000 yr using additional freshwater flux perturbations $\delta F^{(j)}(t)$. In each of these 1000 yr long chunks, the freshwater forcing scenario (j) was selected with $F^{(j)}(t) = F_i(t) + \delta F^{(j)}(t)$ that minimized the cost function

$$J^{(j)}(t) = \int_{t}^{t+\tau} \beta \left(T_s(t') - T_r(t') \right)^2 + \gamma \left(\dot{T}_s(t') - \dot{T}_r(t') \right)^2 dt', \quad (1)$$

within this window $[t, t+\tau]$ with $\tau = 1000$ yr. The simulated temperature evolution for $T_s(t')$ in the integral is a function of the applied freshwater forcing $F^{(j)}(t')$. The resulting concatenated freshwater forcing time series $F(t)$ is shown in Fig. 2 (upper panel). Similar to data-assimilation methods that adjust parameters and/or dynamical variables to reduce the mismatch between observations and models, $F(t)$ has the sole purpose to force LOVECLIM into a realistic trajectory with respect to millennial-scale subtropical North Atlantic SST anomalies during MIS3. We did not choose the Greenland temperature reconstruction as an optimization target, because it shows only very weak differences between DO and Heinrich stadials, in contrast to the North Atlantic SST reconstructions.

3   Results

3.1   Freshwater forcing

The applied North Atlantic freshwater forcing $F(t)$ captures the dominant meltwater pulses associated with Heinrich stadials (Fig. 2). It compares well with a recent freshwater forcing estimate (Jackson et al., 2010) obtained with a North Atlantic box model through Bayesian inversion methods[1]. In both cases stadial–interstadial transitions are triggered by negative forcing anomalies, which increase North Atlantic surface densities and subsequently strengthen the AMOC (Fig. 2, middle panel). Negative freshwater forcing can be interpreted to represent a positive ice-sheet mass balance, which in our modeling framework mimics a reduction of the continental runoff as well as excess evaporation over precipitation over the North Atlantic region. As an independent validation of our freshwater forcing, we compare the time-integral of $F(t)$, which represents the corresponding global sea level changes, with the composite IRD records from the Nordic Seas cores PS2644 and SO82-5 (Fig. 1) on the GICC05 timescale (see Sect. 3.6 for more details on the

---

[1] Some discrepancies between $F(t)$ and the freshwater estimate of Jackson et al. (2010) arise from the different AMOC sensitivities to freshwater perturbations and the different choice of the optimization target (GISP2 for the box model Jackson et al., 2010 and Iberian margin SST for LOVECLIM)

synchronization of the cores and the model results). The rationale of this comparison is that high values of IRD correspond to additional freshwater discharge and sea level rise. Furthermore, rising sea level can amplify iceberg calving through ice-shelf instabilities. Except for the simulated sea level rise associated with Heinrich event 5, which is not captured in these eastern North Atlantic paleorecords, we find a relatively good match between model simulation and reconstruction, thus supporting the realism of the applied freshwater forcing. It should be noted here that the simulated DO-related sea level changes are a factor 2–3 smaller than those reconstructed from the Red Sea during this time (Siddall et al., 2003).

3.2 AMOC response

As a result of the applied anomalous North Atlantic freshwater fluxes, the AMOC weakens and strengthens on millennial timescales. Heinrich stadials correspond to a complete shutdown of the AMOC, whereas DO stadials are associated with a 50 % weakening of the AMOC, relative to the interstadial periods (Fig. 2). The resulting AMOC time series compares reasonably well with a reconstruction of Atlantic bottom currents obtained from mass-normalized anhysteretic remanent magnetization (ARM) data (Kissel et al., 2008) (Fig. 2, middle panel), even though the model and paleoceanography time series are based on different underlying age models (a more detailed discussion of age-scale uncertainties is provided is Sect. 3.6).

3.3 Temperature response

The excellent agreement between simulated and reconstructed SST anomalies in the Iberian margin area (Fig. 2 lower panel) is to be expected, because the latter has been used as the target for the optimization of the freshwater fluxes. One important finding is that the temperature drop in the northeast Atlantic around 36 ka BP (referred to as C7, adopting the Chapman and Shackleton, 1999 terminology) can be obtained in our simulation only by a complete shutdown of the AMOC, which is induced by a prolonged freshwater flux of ~ 0.1 Sv. This is consistent with the presence of an IRD pulse in the Greenland and Irminger seas (see Fig. 1), a drop in sea-surface salinity in SO82-5 (van Kreveld et al., 2000) and changes in benthic $\delta^{18}O$ (Margari et al., 2010). Whereas the IRD pulse is well pronounced in the composite IRD time series (Fig. 1), as well as in records from the Irminger basin (SU 90-24) (Elliot et al., 2002), the southern Gardar Drift (JPC-13) (Hodell et al., 2010) and from the Iberian margin (MD95-2040) (Schönfeld et al., 2003); it appears to be absent in other southwestern Atlantic IRD records (e.g., Grousset et al., 1993; Rashid et al., 2003; Nave et al., 2007).

We move on to a more detailed comparison with other temperature reconstructions from the North Atlantic and

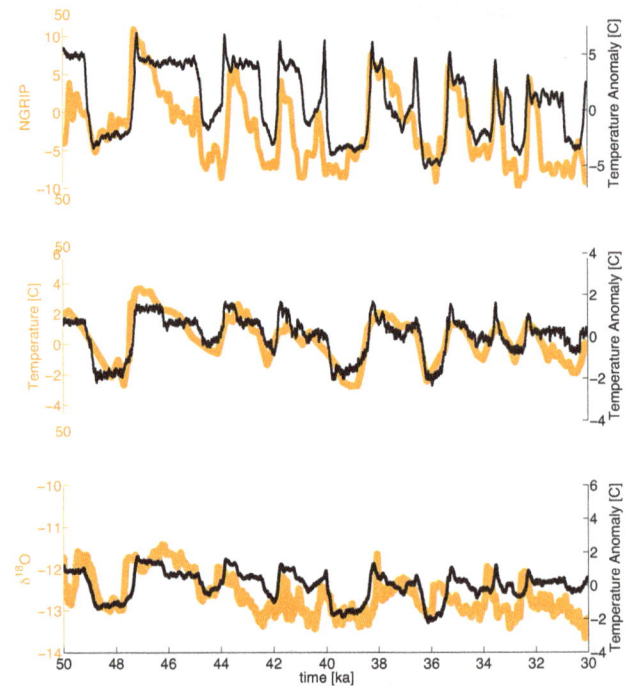

Fig. 3. From top to bottom: time series of simulated northeastern Greenland air temperature anomalies (40° W–10° E, 66–85° N) compared to the NGRIP temperature reconstruction (Huber et al., 2006); simulated SST in the western Mediterranean compared to alkenone-based SST reconstructions from the Alboran Sea ODP hole 161-977A (Martrat et al., 2007) and simulated air temperature anomalies over Turkey (25–46° E, 35–42° N) compared to a speleothem $\delta^{18}O$ record (‰) from Sofular Cave, Turkey (Fleitmann et al., 2009). Model results are in black and paleoproxy records in orange.

Mediterranean realm. Figure 3 shows the comparison between simulated surface temperatures in Greenland and NGRIP temperature reconstructions (Huber et al., 2006) on the SS09 timescale. In accordance with paleodata, the simulated Heinrich stadials in Greenland attain very similar minimum temperatures than the DO stadials. This behavior in Greenland is quite distinct from SST reconstructions (Fig. 2 lower panel and Fig. 3 middle panel), which exhibit a marked difference between Heinrich and non-Heinrich stadials. These results indicate the presence of a nonlinear sea-ice feedback (Li et al., 2005; Deplazes et al., 2013) which saturates when sea ice reaches a certain extent, thus capping cooling over Greenland. Simulated stadial–interstadial transitions attain values of about 9 °C in Greenland, which is smaller than the reconstructed values of up to 16 °C (Capron et al., 2010). Moreover, the simulation does not capture the slow interstadial cooling seen in the reconstructions. Instead interstadial periods have relatively constant Greenland temperatures in the model, except for an initial overshoot. These dynamics clearly differs from the behavior of SSTs in the northeastern Atlantic/Mediterranean (Figs. 2 and 3), which

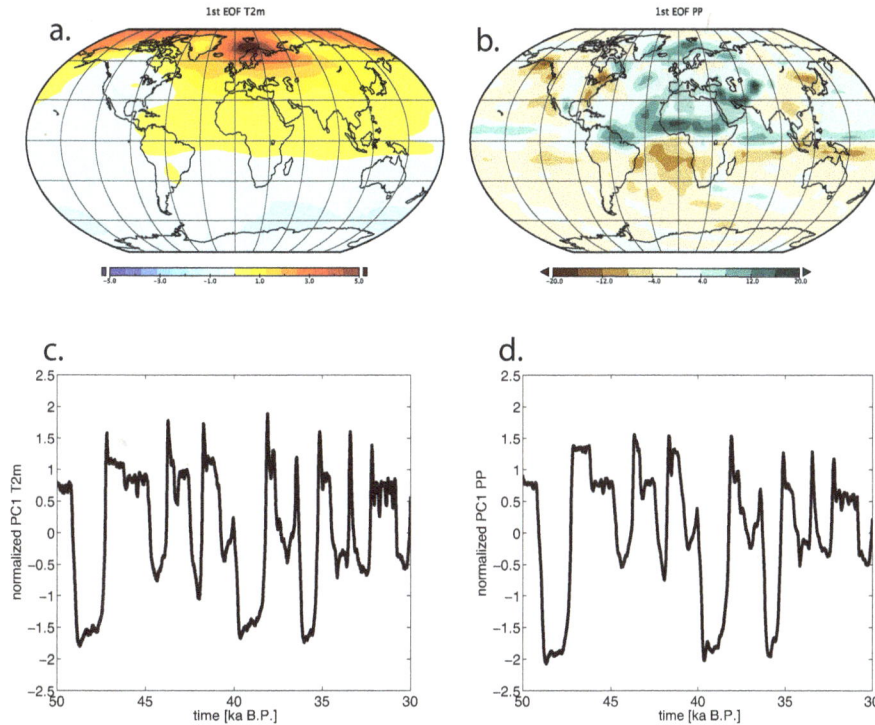

Fig. 4. (a) Pattern of first EOF of detrended 2 m air temperature anomalies (°C); **(b)** pattern of first EOF of detrended precipitation anomalies (cm yr^{-1}); **(c)** normalized principal component of 1st EOF of detrended 2 m air temperature, which explains 64 % of the variance; **(d)** normalized principal component of 1st EOF of detrended precipitation, which explains 16 % of the variance.

tracks the underlying AMOC variability much more accurately. Moving further into the eastern Mediterranean region, we find a reasonable qualitative match between the simulated temperature anomalies in Turkey and the δ^{18}O record from independently dated speleothems from Sofular Cave (Fleitmann et al., 2009), which capture a combined temperature/hydroclimate signal.

A more comprehensive spatial view of the simulated DO/Heinrich dynamics is obtained from an EOF analysis of global surface air temperatures (Fig. 4a). The dominant EOF mode is characterized by a meridional temperature seesaw in accordance with numerous other modeling studies (Stouffer et al., 2006, 2007; Timmermann et al., 2009a; Kageyama et al., 2013) and paleoclimate data sets (Blunier et al., 1998; Barker et al., 2009; Stenni et al., 2011). Interstadial conditions are characterized by northern hemispheric warming with strongest amplitudes over the Greenland, Iceland, and Norway seas and the Arctic Ocean. The warming extends into North Africa, Asia and the western North Pacific. This warming pattern is in general agreement with pollen-derived temperature reconstructions for GI8 (38 ka BP) and GI6 (33 ka BP) (Harrison and Sánchez-Goñi, 2010). The corresponding principal component time series (Fig. 4c) clearly features the enhanced cooling during massive Heinrich stadials (HS5 and HS4) and the C7 stadial, in contrast to the weaker cooling associated with DO stadials. Simulated

southern hemispheric cooling during interstadials is consistent with the presence of a bipolar temperature seesaw (Stocker, 1998; Stocker and Johnsen, 2003).

3.4 Hydroclimate response

As a result of the very strong North Atlantic cooling during Heinrich stadials (HS5, HS4) and during the C7 stadial (Fig. 3), northern hemispheric trade winds intensify by up to 60 %, which leads to a southward shift of the Intertropical Convergence Zones, extending from South America, into the tropical Atlantic, equatorial Africa and the Indian Ocean. This is illustrated by the EOF analysis of simulated precipitation in Fig. 4b and d.

The lower amplitude cooling during DO stadials weakens the trade winds by only 30 %. The corresponding southward shift of the tropical rainbands is less pronounced than for Heinrich stadials as shown by the leading principal component of the rainfall EOF analysis (Fig. 4d). In spite of a high correlation (0.92) between the principal components of temperature and rainfall, there are some notable differences. The precipitation mode exhibits a more pronounced two-step structure for the interstadial DO12 (around 47–46 ka) and a stronger difference between Heinrich stadials and DO stadials (Fig. 4d) than the temperature EOF mode (Fig. 4c). Qualitatively the patterns of simulated temperature and rainfall changes agree with those obtained from coupled general

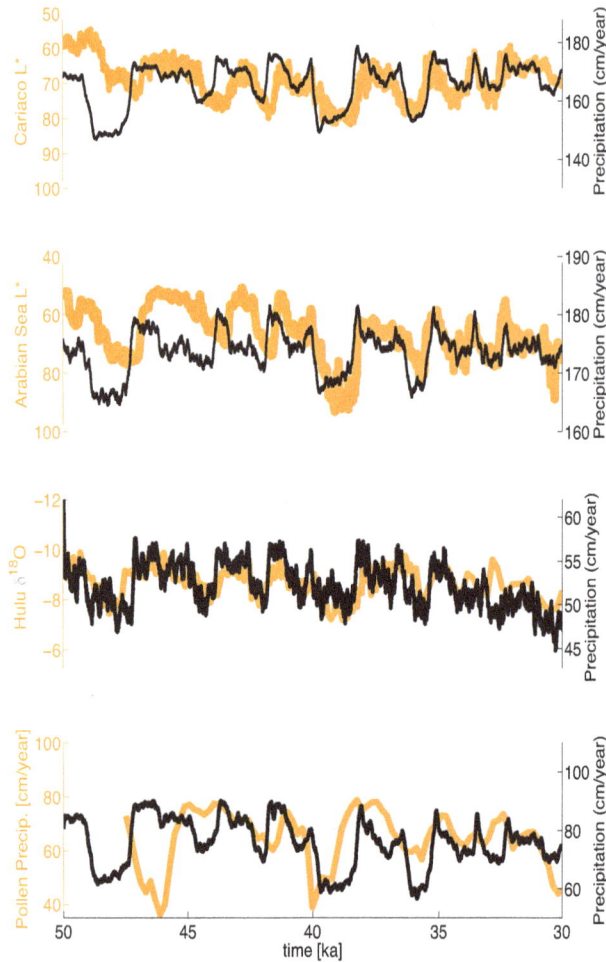

Fig. 5. From top to bottom: time series of simulated annual precipitation anomalies over the Cariaco Basin (60–50° W, 5–20° N) compared to a reflectance record from the Cariaco Basin (Deplazes et al., 2013); time series of simulated Arabian Sea annual precipitation (45–65° E, 5–15° N) compared to a reflectance record (L^*) from the northeastern Arabian Sea (Deplazes et al., 2013); simulated precipitation in eastern China (114–124° E, 28–35° N) compared to a speleothem $\delta^{18}O$ record (‰) from Hulu Cave, China (Wang et al., 2001); time series of simulated Iberian region annual precipitation (10–1° W, 35–40° N) compared to a composite of pollen-derived precipitation estimates from the Iberian margin and the Alboran Sea (Sánchez-Goñi et al., 2002); Model results are in black and paleoproxy records in orange.

circulation models subjected to North Atlantic freshwater perturbations (Broccoli et al., 2006; Timmermann et al., 2007; Kageyama et al., 2013).

Comparing the simulated northern hemispheric rainfall changes on a regional scale with hydroclimate reconstructions for the Mediterranean region (Sánchez-Goñi et al., 2002), the Cariaco Basin (Deplazes et al., 2013), the Arabian Sea (Deplazes et al., 2013), eastern China (Wang et al., 2001) (Fig. 5) and Central America (Hodell et al., 2008) (Fig. 6, upper panel), we find an excellent agreement between model

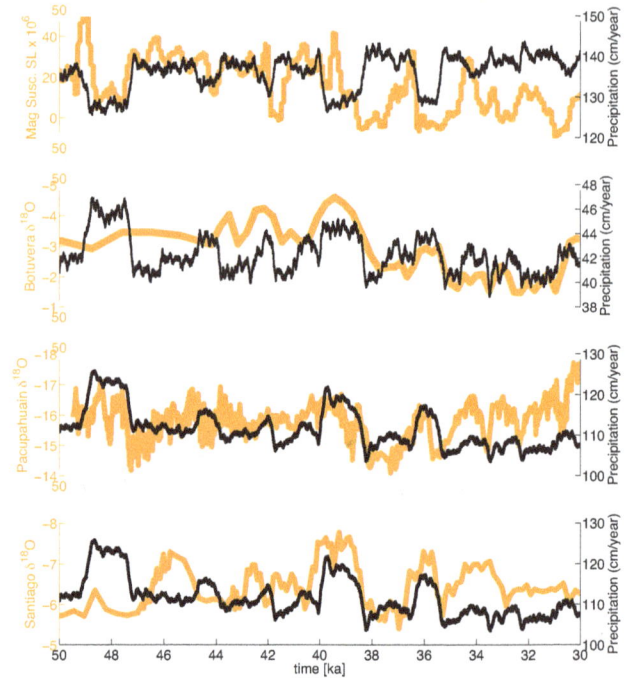

Fig. 6. From top to bottom: time series of simulated annual precipitation anomalies over Guatemala (104–93° W, 12–30° N) compared to a magnetic susceptibility record from Lake Peten Itza, Guatemala (Hodell et al., 2008); Brazil (44–60° W, 20–30° S) compared to $\delta^{18}O$ (‰) of a speleothem record from Botuvera Cave, Brazil (Wang et al., 2007); Peru and Ecuador (85–70° W, 3–15° S) compared to speleothems $\delta^{18}O$ records from Pacupahuain Cave, Peru (Kanner et al., 2013) and Santiago Cave, Ecuador (Mosblech et al., 2012). Model results are in black and paleoproxy records in orange.

and data with stadial (interstadial) conditions corresponding to increased aridity (pluvials). While the pollen-derived precipitation estimate (Sánchez-Goñi et al., 2002) and the model output suggest much drier conditions over the Iberian region during HS5, the chronology of the proxy record is based on a graphic correlation with the GISP2 ice core, which places the HS5 much later than the new GICC05 chronology (Obrochta et al., 2014). It should be noted that reflectance and magnetic susceptibility are indirect hydroclimate proxies and thus cannot give quantitative estimates. In addition, speleothem $\delta^{18}O$ can be potentially affected by other processes such as changes in temperature, soil evaporation and the water vapor sources. Age model uncertainties associated with the Arabian Sea record (Deplazes et al., 2013) could preclude any conclusions with respect to synchronicity with North Atlantic stadials. However, the high level of correspondence between our simulated precipitation changes and the Arabian Sea reflectance record indicates that North Atlantic stadials lead to drier synchronous conditions over the Arabian Sea.

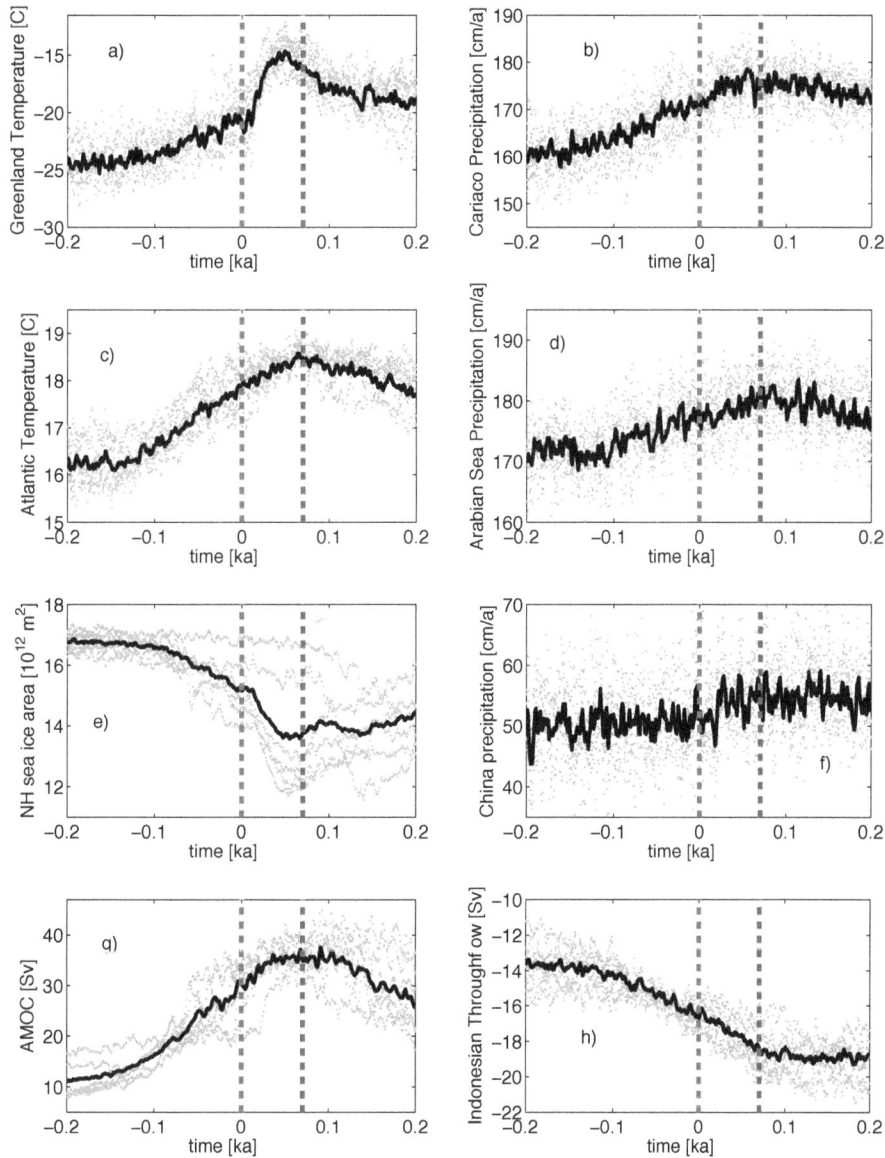

Fig. 7. Composite (thick black line) of DO stadial–interstadial transitions showing different simulated variables relative to the maximum time derivative in simulated Greenland temperatures occurring at 47.25, 43.9, 41.8, 38.3, 35.3, 33.6 and 32.34 ka BP: (**a**) Greenland air temperature, (**b**) Cariaco Basin precipitation, (**c**) North Atlantic temperature, (**d**) Arabian Sea precipitation, (**e**) northern hemispheric sea-ice area, (**f**) eastern China precipitation, (**g**) Maximum of meridional streamfunction in North Atlantic, (**h**) strength of Indonesian Throughflow. The blue and red dashed lines respectively represent the time of the largest positive time derivative of Greenland temperatures and the time when North Atlantic temperatures attain the maximum value. The gray dots represent the individual data points before calculating the composite.

The reverse pattern can be found for southern hemispheric hydroclimate proxies in Brazil (Wang et al., 2007), Peru (Kanner et al., 2013) and Ecuador (Mosblech et al., 2012) as well as for simulated rainfall changes (Fig. 6, lower three panels). In Sect. 3.6 we will try to reconcile some age-model discrepancies by projecting model and proxy data onto the common NGRIP GICC05 timescales (Andersen et al., 2006; Svensson et al., 2006).

3.5 Abruptness of stadial–interstadial transitions

To determine the response time of various climate variables and ocean transport indicators to freshwater changes, we calculate a composite (Fig. 7, black line) based on several stadial–interstadial DO transitions by aligning the model data relative to the maximum temperature derivative of the simulated Greenland temperature (47.25, 43.9, 41.8, 38.3, 35.3, 33.6, 32.34 ka BP). According to this analysis we find that the averaged DO transition takes place within 150 to 200 yr for all the climate variables in Fig. 7. However, for Greenland temperatures (Fig. 7a) and northern hemispheric sea-ice area (Fig. 7e), we see a considerable acceleration and an associated increase of abruptness 100 yr into the transition. This is in agreement with previous estimates of the abruptness of DO stadial–interstadial transitions in Greenland ice cores (∼ 125 yr) (Capron et al., 2010), although much higher rates were reported in atmospheric circulation proxies (Steffensen et al., 2008).

As already demonstrated in Figs. 3 and 5 rainfall changes in the Cariaco Basin area and the Arabian Sea clearly track millennial-scale SST variations in the North Atlantic region. This is further supported by the composite analysis which reveals a very similar time evolution of these variables for the averaged DO stadial–interstadial transition. Rainfall changes in eastern China are less well pronounced owing to a much larger level of simulated rainfall variability that is unrelated to DO cycles.

According to Fig. 7h, changes in the barotropic transport across the Indonesian archipelago occur almost in unison with the AMOC. This surprisingly fast adjustment can be attained by two processes: (i) wind changes in the Pacific (Timmermann et al., 2005b), and (ii) fast oceanic adjustment processes involving wave propagation from the Atlantic into the Indian and Pacific oceans, as discussed in Timmermann et al. (2005a). The former can modulate the Indonesian Throughflow via the island rule (Godfrey, 1989), whereas the latter would have to change the joint effect of baroclinicity and relief (JEBAR) term in the barotropic transport equation (Sarkisyan and Ivanov, 1971; Cane et al., 1998). Irrespective of the relative magnitudes of these terms, our analysis clearly documents that the DO variability has far-reaching fast oceanic impacts that extend also into the other ocean basins.

3.6   Common age scale

To better compare the paleorecords in Figs. 2, 3, 5 and the simulated climate variables, we make an attempt to bring the time series all onto the same age scale. We have chosen the Greenland Ice Core Chronology 2005 (GICC05) (Andersen et al., 2006; Svensson et al., 2006) as the common age model. To project the model simulation (30–50 ka BP) onto this age model, we compare the simulated Greenland temperature with the NGRIP δ^{18}O (Svensson et al., 2006) and identify in a 1800 yr sliding window the lag at which the lag correlation attains its maximum value. Here we allow maximum lags of ±750 yr. To avoid large local discontinuities or reversals in the age model projection, we subsequently filtered the resulting time series of age adjustments for the model simulation using a 500 yr running mean. The resulting Greenland temperature-based age shift is then applied to all other model variables, thus keeping, at least to first order, the lead-lag structure within the model intact. We also project the

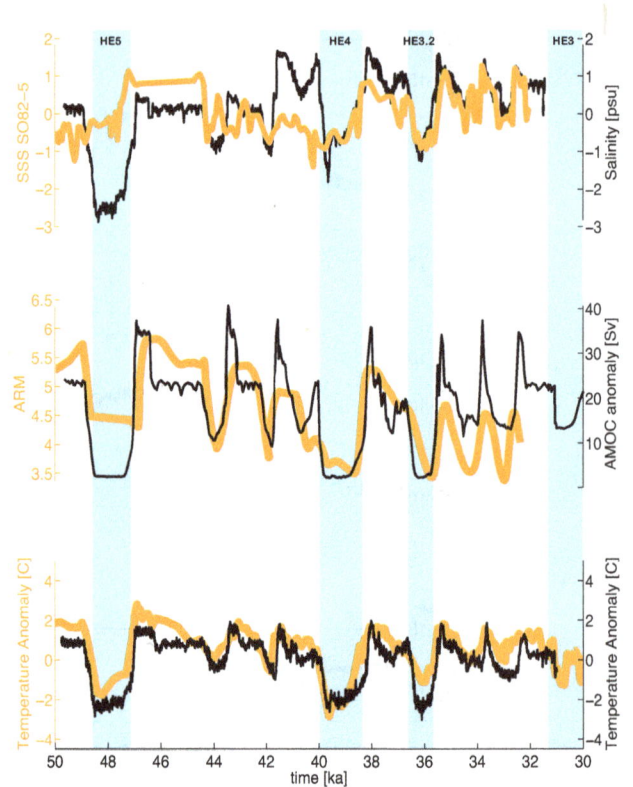

Fig. 8. From top to bottom: time series projected onto GICC05 age scale (Andersen et al., 2006; Svensson et al., 2006) of Nordic Seas salinity from core S082-5 (van Kreveld et al., 2000) and simulated surface salinity anomalies at this location; simulated maximum meridional overturning circulation in the North Atlantic (Sv) compared to North Atlantic marine sediment cores ARM data (Kissel et al., 2008); simulated SST anomalies off the Iberian margin (15–8° W, 37–43° N) compared to alkenone-based SST anomalies from marine sediment core MD01-2444 (Martrat et al., 2007). Model results are in black and paleoproxy records in orange.

sea-surface salinity data from SO82-5 (van Kreveld et al., 2000), the ARM data and the Iberian margin SST (Fig. 8) as well as the Cariaco and Arabian seas color records (Fig. 9), onto the GICC05 timescale. Here we assume that at least in a 1800 yr sliding window, the proxy data varies in synchrony with the Greenland temperature record at zero lag. This assumption is well justified by the model results that show maximum correlation of 0.92 between the principal components of the leading EOFs of temperature (Fig. 4a and c) and precipitation (Fig. 4b and d) at zero lag. Furthermore, our model-based composite analysis of DO stadial–interstadial transitions supports the notion of near synchronicity (within ±100 yr) of the physical variables under consideration.

Having synchronized the model and paleoproxy data with the NGRIP δ^{18}O record on GICC05, we find a much better agreement between model and paleoproxy records, particularly for the period 50–40 ka BP. The ARM data now nicely features marked AMOC weakening during HS5, HS4 and C7 as well as during most of the DO stadials (Fig. 8). In addition,

Table 1. Table showing the timing of stadials HS5, HS4, C7/HS3.2 and HS3 as recorded in high-resolution well-dated paleorecords. We calculate the mean (ka BP) and one standard deviation (ka) of the timing of each stadial as well as the duration (yr) of each stadial and its standard deviation (yr). In the Pacupahuain record, the two peaks around HS5 were treated as a single event. A similar method was used for HS3 in the Hulu Cave record.

Paleoproxy record	HS5 ka BP	HS4 ka BP	C7/HS3.2 ka BP	HS3 ka BP	Ref.
Greenland δ^{18}O (GICC05)	48.8–46.9	39.95–38.25	36.65–35.5	32–28.9	Huber et al. (2006)
Sofular Cave δ^{18}O	48.8–47.7	39.9–38.2	36.6–35.95	31.5–29.5	Fleitmann et al. (2009)
Pacupahuain Cave δ^{18}O	48.7–48.2 48.0–47.45	40.4–38.4	36.4–35.8	30.5–28.9	Kanner et al. (2013)
Santiago Cave δ^{18}O	49–48.2	40.1–38.6	36.6–35.6	–	Mosblech et al. (2012)
Hulu Cave δ^{18}O	48.8–47.6	39.8–38.1	36–35.2	31.3–29.8 28.8–27.9	Wang et al. (2001)
Mean (ka BP)	48.8–47.6	40.0–38.3	36.45–35.6	31.3–28.8	
Standard deviation (ka)	0.1–0.5	0.2–0.2	0.2–0.3	0.5–0.6	
Duration (yr)	1250	1720	840	1820	
Standard deviation (yr)	360	160	210	730	

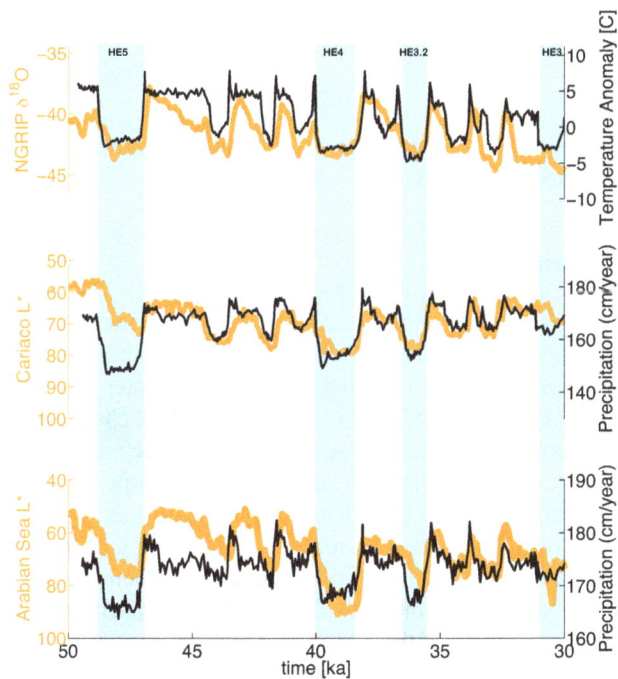

Fig. 9. From top to bottom: time series projected onto GICC05 age scale of simulated NE Greenland air temperature anomalies (40° W–10° E, 66–85° N) and NGRIP δ^{18}O; simulated annual precipitation anomalies over the Cariaco Basin (60–50° W, 5–20° N) compared to a reflectance record from the Cariaco Basin (Deplazes et al., 2013); time series of simulated Arabian Sea annual precipitation (45–65° E, 5–15° N) compared to a reflectance record (L^*) from the northeastern Arabian Sea (Deplazes et al., 2013). Model results are in black and paleoproxy records in orange.

the precipitation records from the Cariaco Basin and the Arabian Sea are now in better agreement with the model, particularly for HS5 (Fig. 9).

We conclude that if forced with a freshwater forcing that leads to simulated salinity anomalies which closely resemble (within the dating uncertainties) paleosalinity reconstructions from the Nordic Seas (Fig. 8), the LOVECLIM model hindcast captures the dominant modes of Heinrich and DO variability found in paleoreconstructions. The model results further support that freshwater forcing triggered changes of the AMOC and North Atlantic SSTs, which subsequently caused the observed hydroclimate shifts across both Hemispheres. This confirms our initial hypothesis that ice-sheet-driven AMOC variations played a crucial role in generating the continuum of millennial-scale DO/Heinrich variability in the North Atlantic during MIS3 (see also Sarnthein et al., 2001). Potential feedbacks of AMOC variability on the mass balance of the major ice sheets will be discussed in Sect. 4.

3.7 Timing and duration of Heinrich stadials

Here we will take the opportunity to revisit the timing of some important climate events during MIS3. We will focus in particular on Heinrich stadials (HS5, HS4 and HS3) and stadial C7 (see Fig. 1) and use high-resolution paleoclimate reconstructions with independent age control that capture Heinrich and DO variability. The timing from NGRIP on GICC05 (Andersen et al., 2006), Sofular Cave, Turkey (Fleitmann et al., 2009), Pacupahuain Cave, Peru (Kanner et al., 2013), Santiago Cave, Ecuador (Mosblech et al., 2012) and Hulu Cave (Wang et al., 2001) is summarized in Table 1. Results from Sect. 3.5 supports the notion of near synchronicity (within ±100 yr) of the physical variables under consideration. Consolidated estimates for the timing of HS5,

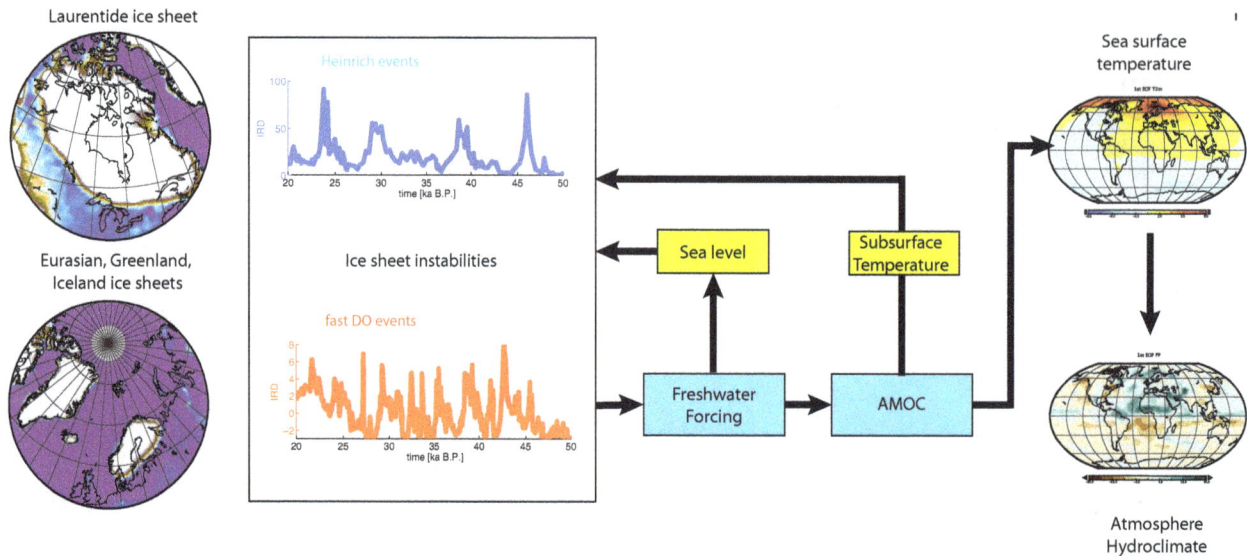

Fig. 10. Schematic illustration of the effect of northern hemispheric ice-sheet instabilities on the AMOC, SST and atmospheric circulation. AMOC changes are likely to provide a positive feedback on ice-sheet instabilities via subsurface temperature anomalies (Alvarez-Solas et al., 2010). Furthermore, sea level changes generated by one ice sheet can trigger ice-shelf instabilities in another ice sheet and subsequent accelerated flow and iceberg calving. Once the ice sheets reach a new mass balance, the freshwater input into the North Atlantic ceases and the AMOC starts its recovery thus initiating a stadial–interstadial DO transition.

HS4, C7 and HS3 are 48.8–47.6, 40.0–38.3, 36.45–35.6, and 31.3–28.8 ka BP, respectively, which agrees well with our model simulation (Figs. 2 and 3). These ranges also agree well with estimates from North Atlantic marine sediment cores for the timing of HS5 and HS4 (50–47 and 40.2–38.3 ka BP, respectively) (Sánchez-Goñi and Harrison, 2010).

The C7 stadial was accompanied by a considerable IRD pulse in the northeastern North Atlantic, as seen for instance in the sediment cores PS2644 (Voelker et al., 2000), SO82-5 (van Kreveld et al., 2000), JPC-13 (Hodell et al., 2010), MD95-2040 (Schönfeld et al., 2003), NA87-22 and SU 90-24 (Elliot et al., 2002). Furthermore, we find strongly reduced surface temperatures in the Atlantic and widespread northern hemispheric aridity (Figs. 2, 3 and 4). In our model simulation and in the paleoproxy data, the C7 stadial shares many common characteristics with the typical response for Heinrich stadials 3–5. In fact in the EOF analysis of the model simulation (Fig. 4), this period is basically indistinguishable from the other prominent Heinrich stadials, both in terms of temperature and rainfall. We therefore propose to introduce the term "Heinrich stadial 3.2" (HS3.2), using the same nomenclature introduced for Heinrich stadial 5.2 (Sarnthein et al., 2001).

Paleorecords as well as the model results display little coherency regarding the amplitude and the timing of HS3. It was suggested (Elliot et al., 1998; Scnoeckx et al., 1999; Grousset et al., 2000) that in contrast to other Heinrich events, HS3 may have originated from the Fennoscandian ice sheet. It is thus possible that iceberg discharges during Heinrich event 3 had a different impact on the AMOC than for other Heinrich events.

4   Conclusions

Here we presented a new transient model simulation that covers the period 30–50 ka BP. This climate model hindcast experiment was designed in such a way that freshwater forcing between 55–10° W, 50–65° N generates AMOC changes and subsequently northeastern Atlantic temperature anomalies that are in agreement with alkenone-based temperature reconstructions from the Iberian margin area. With this weak constraint on model/data agreement, we were able to independently evaluate the model performance with numerous other high-resolution climate proxies from both hemispheres. A marked weakening of the AMOC reduces the oceanic and atmospheric poleward heat transport thus leading to a strong cooling over the North Atlantic region (Kageyama et al., 2013). In our model the cooling is centered on Scandinavia, extends over Greenland and northern Europe and is also simulated over southern Europe, North Africa and Asia. The cooling is the strongest at high latitudes due to sea-ice albedo feedbacks. Such temperature changes lead to a stronger surface temperature gradient over the North Atlantic and therefore to a strengthening of the North Easterly trades. The cooler conditions over the North Atlantic and stronger trades induce a southward shift of the ITCZ over the Atlantic region with drier conditions simulated over Europe, the northern part of South America, North Africa and the Middle East.

The resulting high level of agreement between the model and paleoproxy records provides strong support for our initial hypothesis: namely that Heinrich and DO variability during MIS3 were caused by ice-sheet-driven changes in the strength of the AMOC. The relationship between stadials and IRD records from the North Atlantic region further indicates that northern hemispheric ice-sheet calving and freshwater discharges play a major role in disrupting the AMOC during MIS3 (Fig. 1). Figure 10 summarizes a scenario explaining DO/Heinrich variability, which follows some elements of Sarnthein et al. (2001). We suggest that ice-sheet instabilities of low amplitude, originating mostly from the Eurasian, Iceland and Greenland ice sheets (Bond and Lotti, 1995), are the main driver of the fast DO variability. The corresponding low amplitude freshwater flux perturbations triggered a weakening of the AMOC, but not a complete collapse. We further acknowledge the possibility that DO climate variability may have caused ice-sheet mass imbalances and calving from circum-Atlantic ice sheets (Bond and Lotti, 1995; Grousset et al., 2000, 2001; Marshall and Koutnik, 2006), thus contributing to the DO-synchronized delivery of IRD into the North Atlantic. In contrast, instabilities from the Laurentide ice sheet occurred less frequently but were associated with much larger iceberg and freshwater discharges, leading to complete AMOC shutdown and larger SST and hydroclimate changes in the North Atlantic realm and beyond. Changes in sea level during Heinrich events (Flückiger et al., 2006) and subsurface temperatures (Shaffer et al., 2004; Mignot et al., 2007; Alvarez-Solas et al., 2010; Marcott et al., 2011; Alvarez-Solas et al., 2013) (Fig. 10) may have subsequently triggered marine-ice-sheet instabilities, thus increasing the initial freshwater discharge. Such processes may have played a key role in synchronizing ice-sheet dynamics in the Northern Hemisphere and in prolonging ice-sheet instabilities during Heinrich events. Once the ice sheets reach a new mass balance, the freshwater input into the North Atlantic ceases, salinity increases rapidly to the more saline Arctic/North Atlantic glacial background state and the AMOC starts its recovery thus initiating a stadial–interstadial DO transition.

A more detailed view of the underlying mechanisms is provided in Fig. 11, which shows the time evolution of the composite IRD record (Fig. 1) from cores SO82-5 and PS2644 (orange, upper curve), the salinity reconstruction (blue, upper curve) from Irminger Sea core SO82-5 (van Kreveld et al., 2000), the North Atlantic ARM data (red, middle curve) (Kissel et al., 2008) as a proxy for changes in AMOC strength, summer SST variations from the Irminger Sea (cyan, middle curve) (van Kreveld et al., 2000) and the GISP2 ice-core temperature reconstruction (black, lower curve) (Alley, 2000). All data were interpolated onto the GICC05 timescales (see caption to Fig. 1 and Sect. 3.6 for more details). Here we begin with the high IRD values during HS4, low salinities and cold conditions in the Irminger Sea and a weak AMOC (40–39 ka BP). Around ∼ 39 ka BP the strong freshwater forcing vanishes abruptly. Concomitantly,

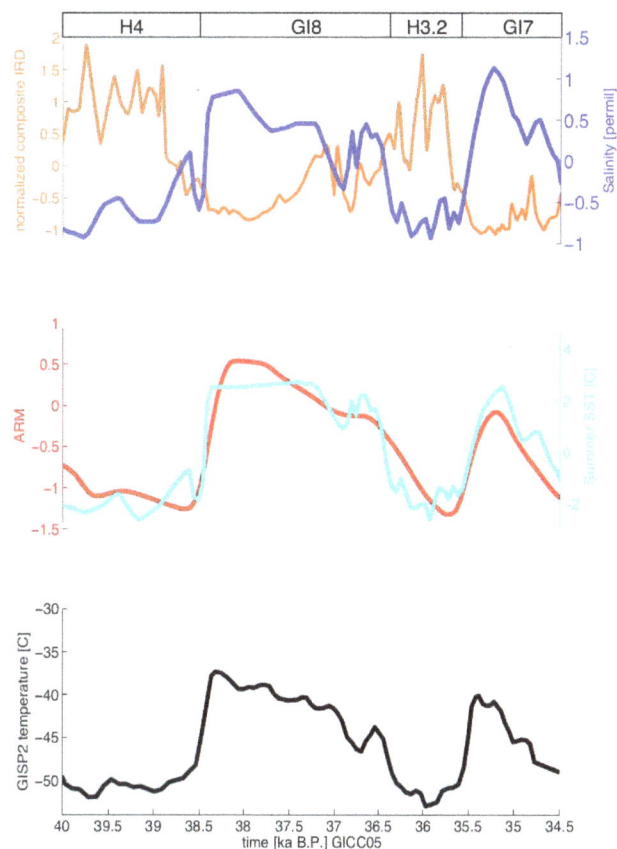

Fig. 11. Top panel: composite IRD record (orange) obtained from the average of the normalized northern North Atlantic IRD records SO82-5 (van Kreveld et al., 2000) and PS2644 (Voelker et al., 2000) (see Fig. 1) and Irminger Sea sea-surface salinities (blue) from SO82-5 (van Kreveld et al., 2000); middle panel: North Atlantic marine sediment cores ARM data (red) (Kissel et al., 2008) and Summer SST anomalies (cyan) from marine sediment core SO82-5 (van Kreveld et al., 2000); bottom panel: GISP2 reconstructed central Greenland temperatures (black) (Alley, 2000). All data are displayed on the GICC05 timescale (see Fig. 1).

sea-surface salinity increases thus initiating the AMOC recovery. The AMOC strengthening leads to Greenland and North Atlantic warming as well northern North Atlantic sea-ice retreat. Greenland interstadial 8 (GI8) is characterized by a very warm initial period which lasts for about 100–200 yr. IRD is at its minimum, Nordic Seas surface salinity is high and so is the strength of the AMOC (Fig. 8). We consider this period of minimum ice-sheet calving a period of positive northern hemispheric ice-sheet mass balance (i.e. growth). Around 37.5 ka BP calving resumes and increases until 37 ka BP. This evolution is briefly interrupted for about 100 yr between 36.9–36.8 ka BP, before a period of rapid iceberg surging and an associated salinity decrease leads into the stadial cooling phase during HS3.2. The stadial iceberg surging period lasts for 1 ka and comes to an end when the ice-sheet calving has exhausted itself. This is the initiation of GI7. The scenario outlined here for a set of paleoproxy data sets is entirely consistent with the modeling-based evidence from Figs. 8 and 9.

Our paper further highlights the different response characteristics of various climate variables to AMOC changes (Fig. 7). The extraordinary abruptness of Greenland temperature changes during the DO stadial–interstadial transition was identified as a regional phenomenon, which is likely induced by sea-ice feedbacks (Li et al., 2010; Deplazes et al., 2013). Given the fact that the North Atlantic temperature and AMOC composite shown in Fig. 7 have already reached 2/3 of their full DO amplitude at zero lag whereas Greenland temperatures have only attained about 50 %, it may appear as if the Greenland record is lagging the other variables. This is merely a reflection of the nonlinearity of the Greenland temperature response. It should be noted here that this feature may impact the synchronization of high-resolution proxy time series with Greenland climate reconstructions.

According to our model and data-based evidence, we conclude that ice-sheet/freshwater-driven AMOC variations and local feedbacks determined the timing and abruptness of DO events as well as their global teleconnections.

Acknowledgements. This project was supported by the Australian Research Council. Model experiments were performed on a computational cluster owned by the Faculty of Science of the University of New South Wales as well as on a cluster from the NCI National Facility at the Australian National University. T. Friedrich and A. Timmermann are supported by NSF grant no. 1010869 and by the Japan Agency for Marine-Earth Science and Technology (JAMSTEC) through its sponsorship of the International Pacific Research Center. M. England acknowledges funding from the ARC Laureate Fellowship program (FL100100214). We thank C. Jackson for providing his optimized freshwater forcing time series as well as M. Elliot and N. Thouveny for providing their data for comparisons. We are also grateful to S. Harrison for organizing a workshop at Macquarie University, that stimulated many interesting discussions on DO events and to M. Sarnthein for insightful discussions.

Edited by: D. Fleitmann

References

Abe-Ouchi, A., Segawa, T., and Saito, F.: Climatic Conditions for modelling the Northern Hemisphere ice sheets throughout the ice age cycle, Clim. Past, 3, 423–438, doi:10.5194/cp-3-423-2007, 2007.

Alley, R.: Ice-core evidence of abrupt climate changes, P. Natl. Acad. Sci. USA, 97, 1331–1334, 2000.

Alvarez-Solas, J., Charbit, S., Ritz, C., Paillard, D., Ramstein, G., and Dumas, C.: Links between ocean temperature and iceberg discharge during Heinrich events, Nat. Geosci., 3, 122–126, doi:10.1038/NGEO752, 2010.

Alvarez-Solas, J., Robinson, A., Montoya, M., and Ritz, C.: Iceberg discharges of the last glacial period driven by oceanic circulation changes, P. Natl. Acad. Sci., 110, 16350–16354, doi:10.1073/pnas.1306622110, 2013.

Andersen, K., Svensson, A., Johnsen, S., Rasmussen, S., Bigler, M., Röthlisberger, R., Ruth, U., Siggaard-Andersen, M.-L., Steffensen, J., Dahl-Jensen, D., Vinther, B., and Clausen, H.: The Greenland Ice Core Chronology 2005, 15–42 ka, Part 1: Constructing the time scale, Quaternary Sci. Rev., 25, 3246–3257, 2006.

Barker, S., Diz, P., Vautravers, M., Pike, J., Knorr, G., Hall, I., and Broecker, W.: Interhemispheric Atlantic seesaw response during the last deglaciation, Nature, 457, 1097–1102, 2009.

Blunier, T., Chappellaz, J., Schwander, J., Dällenbach, A., Stauffer, B., Stocker, T., Raynaud, D., Jouzel, J., Clausens, H., Hammer, C., and Johnsen, S.: Asynchrony of Antarctic and Greenland climate change during the last glacial period, Nature, 394, 739–743, 1998.

Bond, G. and Lotti, R.: Iceberg discharges into the North Atlantic on millennial time scales during the last glaciation, Science, 267, 1005–1010, 1995.

Braun, H., Ditlevsen, P., and Chialvo, D. R.: Solar forced Dansgaard-Oeschger events and their phase relation with solar proxies, Geophys. Res. Lett, 35, L06703, doi:10.1029/2008GL033414, 2008.

Broccoli, A., Dahl, K., and Stouffer, R.: Response of the ITCZ to Northern hemisphere cooling, Geophys. Res. Lett., 33, L01702, doi:10.1029/2005GL024546, 2006.

Brovkin, V., Ganopolski, A., and Svirezhev, Y.: A continuous climate-vegetation classification for use in climate-biosphere studies, Ecol. Model., 101, 251–261, 1997.

Cane, M. A., Kamenkovich, V. M., and Krupitsky, A.: On the Utility and Disutility of JEBAR, J. Phys. Oceanogr., 28, 519–526, 1998.

Capron, E., Landais, A., Chappellaz, J., Schilt, A., Buiron, D., Dahl-Jensen, D., Johnsen, S. J., Jouzel, J., Lemieux-Dudon, B., Loulergue, L., Leuenberger, M., Masson-Delmotte, V., Meyer, H., Oerter, H., and Stenni, B.: Millennial and sub-millennial scale climatic variations recorded in polar ice cores over the last glacial period, Clim. Past, 6, 345–365, doi:10.5194/cp-6-345-2010, 2010.

Chapman, M. and Shackleton, N.: Global ice-volume fluctuations, North Atlantic ice-rafting events, and deep-ocean circulation changes between 130 and 70 ka, Geology, 27, 795—98, doi:10.1130/0091-7613, 1999.

Dansgaard, W., Johnsen, S., and Clausen, H.: Evidence for general instability of past climate from a 250-kyr ice-core record, Nature, 364, 218–220, 1993.

Deplazes, G., Lückge, A., Peterson, L., Timmermann, A., Hamann, Y., Hughen, K., Röh, U., Laj, C., Cane, M., Sigman, D., and Haug, G.: Links between tropical rainfall and North Atlantic climate during the last glacial period, Nat. Geosci., 6, 213–217, 2013.

Dokken, T., Nisancioglu, K., Li, C., Battisti, D., and Kissel, C.: Dansgaard–Oeschger cycles: Interactions between ocean and sea ice intrinsic to the Nordic seas, Paleoceanography, 28, 491–502, 2013.

Elliot, M., Labeyrie, L., Bond, G., Cortijo, E., Turon, J.-L., Tisnerat, N., and Duplessy, J.-C.: Millennial-scale iceberg discharges in the Irminger Basin during the last glacial period: Relationship with the Heinrich events and the environmental settings, Paleoceanography, 13, 433–446, 1998.

Elliot, M., Labeyrie, L., and Duplessy, J.-C.: Changes in North Atlantic deep-water formation associated with the Dansgaard–Oeschger temperature oscillations (60–10 ka), Quaternary Sci. Rev., 21, 1153–1165, 2002.

Fleitmann, D., Cheng, H., Badertscher, S., Edwards, R., Mudelsee, M., Göktürk, O., Fankhauser, A., Pickering, R., Raible, C., Matter, A., Kramers, J., and Tüysüz, O.: Timing and climatic impact of Greenland interstadials recorded in stalagmites from northern Turkey, Geophys. Res. Lett., 36, L19707, doi:10.1029/2009GL040050, 2009.

Flückiger, J., Knutti, R., and White, J.: Oceanic processes as potential trigger and amplifying mechanisms for Heinrich events, Paleoceanography, 21, PA2014, doi:10.1029/2005PA001204, 2006.

Garcin, Y., Williamson, D., Taieb, M., Vincens, A., Mathé P.-E., and Majule, A.: Centennial to millennial changes in maar-lake deposition during the last 45,000 years in tropical Southern Africa (Lake Masoko, Tanzania), Palaeogeogr. Palaeocl., 239, 334–354, 2006.

Godfrey, J.: A Sverdrup model of the depth-integrated flow from the world ocean allowing for island circulations, Geophys. Astrophys. Fluid Dyn., 45, 89–112, 1989.

Goosse, H., Brovkin, V., Fichefet, T., Haarsma, R., Huybrechts, P., Jongma, J., Mouchet, A., Selten, F., Barriat, P.-Y., Campin, J.-M., Deleersnijder, E., Driesschaert, E., Goelzer, H., Janssens, I., Loutre, M.-F., Morales Maqueda, M. A., Opsteegh, T., Mathieu, P.-P., Munhoven, G., Pettersson, E. J., Renssen, H., Roche, D. M., Schaeffer, M., Tartinville, B., Timmermann, A., and Weber, S. L.: Description of the Earth system model of intermediate complexity LOVECLIM version 1.2, Geosci. Model Dev., 3, 603–633, doi:10.5194/gmd-3-603-2010, 2010.

Grousset, F., Labeyrie, L., Sinko, J., Cremer, M., Bond, G., Duprat, J., Cortija, E., and Huon, S.: Patterns of Ice-Rafted Detritus in the Glacial North Atlantic (40–55° N), Paleoceanography, 8, 175–192, 1993.

Grousset, F., Pujol, C., Labeyrie, L., Auffret, G., and Boelaert, A.: Were the North Atlantic Heinrich events triggered by the behavior of the European ice sheets?, Geology, 28, 123–126, 2000.

Grousset, F., Cortijo, E., Huon, S., Hervé, L., Richter, T., Burdloff, D., Duprat, J., and Weber, O.: Zooming in on Heinrich layers, Paleoceanography, 16, 240–259, 2001.

Harrison, S. and Sánchez-Goñi, M.: Global patterns of vegetation response to millennial-scale variability and rapid climate change during the last glacial period, Quaternary Sci. Rev., 29, 2957–2980, 2010.

Heinrich, H.: Origin and consequences of cyclic ice rafting in the northeast Atlantic Ocean during the past 130,000 years, Quatern. Res., 29, 142–152, 1988.

Hemming, S.: Heinrich events: Massive late Pleistocene detritus layers of the North Atlantic and their global climate imprint, Rev. Geophys., 42, RG1005, doi:10.1029/2003RG000128, 2004.

Hodell, D., Anselmetti, F., Ariztegui, D., Brenner, M., Curtis, J., Gilli, A., Grzesik, D., Guilderson, T., Müller, A., Bush, M., Correa-Metrio, A., Escobar, J., and Kutterolf, S.: An 85-ka record of climate change in lowland Central America, Quaternary Sci. Rev., 27, 1152–1165, 2008.

Hodell, D., Evans, H., Channell, J., and Curtis, J.: Phase relationships of North Atlantic ice-rafted debris and surface-deep climate proxies during the last glacial period, Quaternary Sci. Rev., 29, 3875–3886, 2010.

Huber, C., Leuenberger, M., Spahni, R., Flückiger, J., Schwander, J., Stocker, T., Johnsen, S., Landais, A., and Jouzel, J.: Isotope calibrated Greenland temperature record over Marine Isotope Stage 3 and its relation to CH_4, Earth Planet. Sc. Lett., 243, 504–519, 2006.

Jackson, C. S., Marchal, O., Liu, Y., Lu, S., and Thompson, W. G.: A box-model test of the freshwater forcing hypothesis of abrupt climate change and the physics governing ocean stability, Paleoceanography, 25, PA4222, doi:10.1029/2010PA001936, 2010.

Kageyama, M., Merkel, U., Otto-Bliesner, B., Prange, M., Abe-Ouchi, A., Lohmann, G., Ohgaito, R., Roche, D. M., Singarayer, J., Swingedouw, D., and Zhang, X.: Climatic impacts of fresh water hosing under Last Glacial Maximum conditions: a multi-model study, Clim. Past, 9, 935–953, doi:10.5194/cp-9-935-2013, 2013.

Kanner, L., Burns, S., Cheng, H., and Edwards, R. L.: High-Latitude Forcing of the South American Summer Monsoon During the Last Glacial, Science, 335, 570–573, 2013.

Kissel, C., Laj, C., Piotrowski, A., Goldstein, S., and Hemming, S.: Millennial-scale propagation of Atlantic deep waters to the glacial Southern Ocean, Paleoceanography, 23, PA2102, doi:10.1029/2008PA001624, 2008.

Krebs, U. and Timmermann, A.: Tropical air-sea interactions accelerate the recovery of the Atlantic Meridional Overturning Circulation after a major shutdown, J. Climate, 20, 4940–4956, 2007.

Li, C., Battisti, D., Schrag, D., and Tziperman, E.: Abrupt climate shifts in Greenland due to displacements of the sea ice edge, Geophys. Res. Lett., 32, L19702, doi:10.1029/2005GL023492, 2005.

Li, C., Battisti, D., and Bitz, C.: Can North Atlantic Sea Ice Anomalies Account for Dansgaard-Oeschger Climate Signals?, J. Climate, 23, 5457–5475, 2010.

Marcott, S., Clark, P., Padman, L., Klinkhammer, G., Springer, S., Liu, Z., Otto-Bliesner, B., Carlson, A., Ungerer, A., Padman, J., He, F., Cheng, J., and Schmittner, A.: Ice-shelf collapse from subsurface warming as trigger for Heinrich events, P. Natl. Acad. Sci., 108, 13415–13419, doi:10.1073/pnas.1104772108, 2011.

Margari, V., Skinner, L. C., Tzedakis, P. C., Ganopolski, A., Vautravers, M., and Shackleton, N. J.: The nature of millennial-scale climate variability during the past two glacial periods, Nat. Geosci., 3, 127–131, 2010.

Marshall, S. J. and Koutnik, M.: Ice sheet action versus reaction: Distinguishing between Heinrich events and Dansgaard-Oeschger cycles in the North Atlantic, Paleoceanography, 21, PA2021, doi:10.1029/2005PA001247, 2006.

Martrat, B., Grimalt, J., Shackleton, N., de Abreu, L., Hutterli, M., and Stocker, T.: Four climate cycles of recurring deep and surface water destabilizations on the Iberian margin, Science, 317, 502–507, 2007.

Masson-Delmotte, V., Schulz, M., Abe-Ouchi, A., Beer, J., Ganopolski, A., Rouco, J. F. G., Jansen, E., Lambeck, K., Luterbacher, J., Naish, T., Osborn, T., Otto-Bliesner, B., Quinn, T., Ramesh, R., Rojas, M., Shao, X., and Timmermann, A.: Climate Change 2013: The Physical Science Basis. Contribution of Working Group I to the Fifth Assessment Report of the Intergovernmental Panel on Climate Change, in: Information from Paleoclimate Archives, Cambridge Univ. Press., Cambridge, UK and New York, NY, USA, 2013.

Menviel, L., Timmermann, A., Mouchet, A., and Timm, O.: Meridional reorganizations of marine and terrestrial productivity during Heinrich events, Paleoceanography, 23, PA1203, doi:10.1029/2007PA001445, 2008.

Mignot, J., Ganopolski, A., and Levermann, A.: Atlantic subsurface temperatures: Response to a shutdown of the overturning circulation and consequences for its recovery, J. Climate, 20, 4884–4898, 2007.

Mosblech, N., Bush, M., Gosling, W., Hodell, D., Thomas, L., van Calsteren, P., Correa-Metrio, A., Valencia, B., Curtis, J., and vanWoesik, R.: North Atlantic forcing of Amazonian precipitation during the last ice age, Nat. Geosci., 5, 817–820, doi:10.1038/NGEO1588, 2012.

Mouchet, A.: A 3D model of ocean biogeochemical cycles and climate sensitivity studies, Ph.D. thesis, Université de Liège, Lìege, Belgium, 2011.

Nave, S., Labeyrie, L., Gherardi, J., Caillon, N., Cortija, E., Kissel, C., and Abrantes, F.: Primary productivty response to Heinrich events in the North Atlantic Ocean and Norwegian Sea, Paleoceanography, 22, PA3216, doi:10.1029/2006PA001335, 2007.

Obrochta, S., Yokohama, Y., Moren, J., and Crowley, T.: Conversion of GISP2-based sediment core age models to the GICC05 extended chronology, Quatern. Geochronol., 20, 1–7, doi:10.1016/j.quageo.2013.09.001, 2014.

Petersen, S., Schrag, D., and Clark, P.: A new mechanism for Dansgaard-Oeschger cycles, Paleoceanography, 28, 1–7, doi:10.1029/2012PA002364, 2013.

Rashid, H., Hesse, R., and Piper, D.: Evidence for an additional Heinrich event between H5 and H6 in the Labrador Sea, Paleoceanography, 18, 1077, doi:10.1029/2003PA000913, 2003.

Sánchez-Goñi, M. and Harrison, S.: Millennial-scale climate variability and vegetation changes during the Last Glacial: Concepts and terminology, Quaternary Sci. Rev., 29, 2823–2827, 2010.

Sánchez-Goñi, M., Cacho, I., Turon, J.-L., Guiot, J., Sierro, F., Peypouquet, J.-P., Grimalt, J., and Shackleton, N.: Synchroneity between marine and terrestrial responses to millennial scale climatic variability during the last glacial period in the Mediterranean region, Clim. Dynam., 19, 95–105, 2002.

Sarkisyan, A. and Ivanov, V. F.: Joint effect of baroclinicity and bottom relief as an important factor in the dynamics of sea currents, Izv. Acad. Sci. USSR, Atmos. Ocean. Phys., 7, 116–124, 1971.

Sarnthein, M., Eystein, J., Weinelt, M., Arnold, M., Duplessy, J., Erlenkeuser, H., Flatoy, A., Johannessen, G., Johannessen, T., Jung, S., Koc, N., Labeyrie, L., Maslin, M., Pflaumann, U., and Schulz, H.: Variations in Atlantic surface ocean paleoceanography, 50–80° N: A time-slice record of the last 30,000 years, Paleoceanography, 10, 1063–1094, 1995.

Sarnthein, M., Stattegger, K., Dreger, D., Erlenkeuser, H., Grootes, P., Haupt, B., Jung, S., Kiefer, T., ad U. Pflaumann, W. K., Schäfer-Neth, C., Schulz, H., Schulz, M., Seidov, D., Simstich, J., van Kreveld, S., Vogelsang, E., Völker, A., and Weinelt, M.: The Northern North Atlantic: A Changing Environment, in: Fundamental Modes and Abrupt Changes in North Atlantic Circulation and Climate over the last 60 ky – Concepts, Reconstruction and Numerical Modeling, Springer, Berlin, 365–410, 2001.

Schönfeld, J., Zahn, R., and de Abreu, L.: Surface to deep water response to rapid climate changes at the Western Iberian Margin, Global Planet. Change, 36, 237–264, 2003.

Schulz, M., Paul, A., and Timmermann, A.: Relaxation oscillators in concert: A framework for climate change at millennial timescales during the late Pleistocene, Geophys. Res. Lett., 29, 2193, doi:10.1029/2002GL016144, 2002.

Scnoeckx, H., Grousset, F., Revel, M., and Boelaert, A.: European contribution of ice-rafted sand to Heinrich layers H3 and H4, Mar. Geol., 158, 197–208, 1999.

Shaffer, G., Olsen, S., and Bjerrum, C.: Ocean subsurface warming as a mechanism for coupling Dansgaard-Oeschger climate cycles and ice-rafting events, Geophys. Res. Lett., 31, L24202, doi:10.1029/2004GL020968, 2004.

Siddall, M., Rohling, E., Almogi-Labin, A., Hemleben, C., Meischner, D., Schmelzer, I., and Smeed, D.: Sea-level fluctuations during the last glacial cycle, Nature, 423, 853–858, 2003.

Steffensen, J., Andersen, K., Biglcr, M., Clausen, H., Dahl-Jensen, D., Fischer, H., Goto-Azuma, K., Hansson, M., Johnsen, S., Jouzel, J., Masson-Delmotte, V., Popp, T., Rasmussen, S., Röthlisberger, R., Ruth, U., Stauffer, B., Siggaàrd-Andersen, M.-L., Sveinbjörnsdóttir, A., Svensson, A., and White, J.: High-Resolution Greenland Ice Core Data Show Abrupt Climate Change Happens in Few Years, Science, 321, 680–684, 2008.

Stenni, B., Buiron, D., Frezzotti, M., Albani, S., Barbante, C., Bard, E., Barnola, J., Baroni, M., Baumgartner, M., Bonazza, M., Capron, E., Castellano, E., Chappellaz, J., Delmonte, B., Falourd, S., Genoni, L., Iacumin, P., Jouzel, J., Kipfstuhl, S., Landais, A., Lemieux-Dudon, B., Maggi, V., Masson-Delmotte, V., Mazzola, C., Minster, B., Montagnat, M., Mulvaney, R., Narcisi, B., Oerter, H., Parrenin, F., Petit, J., Ritz, C., Scarchilli, C., Schilt, A., Schüpbach, S., Schwander, J., Selmo, E., Severi, M., Stocker, T., and Udisti, R.: Expression of the bipolar see-saw in Antarctic climate records during the last deglaciation, Nat. Geosci., 4, 46–49, 2011.

Stocker, T.: The seesaw effect, Science, 282, 61–62, 1998.

Stocker, T. and Johnsen, S.: A minimum thermodynamic model for the bipolar seesaw, Paleoceanography, 18, 1087, doi:10.1029/2003PA000920, 2003.

Stouffer, R., Yin, J., Gregory, J., Dixon, K., Spelman, M., Hurlin, W., Weaver, A., Eby, M., Flato, G., Hasumi, H., Hu, A., Jungclaus, J., Kamenkovich, I., Levermann, A., Montoya, M., Murakami, S., Nawrath, S., Oka, A., Peltier, W., Robitaille, D., Sokolov, A., Vettoretti, G., and Weber, S.: Investigating the causes of the response of the thermohaline circulation to past and future climate changes, J. Climate, 19, 1365–1387, 2006.

Stouffer, R., Seidov, D., and Haupt, B.: Climate response to external sources of freshwater: North Atlantic versus the Southern Ocean, J. Climate, 20, 436–448, 2007.

Svensson, A., Andersen, K., Bigler, M., Clausen, H., Dahl-Jensen, S. D., Johnsen, S., Muscheler, R., Rasmussen, S., Röthlisberger, R., Steffensen, J., and Vinther, B.: The Greenland Ice Core Chronology 2005, 15–42 ka, Part 2: Comparison to other records, Quaternary Sci. Rev., 25, 3258–3267, 2006.

Timm, O. and Timmermann, A.: Simulation of the last 21,000 years using accelerated transient boundary conditions, J. Climate, 20, 4377–4401, 2007.

Timm, O., Timmermann, A., Abe-Ouchi, A., and Segawa, T.: On the definition of paleo-seasons in transient climate simulations, Paleoceanography, 23, PA2221, doi:10.1029/2007PA001461, 2008.

Timmermann, A., Schulz, M., Gildor, H., and Tziperman, E.: Coherent resonant millennial-scale climate oscillations triggered by massive meltwater pulses, J. Climate, 16, 2569–2585, 2003.

Timmermann, A., An, S. I., Krebs, U., and Goosse, H.: ENSO suppression due to a weakening of the North Atlantic thermohaline circulation, J. Climate, 18, 3122–3139, 2005a.

Timmermann, A., Krebs, U., Justino, F., Goosse, H., and Ivanochko, T.: Mechanisms for millennial-scale global synchronization during the last glacial period, Paleoceanography, 20, PA4008, doi:10.1029/2004PA001090, 2005b.

Timmermann, A., Okumura, Y., An, S.-I., Clement, A., Dong, B., Guilyardi, E., Hu, A., Jungclaus, J., Krebs, U., Renold, M., Stocker, T., Stouffer, R., Sutton, R., Xie, S.-P., and Yin, J.: The influence of shutdown of the Atlantic meridional overturning circulation on ENSO, J. Climate, 19, 4899–4919, 2007.

Timmermann, A., Menviel, L., Okumura, Y., Schilla, A., Merkel, U., Hu, A., Otto-Bliesner, B., and Schulz, M.: Towards a quantitative understanding of millennial-scale Antarctic Warming events, Quaternary Sci. Rev., 29, 74–85, doi:10.1016/j.quascirev.2009.06.021, 2009a.

Timmermann, A., Timm, O., Stott, L., and Menviel, L.: The roles of CO_2 and orbital forcing in driving southern hemispheric temperature variations during the last 21 000 yr, J. Climate, 22, 1626–1640, 2009b.

van Kreveld, S., Samthein, M., Erlenkeuser, H., Grootes, P., Jung, S., Nadeau, M., Pflaumann, U., and Voelker, A.: Potential links between surging ice sheets, circulation changes, and the Dansgaard–Oeschger cycles in the Irminger Sea, 60–18 kyr, Paleoceanography, 15, 425–442, 2000.

Vidal, L., Labeyrie, L., Cortijo, E., Arnold, M., Duplessy, J., Michel, E., Becque, S., and van Weering, T.: Evidence for changes in the North Atlantic Deep Water linked to meltwater surges during the Heinrich events, Earth Planet. Sc. Lett., 146, 13–27, 1997.

Voelker, A., Grootes, P. M., Nadeau, M.-J., and Sarnthein, M.: Radiocarbon levels in the Iceland sea from 25–53 kyr and their link to the Earth's magnetic field intensity, Radiocarbon, 42, 437–452, 2000.

Wang, X., Auler, A., Edwards, R., Cheng, H., Ito, E., Wang, Y., Kong, X., and Solheid, M.: Millennial-scale precipitation changes in southern Brazil over the past 90,000 years, Geophys. Res. Lett., 34, L23701, doi:10.1029/2007GL031149, 2007.

Wang, Y., Cheng, H., Edwards, R., An, Z., Wu, J., Shen, C., and Dorale, J.: A high-resolution absolute-dated Late Pleistocene monsoon record from Hulu Cave, China, Science, 294, 2345–2348, 2001.

Zahn, R., Schönfeld, J., Kudrass, H.-R., Park, M.-H., Erlenkeuser, H., and Grootes, P.: Thermohaline instability in the North Atlantic during meltwater events: Stable isotope and ice-rafted detritus records from core SO75-26KL, Portuguese margin, Paleoceanography, 12, 696–710, 1997.

2

Water mass evolution of the Greenland Sea since late glacial times

M. M. Telesiński[1], R. F. Spielhagen[1,2], and H. A. Bauch[1,2]

[1]GEOMAR Helmholtz Centre for Ocean Research Kiel, Wischhofstrasse 1–3, 24148 Kiel, Germany
[2]Academy of Sciences, Humanities, and Literature, 53151 Mainz, Germany

Correspondence to: M. M. Telesiński (mtelesinski@geomar.de)

Abstract. Four sediment cores from the central and northern Greenland Sea basin, a crucial area for the renewal of North Atlantic deep water, were analyzed for planktic foraminiferal fauna, planktic and benthic stable oxygen and carbon isotopes as well as ice-rafted debris to reconstruct the environmental variability in the last 23 kyr. During the Last Glacial Maximum, the Greenland Sea was dominated by cold and sea-ice bearing surface water masses. Meltwater discharges from the surrounding ice sheets affected the area during the deglaciation, influencing the water mass circulation. During the Younger Dryas interval the last major freshwater event occurred in the region. The onset of the Holocene interglacial was marked by an increase in the advection of Atlantic Water and a rise in sea surface temperatures (SST). Although the thermal maximum was not reached simultaneously across the basin, benthic isotope data indicate that the rate of overturning circulation reached a maximum in the central Greenland Sea around 7 ka. After 6–5 ka a SST cooling and increasing sea-ice cover is noted. Conditions during this so-called "Neoglacial" cooling, however, changed after 3 ka, probably due to enhanced sea-ice expansion, which limited the deep convection. As a result, a well stratified upper water column amplified the warming of the subsurface waters in the central Greenland Sea, which were fed by increased inflow of Atlantic Water from the eastern Nordic Seas. Our data reveal that the Holocene oceanographic conditions in the Greenland Sea did not develop uniformly. These variations were a response to a complex interplay between the Atlantic and Polar water masses, the rate of sea-ice formation and melting and its effect on vertical convection intensity during times of Northern Hemisphere insolation changes.

1 Introduction

The Nordic Seas are an important region for the global oceanic system. First of all, they are the main gateway between the Arctic and North Atlantic oceans (Hansen and Østerhus, 2000). They also play a fundamental role in the overturning circulation being one of the deep water formation regions (Marshall and Schott, 1999). Paleoceanographic studies in this area are crucial to improve our understanding of the pace and amplitude of natural variability during the last glacial–interglacial transition and within the Holocene. While a significant number of detailed studies focuses on the eastern part of the region, along the North Atlantic Current (NAC) flow (e.g., Hald et al., 2007; Risebrobakken et al., 2011), less effort has been devoted to its central and western parts (e.g., Fronval and Janssen, 1997; Bauch et al., 2001). Problems with the accessibility due to the ice cover and low sedimentation rates (Nørgaard-Pedersen et al., 2003; Telesiński et al., 2013), which do not allow high resolution studies, are among the main reasons here.

Recently, Telesiński et al. (2013) presented a new record from the central Greenland Sea that allowed studying the oceanographic changes since the late glacial (22.3 ka) in a relatively high temporal resolution. That study revealed significant variability of the oceanic environment on multicentennial to multimillennial timescales. Although the record was generally in agreement with earlier studies, it also revealed some unusual features such as, e.g., an extreme freshwater-related planktic low-δ^{18}O spike during the deglaciation and microfossil evidence for a late Holocene warming. Here we now correlate and compare that record with three other sediment cores from the northern Greenland Sea and with other paleoceanographic archives from the Nordic Seas to reconstruct the paleoceanography on a larger regional scale. Furthermore, subsurface temperature reconstructions and a first high-resolution benthic stable isotope record from the Greenland Sea are presented and allow assessing the spatial range of variability found in the central Greenland Sea and the history of the overturning circulation in the area.

2 Study area

The Nordic Seas constitute the only deep-water connection between the North Atlantic and the Arctic oceans (Fig. 1). Relatively warm and saline ($T \sim 6$–$11\,^\circ$C, $S > 35$) Atlantic Water (AW) flows north along the Norwegian, Barents Sea and Svalbard continental margins and enters the Arctic through the Fram Strait and Barents Sea. In the west, cold, low-saline ($< 0\,^\circ$C, < 34.4) Polar Water (PW) flows south through the Fram Strait and along the Greenland continental margin to enter the North Atlantic through the Denmark Strait (Rudels et al., 1999). The strong gradient between these two main surface water masses makes the Nordic Seas sensitive to climatic changes. The central part of the Nordic Seas is the domain of Arctic Water (ArW), a result of PW and AW mixing. ArW is separated from PW by the Polar Front and from AW by the Arctic Front (Swift, 1986).

The vertical structure of the water column in the central Greenland Sea consists of three layers. At the surface, there is a thin layer of Arctic Surface Water originating from the East Greenland Current (EGC). Underneath, a layer of Atlantic Intermediate Water exists, which is supplied from the NAC. The weakly stratified Greenland Sea Deep Water, a product of deep convection, is found below (Marshall and Schott, 1999).

The Nordic Seas are one of the areas where deep water convection and the formation of North Atlantic Deep Water (NADW) take place today (e.g., Rudels and Quadfasel, 1991; Marshall and Schott, 1999). The western branches of the NAC and the eastern branches of the EGC create a cyclonic circulation in the Greenland Sea and lead to doming of the upper water layers. As the two water masses mix, they increase their density and sink to the bottom (Hansen and Østerhus, 2000). Subsequently, the water leaves the Nordic Seas as the Denmark Strait and Iceland-Scotland Overflow Waters.

Sea ice plays an important preconditioning role in the Greenland Sea compared to other convectional areas. In early winter, the formation of sea ice leads to brine rejection. The surface layer increases its density and sinks to about 150 m by mid-January. The sea-ice cover forms a wedge (Is Odden) extending far to the northeast, also over the Vesterisbanken area. Preconditioning continues later in the winter, with mixed-layer deepening in the ice-free area (Nord Bukta) to 300–400 m, induced by strong winds blown over the ice. Typically in March, near-surface densities are high enough to develop deep convection (down to > 2000 m) in

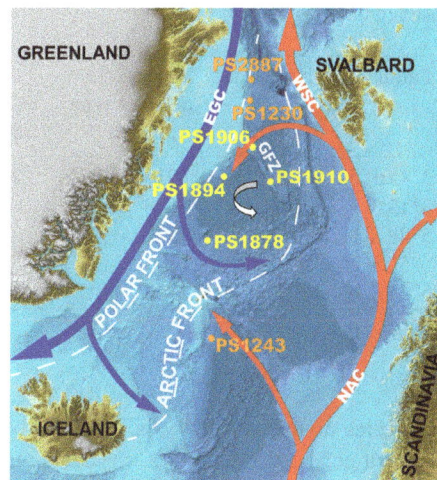

Fig. 1. Present day surface water circulation in the Nordic Seas. Cores used in this study are marked by yellow dots; other cores mentioned in text are marked by orange dots. Red arrows indicate Atlantic Water, blue arrows – Polar Water, white broken lines – oceanographic fronts. White arrow – present-day deep convection (Marshall and Schott, 1999). EGC – East Greenland Current, NAC – North Atlantic Current, WSC – West Spitsbergen Current, GFZ – Greenland Fracture Zone. Bathymetry from The International Bathymetric Chart of the Arctic Ocean (http://www.ibcao.org, 2012).

the Greenland Sea, if the meteorological conditions are favorable (Marshall and Schott, 1999).

At present, the sites investigated in this study are all located within the ArW domain. A detailed description of site PS1878 was given by Telesiński et al. (2013). The three sites from the northern Greenland Sea, PS1894, PS1906 and PS1910, are located on the Greenland continental slope, on the northern and on the southern part of the Greenland Fracture Zone crest, respectively.

3 Material and methods

The sediment cores used in this study were retrieved during the ARK-VII/1 expedition of RV *Polarstern* in 1990 (Fig. 1). Core PS1878 is compiled from a giant box core PS1878-2 and a kasten core PS1878-3 (Telesiński et al., 2013), whereas the three others are giant box cores (Table 1). All cores consisted of brown to olive grey sediments of clay to silty sand. They were sampled continuously every 1 cm. Additionally, surface sediments of cores PS1894, PS1906 and PS1910 were collected. Further preparation included freeze-drying, wet-sieving with deionized water through a 63 μm mesh, and dry-sieving into size fractions using 100, 125, 250, 500 and 1000 μm sieves. Each size fraction was weighed.

In representative splits (> 300 specimens) of the 100–250 μm size fraction planktic foraminifera were counted. Samples containing less than 100 specimens were not used

Table 1. Cores used in the study.

Core	Latitude	Longitude	Water depth (m)	Core type	Core length (cm)
PS1878-2	73°15.1′ N	9°00.9′ W	3038	BC[a]	27
PS1878-3	73°15.3′ N	9°00.7′ W	3048	KC[b]	113
PS1894-7	75°48.8′ N	8°15.5′ W	1992	BC[a]	42
PS1906-1	76°50.5′ N	2°09.0′ W	2990	BC[a]	33
PS1910-1	75°37.0′ N	1°19.0′ E	2448	BC[a]	33

[a] BC – giant box core, [b] KC – kasten core.

for the relative species abundance analysis. The number of planktic foraminifera per 1 g dry sediment was calculated to serve as a semiquantitative proxy for bioproductivity.

Identification and counting of several mineral grain types > 250 μm was used as a proxy for the intensity of ice-rafting and the identification of tephra layers. As ice-rafted debris (IRD) we interpret all lithic grains > 250 μm, except for unweathered volcanic glass. In the high latitudes, such coarse particles can be transported into a deep ocean basin preferentially by icebergs while sea ice mainly transports finer material (Clark and Hanson, 1983; Nürnberg et al., 1994).

For the analysis of stable oxygen and carbon isotopes, specimens of the planktic foraminiferal species *Neogloboquadrina pachyderma* (sin.) (all cores) and two benthic species – the epibenthic *Cibicidoides wuellerstorfi* and the shallow infaunal *Oridorsalis umbonatus* (cores PS1894, PS1910 and PS1878) – were used. Because of departures from isotopic calcite equilibrium, the measured $\delta^{18}O$ values of these two species were corrected by +0.64 and +0.36‰, respectively (cf. Duplessy et al., 1988). Twenty-five specimens were picked from the 125–250 μm (*N. pachyderma* (sin.) and *O. umbonatus*) and 250–500 μm (*C. wuellerstorfi*) size fractions. All stable isotope analyses were carried out in the isotope laboratories of GEOMAR Helmholtz Centre for Ocean Research Kiel and the University of Kiel on Finnigan MAT 251 and Thermo MAT 253 mass spectrometers. Results are expressed in the δ notation referring to the PDB (Pee Dee Belemnite) standard and are given as $\delta^{18}O$ and $\delta^{13}C$ with an analytical accuracy of < 0.06 and < 0.03‰, respectively.

Absolute summer subsurface temperatures (100 m water depth) were calculated at site PS1878 between 15 and 0 ka using transfer functions based on a modern training set from the Arctic (Husum and Hald, 2012) and the C2 software, version 1.7.2 (Juggins, 2011). A weighted average partial least-squares statistical model with three components (WA-PLS C3) and leave-one-out ("jack knifing") cross validation was used. The root mean-squared error of prediction is 0.52 °C. Unlike Husum and Hald (2012), who used the > 100 μm size fraction, we ran the transfer function using the 100–250 μm size fraction. Although the coarser sediments contained relatively few foraminifera, we acknowledge that this might have slightly biased the results. Further,

Table 2. AMS ^{14}C measurements and their calibrated ages for the cores used in the study (BP – before present).

Lab. no.	Depth (cm)	[14]C age ± standard deviation	Calibrated age (yr BP)
Core PS1878-2			
Poz-45376	0.5	775 ± 35	426
Poz-45377	12.5	3300 ± 40	3143
Core PS1878-3			
Poz-45378	11.5	3295 ± 35	3139
Poz-45380	19.5	4525 ± 35	4746
Poz-54381	25.5	5580 ± 50	5961
Poz-54382	30.5	6760 ± 50	7295
Poz-45384	39.5	8410 ± 60	9028
Poz-45385	58.5	11 100 ± 60	12 613
KIA 47284	95.5	16 620 ± 110	19 266
Core PS1894-7			
KIA 7088	0.5	3845 ± 40	3794
KIA 47258	5.5	5390 ± 35	5773
KIA 7089	9.5	5745 ± 40	6174
KIA 47259	16.5	8075 ± 45	8528
KIA 7090	21.5	8910 ± 55	9564
KIA 7091	35.5	14 430 ± 70	17 051
Core PS1906-1			
KIA 7084	4.5	4360 ± 30	4482
KIA 7083	11.5	7965 ± 40	8420
KIA 7082	22.5	17 040 ± 80	19 731
KIA 7081	32.5	19 130 ± 90	22 334
Core PS1910-1			
KIA 44390	0.5	2655 ± 30	2336
Poz-45386	4.5	4820 ± 35	5122
Poz-45387	11.5	6950 ± 50	7457
KIA 44393	17.5	11 340 ± 50	12 794
Poz-45388	30.5	1680 ± 100	19 625

reconstructed temperatures below 2 °C are considered to be uncertain as the modern training set does contain very few data points below 2 °C (Husum and Hald, 2012).

4 Chronology

AMS ^{14}C datings were performed on monospecific samples of *N. pachyderma* (sin.) (Table 2). All radiocarbon ages were corrected for a reservoir age of 400 yr, calibrated using Calib Rev 6.1.0 software (Stuiver and Reimer, 1993) and the Marine09 calibration curve (Reimer et al., 2009) and are given in thousand calendar years before 1950 AD (ka).

The records cover the last ca. 20–23 kyr. The three box cores from the northern Greenland Sea have average

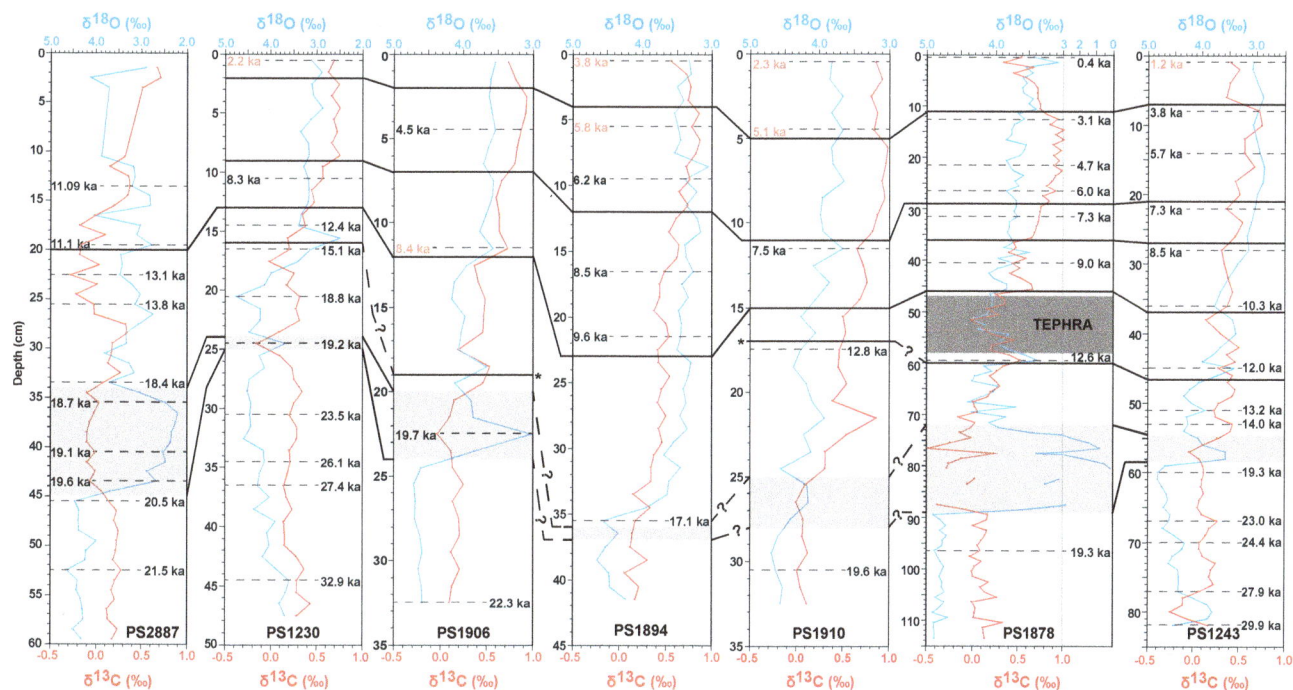

Fig. 2. Planktic oxygen and carbon stable isotope records of cores from the Nordic Seas and suggested correlation. Calibrated AMS ^{14}C dates are shown. Dates excluded from the correlation are marked in pale red. Light grey shadings indicate the light carbon and oxygen isotope excursions interpreted as freshwater discharges, marking the onset of the deglaciation.

sedimentation rates of 1.5–2.0 cm kyr^{-1}. These low rates, together with bioturbation and uncertain reservoir ages, make age models of these records unreliable if based only on ^{14}C datings. This is best illustrated by relatively old ages yielded from the surface samples of these cores (2.3–3.8 ka). However, the surface sample of core PS1878 yielded a younger age (0.426 ka) and contained recent sediments (Telesiński et al., 2013). Therefore we assume that sedimentation in the entire study area did not terminate in the late Holocene. To account for the apparent inaccuracy of part of the AMS ^{14}C dates we attempted to improve the consistency of the age models of these cores by correlating the stable isotope data (and, in a few cases, also other proxies) and using linear interpolation between correlated points and reliable ^{14}C-dated samples. In addition to our own data, we also used three nearby records of comparable sedimentation rates, time range and water depths. These include cores PS2887 (Nørgaard-Pedersen et al., 2003) as well as PS1230 from the western Fram Strait and PS1243 from the SW Norwegian Sea (Bauch et al., 2001). As the base for the correlation we used core PS1878, which has the highest temporal resolution and a reliable chronological framework based on ^{14}C datings in the younger part of the record (Fig. 2). Due to poorer ^{14}C age control and more speculative reservoir ages in the older part of the records, our improved age model is restricted to the last 15 kyr.

5 Results

5.1 Planktic foraminifera, ice-rafted detritus (IRD) and reconstructed subsurface temperatures

Four of our faunal records from the Greenland Sea show significantly different planktic foraminiferal abundances (Fig. 3), most likely due to different sedimentation rates. Therefore, absolute numbers of foraminiferal specimens in individual samples are not a meaningful proxy when cores are compared with one another. The records begin with relatively low abundances of the foraminiferal fauna strongly dominated by *N. pachyderma* (sin.) (Fig. 4, between ca. 23 and 12 ka), a polar species dwelling at water depths of ca. 50–200 m (Carstens et al., 1997). There are, however, a number of prominent, short-lived peaks of high foraminiferal abundance. They are most common and most prominent in core PS1878, supposedly due to its highest time resolution, but they are also noticeable in cores PS1906 and PS1894.

A significant early change among the faunal data is observed in core PS1894. Here, an increase to 20–30 % is found for the subpolar species *N. pachyderma* (dex.) and *Turborotalita quinqueloba* already around 17 ka. In the other cores a similar change is not noted until ca. 12 ka when both the percentages of subpolar species and the total abundance increase. Throughout the remaining part of the records the abundance stays high although significant variability can be observed. The portions of subpolar species remain high for

Planktic foraminifera/g

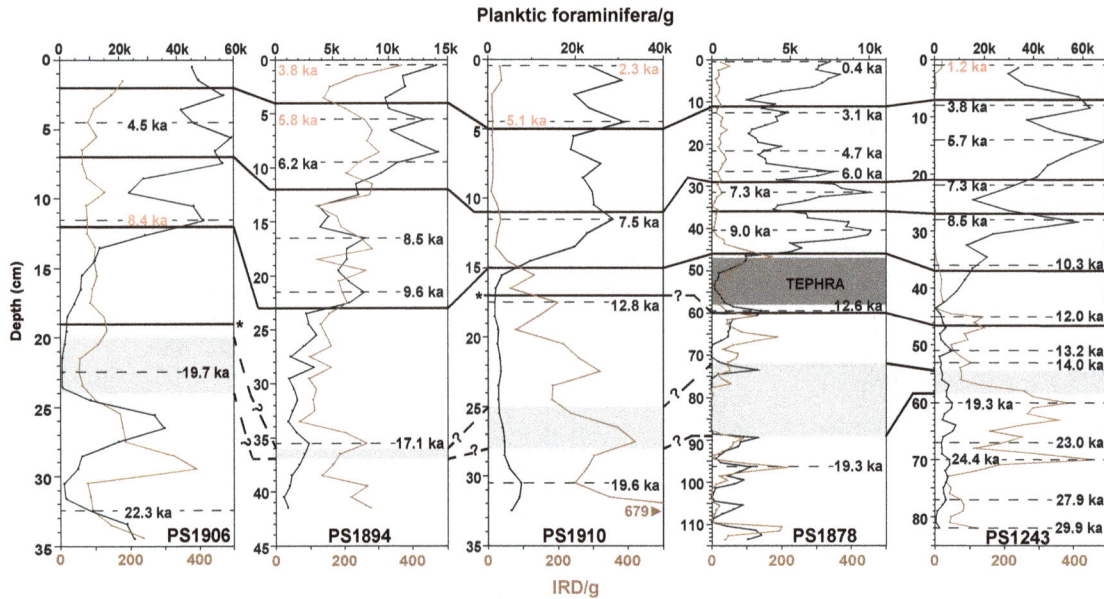

Fig. 3. Planktic foraminifera and IRD abundance (per 1 g dry sediment) of cores used in this study and core PS1243. Correlation and ages as in Fig. 2.

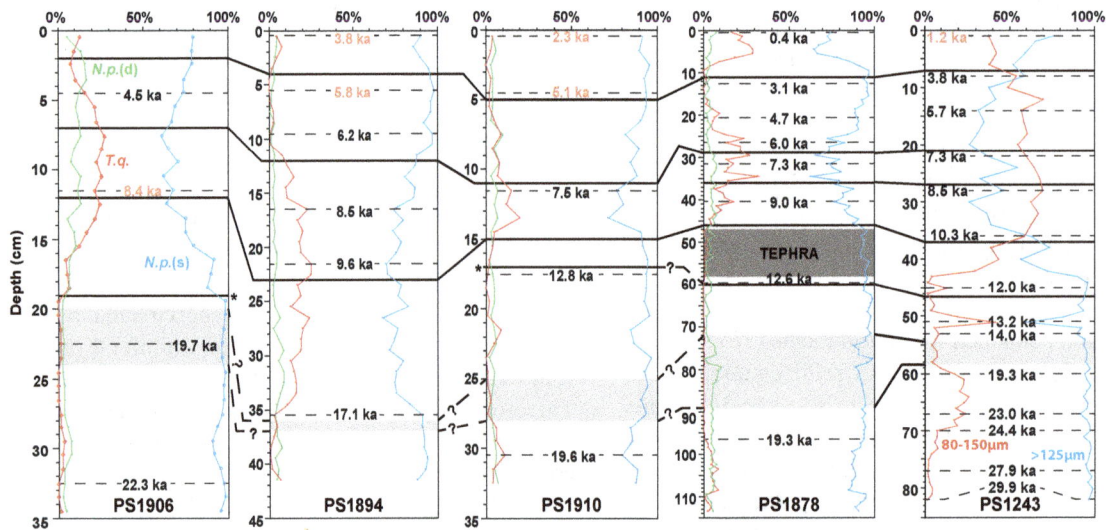

Fig. 4. Relative abundances of the three most common planktic foraminifera species in cores used in this study and core PS1243. *N.p.* (s) – *N. pachyderma* (sin.), *N.p.* (d) – *N. pachyderma* (dex.), *T.q.* – *T. quinqueloba*. Correlation and ages as in Fig. 2. Note the different size fractions used in core PS1243.

a few thousand years and then decrease gradually and un-simultaneously to reach pre-Holocene values (< 10–20 %) again after ca. 5 ka. A second, major increase can be observed after 3 ka in core PS1878 and, less clearly, PS1894. We did not find any significant signs of dissolution in the studied foraminifera. Both tests of robust *N. pachyderma* and more fragile subpolar species are generally well preserved throughout the cores.

As expected, the IRD records show high amounts of coarse lithogenic grains in the glacial part and low numbers during

the Holocene (Fig. 3). Only the IRD content of core PS1894 remains relatively high throughout the entire record with slightly lower values between ca. 17 and 10 ka. In core PS1894, as well as in the lower part of cores PS1906 and PS1878, the IRD content seems to be positively correlated with the foraminiferal abundance, while in core PS1910 and in the upper part of PS1906 and PS1878 these two proxies appear inversely correlated.

The subsurface temperature record of core PS1878 shows values steadily increasing from around 2 °C around 15 ka to a maximum of 3–3.5 °C between 8 and 5.7 ka (Fig. 7). Thereafter it decreases stepwise to values around 2 °C between 3.8 and 2.3 ka. Subsequently the record shows rapidly increasing temperatures with a peak value of ca. 3.5 °C at 1.3 ka and a decrease to ca. 3 °C until today.

5.2 Stable isotopes

The planktic oxygen isotope records start with relatively heavy and stable values of 4.3–4.9‰ (Fig. 2). After ca. 18 ka, sharp peaks of very light values (min. 0.15‰) occur (most pronounced in cores PS1906 and PS1878). Similar peaks are also found in cores PS1230, PS1243 (Bauch et al., 2001) and PS2887 (Nørgaard-Pedersen et al., 2003) that we used for the correlation. A trend towards lower $\delta^{18}O$ values commences thereafter and lasts until the end of the record. A distinct, though irregular, variability can be observed within the trend (Figs. 2, 5).

The oldest part of all planktic carbon isotope records (> 18 ka) exhibits low and stable values around 0.0–0.3‰. Simultaneous with the light $\delta^{18}O$ peaks, the $\delta^{13}C$ values decrease slightly and a trend of increasing values commences thereafter. Around 7 ka the $\delta^{13}C$ values reach a high plateau of 0.7–1.0‰, which lasts until 3 ka and ends with a relatively sudden drop.

Because *O. umbonatus* and *C. wuellerstorfi* were partly absent in the lowermost parts of our cores, the benthic stable isotope records cover only the last 16 kyr (Fig. 6). The oxygen isotope ratios of both benthic species generally show a decreasing trend parallel to the planktic record with values ca. 0.7–1.0‰ heavier than those of *N. pachyderma* (sin.). The epibenthic (*C. wuellerstorfi*) $\delta^{13}C$ data follows the planktic $\delta^{13}C$ records in terms of the main trends, but values are 0.2–1.0 ‰ higher and changes are of lower amplitude. The only major exception is the youngest (< 3 ka) part of record PS1894 in which benthic $\delta^{13}C$ values continue to rise slightly while the planktic record decreases. All data sets are available from http://www.pangaea.de.

6 Discussion

6.1 Last Glacial Maximum (LGM)

The heavy $\delta^{18}O$ values of > 4.5‰ in the Greenland Sea planktic records (Fig. 2) are typical for the late LGM waters in the Nordic Seas and Fram Strait (e.g., Sarnthein et al., 1995; Nørgaard-Pedersen et al., 2003). The low foraminiferal abundance and species diversity (Figs. 3, 4) are evidence of a low biological productivity in the Greenland Sea during the LGM. The latter might be a result of a perennial sea-ice cover that would strongly limit the penetration of sunlight and reduce the growth of phytoplankton that the foraminifera feed on.

Low $\delta^{13}C$ values might suggest that the foraminifera lived in poorly ventilated water (cf. Duplessy et al., 1988), which

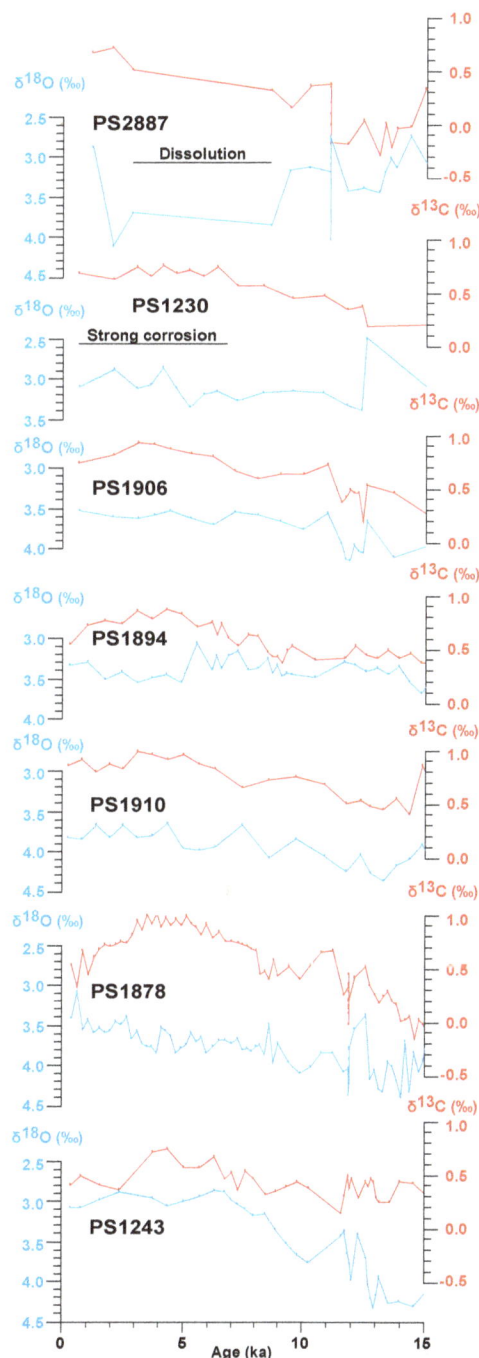

Fig. 5. Planktic oxygen and carbon stable isotope records of cores from the Nordic Seas plotted vs. age (since 15 ka).

seems obvious in a perennially ice-covered ocean. However, relatively high $\delta^{13}C$ values (> 0.7‰) are found at present also in the perennially ice-covered areas of the central Arctic Ocean (Spielhagen and Erlenkeuser, 1994). Therefore, we hesitate to relate the low $\delta^{13}C$ solely to the sea ice and/or strong stratification of the upper water layers. In addition, the carbon cycle in the glacial ocean may have been much

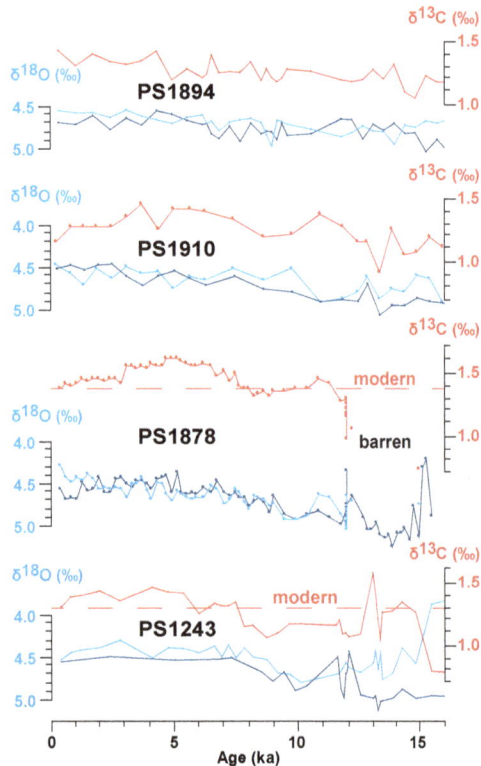

Fig. 6. Benthic oxygen (light and dark blue for *C. wuellerstorfi* and *O. umbonatus*, respectively) and carbon (red, *C. wuellerstorfi*) stable isotope records (in ‰ vs. PDB) of cores PS1894, PS1910, PS1878 and PS1243 vs. age (since 16 ka). Broken lines in PS1878 and PS1243 mark modern (core-top) $\delta^{13}C$ values of *C. wuellerstorfi* from the central Greenland Sea and site PS1243, respectively (Bauch and Erlenkeuser, 2003).

different than at present, which makes it difficult to unambiguously interpret the carbon isotope record in this interval.

The LGM sediments, especially in cores PS1906 and PS1910, contain high amounts of coarse ice-rafted debris if compared to younger layers (Fig. 3). This indicates that numerous icebergs were passing the area and dropping parts of their freight. The IRD concentration is highly variable and marked by numerous prominent peaks. These peaks clearly coincide with foraminiferal abundance peaks in cores PS1894, PS1878 and partly in core PS1906. As already discussed previously at site PS1878, the IRD peaks may represent sporadic and relatively short intervals of somewhat ameliorated conditions during times of decreased seasonal sea ice and slightly warmer surface water that resulted in a higher biological productivity, an increased IRD delivery, and thus, a higher sedimentation rate (Telesiński et al., 2013). The duration of these intervals may be overrepresented in the sediment record, the most compelling example being the IRD and foraminiferal peaks in core PS1906 at ca. 25–30 cm (ca. 20–22 ka). Variable sedimentation rates and the uncertainties in our age models for the LGM make it difficult to say whether

the ameliorated conditions occurred basin wide or had a diachronous nature.

6.2   Deglaciation

Prominent low $\delta^{18}O$ peaks accompanied by low $\delta^{13}C$ values are recorded in the deglacial parts of cores PS1878 (ca. 18 ka) and PS1906 (19.7 ka), as well as PS1230 (19.2 ka, Bauch et al., 2001) and PS2887 (19.6–18.7 ka, Nørgaard-Pedersen et al., 2003). Similar, though more obscure features can be traced in cores PS1894 and PS1910 (Fig. 2). We interpret them as a result of the occurrence of isotopically light freshwater that lowered the regional surface and near-surface water salinity (Sarnthein et al., 1995; Spielhagen et al., 2004; Telesiński et al., 2013). In cores PS1906 and PS1878 the high amplitude of the $\delta^{18}O$ peaks is accompanied by low IRD abundance in the respective intervals, which may suggest that the freshwater originated from catastrophic discharges from remote and/or terrestrial sources (e.g., outbursts from ice-dammed or subglacial lakes) rather than from a delivery by melting icebergs or nearby glaciers.

On the other hand, in the well-dated record from core PS2887 (Nørgaard-Pedersen et al., 2003) $\delta^{18}O$ values remained low for more than 2 kyr and the interpolated age of the spike in PS1878 (18–15 ka) fits well with the duration of the Heinrich stadial 1 (HS1). This may suggest that the freshwater persisted in the Greenland Sea for several thousand years and that the low foraminiferal abundance during this time might be a result of a salinity decrease below the level tolerated by planktic foraminifers. The lack of IRD might then be caused by a decrease in iceberg mobility and melt rate due to a rigid sea-ice cover that is expected to grow on top of a cold and freshened water surface.

We realize that the reservoir ages during the deglaciation, especially in the event when massive freshwater discharges rapidly affected the ocean's surface, remain highly uncertain and may have been considerably larger than at present (Waelbroeck et al., 2001; Hanslik et al., 2010; Stern and Lisiecki, 2013). Although the low sedimentation rates in some of our cores increase the uncertainty of the ^{14}C-based age models, our regional comparison shows that the major deglacial freshwater discharges into the western Nordic Seas were roughly coeval. We consider that these events were likely triggered by the global sea level rise that started around 20 ka (Clark and Mix, 2002) and came from the Greenland Ice Sheet and, perhaps, other circum-Arctic ice sheets (e.g., Sarnthein et al., 1995).

The low carbon isotope ratios during these freshwater events (Fig. 2) might be an indication of a reduced ventilation of the upper water column that was forced by a stable, highly stratified surface water lid (cf. Sarnthein et al., 1995; Spielhagen et al., 2004). If the surface stratification of the Greenland Sea was indeed a basin-wide phenomenon, as shown by our records, it supports the interpretation of a slowdown of the Atlantic Meridional Overturning Circulation

(AMOC) during HS1 (McManus et al., 2004; Stanford et al., 2011; Telesiński et al., 2013). Furthermore, it also gives a rough chronological framework for the onset of the deglaciation (ca. 18 ka).

Although our benthic oxygen isotope records do not cover the initial part of HS1, the $\delta^{18}O$ data of *O. umbonatus* indicate, like the planktic record, a distinct decrease around 15.5–15.0 ka in PS1878 (Fig. 6). Such simultaneously occurring surface and bottom water depletions in $\delta^{18}O$ are often interpreted as a result of brines rejected during sea-ice formation (e.g., Dokken and Jansen, 1999; Hillaire-Marcel and de Vernal, 2008). The likelihood that such brines formed in this way and could sink into intermediate or even much greater depths without significant dilution remains unproven (for a discussion see also Bauch and Bauch, 2001; Rasmussen and Thomsen, 2009). More recently, another scenario was proposed to explain the occurrence of light $\delta^{18}O$ excursions during HS1 (Stanford et al., 2011). It suggests that meltwater loaded with fine sediments entered the Nordic Seas below the sea surface as a hyperpycnal flow. In our record, the negative benthic $\delta^{18}O$ excursion at 15.5–15.0 ka may result from such a mechanism. However, in the record studied by Stanford et al. (2011), the benthic oxygen isotope depletion has an amplitude larger than the planktic record, which is not observed in our record. Stanford et al. (2011) explain that, after losing the sediment load, the remaining relatively fresh, low density and low-$\delta^{18}O$ water rose towards the surface (while strongly mixing with ambient water), resulting in the amplitude difference. Possibly the freshwater event in or close to the Greenland Sea released both a sediment-loaded and a largely sediment-free freshwater plume, which in combination may explain the strong near-surface and weaker bottom water $\delta^{18}O$ decreases. The sediment-loaded plume mechanism may also explain the significant thickness of the layers in cores PS1878 and PS2887 with light $\delta^{18}O$ values. While the plume was losing its load, sedimentation rates likely increased dramatically in the affected areas, resulting in relatively thick fine-grained deposits. The duration of the freshwater outbursts was probably significantly shorter than what appears from the linear age interpolation between the dating points. However, sea ice may have played a role as a further freshwater supplier by extending the range and duration of the freshwater event.

Following the freshwater event(s), planktic $\delta^{18}O$ values increased to $\sim 4‰$ or more (Figs. 2, 5), indicating that the freshwater influence had decreased by this time. Also, the increasing $\delta^{13}C$ values may further suggest that either the ventilation and/or the subsurface water structure with respect to stratification and bioproductivity had changed again.

The gradual and low–amplitude changes in the oxygen isotope record of PS1910 make it likely that the site was not directly influenced by major freshwater discharges. Short-lived freshwater events like those recorded in PS1878 between 15 and 13 ka may have taken place at site PS1910 (as well as PS1906 after the major event) but may be obscured by

the core's low resolution. The generally heavy $\delta^{18}O$ values throughout the deglaciation, as well as later on, do indicate a notable inflow of Atlantic waters to this area.

Site PS1894 is located on the Greenland continental slope, in direct proximity to the EGC and under the sea-ice cover. Thus, the lowest $\delta^{18}O$ values in this record might result from the weakest influence of AW and the lowest salinity, compared to other sites. Today, the salinity at site PS1894 is 1–2 psu (practical salinity units) lower than farther to the east, in the ice-free areas (Thiede and Hempel, 1991). In contrast to the other sites, the main onset of the deglaciation (after 17 ka) seems to be characterized by a warming of the (sub)surface water rather than by a freshwater inflow, as the oxygen isotope ratio decrease is accompanied by the appearance of subpolar foraminiferal species (Figs. 2, 4). It is possible that a minor enhancement of the Atlantic Water inflow into the northwestern Greenland Sea coincided with and probably also contributed to the termination of LGM-type conditions and to the onset of deglacial changes at this site. It might seem counterintuitive that at this site, which is the one most affected by PW today, the subpolar species appeared so early and in such high amounts (around 20 %), especially since even in late Holocene sediments this group constitute less than 20 % of the planktic fauna in this area (Husum and Hald, 2012). However, an occurrence of subpolar species, in particular those of smaller sizes, might indicate the advection of Atlantic waters subducted below stratified and sea-ice covered surface water layers (Bauch et al., 2001). Such a mechanism is confirmed by modern oceanographic measurements on a W–E profile across the Greenland Sea, showing higher subsurface temperatures at stations covered with sea ice than in ice-free areas (Thiede and Hempel, 1991).

Although the PS1894 oxygen isotope record does not indicate any major direct freshwater discharges in this area (Fig. 2), surface water salinity was apparently lower than at the other sites, as indicated by the low $\delta^{18}O$ values, probably as a result of the proximity of the ice margin and the EGC.

6.3 Younger Dryas (YD)

Only core PS1878 contains a clear light $\delta^{18}O$ excursion (12.8–11.9 ka) that, according to our age model, fits into the time span of the YD (12.9–11.7 ka, cf. Broecker et al., 2010). However, less prominent oxygen isotope peaks of the same age can be found in cores PS1906 and PS1910, as well as in PS1230 and PS1243 (Bauch et al., 2001). We associate these peaks also with the YD and used them for the correlation of the cores (Figs. 2, 5). The oxygen isotope record of core PS1894 contains no indications that could be linked to the YD cooling. However, as already mentioned above, this record exhibits generally low $\delta^{18}O$ values ($< 3.5‰$ across the YD interval), often lower than those of the light $\delta^{18}O$ excursions in the other records. It indicates that this site was under a constant influence of relatively fresh PW, which makes the identification of a YD freshwater signal difficult.

In general, the origin and cause of the YD has been a matter of debate for decades now (e.g., Broecker et al., 1989; Teller et al., 2005; Murton et al., 2010; Fahl and Stein, 2012; Fisher and Lowell, 2012; Not and Hillaire-Marcel, 2012). A discharge of large amounts of freshwater from the deglacial Lake Agassiz to the North Atlantic and, in particular, to the areas of deep water convection is still considered the most likely cause for the YD (Broecker et al., 2010). While a rerouting from the Gulf of Mexico to the St. Lawrence River was proposed earlier as one triggering mechanism (Broecker et al., 1989), recent modeling results of Condron and Winsor (2012) indicate that only a freshwater discharge to the Arctic (probably via the Mackenzie Valley; cf. Tarasov and Peltier, 2006) was able to reach the deep water formation regions in the North Atlantic (including our study area) and weaken the AMOC sufficiently to trigger the YD. Our finding of a coeval low $\delta^{18}O$ signal at ~ 13 ka in Fram Strait and Greenland Sea records is in support of hypotheses that suggest the Arctic region (including the East Greenland margin) as the main source area for the freshwater pulse. It seems unlikely that a large-volume freshwater transport occurred from the south, i.e., opposite to the dominant flow direction in the Greenland Sea. Following the modeling results of Condron and Winsor (2012), our data make the hypothesis of an Arctic trigger for the YD cold event more convincing.

6.4 Holocene

Although the onset of the Holocene in our records is expressed by the typical proxy changes for a glacial–interglacial transition, it looks different at the individual sites. In the southern Fram Strait (site PS1906) both the foraminiferal abundance and the percentage of subpolar species increased relatively rapidly around 12 ka. This was possibly related to the onset of enhanced surface flow of the NAC branch along the eastern Nordic Seas following shortly upon the YD (e.g., Sarnthein et al., 2003; Hald et al., 2007; Risebrobakken et al., 2011). Farther south, at sites PS1910 and PS1878, that increase was much more gradual and highest values there were reached between 10 and 8 ka. Subsurface waters at site PS1878 also warmed more slowly reaching $\sim 3\,°C$ only around 8 ka (Fig. 7). This confirms that in the earliest Holocene the influence of the melting Greenland Ice Sheet was strong and acted as a negative feedback to the orbitally forced climatic optimum (cf. Blaschek and Renssen, 2013). The decrease of IRD deposition at three of our sites (PS1906, PS1910, PS1878) indicates that only few icebergs still reached the southeastern Greenland Sea due to a northwestward expansion of the warmer water masses. The decrease in IRD deposition was less prominent in the southern Fram Strait at this time, most probably due to the proximity of the Transpolar Drift which still brought numerous icebergs from the Arctic Ocean into this region. Site PS1894 showed the least significant changes at the onset of the Holocene (Figs. 8, 9). The proxy data indicate that the eastern part of

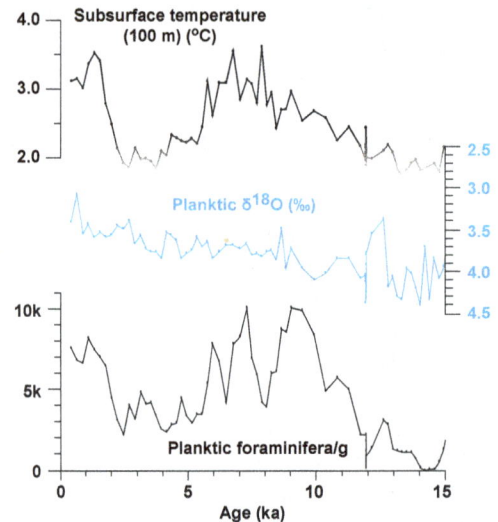

Fig. 7. Absolute subsurface temperatures calculated using the transfer function of Husum and Hald (2012) on planktic foraminifera from core PS1878 and planktic oxygen isotope and total planktic foraminiferal abundance records. Calculated temperatures below $2\,°C$ should be considered uncertain.

the Greenland Sea remained under polar conditions with cold surface water, numerous icebergs and sea-ice cover for most of the time.

For the entire study area it is difficult to determine a coeval thermal maximum, which we define as the interval with the highest percentage of subpolar species (or highest absolute temperatures in core PS1878). Not only the course of the initial warming but also the duration and termination of the warmest interval differed between the individual sites. In the southern Fram Strait (site PS1906) the thermal maximum interval apparently started already around 11.5 ka and ended gradually between 7 and 3 ka. At sites PS1894, PS1910 and PS1878 it was significantly shorter and can be dated to ca. 11–9.5, 10.5–7 and 8–5.5 ka, respectively. This might at least in part be attributed to uncertainties in the correlation between the records, which was mainly based on the isotope records. Nevertheless, the onset of the warmest interval around 11–9 ka accords with many other Nordic Seas records (e.g., Bauch et al., 2001; Sarnthein et al., 2003; Giraudeau et al., 2010; Risebrobakken et al., 2011; Husum and Hald, 2012) where the beginning of the Holocene thermal maximum (HTM) was related to maximum insolation in the high latitudes (e.g., Andersen et al., 2004; Risebrobakken et al., 2011) and the maximum in northward oceanic heat transport by the NAC (Risebrobakken et al., 2011). The late onset of the thermal maximum at site PS1878 might have resulted from the large distance between the site and the core of the NAC (Fig. 1). Since this onset was time-transgressive along the main pathway of the NAC (Hald et al., 2007), a similar development may also be expected westward. In principle, the presence of freshwater in the earliest Holocene (Fig. 5)

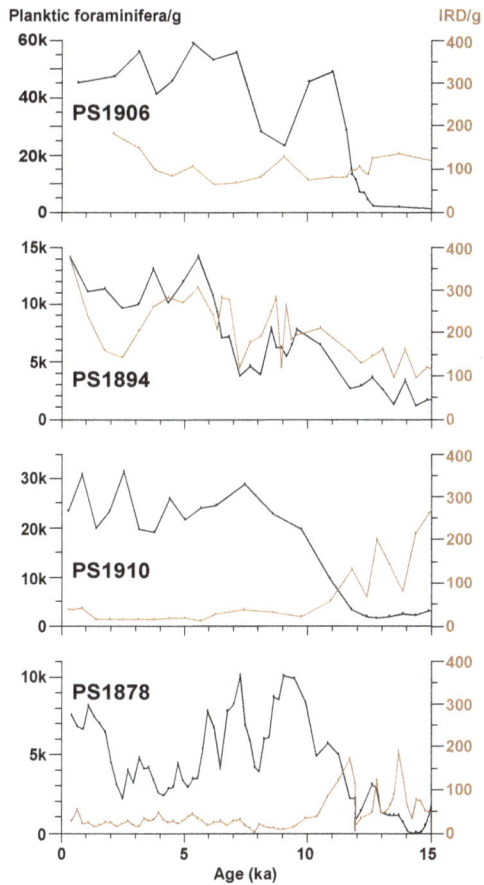

Fig. 8. Planktic foraminifera and IRD abundance (per 1 g dry sediment) of cores used in this study, plotted vs. age (since 15 ka).

Fig. 9. Relative abundance of the three most common planktic foraminifera species in cores used in this study, plotted vs. age (since 15 ka). Abbreviations as in Fig. 4. Asterisks mark the modern (core-top) values (own data).

may have had a cooling effect, but this should have also been the case at the other three sites. Furthermore, the relative proximity of the remnant Greenland Ice Sheet, still delivering cold meltwater, could have acted as a negative feedback for the early Holocene warming (Blaschek and Renssen, 2013). The transfer function yielded temperatures of 3–3.5 °C at 100 m water depth between 8 and 5.5 ka. This is significantly warmer than modern temperatures at this depth in the Vesterisbanken area (max. 2 °C, Thiede and Hempel, 1991) and indicates that the advection of Atlantic waters to the area between 8 (or even 10.5) and 5.5 ka was stronger than at present.

The transition between the thermal maximum and the Neoglacial cooling as found in our records between ca. 6–5 and 3 ka was also not simultaneous and, with the exception of PS1878, was much more gradual than the early Holocene warming (Figs. 7, 9). Although in cores PS1906 and PS1878 relatively late, such a timing is in good general agreement with other studies (e.g., Bauch et al., 2001; Sarnthein et al., 2003; Hald et al., 2007; Giraudeau et al., 2010; Rasmussen and Thomsen, 2010; Husum and Hald, 2012; Werner et al., 2013; for some remarkable exceptions see Risebrobakken et

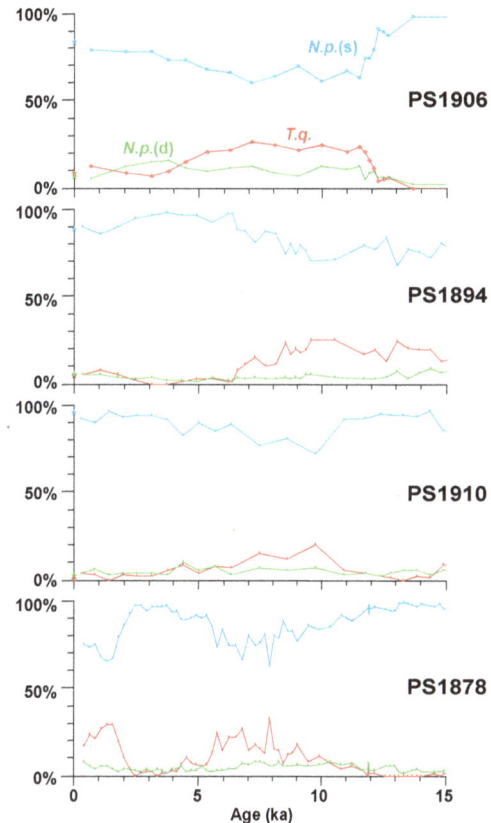

al., 2011). The Neoglacial cooling was very likely forced primarily by decreasing insolation (Andersen et al., 2004), while the regional variations in its timing and scale are a manifestation of the reorganization of the specific water mass configuration in the Nordic Seas. This reorganization involved, e.g., changes in the strength and routing of the individual NAC and EGC branches, the amount of meltwater, and the relocation of the convection centers and eventually resulted in the establishment of a type of overall water mass distribution and circulation as we see it today (Bauch et al., 2001).

The δ^{13}C "plateau" between ca. 7 and 3 ka (Fig. 5) is common in Nordic Seas records (e.g., Vogelsang, 1990; Fronval and Jansen, 1997; Bauch et al., 2001; Sarnthein et al., 2003; Risebrobakken et al., 2011; Werner et al., 2013) and reflects a period of maximum ventilation of subsurface waters, relatively stable and modern-like environmental conditions (Bauch et al., 2001; Sarnthein et al., 2003), and perhaps a significantly changed surface water structure (Bauch and Weinelt, 1997). Its onset also corresponds to the establishment of the modern Iceland–Scotland Overflow Water (Thornalley et al., 2010) and AMOC strengthening (Hall et al., 2004). Our benthic δ^{13}C records (Fig. 6) and other benthic records from the Nordic Seas (Bauch et al., 2001; Sarnthein et al., 2003) also exhibit relatively high values in this interval. This implies good ventilation of the bottom water and suggests that intensive deep water convection took place in the Nordic Seas between 7 and 3 ka. An AMOC intensification after 7 ka would also imply enhanced inflow of AW and PW into the Greenland Sea since the increased convection rate must be compensated by an increased inflow of both saline AW from the south and cold PW from the north. The increasing influence of cold PW amplified the Neoglacial cooling in the area, which might explain the relatively rapid warm–cold transition at site PS1878 at 5.5 ka, similar to what was found in the eastern Fram Strait (Werner et al., 2013). The cooling, in turn, likely enhanced sea-ice formation and strong winds, which opened up ice leads and provoked super-cooling processes further intensifying deep water formation. The bottom water at site PS1878 was particularly well ventilated compared to other Holocene records from the Nordic Seas (Fig. 6, cf. Bauch et al., 2001; Sarnthein et al., 2003). This indicates that deep convection was taking place in the central Greenland Sea, in the proximity of this site, with maximum vigor between 7 and 3 ka.

The planktic δ^{13}C decrease after around 3 ka, observed in all our records (Fig. 5), appears to be a sound stratigraphic time marker in many Nordic Seas records (Bauch and Weinelt, 1997). Moreover, as it occurs all across the Nordic Seas including the Barents Sea (e.g., Vogelsang, 1990; Fronval and Jansen, 1997; Bauch et al., 2001; Sarnthein et al., 2003; Risebrobakken et al., 2011; Werner et al., 2013) this event clearly bears a supraregional implication. A reconstruction of sea-ice conditions in the Fram Strait (Müller et al., 2012) revealed increasing sea-ice coverage since 8 ka. At about 3 ka a further significant expansion of the sea-ice cover occurred and sea-ice conditions became more fluctuating. Although in the record from the East Greenland Shelf (Müller et al., 2012) no increase in sea-ice coverage is observed before 3 ka (perhaps because this area was strongly influenced by sea ice during the entire Holocene), the total sea-ice cover in the Nordic Seas was probably increasing. A similar timing in ice increase is also confirmed for the western Barents Sea slope (Sarnthein et al., 2003). Renssen et al. (2006) indicated that a negative solar irradiance anomaly and associated cooling may cause an expansion of sea ice and a temporary relocation of deep water formation sites in the Nordic Seas. One of the strongest anomalies in the Holocene occurred between 2.85 and 2.6 ka and could have triggered the sudden increase in sea-ice extent, increased the stratification of the upper water layers and decreased the ventilation of the subsurface water. This solar irradiance anomaly may also have triggered the increase in ice rafting in the North Atlantic around that time (Bond et al., 2001; Renssen et al., 2006).

In two of our benthic carbon isotope records (PS1910 and PS1878, Fig. 6) we observe a decrease of values around 3 ka, which paralleled that in the planktic record. This is, however, not generally the case elsewhere (e.g., at site PS1894 or in the central and eastern Nordic Seas; Bauch et al., 2001; Sarnthein et al., 2003; Werner et al., 2013). The decrease in benthic δ^{13}C values suggests that, probably as a result of a more extensive sea-ice cover and a stronger stratification of the upper water layers, deep convection diminished or did not reach down to maximum depth of the basins any longer (Renssen et al., 2006). Sites PS1910 and PS1878 were most likely located closest to the convection center and the decrease in convection rate or depth was recorded here as a benthic δ^{13}C decrease. At other sites that were located farther from the convection center, the bottom waters were not as well ventilated before 3 ka and therefore the relative decrease in ventilation was not large enough to be recorded in the sediment archive.

As described earlier (Telesiński et al., 2013), significant changes are observed in core PS1878 since 3 ka. The total foraminiferal abundance (Fig. 8) and percentage of subpolar species (Fig. 9) increase and planktic carbon and oxygen isotope ratios decrease. These changes were interpreted as evidence of a warming of subsurface waters caused by an NAO-induced increase in AW inflow, amplified by stronger upper water layers stratification (Telesiński et al., 2013). The benthic data from core PS1878 show that the planktic and the two benthic oxygen isotope records, which in the older part of the record ran roughly parallel to each other, diverge after 3 ka (Figs. 5, 6). The planktic values begin to decrease after the stable interval of the Middle Holocene and *O. umbonatus* values start to increase, while *C. wuellerstorfi* oxygen isotope ratios follow the earlier slightly decreasing trend. As a result of the decrease in convection rate and depth, probably not only the surface and bottom waters began to differentiate from each other, but also, at a smaller scale, the epibenthic and infaunal biotopes became more distinct than before due to more stagnant conditions.

In the other records from the Greenland Sea the changes after 3 ka are not as obvious. At site PS1894, strongly affected by PW, the conditions seem to be similar to those at other sites during the LGM, with at least seasonally open water conditions and somewhat warmer upper water layers (Figs. 5, 8, 9; see discussion above). Virtually no indications of warming or increased AW influence can be found at sites PS1906 and PS1910 at that time.

The high-resolution subsurface temperature reconstruction from site PS1878 indicates a warming from ca. 2 °C at 2.5 ka to 3.5 °C at 1.5 ka, confirming that conditions in the central Greenland Sea in the late Holocene were comparable to the early Holocene warm interval (cf. Telesiński et al., 2013). The scale of this warming (1.5 °C) is comparable to that of the modern warming in the Arctic (e.g., Spielhagen et al., 2011) though, of course, on a significantly longer timescale. A comparison with the faunal data from other Greenland Sea cores (Fig. 9) shows that this phenomenon was confined to the central part of the Greenland Sea and may have resulted from the co-occurrence of the stronger water column stratification and the enhanced inflow of Atlantic waters to the site.

7 Summary and conclusions

With the records presented in this study we were able here to reconstruct for the first time a millennial- to multicentennial-scale image of the late glacial and Holocene paleoceanographic evolution in the northern and central Greenland Sea. Despite the low sedimentation rates in the northern part of the study area and the related chronological uncertainties, the correlation and comparison with a high resolution record PS1878 (Telesiński et al., 2013) allowed us to study the spatial and temporal variability of the most important oceanographic processes. The integration of surface, subsurface and bottom water proxies gave an almost complete image.

During the LGM environmental conditions were to a large extent similar across the Greenland Sea. Cold conditions with a dense sea-ice cover, numerous icebergs and low biological productivity prevailed in the area. During the deglaciation the Greenland Sea was affected by freshwater discharges. Although we argue that they were roughly simultaneous (between 18 and 15 ka) and may have had a common trigger mechanism, their sources and character were probably different. During the YD the Greenland Sea was affected by a major deglacial freshwater discharge most probably originating from the Arctic. Our data suggest a thicker but weaker halocline and a deepening of AW.

The onset, duration and decline of the early Holocene warm interval were apparently different in age and scale at each site, reflecting regional differences in the reorganization of the ocean circulation of the area. As peak warming occurred not simultaneously at all sites, the thermal maximum in the central Greenland Sea was not reached until ca. 8 ka, which is relatively late compared to other Nordic Seas records. Maximum subsurface temperatures ($> 3\,°C$) were higher than at present, indicating a strong influence of Atlantic waters. Since 7 ka high $\delta^{13}C$ values, both planktic and benthic, indicate the establishment of the modern ocean circulation system in the Nordic Seas with maximum deep convection in the Greenland Sea. Despite a strong AMOC, decreasing insolation led to the Neoglacial cooling and an increase in sea-ice coverage. At 3–2.8 ka a solar irradiance minimum may have triggered a rapid expansion of the sea-ice cover that led to a stronger stratification of the upper water layers and, subsequently, to a weakening of deep convection in the Greenland Sea and of the AMOC. Eventually, an increase in AW inflow into the Nordic Seas led to subsurface warming in the central Greenland Sea (site PS1878). Probably due to a relatively stable water stratification, as well as increased presence of sea ice (and thus an isolation of the subsurface water from the atmosphere and other water masses), subsurface temperatures rose again to a level comparable with the early Holocene thermal maximum at this site.

Comparison of the Greenland Sea records suggests insolation to be the primary driver controlling the regional paleoceanographic evolution while the routing and intensity of AW inflow seems to control the spatial variability in the area. Other processes – such as sea-ice formation, deep convection, freshwater discharges, etc. – also played an important role in the observed local differences.

Acknowledgements. This work is a contribution to the CASE Initial Training Network funded by the European Community's 7th Framework Programme FP7 2007/2013, Marie Curie Actions, under Grant Agreement no. 238111. We thank reviewers Juliane Müller and Thomas Cronin, as well as Christelle Not and Kirstin Werner for their constructive criticism and suggestions which improved the manuscript. Our gratitude goes to Katrine Husum for her help with performing the transfer function calculations. We are grateful to Lulzim Haxhiaj as well as Helmut Erlenkeuser and his staff for performing the stable isotope measurements and to the Leibniz Laboratory, Kiel University, and the Poznan Radiocarbon Laboratory for the AMS ^{14}C datings.

Edited by: H. Renssen

References

Andersen, C., Koç, N., Jennings, A. E., and Andrews, J. T.: Nonuniform response of the major surface currents in the Nordic Seas to insolation forcing: Implications for the Holocene climate variability, Paleoceanography, 19, PA2003, doi:10.1029/2002PA000873, 2004.

Bauch, D. and Bauch, H. A.: Last glacial benthic foraminiferal $\delta^{18}O$ anomalies in the polar North Atlantic: A modern analogue evaluation, J. Geophys. Res., 106, 9135–9143, 2001.

Bauch, H. A. and Erlenkeuser, H.: Interpreting Glacial-Interglacial Changes in Ice Volume and Climate From Subarctic Deep Water Foraminiferal $\delta^{18}O$, in: Earth's Climate and Orbital Eccentricity: The Marine Isotope Stage 11 Question, Geoph. Monog. Series 137, edited by: Droxler, L. H., Poore, A. W., and Burckle, R. Z., American Geophysical Union, Washington, D.C., 87–102, 2003.

Bauch, H. A. and Weinelt, M. S.: Surface water changes in the Norwegian Sea during last deglacial and Holocene times, Quaternary Sci. Rev., 16, 1115–1124, 1997.

Bauch, H. A., Erlenkeuser, H., Spielhagen, R. F., Struck, U., Matthiessen, J., Thiede, J., and Heinemeier, J.: A multiproxy reconstruction of the evolution of deep and surface waters in the subarctic Nordic seas over the last 30,000 yr, Quaternary Sci. Rev., 20, 659–678, 2001.

Blaschek, M. and Renssen, H.: The Holocene thermal maximum in the Nordic Seas: the impact of Greenland Ice Sheet melt and other forcings in a coupled atmosphere–sea-ice–ocean model, Clim. Past, 9, 1629–1643, doi:10.5194/cp-9-1629-2013, 2013.

Bond, G. C., Kromer, B., Beer, J., Muscheler, R., Evans, M. N., Showers, W., Hoffmann, S., Lotti-Bond, R., Hajdas, I., and Bonani, G.: Persistent solar influence on North Atlantic climate during the Holocene, Science, 294, 2130–2136, doi:10.1126/science.1065680, 2001.

Broecker, W. S., Kennett, J. P., Flower, B. P., Teller, J. T., Trumbore, S., Bonani, G., and Wolfli, W.: Routing of meltwater from the Laurentide Ice Sheet during the Younger Dryas cold episode, Nature, 341, 318–321, 1989.

Broecker, W. S., Denton, G. H., Edwards, R. L., Cheng, H., Alley, R. B., and Putnam, A. E.: Putting the Younger Dryas cold event into context, Quaternary Sci. Rev., 29, 1078–1081, doi:10.1016/j.quascirev.2010.02.019, 2010.

Carstens, J., Hebbeln, D., and Wefer, G.: Distribution of planktic foraminifera at the ice margin in the Arctic (Fram Strait), Mar. Micropaleontol., 29, 257–269, 1997.

Clark, D. L. and Hanson, A.: Central Arctic Ocean sediment texture: A key to ice transport mechanism, in: Glacial-marine sedimentation, edited by: Molnia, B. F., Plenum Press, New York, 301–330, 1983.

Clark, P. U. and Mix, A. C.: Ice sheets and sea level of the Last Glacial Maximum, Quaternary Sci. Rev., 21, 1–7, doi:10.1016/S0277-3791(01)00118-4, 2002.

Condron, A. and Winsor, P.: Meltwater routing and the Younger Dryas, P. Natl. Acad. Sci. USA, 109, 19928–19933, doi:10.1073/pnas.1207381109, 2012.

Dokken, T. M. and Jansen, E.: Rapid changes in the mechanism of ocean convection during the last glacial period, Nature, 401, 458–461, 1999.

Duplessy, J. C., Labeyrie, L. D., and Blanc, P. L.: Norwegian Sea Deep Water Variations over the Last Climatic Cycle: Paleo-Oceanographical Implications, in: Long and Short Term Variability of Climate, edited by: Wanner, H. and Siegenthaler, U. Springer, New York, 83–116, 1988.

Fahl, K. and Stein, R.: Modern seasonal variability and deglacial/Holocene change of central Arctic Ocean sea-ice cover: New insights from biomarker proxy records, Earth Planet. Sc. Lett., 351-352, 123–133, doi:10.1016/j.epsl.2012.07.009, 2012.

Fisher, T. G. and Lowell, T. V.: Testing northwest drainage from Lake Agassiz using extant ice margin and strandline data, Quatern. Int., 260, 106–114, doi:10.1016/j.quaint.2011.09.018, 2012.

Fronval, T. and Jansen, E.: Eemian and early Weichselian (140–60 ka) paleoceanography and paleoclimate in the Nordic seas with comparisons to Holocene conditions, Paleoceanography, 12, 443–462, 1997.

Giraudeau, J., Grelaud, M., Solignac, S., Andrews, J. T., Moros, M., and Jansen, E.: Millennial-scale variability in Atlantic water advection to the Nordic Seas derived from Holocene coccolith concentration records, Quaternary Sci. Rev., 29, 1276–1287, doi:10.1016/j.quascirev.2010.02.014, 2010.

Hald, M., Andersson, C., Ebbesen, H., Jansen, E., Klitgaard-Kristensen, D., Risebrobakken, B., Salomonsen, G. R., Sarnthein, M., Sejrup, H. P., and Telford, R. J.: Variations in temperature and extent of Atlantic Water in the northern North Atlantic during the Holocene, Quaternary Sci. Rev., 26, 3423–3440, doi:10.1016/j.quascirev.2007.10.005, 2007.

Hall, I. R., Bianchi, G. G., and Evans, J. R.: Centennial to millennial scale Holocene climate-deep water linkage in the North Atlantic, Quaternary Sci. Rev., 23, 1529–1536, doi:10.1016/j.quascirev.2004.04.004, 2004.

Hansen, B. and Østerhus, S.: North Atlantic–Nordic Seas exchanges, Prog. Oceanogr., 45, 109–208, doi:10.1016/S0079-6611(99)00052-X, 2000.

Hanslik, D., Jakobsson, M., Backman, J., Björck, S., Sellén, E., O'Regan, M., Fornaciari, E., and Skog, G.: Quaternary Arctic Ocean sea ice variations and radiocarbon reservoir age corrections, Quaternary Sci. Rev., 29, 3430–3441, doi:10.1016/j.quascirev.2010.06.011, 2010.

Hillaire-Marcel, C. and de Vernal, A.: Stable isotope clue to episodic sea ice formation in the glacial North Atlantic, Earth Planet. Sc. Lett., 268, 143–150, doi:10.1016/j.epsl.2008.01.012, 2008.

Husum, K. and Hald, M.: Arctic planktic foraminiferal assemblages: Implications for subsurface temperature reconstructions, Mar. Micropaleontol., 96–97, 38–47, doi:10.1016/j.marmicro.2012.07.001, 2012.

Juggins, S.: C2, Version 1.7.2, Software for Ecological and Palaeoecological Data Analysis and Visualization, http://www.campus.ncl.ac.uk/staff/Stephen.Juggins/index.html, University of Newcastle, Newcastle upon Tyne, UK, 2011.

Marshall, J. and Schott, F.: Open-ocean convection: Observations, theory, and models, Rev. Geophys., 37, 1–64, 1999.

McManus, J. F., Francois, R., Gherardi, J.-M., Keigwin, L. D., and Brown-Leger, S.: Collapse and rapid resumption of Atlantic meridional circulation linked to deglacial climate changes, Nature, 428, 834–837, doi:10.1038/nature02494, 2004.

Müller, J., Werner, K., Stein, R., Fahl, K., Moros, M., and Jansen, E.: Holocene cooling culminates in sea ice oscillations in Fram Strait, Quaternary Sci. Rev., 47, 1–14, doi:10.1016/j.quascirev.2012.04.024, 2012.

Murton, J. B., Bateman, M. D., Dallimore, S. R., Teller, J. T., and Yang, Z.: Identification of Younger Dryas outburst flood path from Lake Agassiz to the Arctic Ocean, Nature, 464, 740–743, doi:10.1038/nature08954, 2010.

Nørgaard-Pedersen, N., Spielhagen, R. F., Erlenkeuser, H., Grootes, P. M., Heinemeier, J., and Knies, J.: Arctic Ocean during the Last Glacial Maximum: Atlantic and polar domains of surface water mass distribution and ice cover, Paleoceanography, 18, 1–19, doi:10.1029/2002PA000781, 2003.

Not, C. and Hillaire-Marcel, C.: Enhanced sea-ice export from the Arctic during the Younger Dryas, Nature Comm., 3, 647, doi:10.1038/ncomms1658, 2012.

Nürnberg, D., Wollenburg, I., Dethleff, D., Eicken, H., Kassens, H., Letzig, T., Reimnitz, E., and Thiede, J.: Sediments in Arctic sea ice: implications for entrainment, transport and release, Mar. Geol., 104, 185–214, 1994.

Rasmussen, T. L. and Thomsen, E.: Stable isotope signals from brines in the Barents Sea: Implications for brine formation during the last glaciation, Geology, 37, 903–906, doi:10.1130/G25543A.1, 2009.

Rasmussen, T. L. and Thomsen, E.: Holocene temperature and salinity variability of the Atlantic Water inflow to the Nordic seas, Holocene, 20, 1223–1234, doi:10.1177/0959683610371996, 2010.

Reimer, P., Baillie, M., Bard, E., Bayliss, A., Beck, J. W., Blackwell, P. G., Bronk Ramsey, C., Buck, C. E., Burr, G. S., Edwards, R. L., Friedrich, M., Grootes, P. M., Guilderson, T. P., Hajdas, I., Heaton, T. J., Hogg, A. G., Hughen, K. A., Kaiser, K. F., Kromer, B., McCormac, F. G., Manning, S. W., Reimer, R. W., Richards, D. A., Southon, J. R., Talamo, S., Turney, C. S. M., van der Plicht, J., and Weyhenmeyer, C. E.: IntCal09 and Marine09 radiocarbon age calibration curves, 0–50,000 years cal BP, Radiocarbon, 51, 1111–1150, 2009.

Renssen, H., Goosse, H., and Muscheler, R.: Coupled climate model simulation of Holocene cooling events: oceanic feedback amplifies solar forcing, Clim. Past, 2, 79–90, doi:10.5194/cp-2-79-2006, 2006.

Risebrobakken, B., Dokken, T., Smedsrud, L. H., Andersson, C., Jansen, E., Moros, M., and Ivanova, E. V.: Early Holocene temperature variability in the Nordic Seas: The role of oceanic heat advection versus changes in orbital forcing, Paleoceanography, 26, PA4206, doi:10.1029/2011PA002117, 2011.

Rudels, B. and Quadfasel, D.: Convection and deep water formation in the Arctic Ocean-Greenland Sea System, J. Mar. Syst., 2, 435–450, doi:10.1016/0924-7963(91)90045-V, 1991.

Rudels, B., Friedrich, H. J., and Quadfasel, D.: The Arctic Circumpolar Boundary Current, Deep Sea-Res. Pt. II, 46, 1023–1062, doi:10.1016/S0967-0645(99)00015-6, 1999.

Sarnthein, M., Jansen, E., Weinelt, M., Arnold, M., Duplessy, J. C., Erlenkeuser, H., Flatøy, A., Johannessen, G., Johannessen, T., Jung, S., Koc, N., Labeyrie, L., Maslin, M., Pflaumann, U., and Schulz, H.: Variations in Atlantic surface ocean paleoceanography, 50°–80° N: A time-slice record of the last 30,000 years, Paleoceanography, 10, 1063–1094, 1995.

Sarnthein, M., van Kreveld, S., Erlenkeuser, H., Grootes, P. M., Kucera, M., Pflaumann, U., and Schulz, M.: Centennial-to-millennial-scale periodicities of Holocene climate and sediment injections off the western Barents shelf, 75° N, Boreas, 32, 447–461, doi:10.1080/03009480310003351, 2003.

Spielhagen, R. F. and Erlenkeuser, H.: Stable oxygen and carbon isotopes in planktic foraminifers from Arctic Ocean surface sediments: Reflection of the low salinity surface water layer, Mar. Geol., 119, 227–250, doi:10.1016/0025-3227(94)90183-X, 1994.

Spielhagen, R. F., Baumann, K.-H., Erlenkeuser, H., Nowaczyk, N. R., Nørgaard-Pedersen, N., Vogt, C., and Weiel, D.: Arctic Ocean deep-sea record of northern Eurasian ice sheet history, Quaternary Sci. Rev., 23, 1455–1483, doi:10.1016/j.quascirev.2003.12.015, 2004.

Spielhagen, R. F., Werner, K., Aagaard-Sørensen, S., Zamelczyk, K., Kandiano, E., Budeus, G., Husum, K., Marchitto, T. M., and Hald, M.: Enhanced Modern Heat Transfer to the Arctic by Warm Atlantic Water, Science, 331, 450–453, doi:10.1126/science.1197397, 2011.

Stanford, J. D., Rohling, E. J., Bacon, S., Roberts, A. P., Grousset, F. E., and Bolshaw, M.: A new concept for the paleoceanographic evolution of Heinrich event 1 in the North Atlantic, Quaternary Sci. Rev., 30, 1047–1066, doi:10.1016/j.quascirev.2011.02.003, 2011.

Stern, J. V. and Lisiecki, L. E.: North Atlantic circulation and reservoir age changes over the past 41,000 years, Geophys. Res. Lett., 40, 3693–3697, doi:10.1002/grl.50679, 2013.

Stuiver, M. and Reimer, P. J.: Radiocarbon calibration program, Radiocarbon, 35, 215–230, 1993.

Swift, J.: The Arctic Waters, in: The Nordic Seas, edited by: Hurdle, B., Springer, New York, 129–151, 1986.

Tarasov, L. and Peltier, W. R.: A calibrated deglacial drainage chronology for the North American continent: evidence of an Arctic trigger for the Younger Dryas, Quaternary Sci. Rev., 25, 659–688, doi:10.1016/j.quascirev.2005.12.006, 2006.

Telesiński, M. M., Spielhagen, R. F., and Lind, E. M.: A high-resolution Late Glacial and Holocene paleoceanographic record from the Greenland Sea, Boreas, doi:10.1111/bor.12045, in press, 2013.

Teller, J. T., Boyd, M., Yang, Z., Kor, P. S. G., and Mokhtari Fard, A.: Alternative routing of Lake Agassiz overflow during the Younger Dryas: new dates, paleotopography, and a re-evaluation, Quaternary Sci. Rev., 24, 1890–1905, doi:10.1016/j.quascirev.2005.01.008, 2005.

Thiede, J. and Hempel, G.: The Expedition ARKTIS-VII/1 of RV "POLARSTERN" in 1990, Ber. Polarforsch., 80, 137 pp., 1991.

Thornalley, D. J. R., Elderfield, H., and McCave, I. N.: Intermediate and deep water paleoceanography of the northern North Atlantic over the past 21,000 years, Paleoceanography, 25, 1–17, doi:10.1029/2009PA001833, 2010.

Vogelsang, E.: Paläo-Ozeanographie des Europäischen Nordmeeres an Hand stabiler Kohlenstoff- und Sauerstoffisotope – Paleoceanography of the Nordic seas on the basis of stable carbon and oxygen isotopes, Berichte aus dem Sonderforschungsbereich 313, Nr. 23, Univ. Kiel, Kiel, 1990.

Waelbroeck, C., Duplessy, J. C., Michel, E., Labeyrie, L., Paillard, D., and Duprat, J.: The timing of the last deglaciation in North Atlantic climate records, Nature, 412, 724–727, doi:10.1038/35089060, 2001.

Werner, K., Spielhagen, R. F., Bauch, D., Christian Hass, H., and Kandiano, E.: Atlantic Water advection versus sea-ice advances in the eastern Fram Strait during the last 9 ka: Multiproxy evidence for a two-phase Holocene, Paleoceanography, 28, 283–295, doi:10.1002/palo.20028, 2013.

3

Multidecadal to millennial marine climate oscillations across the Denmark Strait (∼ 66° N) over the last 2000 cal yr BP

J. T. Andrews and A. E. Jennings

INSTAAR and Department of Geological Sciences, University of Colorado, Boulder, CO 80309, USA

Correspondence to: J. T. Andrews (andrewsj@colorado.edu)

Abstract. In the area of Denmark Strait (∼ 66° N), the two modes of the North Atlantic Oscillation (NAO) and Arctic Oscillation (AO) are expressed in changes of the northward flux of Atlantic water and the southward advection of polar water in the East Iceland current. Proxies from marine cores along an environmental gradient from extensive to little or no drift ice, capture low frequency variations over the last 2000 cal yr BP. Key proxies are the weight% of calcite, a measure of surface water stratification and nutrient supply, the weight% of quartz, a measure of drift ice transport, and grain size. Records from Nansen and Kangerlussuaq fjords show variable ice-rafted debris (IRD) records but have distinct mineralogy associated with differences in the fjord catchment bedrock. A comparison between cores on either side of the Denmark Strait (MD99-2322 and MD99-2269) show a remarkable millennial-scale similarity in the trends of the weight% of calcite with a trough reached during the Little Ice Age. However, the quartz records from these two sites are quite different. The calcite records from the Denmark Strait parallel the 2000 yr Arctic summer-temperature reconstructions; analysis of the detrended calcite and quartz data reveal significant multi-decadal–century periodicities superimposed on a major environmental shift occurring ca. 1450 AD.

1 Introduction

The region of Denmark Strait (Figs. 1 and 2) is the northern end member for calculation of the North Atlantic Oscillation (NAO) index (Dawson et al., 2002; Hurrell et al., 2003; Kwok and Rothrock, 1999; Zhang et al., 2004) and is effected by variations in sea ice and freshwater, exported through Fram Strait which is associated with the Arctic Oscillation

(AO) (Thompson and Wallace, 1998; Wang and Ikeda, 2000; Darby et al., 2012). Our records in this paper cannot detect multi-year oscillations, but they can resolve multidecadal to -millennial periodicities. Denmark Strait (Fig. 1) is one of the main gateways linking the polar and temperate realms. Strong thermal and salinity gradients exist along a series of marine fronts (Belkin et al., 2009) as northward advected Atlantic water comes in contact with southward-flowing polar and Arctic waters. Drift ice (Koch, 1945), in the form of sea ice and icebergs, is a pervasive feature of the East Greenland margin with landfast sea-ice retaining icebergs within the fjords for months or years (Dwyer, 1995). In contrast, the northern shelf of Iceland varies greatly in ice coverage from little or no drift ice to extensive coverage, and which has serious effects on farming and the fishery (Divine and Dick, 2006; Ogilvie et al., 2000; Ogilvie, 1997). In the 1870s for example the whaler, Captain David Gray, showed (1881) Iceland nearly encircled by drift ice. Blindheim et al. (2001) provided an important framework for paleoceanographic studies in the area of Denmark Strait (Fig. 2) by sketching out the major changes in surface oceanography that were associated with NAO positive/negative atmospheric modes (Hurrell et al., 2003; van Loon and Rogers, 1978) (Fig. 2c). This has also been extended to include the impact on intermediate and deep water circulation in the area of the subpolar gyre, with its center south of Denmark Strait (Sarafanov, 2009), and to the extent of sea ice in the region (Parkinson, 2000; Zhang et al., 2004). Yearly indexes (1899–2000 AD) for the NAO and AO are positively correlated but the explained variance is only modest ($r^2 = 0.3$) (Fig. 2d) (CRU, 2004; JISAO, 2004). However, the plot indicates that 70 % of the time the sign of the indexes match indicating that although the strength of the departures may vary the two oscillations are largely in-phase. However, the NAO and AO are a multi-year oscillations (Fig. 2c) whereas in most paleoceanographic records the sampling resolution is decadal at best, hence we should talk about NAO/AO-like intervals. Several recent papers have focused on conditions along the Iceland shelf over the last 1000–2000 cal yr BP using a variety of proxies (Andrews et al., 2009a; Axford et al., 2011; Jennings et al., 2001; Knudsen et al., 2004; Masse et al., 2008; Sicre et al., 2008), but there has been little effort to compare the proxies across the Denmark Strait (Jennings et al., 2002a; Andresen et al., 2013). The goal of this paper is to present proxy data from four areas, namely East Greenland fjord mouths, East Greenland mid-shelf, NW/N Iceland, and SW Iceland (Fig. 1), and assess whether there is any evidence for significant oscillations in our marine proxies over the last 2000 cal yr BP. In particular, changes in the strength and extent of the Irminger current (IC) and the East Iceland current (EIC) (Fig. 2) modulate changes in nutrient availability and in the presence/absence of drift ice. SW Iceland is always under the influence of the IC and is only very rarely affected by sea ice. SW Iceland shelf sediments contain little or no quartz, no presence of the sea-ice biomarker IP25 (Belt et al., 2007), and have high carbonate content (Axford et al., 2011). The opposite end member is represented by the inner NE Greenland shelf, which is always influenced by the East Greenland current (EGC) and extensive landfast and drifting sea ice (Reeh, 2004). During the Little Ice Age (LIA), the northeast Greenland fjords may have been covered by permanent landfast sea ice (Funder et al., 2011; Reeh et al., 2001), thus restricting both sea-ice and iceberg drift and probably curtailing the bloom of sea-ice diatoms (Muller et al., 2011). In such an environment, ice-rafted debris (IRD) is restricted, laminated sediments replace bioturbated muds (Dowdeswell et al., 2000; Jennings and Weiner, 1996), and marine carbonate production would be extremely limited. The IC is present within the deep cross-shelf troughs as an intermediate water mass, which under present-day conditions is being transported into the SE Greenland fjords causing retreat of tidewater glaciers (Andresen et al., 2012; Syvitski et al., 1996; Straneo et al., 2010). Hence the most sensitive area to NAO/AO-like variability is represented by the outer E Greenland shelf and NW/N Iceland (Fig. 3) (Jennings et al., 2011; Olafsdottir et al., 2010; Andresen et al., 2013), where drift ice can impinge on the shelf during NAO positive circulation (Fig. 2c), and thus variable values in IRD and IP25 proxies can be expected (Andrews et al., 2009a).

2 The oceanographic, climatic, and physical background

The present-day velocity of bottom currents through the Denmark Strait is high (Jochumsen et al., 2012) and little sediment accumulation takes place. Core-top dates are no younger than 8000 cal yr BP (Andrews and Cartee-

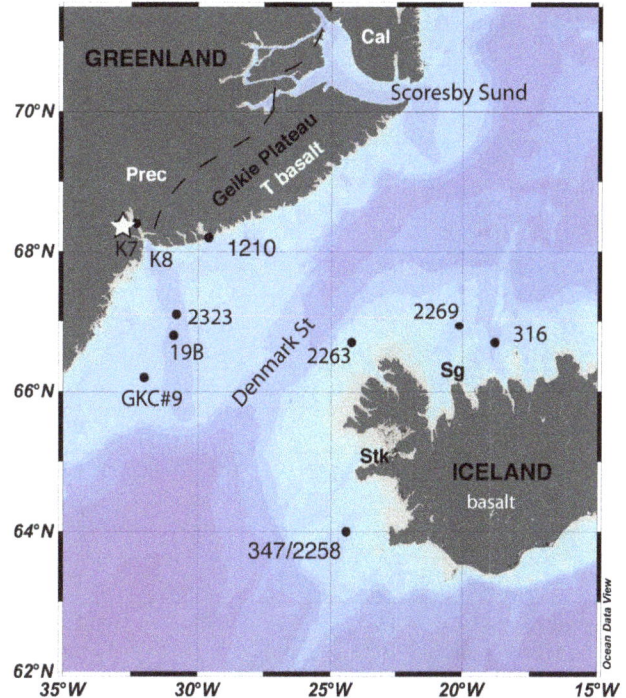

Fig. 1. Bathymetry and core locations – see Table 1 for details. The basalt outcrop of the Geikie Plateau is delimited by a dashed line. The letters Stk refer to the location of the Stykksholmur climate station (Fig. 3a) and Sg to the Siglunes transect (Fig. 3a). Core K14 (see text) is at the same location as core 1210. The large white star in Kangerlussuaq Fjord shows the location of the Tertiary felsic intrusion. North of Scoresby Sund is an extensive outcrop of Caledonide (Cal) sediments and igneous rocks (Higgins et al., 2008). Elsewhere the bedrock is dominated by Precambrian (Prec) basement.

Schoofield, 2003), thus limiting late Holocene proxy records to the adjacent shelves rather than from the Denmark Strait itself.

The East Greenland current (EGC) (Fig. 2) transports polar water, sea ice, and icebergs along the E Greenland margin (Rudels et al., 2012). The East Iceland current (EIC) branches from the main EGC and flows toward the N Iceland shelf where it comes in contact with the North Iceland Irminger current (NIIC), a northward flowing branch of the Irminger current (IC). The IC flows northward along the W Iceland shelf and a branch is directed westward where it moves landward in the Kangerlussuaq Trough as an intermediate water mass (Jennings et al., 2002a, 2011) although the main branch moves southward along the shelf break.

Instrumental and observational data are available from either side of the Denmark Strait for the last one or two centuries (Fig. 3), whereas incursions of sea ice onto the N Iceland coast have a longer record (Wallevik and Sigurjonsson, 1998) (Fig. 3d). The mean annual temperature (MAT) records from opposite sides of the Denmark Strait (Stykksholmur: W Iceland and Ammassalik: SE Greenland) share significant similarities with r^2 values (1895 AD to

Table 1. Core locations (see Fig. 1).

	Core type	Length (cm) 2 cal ka BP	Long.	Lat.	Water depth (m)	Pb210	SAR (mm yr^{-1})	Sampling resolution
			E Greenland					
JM96-1210	GC	240	−29.60	68.20	452	yes	1	10 yr
BS1191-K14	GC	> 162	−29.60	68.19	459	no	1.2	
BS1191-K7	GC	> 241	−32.10	68.26	862		3.2	∼ 15 yr
BS1191-K8	GC	> 234	−31.86	68.13	872		2.6	15 yr
MD99-2322	PC	230	−30.9	66.8	714	no	1.56	17 yr
HU93-019B	BC	48.5	−30.80	67.10	713	yes	1.94	5 yr
PO175GKC#9	GKC	20	−32.00	66.20	313	yes	0.63	16 yr
			NW/N ICELAND					
B997-316	PC and SGC	269 and 20	−18.80	66.70	658	no	2.47	20 yr
MD99-2269	PC	410	−20.85	66.63	365	no	2.03	20 yr
MD99-2263	BC	45.5	−24.20	66.70	235	yes		
			SW ICELAND					
B997-347	PC	34	−24.2	63.2	321	no	0.17	100 yr
MD99-2258	BC	24	−24.4	64	355	yes	0.62	15 yr
BC	box core							
GC and GKC	gravity core							
PC	piston core							
SGC	short gravity core							

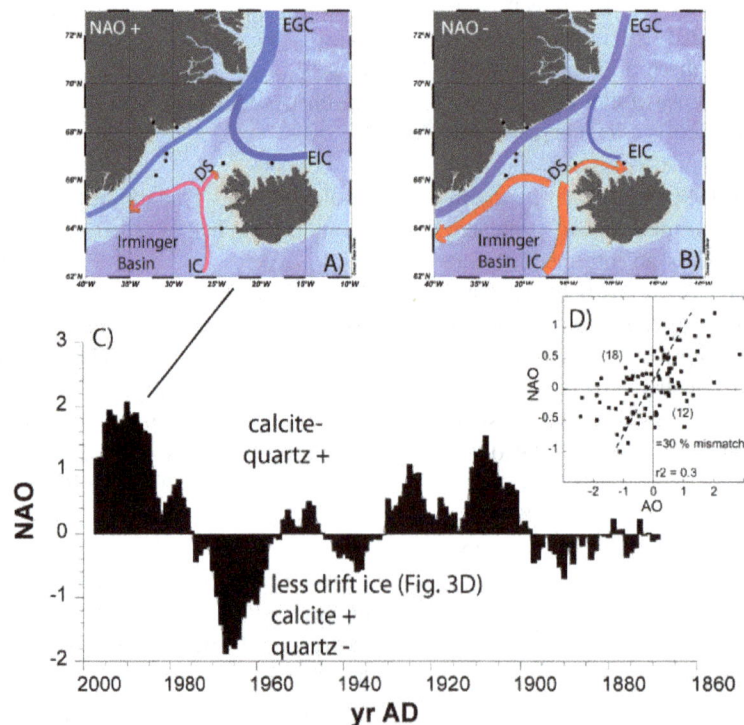

Fig. 2. EGC = East Greenland current; EIC = East Iceland current; IC = Irminger current. Core sites (Fig. 1) are shown as small dots. (**A**) Schematic plot of surface currents in the area of Denmark Strait (DS) under the influence of a NAO positive regime, and (**B**) NAO negative regime (after Blindheim et al., 2001). (**C**) Variations in the NAO winter index – data smoothed with a 10 yr running mean. (**D**) Plot of the annual indexes for the AO and NAO (1899–2000 AD).

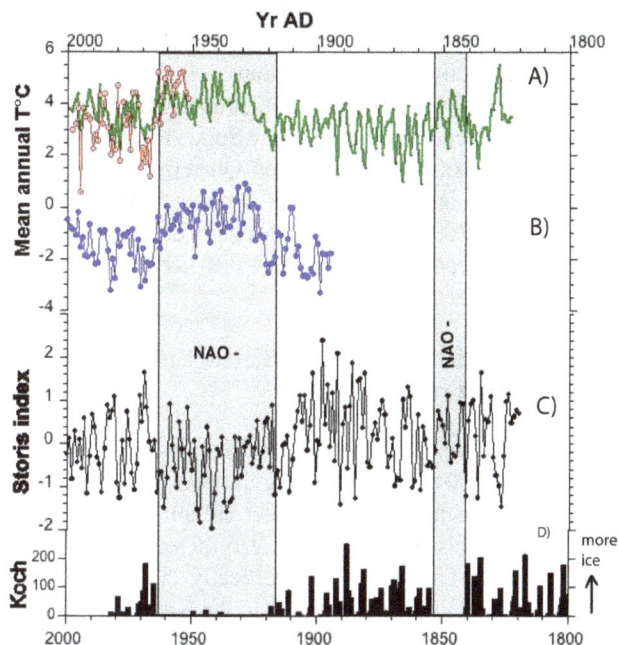

Fig. 3. Historical data from the area of Denmark Strait: (A) mean annual temperature data from W Iceland (green), and the Siglunes (red) marine record (Fig. 1); (B) MAT (blue) from SE Greenland; (C) index of storis (Schmith and Hanssen, 2003); (D) revised Koch sea-ice index. Grey shaded areas represent intervals of reduced drift ice during inferred NAO negative circulation (Fig. 2).

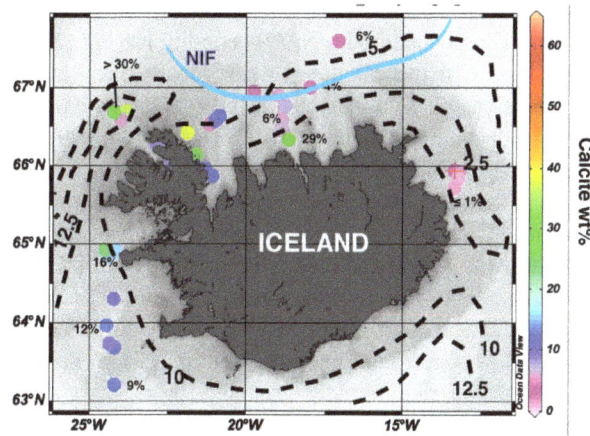

Fig. 4. Total carbonate weight% in the > 2 mm seafloor sediments around Iceland (colored circles and black numbers) (i.e. 12 %), and the index of vertical mixing dashed black lines show the (Stefansson and Olafsson, 1991) (units see text) means for 20 May–15 June, 1972–1984. The heavy blue line shows the average position of the North Iceland Front (NFI)(Belkin et al., 2009).

the present) of 0.65. There is a tight coupling between the land and marine temperatures as can be seen by the parallel trends in the Stykksholmur MAT and the average water column temperatures from the Siglunes hydrographic section (Olafsson, 1999) (Fig. 3); however, direct yearly correlations between the NAO index and the MAT data are low due to the complex underlying causes of the NAO (Zhang and Vallis, 2006). Positive values of the storis index (Schmith and Hanssen, 2003) indicate heavy ice years off SW Greenland, fed by the EGC. Zhang and Vallis (2006) noted lagged correlations between low-frequency data of Great Salinity Anomalies (GSA) (Belkin et al., 1998; Dickson et al., 1988), Iceland sea-ice extent, Labrador Sea sea surface temperature (SST) anomalies, and the NAO index. Indeed they noted (Zhang and Vallis, 2006, p. 476) that over the last century positive Iceland sea-ice anomalies lead (forces?) the positive NAO by 7 yr. Jennings et al. (2002a, p. 56) also drew attention to the "NAO paradox" (Dickson et al., 2000), noting that "[...]an extreme example of the lack of correspondence between the NAO and sea-ice flux is the Great Salinity Anomaly" because this occurred during a NAO-mode. Moreover, Lamb (1979) suggested that the climate generated by the 1969 GSA could be an analog for conditions during the LIA in the North Atlantic.

Stefansson and Olafsson (1991) provided a link between Icelandic oceanography and marine environmental proxies by synthesizing nutrient data, collected on the regular hy-

drographic cruises, from the Iceland Marine Research Institute (Stefansson, 1968). A key argument was the importance of water stratification as a control on productivity. They computed an index of vertical mixing (IVM), the reciprocal of the water column stability, and defined IVM as: $10^{-2}[D\sigma_t/Dz]-1$. Stefansson and Olafsson (1991) noted that the IVM has statistically significant positive correlations with nutrient variables. Maps of IVM values shows a tongue of high values extending northward along the W/NW Iceland shelf corresponding to the flow of the Irminger current (Fig. 4). The contours on the IVM parallel the calcite weight% of surface sediment from the seafloor, with a tongue of increased values heading NNE along the west coast terminating in the Djupall Trough, where calcite weight% (wt%) exceeds 30 wt% (Fig. 4). This broad trend supports the hypothesis that calcite wt% is a measure of water stratification and temperature (Andrews et al., 2001; Mahadevan et al., 2012; Thórdardóttir, 1984). There is a statistically significant relationship ($n = 26$, $p = 0.0002$) between 50 m water temperature and the weight% of carbonate for all sites, but excluding Djupall, which for its temperature, has anomalously high carbonate values. The relationship indicates an increase in carbonate of 1.8 ± 0.4 % for a 1 °C increase. The temperature data specifically relate to late July 1997 (Helgadottir, 1997).

The major component of the biogenic calcite is the Coccolithophorids, which bloom along the Iceland and East Greenland margins (Balestra et al., 2004; Thórdardóttir, 1984; Dylmer et al., 2013) – an exception is NW Iceland where foraminifera are a significant component. Icebergs may also act to disturb the water column, break down stratification and increase nutrient supply (Burton et al., 2012). Increased stratification is linked to excursions of fresh polar/Arctic waters onto the Iceland shelf (Olafsson, 1999), which in turn is associated with NAO+ conditions (Fig. 2), and thus changes in the calcite wt%, or ideally the flux, might serve as an index to changes in the NAO/AO or the Atlantic Multidecadal Oscillation (AMO) (Hakkinen, 1999; Mauritzen and Hakkinen, 1997).

Changes in sea-ice extent are considered a major climatic forcing for changes in the Holocene (Jennings et al., 2002a; Smith et al., 2003) and modelling indicates that such changes might be forced by volcanism (Schleussner and Feulner, 2013; Zhong et al., 2011). Sea ice is a pervasive factor along the East Greenland margin (Melling, 2012). Sea-ice charts (www.dmi.dk/dmi/index/gronland/iskort/iskort_-_ostgronland.htm) for the last few years show a narrow band of fast ice exists from 67° N northward and extends into the fjords, whereas offshore from this ice concentrations range from < 9/10th to the ice marginal zone (< 20 %), which from December through the spring lies close to the shelf break (Alonso-Garcia et al., 2013). In NAO/AO + years, drift ice is frequently transported in the EIC toward Iceland; 20th Century observations and historical sources indicate significant variability on decadal timescales (Fig. 3c, d) (Polyakova et al., 2006).

Kangerlussuaq Fjord, the Geikie Plateau, and Scoresby Sund all contain fast-flowing tidewater glaciers that contribute icebergs from their tidewater margins (Dwyer, 1993; Nuttall, 1993; Seale et al., 2011). Several of the fast-flowing ice streams are buttressed by a sikussuaq, a mélange of sea ice, bergy bits, and icebergs (Mugford and Dowdeswell, 2010; Reeh, 2004; Reeh et al., 2001; Syvitski et al., 1996). Sikussuaq and landfast ice within the fjord, restrict the movement of icebergs, and consequentially impact their ability to transport and release ice-rafted debris (IRD) to the adjacent shelf. Icebergs are frequent in the area south of Scoresby Sund (Dowdeswell et al., 1992) and can cause scouring and remobilization of sediment on the sea floor (Syvitski et al., 2001). Retreat of the tidewater glaciers has been linked to the introduction of "warm" Atlantic Intermediate water into the fjords (Andresen et al., 2011; Straneo et al., 2010; Howat et al., 2008; Joughin et al., 2008; Syvitski et al., 1996). Coarse (> 2 mm) IRD decreases rapidly seaward from the E Greenland fjords following a power law (Andrews et al., 1997) and is thus only rarely noted in late Holocene Icelandic marine sediments.

The bedrock geology map (www.geus.dk) for Nansen Fjord shows Achaean gneisses and granites overlain in places by K–TK shales, siltstones and sandstones, which are capped by an extensive early Tertiary basalt outcrop forming the Geikie Plateau (Larsen et al., 1999) (Fig. 1). The bedrock geology of Kangerlussuaq Fjord is complex (Map sheet 13, 1 : 250 000 Greenland Geological Survey) and differs from the Nansen Fjord in that outcrops of KT sediments and flood basalts are rare. Archean gneisses and granites dominate the bedrock geology. A massive felsic Tertiary intrusion outcrops on the south flank of the fjord (Fig. 1). North of Scoresby

Sund there is an extensive coast-parallel outcrop of Caledonian sediments and igneous rocks (Higgins et al., 2008), which have a distinctive mineral signature in their fjord and shelf sediments (Andrews et al., 2010), although this signature is severely muted south of Scoresby Sund. The bedrock of Iceland consists of late Tertiary and Quaternary basalts and other volcanic facies with minor sediment inclusions (Hardarson et al., 2008). Quartz is virtually absent in the bedrock and is transported to Iceland in drift-ice (Eiriksson et al., 2000; Moros et al., 2006).

3   A working model for detecting climate change in the Denmark Strait region

Malmberg (1969) suggested that the reduction in the salinity of surface (upper 200 m) waters during the time of the GSA would increase the stability of the water column, hence reduce convection. This process would also increase the formation of sea ice. This thesis was critically evaluated by Marsden et al. (1991) who showed that in the Greenland and Labrador Seas between 1953 and 1980, sea-ice anomalies indeed lagged salinity anomalies. We link incursions of sea ice to an increase in the transport of "foreign" minerals onto the Iceland shelf and concomitantly to a decrease in marine productivity. These changes may thus provide a proxy for the Thermohaline Circulation (THC) (Broecker, 1997). Axford et al. (2011) compiled marine and lake climate proxies from Iceland at a 100 yr sampling resolution for AD 300–1900 cal yr. Further analysis (in our paper) of 18 variables from the 5 marine cores by Principal Component Analysis (PCA) (Davis, 1986) indicated that the 1st PC axis explained 68 % of the variance, with changes in quartz, the sea-ice biomarker IP25, and δ^{18}O of foraminifera being negatively correlated with changes in calcite wt%. The scores on the 1st PC (see later) represent a 1st-order climate signal for the eastern side of the Denmark Strait. This "seesaw" of proxies fits well with the schematic picture of NAO variations depicted on Fig. 2, such that with incursions of sea ice and reduction in the IC during positive NAO/AO conditions off Iceland, we would expect an increase in quartz, an increase in the sea-ice biomarker IP25, and a reduction in calcite production.

We employ three major proxies to capture changes in marine conditions. These are (1) changes in sediment mineralogy, especially the variations in quartz weight% (wt%), which represents an external input to the Iceland bedrock signatures (Andrews et al., 2009b; Eiriksson et al., 2000; Moros et al., 2006); (2) variations in IRD, defined as clasts > 2 mm or very coarse sand; and (3) variations in the calcite content. In terms of our proxies, changes in stratification would result in changes in coccolithophorid accumulation rates associated with changes in the timing of the extent and nature of the sea-ice margin (Balestra et al., 2004; Giraudeau et al., 2004, 2010).

3.1 Cores, chronology, and methods

We present data from cores (Table 1, Fig. 1) retrieved during cruises in 1991, 1993, 1996, 1997, and 1999 (e.g. Helgadottir, 1997; Labeyrie et al., 2003). All the radiocarbon dates on the cores have been published in INSTAAR Date Lists (Dunhill et al., 2004; Quillmann et al., 2009; Smith and Licht, 2000), apart from new dates obtained on cores HU93030-19B, MD99-2322, JM96-1210GGC, the three PO175GKC cores (Table 2), and 4 dates from B997-316PC (Jonsdottir, 2001). ^{210}Pb and ^{137}Cs measurements have been carried out (Table 2) on a GKC core and several box cores (Alonso-Garcia et al., 2013; Andrews et al., 2009a; Smith et al., 2002).

An important issue in radiocarbon-based chronologies is the value(s) for the ocean reservoir correction on either side of the Denmark Strait (Eiriksson et al., 2010; Hjort, 1973; Jennings et al., 2002b, 2011). The presence of well-dated tephras in many cores enables some insights into the problem (Jennings et al., 2002b), but no corrections have been determined for NW/W Iceland over the last 2000 cal yr. Our value for ΔR of 0, and usage in radiocarbon calibration programs (Reimer et al., 2002) means that our ^{14}C age calibrated ages may be too old. OxCal was used to calibrate the radiocarbon dates (Blockley et al., 2007). Core lengths accumulated over the last 2000 cal yr, sediment accumulation rates (SAR), and sampling intervals are listed on Table 1. These data show that there is the potential to reconstruct conditions across the Denmark Strait at multidecadal to millennial resolution, although the impact of bioturbation is always an issue (Anderson, 2001). However, the high SARs and the preservation of discrete tephra events in several of the cores (e.g. Jennings et al., 2002b) suggest only a limited impact. The age scales on subsequent figures are based on interpolation of the depth–age relationships between each date in a core (Table 1), rather than fitting them to a mathematical model. As noted in Andrews et al. (1999) the errors on interpolated ages are ≥ the error estimates on the bounding dates (i.e. Table 2). We do this with the clear knowledge that "All age-depth models are wrong: but how badly?" (Telford et al., 2003). Thus, as we will discuss later, potential errors in depth–age models makes it uncertain as to whether there are leads and lags in responses between sites, or whether the differences are associated with imperfect age models. These issues can be mitigated by wiggle matching, based on a high density of dates (e.g. Telford et al., 2003; Sejrup et al., 2010), but are not financially practical for a large regional study such as ours.

Our proxies have been measured using methodologies that have been well documented in the literature. The mineralogy of the sediments has been estimated by quantitative X-ray diffraction analysis using ZnO as a calibration standard (Eberl, 2003; McCarty, 2002; Omotoso et al., 2006) on the < 2 mm sediment matrix (Andrews et al., 2010). We use the downcore variations in mineralogy to seek an understanding of changes in sediment provenance using an Excel macro program called SedUnMix (Andrews and Eberl, 2012). Cal-

cite and total carbonate (as there is little/no dolomite in the samples calcite ~ carbonate%s) have been measured by XRD, coulometer, or by a WHOI carbonate device – comparisons of duplicate runs indicated excellent reproducibility between the different methods. In addition to sediment samples, large IRD clasts from the JM96-1210, GKC#9, and MD99-2262 (W Iceland) cores were pulverized for qXRD in order to gain further insights into sediment source mineralogy. We have taken surface samples and cores from Kangerlussuaq Fjord in 1988, 1991, and 1993 (Smith and Andrews, 2000; Syvitski et al., 1996), and surface samples were also taken on the Sir James Clark Ross cruise JR106 in 2004. It is appropriate at this stage to note that we are often dealing with percentage data, i.e. the non-clay and clay mineral weight% (wt%) sum to 100, and this poses problems in the interpretation of changes in percentage data (Aitchison, 1986). On X-radiographs, clasts > 2 mm were counted continuously in 2 cm depth increments (Andrews et al., 1997; Grobe, 1987; Pirrung et al., 2002). If such data are not available we use the percentage of sand-size grains in fractions > 1 mm as an index of iceberg rafting (Jennings et al., 2011), as sea ice derived from the Arctic Ocean contains predominantly silt and clay-size particles (Dethleff, 2005; Dethleff and Kuhlmann, 2010).

4 Results and interpretation

We present data from our study areas beginning with the heavily glaciated fjords of E Greenland and ending with the the ice-free waters off SW Iceland – the impact of low-frequency variations will vary along such a transect. The least affected areas will be the ice-free SW Iceland shelf, whereas the impact on the numerous E Greenland tidewater glaciers will be very significant if the NAO-mode (Fig. 2) is sufficiently persistent to increase the flow of Atlantic water into the fjords (Straneo et al., 2010; Andresen et al., 2012). This would increase glacier calving and IRD deposition in the fjords and inner shelf, but the increased rate of iceberg melting would limit the area over which IRD could be delivered.

4.1 Fjords, East Greenland

4.1.1 Geikie Plateau

Core JM96-1210GGC was taken near the mouth of Nansen Fjord close to the site of BS1191-K14 (Jennings and Weiner, 1996) (Fig. 1). Local IRD basalt clasts contain some quartz from crustal contamination. The Christian IV Gletscher at the head of the fjord has an estimated calving rate of ~ 2 km^3 yr^{-1} (Andrews et al., 1994). CTD casts in 1996 indicated that water temperatures below 150 m were between 1.5 and 2 °C, thus ensuring rapid melt of icebergs exiting the fjord. The depth–age model is based on 3 ^{14}C dates (Table 2) and ^{210}Pb and ^{137}Cs measurements (Smith et al., 2002); the sediment accumulation rate (SAR) averaged 0.11 cm yr^{-1}.

The average sampling interval over the last 2000 cal yr is 20 yr sample^{-1}.

The grain-size spectra (Fig. 5a), multi-modal peaks in the > 2 mm fraction, are clearly ice-rafted (Fig. 5b). Sand content is relatively uniform (Fig. 5b) but the largest fraction is medium to coarse silt reflecting glacial abrasion. The calcite wt% is at the detection limit; an interpretation is complicated by the presence of marine mudstones within the drainage (Larsen et al., 1999). To express the qXRD data in terms of provenance, the minerals have been grouped (e.g. Na and Ca-feldspars grouped into plagioclase) and smoothed with a 5-point moving average. The 2000 yr record of changes in mineralogy in part reflects changes in grain size with a fining-upward trend being mimicked by a decrease in quartz. The plagioclase and quartz percentages tend to parallel each other and both show significant oscillations. In Fig. 5d we have used SedUnMix to estimate the downcore variability of the mineral signal by comparing each sample with the uppermost 5 samples in the core (i.e. the recent sediment sources) (Andrews and Eberl, 2012). As might be expected there is not a great deal of variation (red line, Fig. 5d) and the average absolute difference in mineralogy between the observed and SedUnMix calculated wt% is generally < 1 wt% (Fig. 5d, blue line). When compared with the foraminifera-based climate divisions from the near by core BS1191-K14 (Table 1) (Jennings and Weiner, 1996) there are some associations and the Little Ice Age (LIA) appears to be represented by an increase in quartz and plagioclase but larger peaks occur between the LIA and the Medieval Warm Period (MWP). Wanner et al. (2011) suggested two cold intervals within the last 2000 cal yr at 500 and 1500 AD – the former event (the so-called Dark Ages cold event) is marked by maximum wt% values of quartz and plagioclase and relatively large deviations in mineralogy.

4.1.2 Kangerlussuaq Fjord

Relative to Nansen Fjord sediments from Kangerlussuaq Fjord are enriched in quartz and the K- and Na-feldspars, but depleted in the Ca-feldspars, pyroxene, maghemite, saponite, and Fe-chlorite. The lithofacies, grain-size and IRD (> 2 mm) records from cores within Kangerlussuaq Fjord were presented earlier (Smith, 1997; Smith and Andrews, 2000). The sediments consisted predominantly of weakly stratified fine-grained muds with IRD (Smith, 1997, p. 78). Here we reproduce the IRD records and new qXRD data for cores BS1191-K7 and K8 (Table 1, Figs. 1 and 7), which although poorly dated (Table 2) cover the last 800 yr, and include the local Little Ice Age (Geirsdottir et al., 2000). The SAR for K8 is 0.26 and 0.11 cm yr^{-1} for K7 – using ^{210}Pb, Smith et al. (2002) determined a SAR over the last 100 yr of 0.42 cm yr^{-1} at K8. Thus the interpolated age estimates for K7 and K8 have unknown errors on the age estimates but are probably in the range of ± 150 yr; for this reason we do not correlate specific multidecadal events between the fjord and trough prox-

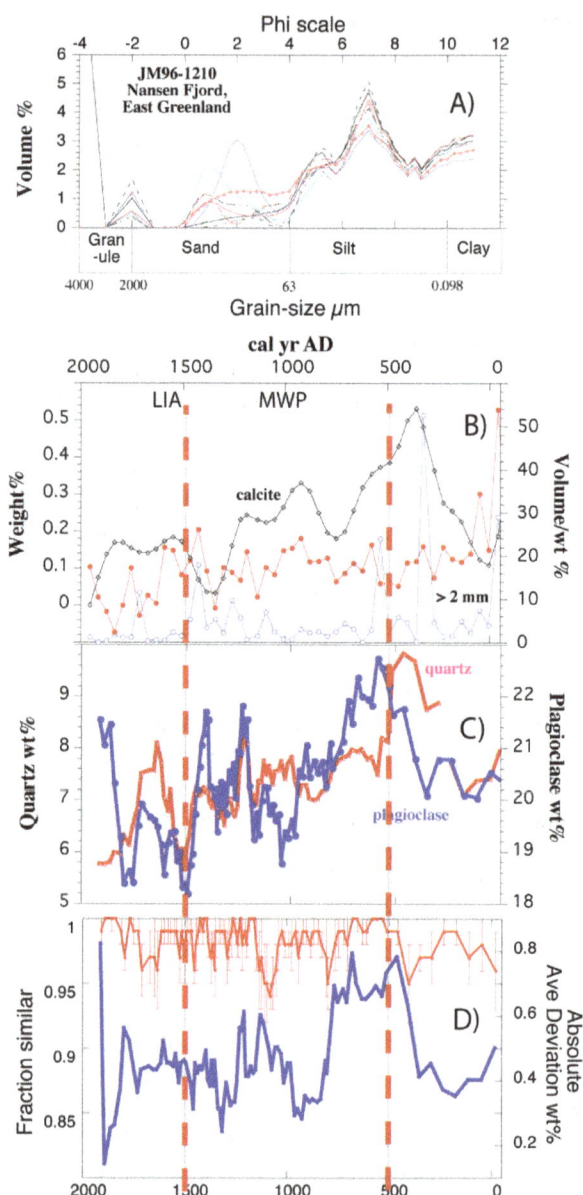

Fig. 5. Data from JM96-1210 and BS1191-K14, Nansen Fjord (Fig. 1). (**A**) Grain-size spectra from JM96-1210; (**B**) plots of the > 2 mm sediment (blue), sand (red), and wt% calcite (smoothed); (**C**) smoothed wt% quartz and plagioclase; (**D**) downcore variations and error in the estimated similarity from the uppermost 5 qXRD sample mineralogy (red) and the absolute average deviation (wt%) from the 9 mineral groups. The two vertical dashed red lines represent the timing of the latest Holocene cold events (Wanner et al., 2011).

ies. The IRD records (Fig. 6) show pulses of ice-rafting with a notable absence $\sim$ 1500 AD. In K7 a low persistent level of IRD is evident in the last 400 yr, whereas in K8, closer to the fjord mouth, no visible IRD > 2 unitmm was captured in the core for the last 300 yr. In K7 higher levels of quartz in the < 2 mm fraction appear to coincide with reduced IRD.

Table 2. Radiocarbon dates.

Core ID	Depth (cm)	[14]C date	Error	From	To	Median cal BP	Sigma
JM96-1210GGC	49	785	15	486	374	438	20
	149.5	1275	15	893	756	821	37
	332	3050	80	3051	2685	2833	94
BS1191-K14	50	855	60	521	305	436	58
	115.5	1440	70	1107	771	938	82
BS1191K7	262.5	1310	60	920	680	805	64
BS1191-K8	107.5	1155	56	768	550	664	53
	226	1391	55	1008	745	881	64
HU93030-019B	55.5	835	20	489	359	436	32
MD99-2322	2.3	675	30	359	145	273	42
	34	693	38	413	148	290	51
	101.5	1267	44	875	670	758	54
	150	1627	46	1254	1021	1137	59
	232.5	2415	20	2081	1908	1987	42
MD99-2263	7	595	15	288	145	253	37
	10	600	15	294	146	257	34
	25	850	15	514	445	482	17
	38.5	1620	15	1253	1127	1192	32
	45	2165	15	1824	1690	1759	35
MD99-2269	1	72	37	55	10	36	11
	42.5	680	30	413	257	328	41
	131	1010	30	646	530	589	31
	177.5	1226	25	858	688	762	41
	265.5	1693	42	1334	1159	1250	43
	412	2396	47	2150	1890	2028	67
B997-316	18.5	402	38	121		44	38
	41.75	755	35	471	305	396	46
	118	1020	40	660	525	595	36
	219.25	1090	90	830	505	655	81
	249.5	1550	120	1342	862	1102	121
MD99-2258	27.5	725	15	429	301	367	35
	31.25	1937	36	1587	1380	1482	52
	70.5	3180	20	3068	2877	2975	48
B997-347	3	770	65	504	285	403	61
	61	3110	45	3020	2760	2886	66

5 Kangerlussuaq Trough

Icebergs with keel depths < 400 m can exit the fjord and icebergs of this size have been observed (Dowdeswell et al., 1992) and inferred based on iceberg scours on the adjacent seafloor (Syvitski et al., 2001). High-resolution seismic surveys of the inner Kangerlussuaq Trough in 1993 and 1996 (Jennings et al., 2002b, 2006, 2011) indicates that cores HU93030-019B and MD99-2322 (henceforth 2322) (Fig. 1) are stratigraphically from the same site and are well below the limit for iceberg scouring. The ^{210}Pb profile and ^{14}C dates from 019B (Smith et al., 2002; Table 2) are conformable in contrast to the results from PO175GKC#9 (Alonso-Garcia et al., 2013), farther down the trough but in only 300 m of water, where Cold Room storage (calcite dissolution) and possible reworking led to the ^{14}C dates being rejected. The calibrated dates from the base of 19B box core and the 2322 Calypso core overlap (Table 2). The SAR for 19B is ~0.127 and 0.148 cm yr^{-1} for 2322; 19B was processed in 1 cm intervals and 2322 averaged 2 cm intervals. The average resolution is one sample every 15 yr.

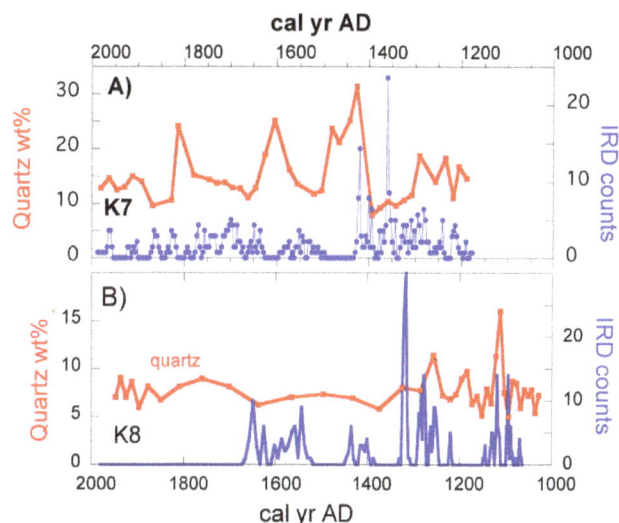

Fig. 6. IRD (blue) and quartz wt% data (red) from BS1191-K7 (**A**) and -K8, outer Kangerlussuaq Fjord (**B**) (Tables 1 and 2).

The sediment in the upper 2000 yr of 2322 is generally silty clay with scattered IRD clasts and coarse sand (Jennings et al., 2011). The mineralogy of the < 2 mm fraction was determined by qXRD and the sampling for the last 2000 cal yr was increased from an earlier study (Andrews et al., 2010). The inner Kangerlussuaq Trough is positioned to receive sediment from the Kangerlussuaq Fjord tidewater glaciers, and from the numerous tidewater glaciers that drain the Geikie Plateau (Fig. 1). A key question is whether the sediment in Kangerlussuaq Trough is being contributed from the erosion of felsic rocks (e.g. granites, gneisses, felsic intrusions – Kangerlussuaq Fjord), from erosion of the complex basalt outcrop (Geikie Plateau), or some fraction from each. A 5-point moving average has been applied to the data to enhance the signal (Burroughs, 2003).

The SedUnMix program allows for 5 samples per end member or source with a maximum of six sources; details of the program are given in Andrews and Eberl (2012). We initially evaluate downcore variability in sediment composition by comparing the downcore sediment samples with the 5 uppermost samples from 019B and computing the standard error based on 10 iterations of the SedUnMix algorithm. The results indicate that in the last ~500 yr there has been little variation in sediment provenance, although commencing ca. 1500 AD there were consistent deviations (Fig. 7a).

The next question is whether we can assign sediment sources and unmix the sediment composition. We represent the Tertiary basalt source by the qXRD mineralogy from JM96-1210 and the Kangerlussuaq Fjord source by surface samples from core BS1191-K7 (Alonso-Garcia et al., 2013; Andrews et al., 2010). However, we note that interpretation is complicated by the closed array problem (Chayes, 1971) so that as the contribution from one source increases there is a tendency ($r = -0.58$) for the other source to decrease

(Fig. 7). The average difference between the measured and calculated weight% for each of the non-clay and clay minerals used in the analyses is 3.3 ± 0.3 wt%, indicating a reasonably good fit between the measured (sources) and calculated (2322) mineral wt%s. The results (Fig. 7b) indicate that Geikie Plateau has been the dominant sediment source with an average of 45 ± 7 % of the sediment being assigned to a Nansen Fjord-like source, compared to 22 ± 11 % from Kangerlussuaq Fjord. There is thus an unaccounted source(s) averaging ~ 33 % of the composition (Fig. 7c). However, our estimates also include a calculation of the standard deviations of the estimate based on iterations of the 2×5 matrix (Andrews and Eberl, 2012); thus, we could account for nearly 100 % of the sediment composition by specific sample selection from the two source areas. Additional sediment sources would certainly be derived from icebergs that originated in Scoresby Sund and northeast Greenland and sediment entrained in sea ice from the Arctic Ocean (Andrews, 2011; Bigg, 1999; Darby and Bischof, 2004; Darby et al., 2011; Seale et al., 2011). Over the 100 yr interval ending ca. 1993 AD, the relative contribution from the Geikie Plateau has increased, whereas the data indicate a decrease in the more felsic source(s) associated with Kangerlussuaq Fjord (Fig. 7b).

The last 2000 yr covers several important climate events, such as the Medieval Warm Period (MWP) and Little Ice Age (LIA) (Fig. 8) although their time boundaries are probably not globally synchronous nor agreed on (Hughes and Diaz, 1994; Broecker, 2001: Mann et al., 2009). For the purpose of this paper we plot the boundaries for the MWP between 950 and 1250 AD, and the LIA from 1250 to 1910 AD. A 15 yr equi-spaced integrated time series was generated for 2322 data using AnalySeries (Paillard et al., 1996). Figure 8a graphs the PC 1 scores from the Iceland marine data set (Axford et al., 2011) versus the calcite wt% from 2322. The 2322 calcite data parallel the lower-resolution PC scores, suggesting a broadly parallel climate evolution on both sides of the Denmark Strait, in keeping with the climate data for the last 100–150 yr (Fig. 3). We use a regime shift indicator (RSI) algorithm (Rodionov, 2004, 2006) to ascertain whether the records show marked regime shifts (Fig. 8b and c), and whether these shifts have any correspondence with the 14C probability distributions of vegetation kill/ice growth on Baffin Island, Canada (Miller et al., 2012) (Fig. 8b). Such a comparison is appropriate, as the changes in the small Baffin Island ice caps have been linked to intervals of persistent volcanism which force variations in the extent of sea ice (Miller et al., 2012; Schleussner and Feulner, 2013). The peaks in the ^{14}C probability distribution reflect intervals when vegetation was able to grow on the uplands of Baffin Island, and the absence of dates represents intervals of persistent ice/snow cover (Williams, 1978). Lowell et al. (2013) reported dates on dead vegetation with similar ages to the north of our sites (Istorvet Ice Cap $\sim 71°$ N) (Miller et al., 2013). Particularly noteworthy in our data is the calcite minimum ~ 1600 AD

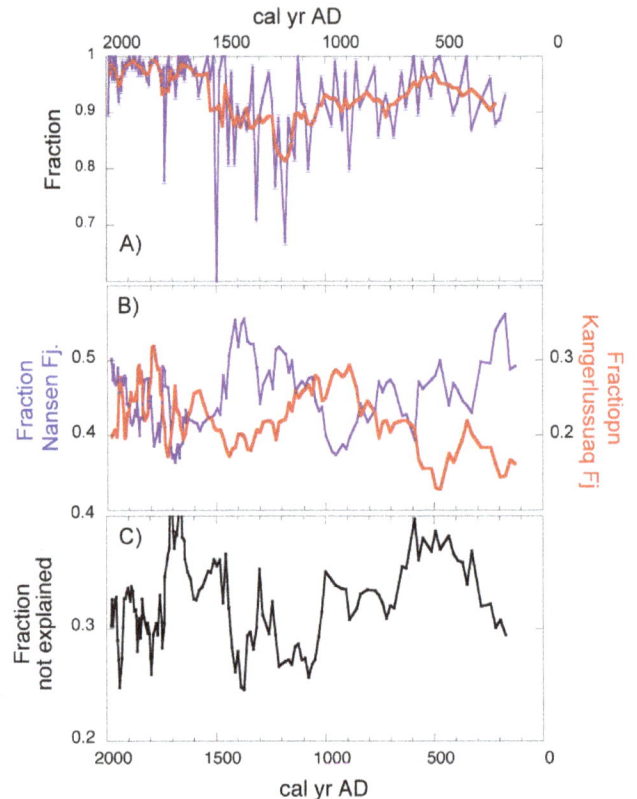

Fig. 7. (A) Changes in the fraction of sediment with a similar mineralogy to the uppermost 5 samples (last 100 yr) at site 2322 (blue) (Fig. 1) and 5 point running mean (red). **(B)** Estimated fractions of sediment to site 2322 from the Nansen Fjord (blue) or Kangerlussuaq Fjord (red) based on mineralogy. **(C)** Changes in the average fraction of sediment not attributable to either source (see Fig. 7b).

which coincides with the abrupt reduction of dates on vegetation emerging from the retreating ice patches, although lagging the vegetation reduction by ~ 100 yr, whereas the older interval, which terminated ~ 1000 AD, occurred ca. 150 yr prior to the regime shift in calcite (Fig. 8b).

The quartz wt% data has a trend opposite to the calcite wt% (Fig. 8c) with an increase starting ca. 800 AD and continuing until 1500 AD when there is a decrease, although there are sharp peaks, which project above the regime shift mean values (Fig. 8c).

5.1 N/NW Iceland: none to variable sea ice, variable IRD

N/NW Iceland sites (Figs. 1 and 2) should be sensitive to low-frequency multidecadal-/century changes, as indeed was demonstrated by Axford et al. (2011) and Andresen et al. (2005). X-radiographs from the numerous cores from the N Iceland shelf indicate that IRD clasts > 2 mm are extremely rare or absent, but quartz is present in the < 2 mm sediment fraction (Moros et al., 2006). The quartz and quartz + k-feldspar data from 16 cores from the Iceland shelf have

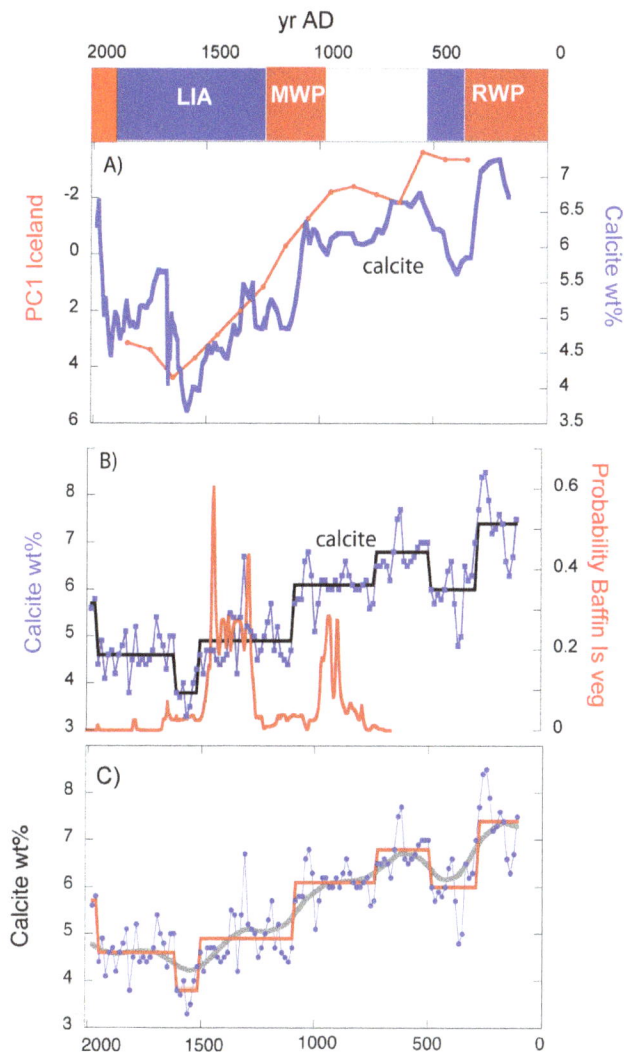

Fig. 8. (A) Climate intervals of the last 2000 yr and a graph of the 1st PC scores from the N Iceland proxy data (red) (Axford et al., 2011) versus the smoothed (5-point average) weight% calcite (blue) from the combined HU93030-019B and MD99-2322; **(B)** 15 yr calcite wt% (blue) and regime shift mean values versus the radiocarbon calibrated probability distribution for Baffin Island vegetation (Miller et al., 2012) (red); **(C)** quartz wt% (grey line with filled circles), regime shifts (black line) and the trend (heavy grey line).

been presented at 250 and 100 yr resolution (Andrews, 2009; Andrews et al., 2009b). We focus on three high-resolution records that span the last 2000 cal yr, namely MD99-2263 (Andrews et al., 2009a; Olafsdottir et al., 2010), MD99-2269 (Andrews et al., 2003; Moros et al., 2006; Stoner et al., 2007), and B997-316PC3 (Jonsdottir, 2001). Olafsdottir et al. (2010, p. 116) noted, "The characteristics for the LIA in both cores (MD99-2256, SW Iceland and MD99-2263/2264) are high amplitude fluctuations, not at all a stable continuous cold bottom water period". The three cores span an oceanographic gradient (Smith et al., 2005; Stefansson, 1962) with MD99-2263 dominated by the NIIC, whereas B997-316PC3

(Jonsdottir, 2001) occurs much further offshore at or near the North Iceland Marine Front (Fig. 4) (Belkin et al., 2009) with bottom water temperatures < 0°. This variation in oceanographic setting raises the question of differences in the ocean reservoir correction during the late Holocene (Eiriksson et al., 2010); however, the radiocarbon-based interpolated age at 105.5 cm in B997-316PC3 is 1433 AD, compared to an age for the Veiðovötn tephra of 1477 AD, identified at this depth (Jonsdottir, 2001; Larsen, 2005), indicates that our suggested reservoir correction is not unreasonable.

The calcite data (Fig. 9a) show large differences in wt% with very high values in MD99-2263 and lower values to the north and east, consistent with the suggestion of Stefansson and Olafsson (1991) and the distribution of calcite in Icelandic seafloor sediments (Fig. 4). There are considerable variations in wt% at each site and there is a marked decrease during the Iceland LIA (Geirsdottir et al., 2009; Grove, 2001; Ogilvie and Jónsson, 2001) and a consistent rise in values in recent decades. In detail (e.g. Fig. 9c), there is a remarkable similarity between the calcite wt% data from N Iceland (316PC3) and our data from the Greenland side of the Denmark Strait (Fig. 8) over the last 2000 yr. This will be examined more fully in Sect. 5 (below).

The quartz wt% time series (Fig. 9b) have values <wt 2 % for 2263, partly because of the very high calcite values at this site, compared to 2269 and 316PC3. The results for 2263 and 316PC show highly variable inputs over the last 1100 cal yr superimposed on a general increase toward the present day. This variability is in keeping with IRD deposition, which by its very nature is spatially and temporally "noisy".

5.2   SW Iceland: no sea ice and no IRD

The clockwise pattern of drift ice around the Iceland coast implies that the SW shelf is only occasionally impacted by drift ice (Divine and Dick, 2006), although Jennings et al. (2001) suggested that polar water may have impacted the area. Calcite is a major component of the sediment and reflects significant biological productivity (Stefansson and Olafsson, 1991). In Axford et al. (2011) it was shown that quartz and the sea-ice biomarker IP25 have values close to the detection limits.

Studies of MD99-2258 (box core) and the MD99-2259 piston core indicates a hiatus between ~1600 and 500 AD (Table 2). However, nearby core B997-347 (Fig. 1, Table 1) has an estimated age of the top-most sample ~1480 AD. A combined plot of calcite wt% from 2258 and 347 (Fig. 9d) shows sharp oscillations in calcite over the last 2000 yr. When compared with data from Kangerlussuaq Trough (Fig. 9d), the SW Iceland record leads by several decades. It is unclear whether this is an artifact of dating uncertainties or not, especially with the regional issue of the ocean reservoir correction (Sejrup et al., 2010). The quartz wt% in 347PC is close to zero, and it only rises above 1 % in 2258 on one occasion (~1790 AD) (Fig. 9e).

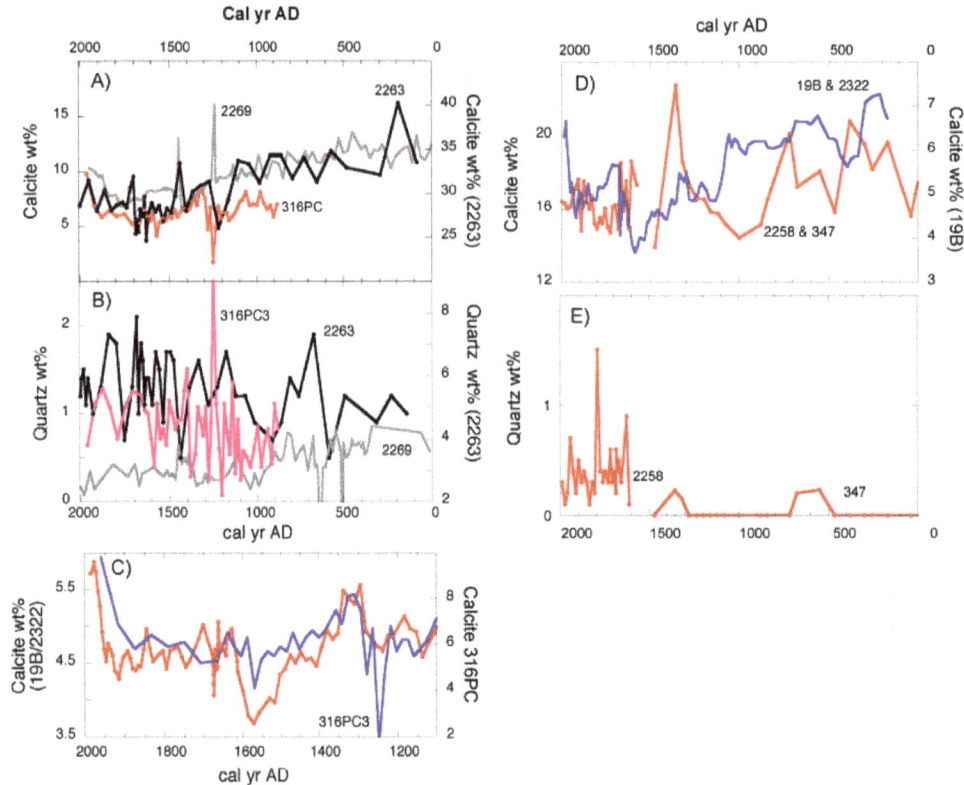

Fig. 9. (A) Calcite wt% data for sites MD99-2263 (left y axis), MD99-2269, and B997-316PC3 (right y axis) from N Iceland (Fig. 1). **(B)** Quartz wt% from same sites and same y axes. Note the difference in y axis values from the left to right axes. **(C)** Calcite wt% data from 316PC versus 19B/2322 over the last 1300 yr. **(D)** Calcite wt% data for cores from SW Iceland (Fig. 1) compared to 2322. **(E)** Quartz wt% data from SW Iceland.

6 Discussion

The common thread in our sites is provided by the variations in flow of the Irminger current (IC) and its conceptual impact on changes in calcite and quartz (Fig. 2) – the IC is a surface current off SW Iceland, a surface or intermediate current off N Iceland, and an intermediate water mass in the deep troughs and fjords of E and SE Greenland – variations in the latter are a prime cause for changes in the tidewater glacier dynamics of the area (Syvitski et al., 1996; Andresen et al., 2011). A number of Arctic/North Atlantic climate reconstructions have been developed, including a 2000 yr estimate of Arctic summer temperatures (Kaufman et al., 2009) (Fig. 11a), 1400 yr Arctic temperature (Shi et al., 2012a, b) and Arctic sea-ice records (Kinnard et al., 2011), and a 700 yr long AMO reconstruction (Mann et al., 2009; Gray et al., 2004). There is a strong association between the Arctic summer temperature time series and the 2322 calcite wt% data (Fig. 10a). On their initial chronologies the calcite and quartz data and the AMO record (Fig. 10b) are only poorly correlated. However, the fit between the AMO and calcite can easily be improved using AnalySeries (Paillard et al., 1996) (7 tie-points) to improve the association with $r = 0.76$ (Fig. 10c). The correction to the calcite time series was such

that an increase in the ocean reservoir correction from 0 to ± 30 yr would accommodate much of the suggested correction. Such a correction is certainly reasonable (Sejrup et al., 2010) but difficult to validate. The adjusted quartz time series had no obvious association with the AMO (not shown).

These data raise the question as to whether there are common trends and periodicities in calcite and quartz across Denmark Strait over the last 2000 cal yr. Stoner et al. (2007) constructed a composite Holocene age model from either side of the Denmark Strait (MD99-2269 and -2322, 440 km apart, Fig. 1) based on paleomagnetic secular variations (PSV), and an age control based on 47 ^{14}C dates. This PSV record has been shown to have regional chronological application (Olafsdottir et al., 2013). The composite chronology showed that the carbonate data from these two sites were remarkably similar (Stoner et al., 2007). We pursue this issue by examining these sites over the last 2000 yr. We do this by producing 30 yr equi-spaced data (Paillard et al., 1996) from 110 to 1970 AD for the two sites (Fig. 11) and 15 yr spaced data for 2322. Trends in the data were detected by using Singular Spectra Analysis (SSA) (Ghil et al., 2002), and the residuals from the trends were processed using SSA and Multi-Taper Method (MTM) analyses (Mann and Lees, 1996; Kondrashov et al., 2005). Given the warnings about age models

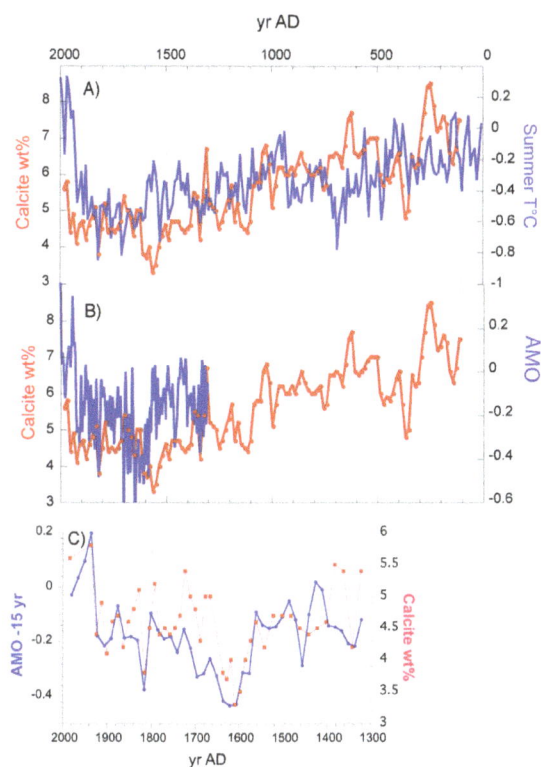

Fig. 10. (A) Plot of the calcite wt% from 19B/2322 (red) versus the summer temperature estimates (Kaufman et al., 2009) (blue). **(B)** Plot of the calcite wt% from 2322 (red) versus the AMO estimates for the last 700 yr (blue) (Mann et al., 2009). **(C)** Plot of the AMO estimates (Man et al., 2009) (blue) and corrected calcite wt% chronology (red).

(Telford et al., 2003) the 30 yr equi-spaced time series are at the limit of acceptance, although the Nyquist frequency essentially restricts the recognition of meaningful periodicities to a century or more. The 15 yr resolution series for 2322 is cautiously accepted given the high and nearly monotonic SAR but any significant period in the series will be multidecadal to century in length.

The trends in calcite from the two sites (Fig. 11a) explain 80.2 and 87.5 % of the data and are very similar to each other ($r^2 = 0.85$) on their respective time series. Both series show a progressive decrease in calcite commencing 500–600 AD, which reaches a minimum between 1550 and 1750 AD, and which is then followed by a modest increase. If this is indeed a reliable index of water stratification (Marsden al., 1991) then it suggests a persistent freshening of the surface water from Roman times on, and by inference a slow down in the THC. There is no obvious break in the data that can be associated with the MWP.

The residuals from the trend for 2322 show significant (95 %) oscillatory modes (Vautard and Ghil, 1989; Kondrashov et al., 2005) at 192 yr (175 yr MTM) and weaker modes at 110 and 76 yr. The reconstructed time series based on these modes (Fig. 11b) explains 82 % of the residual vari-

ance. A oscillatory mode (Vautard and Ghil, 1989) was also dominant on the 2269 calcite residuals (Fig. 11c), with a period of 114 yr (128 yr MTM) and weaker modes at 292 yr (370 yr MTM), and 67 yr (86 yr MTM). The reconstructed time series based on these modes explains 91 % of their variance. The oscillatory modes at these two sites, some 440 km apart (Fig. 1), are similar but not identical; however, the reconstructed time series from the two sites (Fig. 11a and b) are not correlated ($r^2 = 0.02$). The 15 yr spaced calcite and quartz data from 2322 were detrended which delimited two significant multidecadal oscillatory modes with periods of 90, and 62 yr compared to a single mode of 68 yr in quartz. The oscillations lie within range usually ascribed to the AMO.

In contrast, the trends in the quartz wt% across Denmark Strait (Fig. 11d) are very different and explain 44 % (2322) and 72 % (2269) of the variance. The 2269 trend is to a degree a mirror image of the calcite data (Fig. 11a) with quartz wt% increasing during the LIA with a low interval during the MWP. On the East Greenland shelf the quartz wt% reaches a maximum ca. 1200 AD and has a distinct low during the LIA (Fig. 11d). This may indicate that the IRD export of sediment from the quartz-rich areas on either side of the Geikie Plateau (Fig. 1) was restricted by pervasive landfast sea ice during the LIA (Reeh, 2004). Residuals from the SSA East Greenland quartz trend (Fig. 11e) had two dominant oscillatory modes with periods of 95 and 190 yr (104 and 296 yr MTM), which explained 78 % of the residual variance, whereas 50 % of the residual variance in the 2269 quartz data are associated with a mode of 155 yr (128 yr MTM). The coherence between the residual calcite and quartz records from these two sites (Fig. 1) was evaluated using MTM; this resulted in some what similar multicentury periods of 440 and 133 yr at 2322 compared to 416 and 181 yr for 2269.

The link between the NAO and ocean circulation and conditions in the Denmark Strait region (Figs. 2 and 3) suggests it is important to see if our records (Figs. 5–9) have any obvious association with a NAO reconstruction (Cook et al., 2002; Trouet et al., 2009; Kinnard et al., 2011). Departures from the mean for the period 1990–1100 AD were derived for the calcite and quartz wt% data for 2322 and 2269, respectively, and compared with the integrated 15 yr NAO index (Fig. 12c). Between 1100 and 1450 AD a prevailing low frequency NAO/AO-like positive index is associated with largely positive departures for both quartz and calcite off East Greenland, whereas on the N Iceland shelf the departures are positive for calcite and negative for quartz (Fig. 12d and e). At around 1450 AD there is a fundamental switch in all records and the reconstructed oscillatory NAO signal is matched by the calcite departures at 2322 (Fig. 12a) whereas the other records (Fig. 12b, d, and e) switch sign with a return to positive calcite departures in the last several decades. The reduction in quartz over the last 900 yr at 2322 may reflect a reduction in iceberg drift because of the development of pervasive land-fast sea ice along the NE Greenland shelf

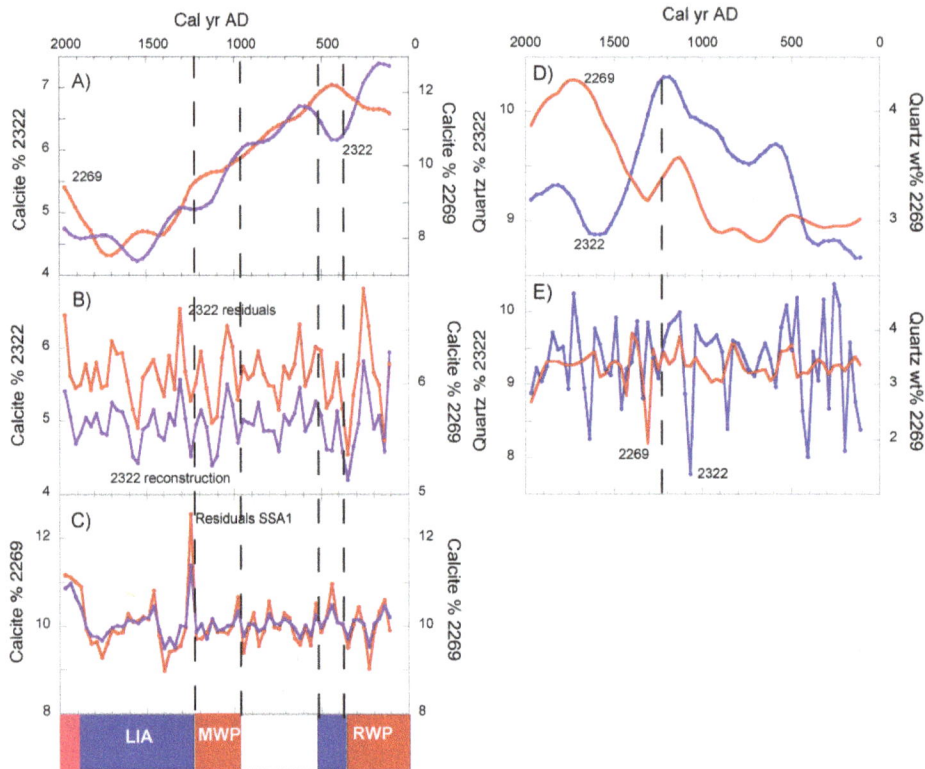

Fig. 11. (A) The SSA1 detrended calcite data for sites MD99-2322 and -2269 (Fig. 1), some 440 km apart. **(B)** Residuals from the SSA1 trend for MD99-2322 (red) and the reconstructed time series (blue). **(C)** Residuals from the SSA1 trend for MD99-2269 (red) and the reconstructed time series (blue). **(D)** SSA trends on quartz wt%. **(E)** residuals from the quartz trends (Fig. 11e). The lower left panel shows the same climate intervals as in Fig. 8.

(Reeh, 2004; Funder et al., 2011). For the interval 1100–1450 the reconstructed NAO (Fig. 12c) (Cook et al., 2002; Trouet et al., 2009; Kinnard et al., 2011) does not accord with our conceptual NAO model (Fig. 2), which links positive calcite departures with a NAO negative circulation. We are unsure as to why this is the case (see also Lehner et al., 2012).

7   Conclusions

Initial and reviewed drafts of this paper strongly focused on "NAO-like" variations as an explanation for our proxies. There is, however, the question as to how far a multi-year atmospheric forcing is a reasonable analog for multi-century to millienial variations (e.g. Figs. 10–12). This might be especially so for our region which lies at the northern margin of the influence of the Icelandic Low. A reviewer suggested that our records might be better considered as representing changes in surface water salinities, which implies a link with variations in the flux of sea ice through Fram Strait (Kwok and Rothrock, 1999; Schmith and Hanssen, 2003), hence the association between the NAO and AO (Fig. 2d) needs to be considered. Several recent papers have evaluated a variety of high-resolution marine proxies from off Iceland and adjacent areas and associated them with known climatic modes, such as the NAO, AO, and AMO (Jennings et al., 2002a; An-

drews et al., 2003; Darby and Bischof, 2004; Jiang et al., 2005; Knudsen et al., 2011; Miettinen et al., 2011; Darby et al., 2012; Alonso-Garcia et al., 2013; Zhang et al., 2007), although such oscillations are superimposed on both long-term insolation trends and abrupt events, such as volcanic eruptions. The underlying controls on our proxies are the variations in the landfast and drift ice and the extent and presence of Atlantic water transported via the Irminger current (Fig. 2). Despite potential problems associated with dissolution the calcite content, cores on both sides of the Denmark Strait share a considerable number of features in common (Figs. 8–12) – this is true not only for the last 2000 cal yr BP but for the Holocene (Andrews et al., 2001; Stoner et al., 2007). The trough of the LIA is well defined (Fig. 11a) but there is no distinct MWP. On the 2000 yr timescale the calcite data show a similar long-term trend to the reconstructed Arctic summer temperatures (Kaufman et al., 2009) (Fig. 10a) and have an association with the AMO (Fig. 10c). The abrupt changes in calcite wt% are temporally associated (Figs. 8b, 12a, and d) with the expansion of small ice caps on Baffin Island and East Greenland (Miller et al., 2012; Lowell et al., 2013). Zhong et al. (2011) modeled the climatic impact of a sustained interval of volcanism in the late 13th and middle 15th centuries and indicated that such a history could

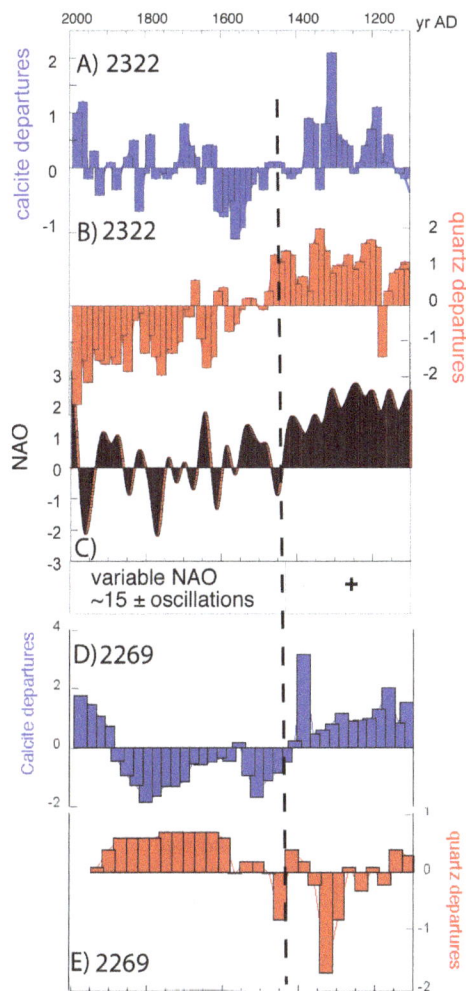

Fig. 12. Departures from the mean for the last 900 yr for calcite (**A** and **D**) and quartz (**A** and **D**) and quartz (**B** and **E**) at sites MD99-2322 and MD99-2269 (Fig. 1) against the NAO index (**C**) (Trouet et al., 2009; Kinnaid et al., 2011). The vertical dashed line ~ 1450 AD represents the time of a major transition.

sidereal years, changes in the rate of sediment accumulation, and the ocean reservoir correction. Thus it is difficult to know whether the offset between the calcite data and AMO is a true lag or associated with the dating (Fig. 10c). The problem could be addressed by a very high density of radiocarbon dates (Sejrup et al., 2010) but this is not practical (because of the cost) in most situations.

Acknowledgements. Support for involvement in cruises in 1991, 1993, 1996, 1997, and 1999 was provided to J. T. Andrews and A. E. Jennings by grants from the National Science Foundation. This paper is a contribution to grants NSF-OCE-0823535 and ANS-1107761. J. T. Andrews thanks C.A.S.E. and Jacques Giraudeau for involving him in the project. The calcite and quartz XRD data for MD99-2269 were provided by Matthias Moros (available in NOAA data base). Mike Mann and Anne de Vernal kindly provided data from published climate reconstructions. We greatly appreciate the comments and advice advanced by three reviewers, and which have led to significant changes in the final version, including a change in the title and the suggestion to add more weight to changes in salinity. We appreciate the Editorial advice and guidance of Hans Renssen. The data generated for this paper will be submitted to the NOAA Paleoclimate Data Base for archiving http://www.ncdc.noaa.gov/paleo/paleo.html.

Edited by: H. Renssen

References

Aitchison, J.: The Statistical Analysis of Compositional Data, Chapman and Hall, London, UK, 1986.

Alonso-Garcia, M., Andrews, J. T., Belt, S. T., Cabedo-Sanz, P., Darby, D., and Jaeger, J.: A multi-proxy and multi-decadal record (to AD 1850) of environmental conditions on the East Greenland shelf (~ 66° N), The Holocene, 23, 1672–1683, 2013.

Anderson, D. M.: Attenuation of millennial-scale events by bioturbation in marine sediments, Paleoceanography, 16, 352–357, 2001.

Andresen, C. S., Bond, G., Kuijpers, A., Knutz, P., and Bjorck, S.: Holocene climate variability at multi-devadal time-scales detected by sedimentological indicators in a shelf core NW off Iceland, Mar. Geol., 214, 323–338, 2005.

Andresen, C. S., McCarthy, D. J., Dylmer, C. V., Seidenkrantz, M. S., Kuijpers, A., and Lloyd, J. M.: Interaction between subsurface ocean waters and calving of the Jakobshavn Isbrae during the late Holocene, Holocene, 21, 211–224, 2011.

Andresen, C. S., Straneo, F., Ribergaard, M. H., Bjørk, A. A., Andersen, T. J., Kuijpers, A., Nørgaard-Pedersen, N., Kjær, K. H., Schjoth, F., Weckström, K., and Ahlstrøm, A. P.: Rapid response of Helheim Glacier in Greenland to climate variability over the past century, Nat. Geosci., 5, 37–41, 2012.

Andresen, C. S., Hansen, M. J., Seidenkrantz, M. S., Jennings, A. E., Knudsen, M. F., Norgaard-Pedersen, N., Larsen, N. K., Kuijpers, A., and Pearce, C.: Mid- to late-Holocene oceanographic variability on the southeast Greenland shelf, Holocene, 23, 167–178, 2013.

result in an expansion of sea ice in the Arctic Ocean and its marginal seas. A notable feature of the LIA reconstructions is the highly variable NAO (Fig. 12c), which is matched in the calcite data from 2322. Our simple conceptual model of the inverse association between the variations in calcite and quartz, based on data from Iceland (Axford et al., 2011) (Fig. 12d and e), is not mimicked by the quartz record on the East Greenland side of the Denmark Strait (Fig. 12a and b). Neither does our current understanding of NAO variations (Fig. 2) and associated changes in our proxies match a reconstructed persistent positive NAO-like scenario between 1100 and 1450 AD (Fig. 12c).

We conclude by noting that there is a fundamental "uncertainty principle" to the chronologies of high-resolution Holocene marine records and the delimitation of episodes such as the MWP and LIA (Fig. 8). This uncertainty is based on: the ± errors on the radiocarbon dates, the calibration to

Andrews, J. T.: Seeking a Holocene drift ice proxy: non-clay mineral variations from the SW to N-central Iceland shelf: trends, regime shifts, and periodicities, J. Quaternary Sci., 24, 664–676, 2009.

Andrews, J. T.: Unravelling sediment transport along glaciated margins (the northwestern Nordic Seas) using quantitative X-ray diffraction of bulk (< 2 mm) sediment, in: Sediment Transport – Flow and Morphological Processes, edited by: Bhuiyan, A. F., InTech, 225–248, doi:10.5772/21795, 2011.

Andrews, J. T. and Cartee-Schoofield, S.: Late Quaternary lithofacies, provenance, and depositional environments (~ 12 to 30 cal ka), north and south of the Denmark Strait, Mar. Geol., 199, 65–82, 2003.

Andrews, J. T. and Eberl, D. D.: Determination of sediment provenance by unmixing the mineralogy of source-area sediments: the "SedUnMix" program, Mar. Geol., 291, 24–33, 2012.

Andrews, J. T., Milliman, J. D., Jennings, A. E., Rynes, N., and Dwyer, J.: Sediment thicknesses and Holocene glacial marine sedimentation rates in three East Greenland fjords (ca. 68° N), J. Geol., 102, 669–683, 1994.

Andrews, J. T., Smith, L. M., Preston, R., Cooper, T., and Jennings, A. E.: Spatial and temporal patterns of iceberg rafting (IRD) along the East Greenland margin, ca. 68° N, over the last 14 cal ka, J. Quaternary Sci., 12, 1–13, 1997.

Andrews, J. T., Barber, D. C., and Jennings, A. E.: Errors in generating time-series and dating events at late Quaternary (radiocarbon) time-scales: Examples from Baffin Bay, Labrador Sea, and East Greenland, in: Mechanisms of Global Climate Change at Millennial Time Scales, edited by: Clark, P. U., Webb, R. S., and Keigwin, L. D., American Geophysical Union, Washington, DC, USA, 1999.

Andrews, J. T., Helgadottir, G., Geirsdottir, A., and Jennings, A. E.: Multicentury-scale records of carbonate (hydrographic?) variability on the N, Iceland margin over the last 5000 yr, Quaternary Res., 56, 199–206, 2001.

Andrews, J. T., Hardardottir, J., Stoner, J. S., Mann, M. E., Kristjansdottir, G. B., and Koc, N.: Decadal to millennial-scale periodicities in North Iceland shelf sediments over the last 12 000 cal yr: long-term North Atlantic oceanographic variability and Solar forcing, Earth Planet. Sc. Lett., 210, 453–465, 2003.

Andrews, J. T., Belt, S. T., Olafsdottir, S., Masse, G., and Vare, L.: Sea ice and marine climate variability for NW Iceland/Denmark Strait over the last 2000 cal yr BP, Holocene, 19, 775–784, 2009a.

Andrews, J. T., Darby, D. A., Eberl, D. D., Jennings, A. E., Moros, M., and Ogilvie, A.: A robust multi-site Holocene history of drift ice off northern Iceland: implications for North Atlantic climate, Holocene, 19, 71–78, 2009b.

Andrews, J. T., Jennings, A. E., Coleman, C. G., and Eberl, D.: Holocene variations in mineral and grain-size composition along the East Greenland glaciated margin (ca. 67–70° N): local versus long-distant sediment transport, Quaternary Sci. Rev., 29, 2619–2632, 2010.

Andrews, J. T., Bigg, G. R., and Wilton, D. J.: Holocene sediment transport from the glaciated margin of East/Northeast Greenland (67–80° N) to the N Iceland shelves: detecting and modeling changing sediment sources, Quaternary Sci. Rev., in press, doi:10.1016/j.quascirev.2013.08.019, 2014.

Axford, Y., Andresen, C., Andrews, J. T., Belt, S. T., Geirsdottir, A., Masse, G., Miller, G. H., Olafsdottir, S., and Vare, L. L.:

Do paleoclimate proxies agree?, statistical comparison of climate and sea-ice reconstructions from Icelandic marine and lake sediments, 300–1900 AD, J. Quaternary Sci., 26, 645–656, 2011.

Balestra, B., Ziveri, P., Monechi, S., and Troelstra, S.: Coccolithophorids from the Southeast Greenland Margin (northern North Atlantic): production, ecology and the surface sediment record, Micropaleontology, 50, 23–34, 2004.

Belkin, I. M., Levitus, S., Antonov, J., and Malmberg, S.-A.: "Great Salinity Anomalies" in the North Atlantic, Prog. Oceanogr., 41, 1–68, 1998.

Belkin, I. M., Cornillon, P. C., and Sherman, K.: Fronts in large marine ecosystems, Prog. Oceanogr., 81, 223–236, 2009.

Belt, S. T., Masse, G., Rowland, S. J., Poulin, M., Michel, C., and LeBlanc, B.: A novel chemical fossil of palaeo sea ice: IP25, Org. Geochem., 38, 16–27, 2007.

Bigg, G. R.: An estimate of the flux of iceberg calving from Greenland, Arct. Antarct. Alp. Res., 31, 174–178, 1999.

Blindheim, J., Toresen, R., and Loeng, H.: Fremtidige klimatiske endringer og betydningen for fiskeressursene, Future climatic changes and their influence on fishing ressources, Environment of the Sea Havets miljø: Norwegian Institute of Marine Research, 73–78, 2001.

Blockley, S. P. E., Blaauw, M., Ramsey, C. B., and van der Plicht, J.: Building and testing age models for radiocarbon dates in lateglacial and early Holoicene sediments, Quaternary Sci. Rev., 26, 1915–1926, 2007.

Broecker, W. S.: Thermohaline Circulation, the Achilles Heel of Our Climate System: Will Man-Made CO_2 Upset the Current Balance?, Science 278, 1582–1588, doi:10.1126/science.278.5343.1582, 1997.

Broecker, W. S.: Was the Medieval Warm Period global?, Science, 291, 1497–1499, 2001.

Burroughs, W. J.: Weather Cycles Real or Imaginary, 2nd Edn., Cambridge University Press, Cambridge, UK, 317 pp., 2003.

Burton, J. C., Amundson, J. M., Abbot, D. S., Boghosian, A., Cathles, L. M., Correa-Legisos, S., Darnell, K. N., Guttenberg, N., Holland, D. M., and MacAyeal, D. R.: Laboratory investigations of iceberg capsize dynamics, energy dissipation and tsunamigenesis, J. Geophys. Res., 117, F01007, doi:10.1029/2011jf002055, 2012.

Chayes, F.: Ratio Correlation, University of Chicago Press, Chicago, 1971.

Cook, E. R., D'Arrigo, R. D., and Mann, M. E.: A well-verified, multiproxy reconstruction of the winter North Atlantic Oscillation index since AD 1400, J. Climate, 15, 1754–1764, 2002.

CRU (Climatic Research Unit): North Atlantic Oscillation (NAO) Data, UEA (University of East Anglia), available at: http://www.cru.uea.ac.uk/cru/data/nao/, 2004.

Darby, D. A. and Bischof, J.: A Holocene record of changing Arctic Ocean ice drift analogous to the effects of the Arctic Oscilation, Palaeoceanography, 19, PA1027, doi:10.1029/2003PA000961, 2004.

Darby, D. A., Myers, W. B., Jakobsson, M., and Rigor, I.: Modern dirty sea ice characteristics and sources: the role of anchor ice, J. Geophys. Res., 116, C09008, doi:10.1029/2010JC006675, 2011.

Darby, D. A., Ortiz, J. D., Grosch, C. E., and Lund, S. P.: 1,500-year cycle in the Arctic Oscillation identified in Holocene Arctic sea-ice drift, Nat. Geosci., 5, 897–900, 2012.

Davis, J. C.: Statistics and Data Analysis in Geology, John Wiley & Sons, New York, 1986.

Dawson, A. G., Hickey, K., Holt, T., Elliot, L., Dawson, S., Foster, I. D. L., Wadhams, P., Jonsdottir, I., Wilkinson, J., McKenna, J., Davis, N. R., and Smith, D. E.: Complex North Atlantic Oscillation (NAO) index signal of historic North Atlantic storm-track changes, Holocene, 12, 363–369, 2002.

Dethleff, D.: Entrainment and export of Laptev Sea ice sediments, Siberian Arctic, J. Geophys. Res., 110, C07009, doi:10.1029/2004JC002740, 2005.

Dethleff, D. and Kuhlmann, G.: Fram Strait sea-ice sediment provenance based on silt and clay compositions identify Siberian and Kara and Laptev sxeas as main source reegions, Polar Res., 29, 265–282, 2010.

Dickson, R. R., Meincke, J., Malmberg, S., and Lee, A.: The "Great Salinity Anomaly" in the northern North Atlantic 1968–1982, Prog. Oceanogr., 20, 103–151, 1988.

Dickson, R. R., Osborn, T. J., Hurrell, J. W., Meincke, J., Blindheim, J., Adlandsvik, B., Vinje, T., Alekseev, G., and Maslowski, W.: The Arctic Ocean response to the North Atlantic oscillation, J. Climate, 13, 2671–2696, 2000.

Divine, D. V. and Dick, C.: Historical variability of the sea ice edge position in the Nordic Seas, J. Geophys. Res., 111, C01001, doi:10.1029/2004JC002851, 2006.

Dowdeswell, J. A., Whittington, R. J., and Hodgkins, R.: The sizes, frequencies, and freeboards of East Greenland icebergs observed using ship radar and sextant, J. Geophys. Res., 97, 3515–3528, 1992.

Dowdeswell, J. A., Whittington, R. J., Jennings, A. E., Andrews, J. T., Mackensen, A., and Marienfeld, P.: An origin for laminated glacimarine sediments through sea-ice build-up and suppressed iceberg rafting, Sedimentology, 47, 557–576, 2000.

Dunhill, G., Andrews, J. T., and Kristjansdottir, G. B.: Radiocarbon Date List X: Baffin Bay, Baffin Island, Iceland, Labrador, and the northern North Atlantic, Occasional Paper No. 56, Institute of Arctic and Alpine Research, University of Colorado, Boulder, USA, 77 pp., 2004.

Dylmer, C. V., Giraudeau, J., Hanquiez, V., and Husum, K.: The coccolithophores Emiliania huxleyi and Coccolithus pelagicus: extant populations from the Norwegian-Iceland Sea and Fram Strait, Biogeosciences Discuss., 10, 15077–15106, doi:10.5194/bgd-10-15077-2013, 2013.

Dwyer, J. L.: Monitoring characteristics of glaciation in the Kangerdlugssuaq Fjord region, East Greenland, using digital LANDSAT MSS and TM Data, University of Colorado, Boulder, USA, 1993.

Dwyer, J. L.: Mapping tide-water glacier dynamics in East Greenland using Landsat data, J. Glaciol., 41, 584–596, 1995.

Eberl, D. D.: User guide to RockJock: a program for determining quantitative mineralogy from X-ray diffraction data, Open File Report 03-78, United States Geological Survey, Washington, DC, USA, 40 pp., 2003.

Eiriksson, J., Knudsen, K. L., Haflidason, H., and Henriksen, P.: Late-glacial and Holocene paleoceanography of the North Iceland Shelf, J. Quaternary Sci., 15, 23–42, 2000.

Eiriksson, J., Knudsen, K. L., Larsen, G., Olsen, J., Heinemeier, J., Bartels-Jonsdottir, H. B., Jiang, H., Ran, L. H., and Simonarson, L. A.: Coupling of palaeoceanographic shifts and changes in marine reservoir ages off North Iceland through the last millennium, Palaeogeogr. Palaeocl., 302, 95–108, 2010.

Funder, S., Goosse, H., Jepsen, H., Kaas, E., Kjær, K. H., Korsgaard, N. J., Larsen, N. K., Linderson, H., Lysa, A., Möller, P., Olsen, J., and Willerslev, E.: A 10,000-Year Record of Arctic Ocean Sea-Ice Variability – View from the Beach, Science, 333, 747–750, 2011.

Geirsdottir, A., Hardardottir, J., and Andrews, J. T.: Late Holocene terrestrial geology of Miki and I. C. Jacobsen Fjords, East Greenland, Holocene, 10, 125–134, 2000.

Geirsdottir, A., Miller, G. H., Axford, Y., and Olafsdottir, S.: Holocene and latest Pleistocene climate and glacier fluctuations in Iceland, Quaternary Sci. Rev., 28, 2107–2118, 2009.

Ghil, M., Allen, M. R., Dettinger, M. D., Ide, K., Kondrashov, D., Mann, M. E., Roberston, A. W., Saunders, A., Tian, Y., Varadi, F., and Yiou, P.: Advanced spectral methods for climatic time series, Rev. Geophys., 40, 1003, doi:10.1029/2000RG000092, 2002.

Giraudeau, J., Jennings, A. E., and Andrews, J. T.: Timing and mechanisms of surface and intermediate water circulation changes in the Nordic Sea over the last 10,000 cal years: a view from the North Iceland shelf, Quaternary Sci. Rev., 23, 2127–2139, 2004.

Giraudeau, J., Solignac, S., Moros, M., Andrews, J. T., and Jansen, E.: Millennial-scale variability in Atlantic water advection to the Nordic Seas derived from coccolith contration records, Quaternary Sci. Rev., 29, 1276–1287, 2010.

Gray, D.: Ice chart of the Arctic Ocean between Greenland and Spitsbergen, From observations by Capt. David Gray, Royal Geographical Society, Control # 503747, Scale 503741:503746,503750,503000, 1881.

Gray, S. T., Graumlich, L. J., Betancourt, J. L., and Pederson, G. T.: A tree-ring based reconstruction of the Atlantic Multidecadal Oscillation since 1567 AD, Geophys. Res. Lett., 31, L12205, doi:10.1029/2004gl019932, 2004.

Grobe, H.: A simple method for the determination of ice-rafted debris in sediment cores, Polarforschung, 57, 123–126, 1987.

Grove, J. M.: The initiation of the "Little Ice Age" in regions round the North Atlantic, Climatic Change, 48, 53–82, 2001.

Hakkinen, S.: A simulation of thermohaline effects of a great salinity anomaly, J. Climate, 12, 1781–1795, 1999.

Hardarson, B. S., Fitton, J. G., and Hjartarson, A.: Tertiary volcanism in Iceland, Jokull, 58, 161–178, 2008.

Helgadottir, G.: Paleoclimate (0 to > 14 ka) of W and NW Iceland: an Iceland/USA contribution to P. A. L. E., Cruise Report B9–97, Marine Research Institute of Iceland, Reykjavik, 1997.

Higgins, A. K., Gilotti, J. A., and Smith, P. M.: The Greenland Caledonides. Evolution of the Northeast margin of Laurentria, Geological Society of America, Boulder, CO, USA, 368 pp., 2008.

Hjort, C.: A sea correction for East Greenland, Geologiska Foreningen i Stockholm Forhandlingar, 95, 132–134, 1973.

Howat, I. M., Joughin, I., Fahnestock, M., Smith, B. E., and Scambos, T. A.: Synchronous retreat and acceleration of southeast Greenland outlet glaciers 2000–06: ice dynamics and coupling to climate, J. Glaciol., 54, 646–660, 2008.

Hughes, M. K. and Diaz, H. F.: Was there a "Medieval Warm Period", and if so, where and when?, in: Medieval Warm Period, edited by: Hughes, M. K. and Diaz, H. F., Kluwer, Boston, 109–142, 1994.

Hurrell, J. W., Kushnir, Y., Ottersen, G., and Vibeck, M. (Eds.): The North Atlantic Oscillation: Climatic Significance and Environmental Impact, American Geophysical Union, Washington, DC, Geoph. Monog. Series, 134, 279 pp., 2003.

Jennings, A. E. and Weiner, N. J.: Environmental change on eastern Greenland during the last 1300 yr: Evidence from Foraminifera and Lithofacies in Nansen Fjord, 68° N, Holocene, 6, 179–191, 1996.

Jennings, A. E., Hardardottir, J., Stein, R., Ogilvie, A. E. J., and Jonsdottir, I.: Oceanographic change and terrestrial human impacts in a post 1400 AD record from the southwest Iceland Shelf, Climatic Change, 48, 83–100, 2001.

Jennings, A. E., Knudsen, K. L., Hald, M., Hansen, C. V., and Andrews, J. T.: A mid-Holocene shift in Arctic sea ice variability on the East Greenland shelf, Holocene, 12, 49–58, 2002a.

Jennings, A. E., Gronvold, K., Hilberman, R., Smith, M., and Hald, M.: High resolution study of Icelandic tephras in the Kangerlussuaq Trough, southeast Greenland, during the last deglaciation, J. Quaternary Sci., 17, 747–757, 2002b.

Jennings, A. E., Hald, M., Smith, L. M., and Andrews, J. T.: Freshwater forcing from the Greenland Ice Sheet during the younger dryas: evidence from southeastern Greenland shelf cores, Quaternary Sci. Rev., 25, 282–298, 2006.

Jennings, A. E., Andrews, J. T., and Wilson, L.: Holocene environmental evolution of the SE Greenland Shelf north and south of the Denmark Strait: Irminger and East Greenland current interactions, Quaternary Sci. Rev., 30, 980–998, 2011.

Jiang, H., Eiriksson, J., Schulz, M., Knudsen, K.-L., and Seidenkrantz, M.-S.: Evidence for solar forcing of sea-surface temperature on the North Icelandic Shelf during the late Holocene, Geology, 33, 73–76, 2005.

JISAO (Joint Institute for the Study of the Atmosphere and Ocean): Arctic Oscillation (AO) time series, 1899–June 2002, available at: http://jisao.washington.edu/data/aots/aojfm18992002.asciiTS1, 2004.

Jochumsen, K., Quadfasel, D., Valdimarsson, H., and Jonsson, S.: Variability of the Denmark Strait overflow: moored time series from 1996–2011, J. Geophys. Res., 117, C12003, doi:10.1029/2012jc008244, 2012.

Jonsdottir, H. B. B.: Late Holocene climatic changes on the North Iceland shelf, Department of Marine Geology, University of Aarhus, Denmark, Aarhus, 124 pp. plus Appendixes, 2001.

Joughin, I., Howat, I., Alley, R. B., Ekstrom, G., Fahnestock, M., Moon, T., Nettles, M., Truffer, M., and Tsai, V. C.: Ice-front variation and tidewater behavior on Helheim and Kangerdlugssuaq Glaciers, Greenland, J. Geophys. Res., 113, F01004, doi:10.1029/2007JF000837, 2008.

Kaufman, D. S., Schneider, D. P., McKay, N. P., Ammann, C. M., Bradley, R. S., Briffa, K. R., Miller, G. H., Otto-Bliesner, B. L., Overpeck, J. T., and Vinther, B. M.: Recent Warming Reverses Long-Term Arctic Cooling, Science, 325, 1236–1239, 2009.

Kinnard, C., Zdanowicz, C. M., Fisher, D. A., Isaksson, E., de Vernal, A., and Thompson, L. G.: Reconstructed changes in Arctic sea ice over the past 1450 yr, Nature, 479, 509–512, doi:10.1038/nature10581, 2011.

Knudsen, K. L., Eirikson, J., Jansen, E., Jiang, H., Rytter, F., and Gudmundsdottir, E. R.: Paleoceanographic changes off North Iceland through the last 1200 yr: foraminifera, stable isotopes,

diatomes and ice rafted debris, Quaternary Sci. Rev., 23, 2231–2246, 2004.

Knudsen, M. F., Seidenkrantz, M. S., Jacobsen, B. H., and Kuijpers, A.: Tracking the Atlantic Multidecadal Oscillation through the last 8000 yr, Nat. Commun., 2, 178, doi:10.1038/ncomms1186, 2011.

Koch, L.: The East Greenland Ice, Meddelelser om Gronland, 130, 346 pp., 1945.

Kondrashov, D., Feliks, Y., and Ghil, M.: Oscillatory modes of extended Nile River records (A.D. 622–1922), Geophys. Res. Lett., 32, L10702, doi:10.1029/2004GL022156, 2005.

Kwok, R. and Rothrock, D. A.: Variability of Fram Strait ice flux and North Atlantic Oscillation, J. Geophys. Res., 104, 5177, doi:10.1016/j.epsl.2008, 1999.

Labeyrie, L., Jansen, E., and Cortijo, E.: Les rapports de campagnes a la mer MD114/IMAGES V, Institut Polaire Francais Paul-Emile Victor, Brest, 2003.

Lamb, H. H.: Climatic variations and changes in the wind and ocean circulation: the Little Ice Age in the northeast Atlantic, Quaternary Res., 11, 1–20, 1979.

Larsen, G.: Explosive Volcanism in Iceland: Three Examples of Hydromagmatic Basaltic Eruptions on long Volcanic Fissures within the past 1200 Years, Geophysical Research Abstracts, 7, 10158, doi:10.1016/j.epsl.2008, 2005.

Larsen, M., Hamberg, L., Olaussen, S., Norgaard-Pedersen, N., and Stemmerik, L.: Basin evolution in southern East Greenland: an outcrop analog for Cretaceous-Paleogene basins on the North Alantic volcanic margins, AAPG Bull., 83, 1236–1261, 1999.

Lehner, F., Raible, C. C., and Stocker, T. F.: Testing the robustness of a precipitation proxy-based North Atlantic Oscillation reconstruction, Quaternary Sci. Rev., 45, 85–94, 2012.

Lowell, T. V., Hall, B. L., Kelly, M. A., Bennike, O., Lusas, A. R., Honsaker, W., Smith, C. A., Levy, L. B., Travis, S., and Denton, G. H.: Late Holocene expansion of Istorvet ice cap, Liverpool Land, east Greenland, Quaternary Sci. Rev., 63, 128–140, 2013.

Mahadevan, A., D'Asaro, E., Lee, C., and Perry, M. J.: Eddy-driven stratification initiates North Atlantic spring phytoplankton blooms, Science, 337, 54–58, 2012.

Malmberg, S.-A.: Hydrographic Changes in the Waters Between Iceland and Jan Mayen in the Last Decade, Jokull, 19, 30–43, 1969.

Mann, M. E. and Lees, J. M.: Robust estimation of background noise and signal detection in climatic time series, Climatic Change, 33, 409–445, 1996.

Mann, M. E., Zhang, Z. H., Rutherford, S., Bradley, R. S., Hughes, M. K., Shindell, D., Ammann, C., Faluvegi, G., and Ni, F. B.: Global signatures and dynamical origins of the Little Ice Age and medieval climate anomaly, Science, 326, 1256–1260, 2009.

Marsden, R. F., Mysak, L. A., and Myers, R. A.: Evidence for stability enhancement of sea ice in the Greenland and Labrador seas, J. Geophys. Res., 96, 4783–4789, 1991.

Masse, G., Rowland, S. J., Sicre, M.-A., Jacob, J., and Belt, S. T.: Abrupt climate changes for Iceland during the last millenium: evidence from high resolution sea ice reconstructions, Earth Planet. Sc. Lett., 269, 564–568, doi:10.1016/j.epsl.2008, 2008.

Mauritzen, C. and Hakkinen, S.: Influence of sea ice on the thermohaline circulation in the Arctic-North Atlantic Ocean, Geophys. Res. Lett., 24, 3257–3260, 1997.

McCarty, D. K.: Quantitative mineral analysis of clay-bearing mixtures: the "Reynolds Cup" contest, Newsletter No. 27, International Union of Crystallography, Chester, England, 12–16, 2002.

Melling, H.: Sea-ice observations: advances and challenges, in: Arctic Climate Change, The ACSYS Decade and Beyond, edited by: Lemke, P. and Jacoobi, H.-W., Springer, New York, 27–116, 2012.

Miettinen, A., Divine, D., Koc, N., Godtliebsen, F., and Hall, I. R.: Multicentennial variability of the sea surface temperature gradient across the subpolar North Atlantic over the last 2.8 kyr, J. Climate, 25, 4205–4219, 2012.

Miller, G. H., Geirsdottir, A., Zhong, Y. F., Larsen, D. J., Otto-Bliesner, B. L., Holland, M. M., Bailey, D. A., Refsnider, K. A., Lehman, S. J., Southon, J. R., Anderson, C., Bjornsson, H., and Thordarson, T.: Abrupt onset of the Little Ice Age triggered by volcanism and sustained by sea-ice/ocean feedbacks, Geophys. Res. Lett., 39, L02708, doi:10.1029/2011gl050168, 2012.

Miller, G. H., Briner, J. P., Refsnider, K. A., Lehman, S., Geirsdóttir, A., Larsen, D. J., and Southon, J. R.: Substantial agreement on the timing and magnitude of Late Holocene ice cap expansion between East Greenland and the eastern Canadian Arctic: a commentary on Lowell et al., Quaternary Sci. Rev., 77, 239–245, doi:10.1016/j.quascirev.2013.04.019, 2013.

Moros, M., Andrews, J. T., Eberl, D. D., and Jansen, E.: Holocene history of drift ice in the northern North Atlantic: evidence for different spatial and temporal modes, Palaeoceanography, 21, PA2017, doi:10.1029/2005PA001214, 2006.

Mugford, R. I. and Dowdeswell, J. A.: Modeling iceberg-rafted sedimentation in high-latitude fjord environments, J. Geophys. Res., 115, F03024, doi:10.1029/2009jf001564, 2010.

Muller, J., Wagner, A., Fahl, K., Stein, R., Prange, M., and Lohmann, G.: Towards quantitative sea ice reconstructions in the northern North Atlantic: a combined biomarker and numerical modelling approach, Earth Planet. Sc. Lett., 306, 137–148, 2011.

Nuttall, A.-M.: Glaciological Investigations in East Greenland using Digital LANDSAT imagery, MA thesis, University of Cambridge, Cambridge, UK, 107 pp., 1993.

Ogilvie, A. E. J.: Fisheries, climate and sea ice in Iceland: an historical perspective, in: Marine Resources and Human Societies in the North Atlantic Since 1500, edited by: Vickers, D., Institute of Social and Economic Research, Memorial University, St Johns, Canada, 69–87, 1997.

Ogilvie, A. E. J. and Jónsson, T.: "Little Ice Age" research: a perspective from Iceland, Climatic Change, 48, 9–52, 2001.

Ogilvie, A. E. J., Barlow, L. K., and Jennings, A. E.: North Atlantic climate ca. AD 1000: millennial reflections on the Viking Discoveries of Iceland, Greenland and North America, Weather, 55, 34–45, 2000.

Olafsdottir, S., Jennings, A. E., Geirsdottir, A., Andrews, J., and Miller, G. H.: Holocene variability of the North Atlantic Irminger current on the south- and northwest shelf of Iceland, Mar. Micropaleontol., 77, 101–118, 2010.

Olafsdottir, S., Geirsdottir, A., Miller, G. H., Stoner, J. S., and Channell, J. E. T.: Synchronizing Holocene lacustrine and marine sediment records using paleomagnetic secular variation, Geology, 41, 535–538, 2013.

Olafsson, J.: Connections between oceanic conditions off N-Iceland, Lake Myvatn temperature, regional wind direction vari-

ability and the North Atlantic Oscillation, Rit Fiskideildar, 16, 41–57, 1999.

Omotoso, O., McCarty, D. K., Hillier, S., and Kleeberg, R.: Some successful approaches to quantitative mineral analysis as revealed by the 3rd Reynolds Cup contest, Clay. Clay Miner., 54, 748–760, 2006.

Paillard, D., Labeyrie, L., and Yiou, P.: Macintosh program performs time-series analysis, EOS T. Am. Geophys. Un., 77, p. 379, doi:10.1029/96EO00259, 1996.

Parkinson, C. L.: Recent trend reversals in Arctic sea ice extents: possible connection to the North Atlantic oscillation, Polar Geogr., 24, 1–12, 2000.

Pirrung, M., Futtere, D., Grobe, H., Matthiessen, J., and Niessen, F.: Magnetic susceptibility and ice-rafted debris in surface sediments of the Nordic Seas: implications for isotope stage 3 oscillations, Geo-Mar. Lett., 22, 1–11, 2002.

Polyakova, E. I., Journel, A. G., Polyakov, I. V., and Bhatt, U. S.: Changing relationship between the North Atlantic Oscillation and key North Atlantic climate parameters, Geophys. Res. Lett., 33, L03711, doi:10.1029/2005gl024573, 2006.

Quillmann, U., Andrews, J. T., and Jennings, A. E.: Radiocarbon Date List XI: East Greenland shelf, West Greenland Shelf, Labrador Sea, Baffin Island shelf, Baffin Bay, Nares Strait, and Southwest to Northwest Icelandic shelf, Occasional Paper No. 59, INSTAAR, University of Colorado, Boulder, USA, 68 pp., 2009.

Reeh, N.: Holocene climate and fjord glaciations in Northeast Greenland: implications for IRD deposition in the North Atlantic, Sediment. Geol., 165, 333–342, 2004.

Reeh, N., Thomsen, H. H., Higgins, A. K., and Weidick, A.: Sea ice and the stability of north and northeast Greenland floating glaciers, in: Annals of Glaciology, edited by: Jeffries, M. O. and Eicken, H., International Glaciological Society, 33, 474–480, 2001.

Reimer, P. J., McCormac, F. G., Moore, J., McCormick, F., and Murray, E. V.: Marine radiocarbon reservoir corrections for the mid- to late Holocene in the eastern subpolar North Atlantic, Holocene, 12, 129–136, 2002.

Rodionov, S. N.: A sequential algorithm for testing climate regime shifts, Geophys. Res. Lett., 31, L09204, doi:10.1029/2004GL019448, 2004.

Rodionov, S. N.: Use of prewhitening in climate regime shift detection, Geophys. Res. Lett., 33, 025901, doi:10.1029/2006GL025904, 2006.

Rudels, B., Anderson, L., Eriksson, P., Fahrbach, E., Jakobsson, M., Jones, P., Melling, H., Prinsenberg, S., Schauer, U., and Yao, T.: Observations in the Ocean, in: Arctic Climate Change, The ACSYS Decade and Beyond, edited by: Lemke, P. and Jacobi, H.-W., Springer, 117–198, 2012.

Sarafanov, A.: On the effect of the North Atlantic Oscillation on temperature and salinity of the subpolar North Atlantic intermediate and deep waters, Ices J. Mar. Sci., 66, 1448–1454, 2009.

Schleussner, C. F. and Feulner, G.: A volcanically triggered regime shift in the subpolar North Atlantic Ocean as a possible origin of the Little Ice Age, Clim. Past, 9, 1321–1330, doi:10.5194/cp-9-1321-2013, 2013.

Schmith, T. and Hanssen, C.: Fram Strait ice export during the nineteenth and twentieth centuries reconstructed from a multiyear sea ice index from southwestern Greenland, J. Climate, 16, 2782–2791, 2003.

Seale, A., Christoffersen, P., Mugford, R. I., and O'Leary, M.: Ocean forcing of the Greenland Ice Sheet: calving fronts and patterns of retreat identified by automatic satellite monitoring of eastern outlet glaciers, J. Geophys. Res., 116, F03013, doi:10.1029/2010jf001847, 2011.

Sejrup, H. P., Lehman, S. J., Haflidason, H., Noone, D., Muscheler, R., Berstad, I. M., and Andrews, J. T.: Response of Norwegian Sea temperature to solar forcing since 1000 AD, J. Geophys. Res., 115, C12034, doi:10.1029/2010jc006264, 2010.

Shi, F., Yang, B., Ljungqvist, F. C., and Yang, F.: Arctic 1400 Year Multiproxy Summer Temperature Reconstruction, IGBP PAGES/World Data Center for Paleoclimatology Data Contribution Series # 2012-151, NOAA/NCDC Paleoclimatology Program, Boulder CO, USA, 2012a.

Shi, F., Yang, B., Charpentier Ljungqvist, F., and Yang, F.: Multiproxy reconstruction of Arctic summer temperatures over the past 1400 yr, Clim. Res., 54, 113–128, doi:10.3354/cr01112, 2012b.

Sicre, M.-A., Jacob, J., Ezat, U., Rousse, S., Kissel, C., Yiou, P., Eiriksson, J., Knudsen, K.-L., Jansen, E., and Turon, J.-L.: Decadal variability of sea surface temperatures off North Iceland over the last 2000 years, Earth Planet. Sc. Lett., 268, 137–142, 2008.

Smith, L. M.: Late Quaternary glacial marine sedimentation in the Kangerlussuaq Region, East Greenland, 68° N, University of Colorado, Boulder, USA, 190 pp., 1997.

Smith, L. M. and Andrews, J. T.: Sediment characteristics in iceberg dominated fjords, Kangerlussuaq region, East Greenland, Sediment. Geol., 130, 11–25, 2000.

Smith, L. M. and Licht, K. J.: Radiocarbon Date List IX: Antarctica, Arctic Ocean, and the Northern North Atlantic, INSTAAR Occasional paper No. 54, University of Colorado, Boulder, CO, 138 pp., 2000.

Smith, L. M., Alexander, C., and Jennings, A. E.: Accumulation in East Greenland Fjords and on the continental shelves adjacent to the Denmark Strait over the last century based on ^{210}Pb geochronology, Arctic, 55, 109–122, 2002.

Smith, L. M., Miller, G. H., Otto-Bliesner, B., and Shin, S.-I.: Sensitivity of the Northern Hemisphere Climate System to extreme changes in Arctic sea ice, Quaternary Sci. Rev., 22, 645–658, 2003.

Smith, L. M., Andrews, J. T., Castañeda, I. S., Kristjánsdóttir, G. B., Jennings, A. E., and Sveinbjörnsdóttir, A. E.: Temperature reconstructions for SW and N Iceland waters over the last 10 cal ka based on δ^{18}O records from planktic and benthic Foraminifera, Quaternary Sci. Rev., 24, 1723–1740, doi:10.1016/j.quascirev.2004.07.025, 2005.

Stefansson, U.: North Icelandic Waters, Rit Fiskideildar III. Bind, 3, 269 pp., 1962.

Stefansson, U.: Dissolved nutrients, oxygen, and water masses in northern Irminger Sea, Deep-Sea Res., 15, 541–555, 1968.

Stefansson, U. and Olafsson, J.: Nutrients and fertility of Icelandic waters, Rit Fiskideildar, XII, 1–56, 1991.

Stoner, J. S., Jennings, A. E., Kristjansdottir, G. B., Andrews, J. T., Dunhill, G., and Hardardottir, J.: A paleomagnetic approach toward refining Holocene radiocarbon based chronostratigraphies: Paleoceanographic records from North Iceland (MD99-2269)

and East Greenland (MD99-2322) margins, Palaeoceanography, 22, PA1209, doi:10.1029/2006PA001285, 2007.

Straneo, F., Hamilton, G. S., Sutherland, D. A., Stearns, L. A., Davidson, F., Hammill, M. O., Stenson, G. B., and Rosing-Asvid, A.: Rapid circulation of warm subtropical waters in a major glacial fjord in East Greenland, Nat. Geosci., 3, 182–186, 2010.

Syvitski, J. P. M., Andrews, J. T., and Dowdeswell, J. A.: Sediment deposition in an iceberg-dominated glacimarine environment, East Greenland: basin fill implications, Global Planet. Change, 12, 251–270, 1996.

Syvitski, J. P. M., Stein, A., Andrews, J. T., and Milliman, J. D.: Icebergs and seafloor of the East Greenland (Kangerlussuaq) continental margin, Arct. Antarct. Alp. Res., 33, 52–61, 2001.

Telford, R. J., Heegaard, E., and Birks, H. J. B.: All age-depth models are wrong: but how badly?, Quaternary Sci. Rev., 23, 1–5, doi:10.1016/j.quascirev.2003.11.003, 2003.

Thompson, D. W. J. and Wallace, J. M.: The Arctic oscillation signature in wintertime geopotential height and temperature fields, Geophys. Res. Lett., 25, 1297–1300, 1998.

Thórdardóttir, T.: Primary Production North of Iceland in relation to Water Masses in May–June 1970–1980, Council for the Exploration of the Sea, C.M. 1984/L20, 1–17, 1984.

Trouet, V., Esper, J., Graham, N. E., Baker, A., Scourse, J. D., and Frank, D. C.: Persistent positive North Atlantic Oscillation mode dominated the Medieval Climate Anomaly, Science, 324, 78–80, 2009.

van Loon, H. and Rogers, J. C.: The Seesaw in Winter Temperatures between Greenland and Northern Europe. Part I: General Description, Mon. Weather Rev., 106, 296–310, 1978.

Vautard, R. and Ghil, M.: Singular spectrum analysis in nonlinear dynamics, with applications to paleoclimatic time-series, Physica D, 35, 395–424, 1989.

Wallevik, J. E. and Sigurjonsson, H.: The Koch index: formulation, corrections and extension, Icelandic Meteorological Office Report, Reykjavik, Iceland, 1998.

Wang, J. and Ikeda, M.: Arctic oscillation and Arctic sea-ice oscillation, Geophys. Res. Lett., 27, 1287–1290, 2000.

Wanner, H., Solomina, O., Grosjean, M., Ritz, S. P., and Jetel, M.: Structure and origin of Holocene cold events, Quaternary Sci. Rev., 30, 3109–3123, 2011.

Williams, L. D.: The Little Ice Age glaciation level on Baffin Island, Arctic Canada, Palaeogeogr. Palaeocl., 25, 199–207, 1978.

Zhang, J. L., Steele, M., Rothrock, A. D., and Lindsay, R. W.: Increasing exchanges at Greenland–Scotland Ridge and their links with the North Atlantic Oscillation and Arctic sea ice, Geophys. Res. Lett., 31, L09307, doi:10.1029/2003GL019304, 2004.

Zhang, R. and Vallis, G. K.: Impact of great salinity anomalies on the low-frequency variability of the North Atlantic climate, J. Climate, 19, 470–482, 2006.

Zhang, R., Delworth, T. L., and Held, I. M.: Can the Atlantic Ocean drive the observed multidecadal variability in Northern Hemisphere mean temperature?, Geophys. Res. Lett., 34, 022007, doi:10.1029/2006GL028683, 2007.

Zhong, Y., Miller, G. H., Otto-Bliesner, B. L., Holland, M. M., Bailey, D. A., Schneider, D. P., and Geirsdottir, A.: Centennial-scale climate change from decadally-paced explosive volcanism: a coupled sea ice-ocean mechanism, Clim. Dynam., 37, 2373–2387, 2011.

4

Sediment transport processes across the Tibetan Plateau inferred from robust grain-size end members in lake sediments

E. Dietze[1], F. Maussion[2], M. Ahlborn[3], B. Diekmann[4], K. Hartmann[5], K. Henkel[3], T. Kasper[3], G. Lockot[5], S. Opitz[6], and T. Haberzettl[3]

[1]Section 5.2 Climate Dynamics and Landscape Evolution, GFZ German Research Centre for Geosciences, Potsdam, Germany
[2]Chair of Climatology, Technische Universität Berlin, Berlin, Germany
[3]Physical Geography, Institute of Geography, Friedrich-Schiller-Universität Jena, Jena, Germany
[4]Alfred Wegener Institute for Polar and Marine Research, Research Unit Potsdam, Potsdam, Germany
[5]Institute of Geographical Sciences, EDCA, Freie Universität Berlin, Berlin, Germany
[6]Institute for Earth and Environmental Sciences, Universität Potsdam, Potsdam, Germany

Correspondence to: E. Dietze (edietze@gfz-potsdam.de)

Abstract. Grain-size distributions offer powerful proxies of past environmental conditions that are related to sediment sorting processes. However, they are often of multimodal character because sediments can get mixed during deposition. To facilitate the use of grain size as palaeoenvironmental proxy, this study aims to distinguish the main detrital processes that contribute to lacustrine sedimentation across the Tibetan Plateau using grain-size end-member modelling analysis. Between three and five robust grain-size end-member subpopulations were distinguished at different sites from similarly–likely end-member model runs. Their main modes were grouped and linked to common sediment transport and depositional processes that can be associated with contemporary Tibetan climate (precipitation patterns and lake ice phenology, gridded wind and shear stress data from the High Asia Reanalysis) and local catchment configurations. The coarse sands and clays with grain-size modes $> 250\,\mu m$ and $< 2\,\mu m$ were probably transported by fluvial processes. Aeolian sands ($\sim 200\,\mu m$) and coarse local dust ($\sim 60\,\mu m$), transported by saltation and in near-surface suspension clouds, are probably related to occasional westerly storms in winter and spring. Coarse regional dust with modes $\sim 25\,\mu m$ may derive from near-by sources that keep in longer term suspension. The continuous background dust is differentiated into two robust end members (modes: 5–10 and 2–5 μm) that may represent different sources, wind directions and/or sediment trapping dynamics from long-range, upper-level westerly and episodic northerly wind transport. According to this study grain-size end members of only fluvial origin contribute small amounts to mean Tibetan lake sedimentation (19 ± 5 %), whereas local to regional aeolian transport and background dust deposition dominate the clastic sedimentation in Tibetan lakes (contributions: 42 ± 14 % and 51 ± 11 %). However, fluvial and alluvial reworking of aeolian material from nearby slopes during summer seems to limit end-member interpretation and should be cross-checked with other proxy information. If not considered as a stand-alone proxy, a high transferability to other regions and sediment archives allows helpful reconstructions of past sedimentation history.

1 Introduction

The Tibetan Plateau is the world's largest mountain plateau with an average elevation of 4300 m above sea level (a.s.l.). It plays an important role for global climate and continent-wide water supply (Immerzeel et al., 2010). Various circulation systems, including the Asian monsoons and the westerlies, interact along a climate gradient from south-east to north-west (e.g. Kutzbach et al., 1993; Böhner, 2006). Summers are dominated by moist summer monsoonal air masses, local convection and cyclonic circulation from plateau heating. Winters are dry and cold, but have much stronger, continuous winds due to the influence of the westerlies (Maussion et al., 2013). Due to the high altitude of the plateau, the wind patterns are strongly related to the mid-troposphere circulation, but are also influenced by orography and seasonal plateau heating (Maussion et al., 2013). The spatio-temporal variability of these circulation patterns and their respective influence on regional climate are still under debate (Xu et al., 2007; Liu et al., 2009; Immerzeel et al., 2010).

The Tibetan Plateau hosts around 6880 lakes larger than 0.1 km^2 and 7 lakes larger than 500 km^2 (Kropáček et al., 2013). Lakes are fed mainly by summer precipitation, glacial melt and to a minor extent by melting permafrost (Liu et al., 2009). These lakes can contain highly resolved, continuous sediment records that can be used to reconstruct past circulation patterns using a variety of biological, geochemical-mineralogical, and sedimentological proxy information (e.g. Mischke et al., 2009; Kasper et al., 2012; Opitz et al., 2012). However, to decipher the complex interactions of regional climate and other drivers of environmental change it is necessary to disentangle the associated processes that translate environmental change into properties, which can be stored in archives over longer times.

Detrital particles reach lakes, which are often the final depocentres, from multiple sediment sources and via different transport processes (fluvial, aeolian, glacial) along local or regional sediment cascades (Stauch and Lehmkuhl, 2003). The size distribution of detrital particles is one of the standard parameters used for sedimentary facies description and discrimination of sediment sources (e.g. Folk and Ward, 1957; Torres et al., 2005; Flemming, 2007). Natural grain-size distributions are composed of dynamic subpopulations that were sorted by different sediment production, transport and accumulation processes (Weltje and Prins, 2007). These processes sort sediments in a characteristic way depending on properties of sediment sources, availability, local surface roughness and transport energies (wind strength, flow and shear velocities; Tsoar and Pye, 1987).

If sediments were mixed, for example in the water column, by biotic activity and/or geochemical alterations (Cohen, 2003) or sampling procedure, mean grain-size information were often interpreted in terms of sediment core position relative to the shore and related to lake level changes (Kasper et al., 2012; Opitz et al., 2012; Dietze et al., 2013; Doberschütz et al., 2013). However, the often used deterministic approaches of descriptive moments of single samples (Folk and Ward, 1957; Blott and Pye, 2001) are limited and rather inappropriate for multimodal grain-size distributions derived from high-dimensional state-of-the-art laser-diffraction analysis. The statistical decomposition of grain-size data into their original subpopulations offers more proper and detailed information on sedimentation dynamics (Dietze et al., 2012).

Decomposition currently includes the fitting of theoretical functions to individual samples (Sun et al., 2002; Bartholdy et al., 2007), where often some prior knowledge of the natural subpopulation is needed (Flemming, 2007) and a geo-

scientific understanding and interpretation of the fitting parameters is rather complicated. Another approach uses all available samples from a certain archive and unmixes the detrital subpopulations with end-member modelling analysis (EMMA; Weltje, 1997; Dietze et al., 2012). EMMA contributes to a better "operationalism" in particle size analysis (Hartmann, 2007) and is based on principles of numerical inversion to reduce redundancy. It considers compositional data constraints and provides sedimentologically interpretable grain-size end members (Weltje, 1997; Dietze et al., 2012). EMMA has been successfully applied to the multimodal grain-size distributions of marine, lacustrine, and aeolian sediments mainly in the Atlantic ocean and the Asian continent (e.g. Stuut et al., 2002; Vriend and Prins, 2005; Weltje and Prins, 2007; IJmker et al., 2012; Dietze et al., 2012, 2013), where different end members were interpreted as proxies of sediment transport processes.

This contribution aims to disentangle sediment transport and deposition signals from lake sediments exhibiting complex grain-size distributions to facilitate a more detailed interpretation of grain-size proxies in future palaeoenvironmental reconstructions. Using end-member modelling analysis it studies grain-size distributions and the robust end-member subpopulations of lacustrine sediment records from six sites across the Tibetan Plateau. The records were chosen to cover different lake-catchment systems, site conditions, and timescales integrating new gridded contemporary wind-shear data. This will allow the general characterisation of the most important and typical sediment transport processes that lead to deposition of clastic sediment in Tibetan lakes. However, it is not the scope of this study to disentangle local effects of sedimentation or different local contributions of sediment transport processes at a certain time. This will follow in future detailed spatial and temporal comparisons.

2   Study sites

Sediments from five lakes of different surface areas (0.0145 to ~ 2000 km^2) and one peat bog were studied, which are located between 4090 and 4720 m a.s.l. at the north-eastern Tibetan Plateau (i.e. lake Donggi Cona, Qinghai province) and across the southern Tibetan Plateau (all other sites, Xizhang province, China; Fig. 1; Table 1). The lakes were chosen to have different maximum water depths (9–220 m) and varying catchment sizes (4.6 to ~ 10000 km^2; Table 1). Sediment cores were retrieved at water depths between 8.4 and 220 m at different distances to the shore and at one peat bog site that was a lake system in the past (Table 2). Few perennial streams and several episodically filled minor channels feed the lakes with water derived mainly from summer monsoonal precipitation and to a minor part from snow melt. Lakes Nam Co, Taro Co and Tangra Yumco further receive glacial melt-waters (Wrozyna et al., 2010; Long et al., 2012; Fig. 1). Only lake Donggi Cona has an outflow since ca. 4 cal ka BP (Opitz

Table 1. Information on lake systems used in this study.

Lake system*	Latitude [° N]	Longitude [° E]	Present altitude [m a.s.l.]	Lake size [km²]	Catchment size [km²]	Salinity	References on lake systems and catchments
Targo Xian	30.77	86.67	4700	0	11.8	fresh	Miehe et al. (2013)
TT Lake	31.10	86.57	4750	0.015	4.6	fresh	unpublished data
Donggi Cona	35.26	98.60	4090	235	3028	fresh	Dietze et al. (2010), Opitz et al. (2012), Stauch et al. (2012)
Taro Co	31.17	84.20	4570	474	7423	fresh	unpublished data
Tangra Yumco	31.25	86.72	4550	824	9893	brackish	Wang et al. (2010)
Nam Co	30.74	90.79	4720	1966	8952	brackish	Wang et al. (2009), Wrozyna et al. (2010), Kasper et al. (2012, 2013)

* See Fig. 1 for location.

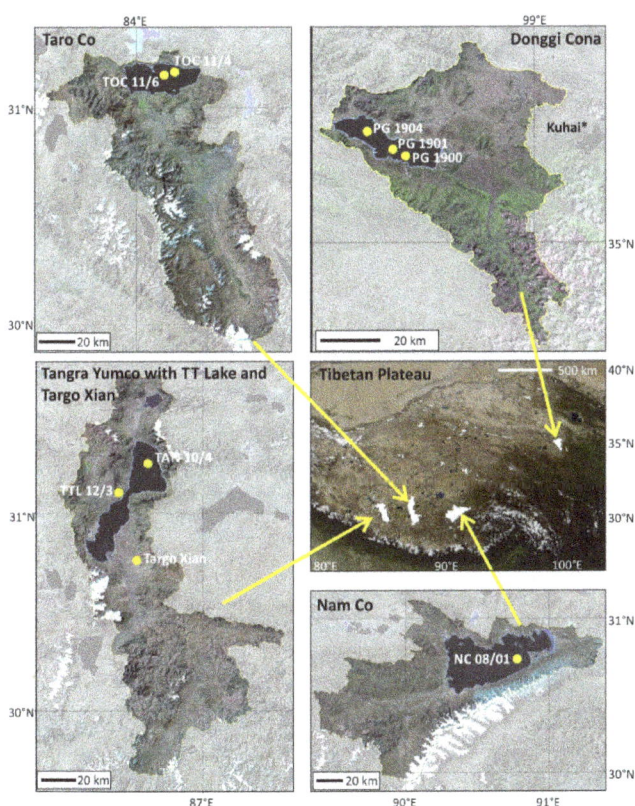

Fig. 1. Studied lakes and their catchments on the Tibetan Plateau, China (for references see Table 1). Map sources: Blue Marble Next Generation, NASA Earth Observatory, 2013 (overview); Landsat ETM+ and SRTM-3, NASA Earth Observatory and USGS, 2001 (catchments); white glaciers from GLIMS Glacier Database, NSIDC, 2013. (*) Lake Kuhai near lake Donggi Cona was studied by Mischke et al. (2009). See kmz-file in Appendix for further information.

et al., 2012) and Targo Xian was part of Lake Tangra Yumco in the past (Miehe et al., 2013). All other lakes are closed lake systems today.

Vegetation in the catchments is composed of semi-arid short-grass steppe, mat-forming *Kobresia*, *Artemisia* dwarf-shrubs and intercalated Cyperaceae-swamps (Kürschner et al., 2005; Miehe et al., 2013) with lower vegetation cover and open soil patches towards the westernmost lake Taro Co (Fig. 1). All lakes are ice covered in wintertime during most of the years, although a complete ice cover may only be obtained at the large lakes during cold, but calm winters (Kropáček et al., 2013). Geological, geomorphological and pedological inventories vary strongly within and among the catchments but were studied preliminarily only for lakes Nam Co (Keil et al., 2010) and Donggi Cona (Dietze et al., 2010; IJmker et al., 2012; Stauch et al., 2012).

3 Methods

3.1 Laboratory analysis

Total (in-)organic matter contents varied below (7 and 45 %) 6 and 12 % in all the cores (Opitz et al., 2012; Ahlborn, unpublished data) with peat phases at Targo Xian containing up to 17 % (Miehe et al., 2013). To study the detrital fraction of the lake sediment composition all sediments were pretreated with hydrogen peroxide (H_2O_2, 15 %, 30 %) until no further reaction occurred indicating that all organic matter was removed. To remove carbonates sediments analysed at the Friedrich Schiller University of Jena (Nam Co, Tangra Yumco, TT Lake, Targo Xian, Taro Co) were treated with hydrochloric acid (HCl, 15 %), whereas the Donggi Cona sediments analysed at Alfred Wegener Institute Potsdam were pretreated with acetic acid (CH_3COOH, 10 %). Each treatment leads to comparable results (P. Schulte, RWTH Aachen, unpublished data). After adding sodium pyrophosphate ($Na_2HPO_4 \cdot 10H_2O$) samples were put in an overhead shaker for 12 h. After 10 s of ultra-sonic bath samples were measured in several cycles until a reproducible signal was obtained using the Fraunhofer optical model in Beckman Coulter laser diffraction particle size analysers LS 13 320

Table 2. Information on lake sediment records used in this study.

Lake system*	Sediment core name	Latitude [°N]	Longitude [°E]	Core length [m]	Time period [ka]	Water depth [m]	Distance to shore [km]	Reference
Targo Xian	Targo Xian core	30.7667	86.6667	3.6	~ 11	0	0	Miehe et al. (2013)
TT Lake	TTL 12/3	31.1029	86.5734	0.89	~ 1.5	8.4	0.02	unpublished data
Donggi Cona	PG 1900	35.2601	98.6027	4.23	~ 14	37.6	2.34	Opitz et al. (2012)
Donggi Cona	PG 1904	35.3336	98.4824	4.46	~ 12	39.5	3.45	Opitz et al. (2012)
Donggi Cona	PG 1901	35.2786	98.5621	5.72	~ 20	40.0	2.67	Opitz et al. (2012)
Taro Co	TOC11/04	31.1689	84.2019	1.22	~ 7.5	68.0	2.866	unpublished data
Taro Co	TOC11/06	31.1561	84.1493	0.9	~ 3	123.0	5.2	unpublished data
Tangra Yumco	TAN 10/04	31.2526	86.7228	1.62	~ 3	220.0	4.683	unpublished data
Nam Co	NC 08/01	30.7374	90.7903	1.15	~ 4	93.0	7.525	Kasper et al. (2012)

* See Fig. 1 for location.

(Jena) and LS 200 (Potsdam). No systematic measurement bias is to be expected in grain-size data sets measured in Jena and in Potsdam (J. Stucki, Beckman Coulter GmbH Germany, personal communication, 2013).

3.2   End-member modelling analysis (EMMA)

Grain-size data was analysed separately for each lake system. All particle size data from Donggi Cona cores PG-1900, PG-1901, and PG-1904 retrieved in similar water depths were compiled in a matrix for a joint evaluation of the same lake system. For Taro Co (TOC) EMMA was performed separately for each core and combined thereafter to see if water depth and distance to the shore affected the results. Event layers (differentiated by fining-upwards facies and distinct colour changes) are difficult to assess with EMMA, as they represent a single process affecting all grain-size fractions in contrast to the size-sorting processes that are of interest here. It was attempted to exclude these layers as best as possible (i.e. ~ 16 and 40 % of the samples were removed from Tangra Yumco and TT Lake). Generally, grain-size classes containing zero values in all samples were ignored to avoid numerical instabilities (Table 3). All samples were recalculated to the constant sum of 100 % (Dietze et al., 2012).

EMMA extracts end members from the eigenspace of a data set. An end member consists of loadings, which are the representation of the end member in the variable space (here: grain-size class) and scores, the composition in the sample space. The loadings were scaled to be genetically meaningful (scale and unit according to original data; Weltje, 1997) using a weight transformation after Klovan and Imbrie (1971). The scores will be discussed elsewhere in the frame of multi-proxy palaeoenvironmental reconstructions. Grain-size end members were calculated after Dietze et al. (2012) using extensions implemented in the R package EMMAgeo (Dietze and Dietze, 2013).

Dietze et al. (2012) showed that from different similarly–likely end-member models some end members had modes at almost identical grain-size fractions. The less an end-member loading varies and the more often it occurs, when using different similarly–likely model runs, the more robust it is. For the similarly–likely models the number of end members q can be varied between q_{min} and q_{max} (Dietze et al., 2012). Similarly, the quantile range (l) ("percentile" in Dietze et al., 2012) included in the weight transformation may range between zero (Miesch, 1976) and a maximum quantile (l_{max}) defined as the highest possible quantile allowing numerically stable weight transformations. The weight transformation reduces scale and outlier effects by standardizing to a certain range of the data. The ranges of q and l can result in potentially infinite numbers of end-member models. For computation efficiency we used a sequence of 50 l values. From combinations of l and the likely range of q only robust and interpretable end-member models were selected. Robust end members (rEM) are defined using the following criteria: (a) unmixing of subpopulation had to be effective and was measured by calculating the overlap of the modes in the end-member loadings. Only models with non-overlapping, interpretable end-member loadings were included and (b) similar end-member loadings should occur in most of the similarly–likely model runs (high frequency of modes in the same or directly adjacent grain-size classes) as determined from histogram-plots (Figs. 2a).

The mean total variance of each end member was calculated to assess the relative importance of each end member in each data set. Furthermore the best-possible parametrisation of an end-member model is given with an optimal weight quantile (l_{opt}), which resulted in the highest explained variance compared to all other l and an optimal number of EM (q_{opt}) at the inflection point within the Q-R^2_{mean} plot of the l_{opt} model in consensus with previous works (Weltje and Prins, 2007).

Table 3. Input and boundary parameters for similarly–likely end-member model runs for robust end-member (rEM) calculation and optimal EMMA (l_{max} and l_{opt} refer to maximum and optimal weight quantiles; q_{opt} indicates the optimal number of end members; see text and Dietze et al., 2012).

Site/core	No. of samples	No. of non-zero grain size classes	l_{max}	l_{opt}	No. of included models*	No. of rEM	Total R^2_{mean} *	q_{opt}
Targo Xian	86	92	0.052	0.031	129	5	0.850	5
TT Lake	53	89	0.028	0.005	241	5	0.889	4
Donggi Cona	128	85	0.012	0.008	225	4	0.841	4
TOC11/4	122	71	0.061	0.010	91	5	0.891	5
TOC11/6	90	72	0.028	0.014	119	3	0.837	3
TOC all	212	72	0.012	0.010	249	5	0.876	5
Tangra Yumco	136	74	0.011	0.004	164	4	0.908	4
Nam Co	92	66	0.016	0.016	128	3	0.860	2

* Mean robust EM model.

3.3 Wind data

To interpret the observed end members in the context of the recent climate, data from the High Asia Reanalysis was used (HAR, Maussion et al., 2011, 2013; http://www.klima.tu-berlin.de/HAR, available for the time period 2001–2011). To consider aeolian sediment transport pathways and energies, the hourly surface wind fields at 10 m height (wind speed and direction) and the wind shear velocity field were extracted from the 10 km resolution products. They were averaged for for the seasons December-January-February (DJF), March-April-May (MAM), June-July-August (JJA), and September-October-November (SON).

4 Results

4.1 EMMA statistics

All sediment records showed multimodal grain-size distributions with different modes or shoulders in the individual samples (Fig. 2b). Calculating similarly–likely EM models resulted in a distinct pattern of EM loadings (i.e. grain-size subpopulations) for each data set (Fig. 2b). In general, histograms indicated between three and five highly frequent primary modes of the loadings at certain grain-size classes (Fig. 2a). Compared to the original multimodal distributions, these rEM subpopulations were clearly unmixed. The mean and standard deviations of these robust end-member loadings, numbered according to the position of their primary mode from coarse to fine grain-size fractions, are shown in Fig. 2b.

As an example, Fig. 3 shows illustrative plots for the Nam Co data set using a sequence of $10 l$ indicating how q_{max}, l_{max}, l_{opt}, and q_{opt} were determined (automatically provided in R package EMMAgeo, Dietze and Dietze, 2013; for q_{min} see Dietze et al., 2012) and which models resulted in over-lapping end-member loadings. Table 3 summarizes these parameters for all data sets. These boundary parameters are not associated, which means the total number of samples, non-zero grain-size classes, optimal number of EM (q_{opt}) or number of included similarly–likely EM models does not affect the number of robust EM or the mean total variance they explain.

Some sites (e.g. Tangra Yumco) showed rather clear rEM loading patterns with very small deviations from the mean indicating a high robustness of the resulting end members. Some sites had rEM that were very robust (e.g. Targo Xian rEM 3–5, Fig. 2). Others were less robust as these may explain only a small part of variance (rEM1 of TOC11/6 and both TOC cores, Fig. 2).

Important features are the smaller secondary modes observed in all loadings independent of laboratory pretreatment. In all cases except for rEM2 of the Nam Co data set they represent numerical artefacts as they appear below the primary modes of other end members. We assume that the secondary modes arise from the orthogonality and linearity constraint in EMMA, because due to measurement procedures and the sedimentary processes itself the eigenspace is likely non-linear. However, Dietze et al. (2012) found that oblique rotations resulted in less interpretable end members, which is why orthogonal constraints were kept. Here, the secondary mode artefacts were masked out to allow a better comparison and discussion of the major modes (Fig. 4).

4.2 Robust end-member loadings

EMMA of the TOC11/4-core and the NC08-1 core produced three rEM explaining 83.7 and 86.0 % of the mean total variance (Table 2). Four rEM occurred in the TT Lake, Donggi Cona and Tangra Yumco data sets explaining 81.4, 84.1, and 90.8 % of the mean total variance. The most frequent end member in the TT Lake data set in the coarse

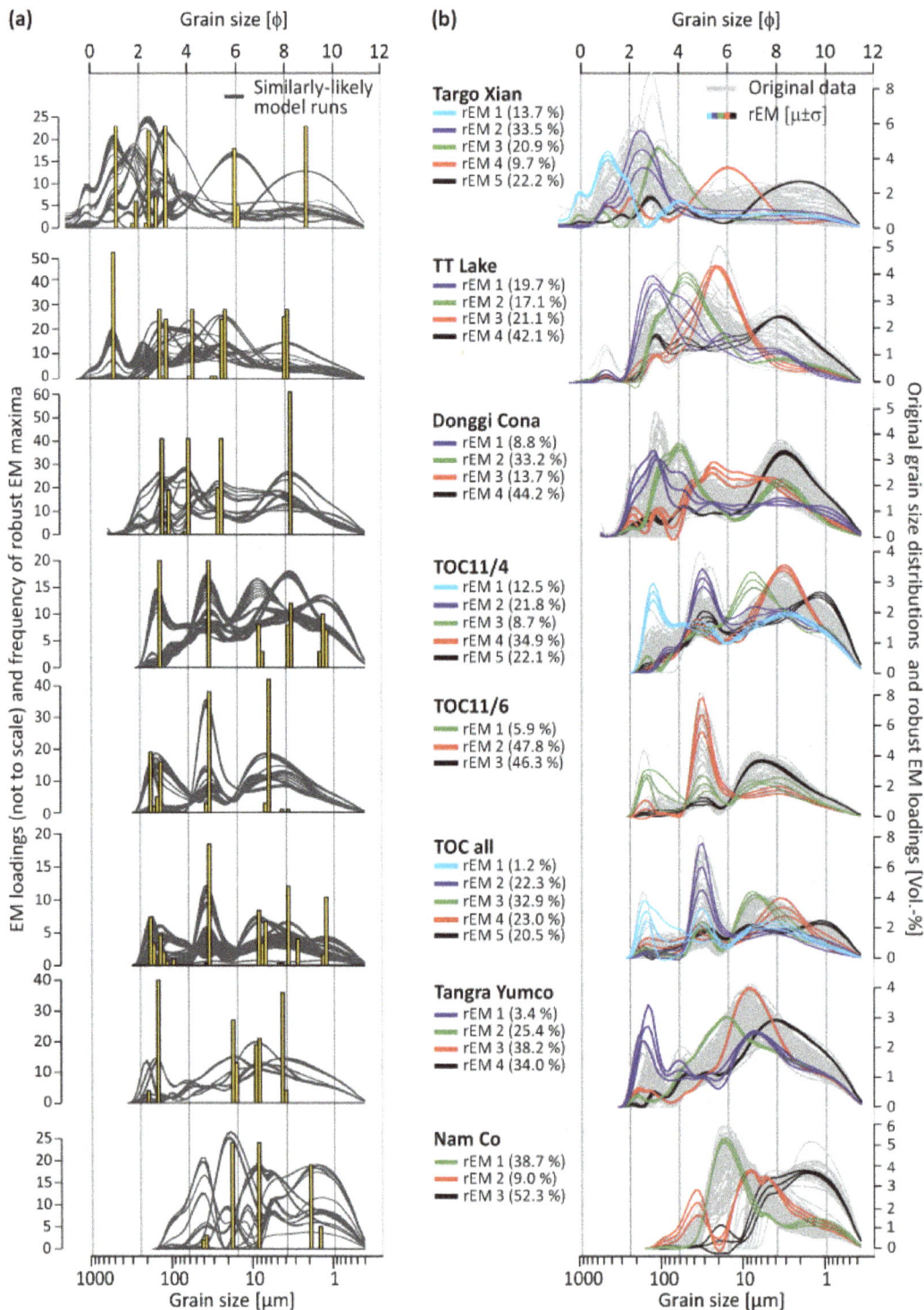

Fig. 2. Results of robust end-member modelling for the studied lake sites (sorted according to lake size). (**a**) End-member (EM) loadings for more than 90 similarly–likely EM model runs (grey graphs) including the frequency of their primary modes (yellow bars) indicating the position of the robust end members. (**b**) Robust end-member (rEM) loadings (coloured, $\mu \pm \sigma$, in brackets: explained variances of each EM) compared to the original grain-size distributions (grey graphs in background).

fraction, however, was not integrated in the mean robust end-member compilation (compare Fig. 2a and b), because it was only present in two samples explaining only 1 % of the mean

total variance, which furthermore resulted in a poorer unmixing of rEM1 and rEM2 modes. Adding the set of modern lake surface samples (Dietze et al., 2012) to the Donggi Cona data

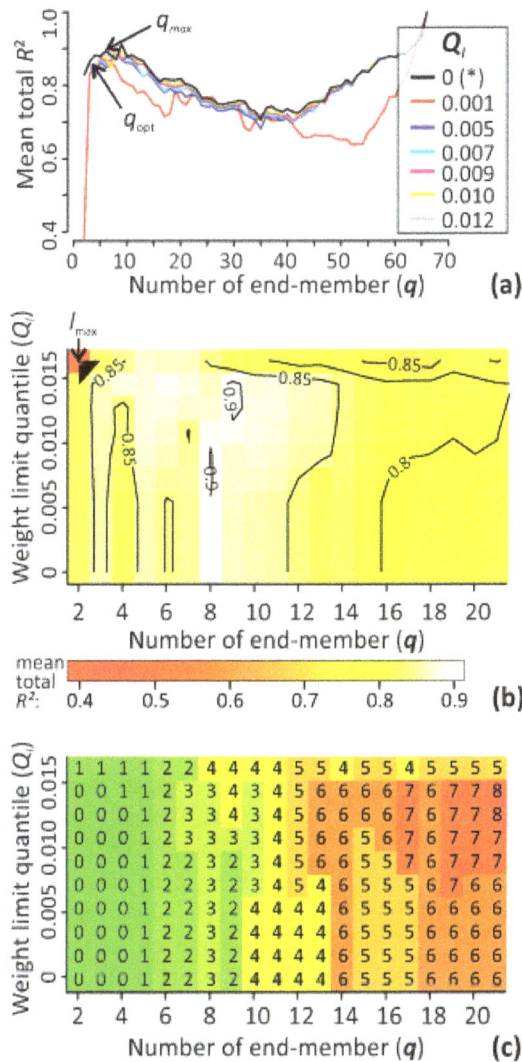

Fig. 3. Example how boundary parameters were defined for all similarly–likely model runs using the Nam Co core grain-size data set. **(a)** $Q - R^2_{mean}$ plot after Dietze et al. (2012) for selected weight quantiles (l) and numbers of end members (q) (q_{opt} at the inflection point after Weltje and Prins (2007); q_{max} at the first local maximum); **(b)** a 3-D pattern of the $Q - R^2_{mean}$ plot including an example sequence of 10 l between 0 (* after Miesch, 1976, corresponds here to l_{opt}) and the numerical maximum of l and the first 20 q that resulted in different end-member models, for which **(c)** shows the number of overlapping modes (only zero overlaps were taken into account for robust similarly–likely model runs).

set resulted in similar end-member composition for the core data indicating the reliability and robustness of the results (not shown).

The grain-size distributions of TOC11/4, TOCall and Targo Xian were composed of five rEM explaining 89.1, 87.6, and 85.0 %, respectively, of the mean total variance of the original data sets. The rEM modes of the individual Taro Co cores were also present at the same grain-size fractions and with similar shape in the rEM of both cores (Fig. 2,

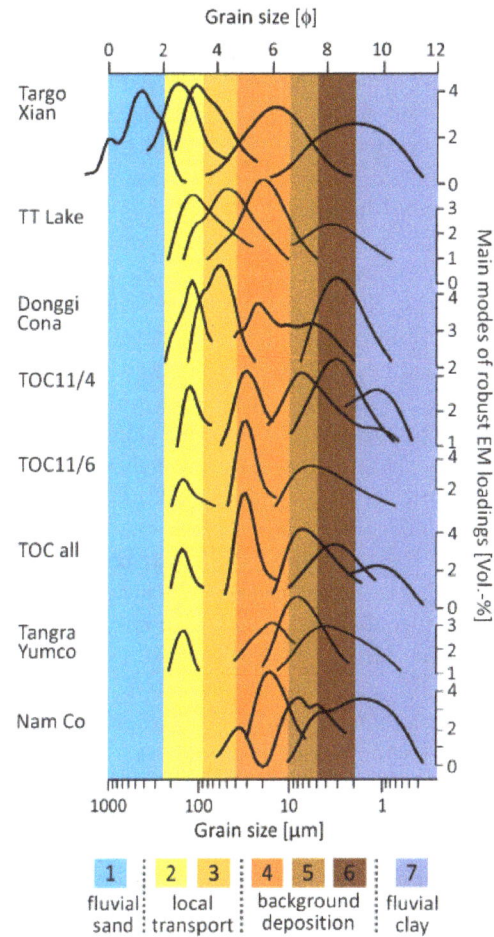

Fig. 4. Main modes of mean robust end-member loadings grouped as typical sediment transport modes. Numbers indicate sediment types: 1 – fluvial sand, 2 – aeolian sand, 3 – coarse local dust, 4 – coarse regional dust, 5/6 – remote dust, 7 – fluvial clays (see text for details).

TOC11/4 and TOC11/6 vs. TOC all). However, the location of the Taro Co core sites played a role in the number of rEM (q) with higher q at TOC11/4 compared to the distal site of TOC11/6 (Table 3).

5   Characteristics and origin of the robust grain-size end members

5.1   Preliminary considerations

Modelled robust end members showed distinct grain-size modes that can be grouped across the Tibetan Plateau, independent of the length of the individual sedimentary sequence. Similar modes were found as typical components in multimodal grain-size data of sediments in the Donggi Cona catchments (Stauch et al., 2012; Dietze et al., 2013). They reflect known distributions of fluvial and aeolian sediment types similar to those investigated at their characteristic depositional environments (Sun et al., 2002, 2007; Vandenberghe, 2013). To better understand the relation between grain sizes, transport mode and depositional environment some climatic parameters, the mean wind velocity, direction and shear velocity across the Tibetan Plateau for the four seasons are presented in Fig. 5 and Table 4. The seasonality in the wind patterns is evident: high velocities and spatially consistent patterns are observed in winter, whereas in summer, a marked cyclonic circulation (the "Tibetan Low") characterizes the surface wind fields, with larger differences between the studied lakes. Averaged across all areas above 3000 m a.s.l. highest shear velocities ($> 0.8\,\mathrm{m\,s^{-1}}$) occur in winter and spring, whereas autumn is the calmest time of the year (Fig. 5, Table 4a). Regionally, highest wind velocities occur more to the west and south of the Tibetan Plateau along the main track of the westerlies and the highest mountain ranges. Hence, more wind shear velocity extreme events ($> 1.2\,\mathrm{m\,s^{-1}}$) that could transport coarser grains occur along the climatic gradient from the westernmost lake Taro Co to Nam Co and are less frequent at lake Donggi Cona (Table 4b).

For the interpretation in terms of typical Tibetan-wide sediment transport and depositional processes, robust grain-size end members are associated with climatic conditions including Tibet-wide lake ice phenology (Kropáček et al., 2013), precipitation and snow/glacial melt (Maussion et al., 2013), the latter being the main triggers of fluvial sediment transport. For simplicity and following the experimental results of Tsoar and Pye (1987) for quartz sphere transport at mean Tibetan shear velocities rEMs were grouped in terms of transport modes with modes $> 90\,\mu m$ being primarily transported locally by saltation. Grain-size rEM between 40 and 90 μm represent local dust sediment transport, whereas grains $<40\,\mu m$ represent regional plateau-wide and remote dust flux in long-term suspension. The finest rEM with modes $< 2\,\mu m$ as a special case might represent local fluvial clays.

5.2 The "local transport" end members

5.2.1 Fluvial sands

At the terrestrial site Targo Xian the coarsest rEM of all sites was found with a mode between 250 and 600 μm in the medium to coarse sand. It had a minor mode at 1 mm and a shoulder at ~ 280 μm with a maximum at 450 μm. It contributed 13.7 % to the mean grain-size variance (Fig. 2). Its scores were high in the Targo Xian lithological unit, where also gravels were found. Tentatively, this rEM also contributed to the small TT Lake but there it explained only 1 % of the grain size.

This rEM indicates a general transport mechanism delivering particles to the site by rolling and saltation in high-energy fluvial suspension (Tsoar and Pye, 1987; Fig. 4). Flu-

Table 4. Frequency on an hourly basis of wind shear velocities events from the High Asia Reanalysis (HAR) calculated as (a) average across the Tibetan Plateau for different seasons (winter, DJF, spring, MAM, summer, JJA, and autumn, SON) and (b) all-year average across the four largest lake catchments.

(a) Seasonal frequency of wind shear velocities events				
U^* [m s^{-1}]	DJF	MAM	JJA	SON
≥ 1.2	0.17 %	0.20 %	0.01 %	0.05 %
[0.8, 1.2]	3.55 %	4.34 %	1.05 %	1.70 %
[0.5, 0.8]	20.64 %	23.10 %	17.99 %	16.02 %
[0.2, 0.5]	46.72 %	45.83 %	54.56 %	46.92 %
< 0.2	28.92 %	26.53 %	26.39 %	35.31 %

(b) Regional frequency of wind shear velocities events				
U^* [m s^{-1}]	Donggi Cona	Nam Co	Tangra Yumco	Taro Co
≥ 1.2	0.04 %	0.20 %	0.31 %	0.33 %
[0.8, 1.2]	3.12 %	5.90 %	5.00 %	6.94 %
[0.5, 0.8]	24.75 %	31.60 %	24.86 %	28.97 %
[0.2, 0.5]	54.56 %	46.72 %	45.83 %	46.92 %
< 0.2	25.56 %	17.59 %	21.69 %	17.74 %

vial sediments can be of multimodal character depending on sediment load and current velocities, typically showing coarse modes between 200 and 400 μm (Sun et al., 2002) or even coarser. Compared to sediment delivered through aeolian transport fluvial deposits are generally coarser and poorly sorted (Tsoar and Pye, 1987). In modern surface sediments and in on-shore high-stand sediments of lake Donggi Cona this mode was also present in small amounts (~ 6–10 %) with a similar mode shape. There, it was interpreted as fluvial and littoral transport, because it appeared only in the fluvial and wave-sorted high-stand facies (Dietze et al., 2012, 2013). Furthermore, it was the prominent mode of typical alluvial fan deposits in an aeolian-fluvial sediment–soil sequence in other environments (e.g. in Europe, Pendea et al., 2009). The absence of fluvial sands in the lacustrine sediments of the larger lakes might be related to the distance to the shore (and to inflows) and the immediate settling, when fluvial transport energy drops at the lake shore. Fluvial transport occurs in spring and summer, when the ground has thawed and most precipitation occurs (Maussion et al., 2013).

5.2.2 Aeolian sands

The coarsest rEM in the larger lake settings had a mode in the fine to very fine sand between 90 and 250 μm. It had a very similar mean shape and maximum position at ~ 150 μm at Taro Co and at Tangra Yumco (Fig. 2a). The mode was finer at Donggi Cona and TT Lake (maximum at 120 μm). It explained ~ 20 % of variance at the small TT Lake and 12.5 % at the TOC11/4 site. It contributed much less to the sediment

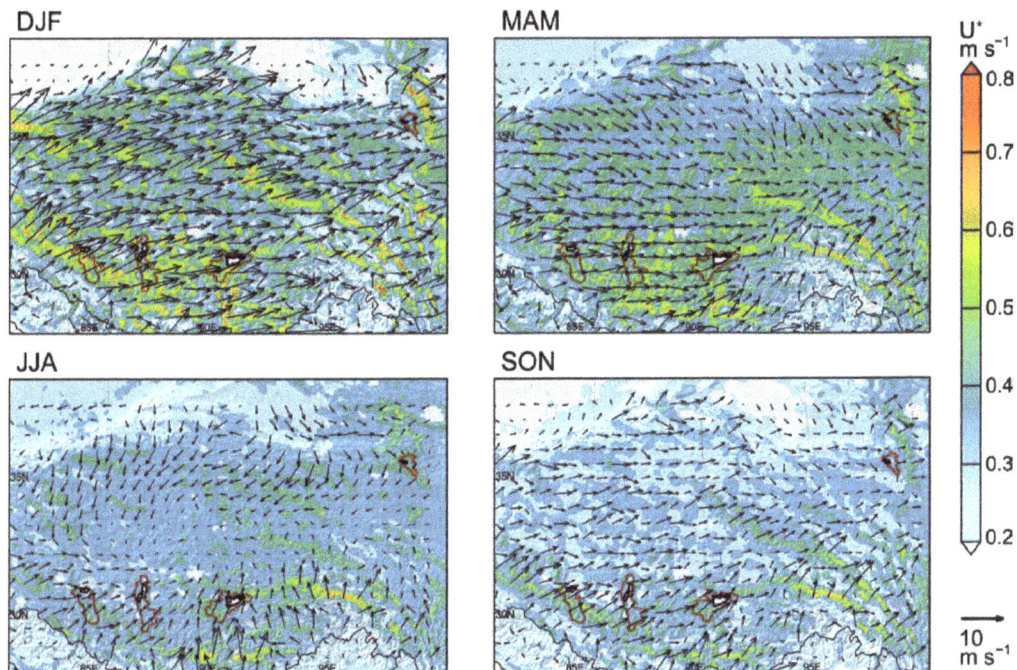

Fig. 5. Mean shear velocities (U^*, colour scale) and wind direction and speed (arrows, plotted every 6th grid point) for winter (DJF), spring (MAM), summer (JJA) and autumn (SON) extracted from the High Asia Reanalysis (HAR). The studied catchments are delimited in red.

at the more distal, deep water positions of Tangra Yumco and Taro Co (TOC11/6). At Targo Xian there were two rEM modes in this range (Figs. 2b and 4). One was well-sorted and coarser between 250 and 100 μm with its maximum at 194 μm. It made up one third of the site's grain-size variance. The second was a relatively broad, but very robust mode in the very fine sand between 50 and 150 μm with a maximum at 92 μm and a secondary shoulder at 60 μm. It explained 21 % of the Targo Xian grain-size variance (Fig. 2). In general, this mode was relatively heterogeneous and less robust (large standard deviation, Fig. 2b) than other rEM modes.

This rEM may be associated with a local transport of rolling and saltation by strong surface winds. The same size fractions dominated in dark laminae of lake Kuhai (Fig. 1, Mischke et al., 2009) and were composed of subangular mineral grains indicative of wind transport. In modern Donggi Cona surface sediments this end member occurred randomly in all samples and was interpreted as sands moving on the frozen lake surface (Dietze et al., 2012). Hence, a (partial) freezing of the lake from the shore is required to contribute sand to more distal positions, also by drifting ice containing trapped sediment. However, at larger lakes only little sandy material can reach the more distal locations because ice traps coarse particles in cracks and occasional snow cover (Nedell et al., 1987), which would explain the lack of this end member in the Nam Co core.

The transport of "aeolian sands" (Sun et al., 2002) on frozen lakes can be associated with very strong, episodic winds that are dominant especially during winter and early spring times on the Tibetan Plateau from westerly and northerly directions (Fig. 5). This transport is most effective during winter and spring storms with mean shear velocities of $> 0.8\,\mathrm{m\,s^{-1}}$ and grains $> 90\,\mu$m could even go into short-term suspension (Tsoar and Pye, 1987; Table 4). A brownish stained ice surface was observed several times at Donggi Cona and lake Kuhai (Fig. 6); especially during spring ice break-up (Mischke et al., 2009). Its mode heterogeneity might be related to different grain inventories at the littoral sources, for example, dune fields and local river beds (Stauch et al., 2012). It fits well with the aeolian sediment type 1.a described by Vandenberghe (2013), who pointed out its local character.

At the terrestrial Targo Xian site the subdivision into two rEM (Fig. 2a) that occurred in samples of different age (not shown) could indicate a systematic change in transported material and/or wind field. The well-sorted coarser rEM can be related to dune formation (Tsoar and Pye, 1987; Sun et al., 2002). However, the finer rEM was finer than typical dune sediment and it might even point to a different process: reworking of sands of different sources by low-energy unconfined alluvial flow (sensu North and Davidson, 2012) as found in Donggi Cona high-stand sediments (Dietze et al., 2013).

Fig. 6. Sediment trapped in lake ice at lake Donggi Cona (Photo: B. Diekmann, February 2007).

5.2.3   Coarse local dust

At the boundary of very fine sand and coarse silt a further rEM had a mode between 40 and 90 μm with maxima in the TT Lake and Donggi Cona data sets at 53 and 63 μm, respectively. It was only present at these two sites and had a secondary shoulder $\sim$ 100 μm (Fig. 2) indicating that saltation and/or high shear velocities might still be important. This rEM was similar to the local dust end member in the lake Donggi Cona surface samples, where it contributed $\sim$ 17 % to the total sediment composition (Dietze et al., 2012) – the same contribution as to TT Lake sediments.

In the catchment of lake Donggi Cona it was the dominant grain-size fraction in sandy loess (Stauch et al., 2012) and constituted an end member in onshore lake high-stand sediments interpreted as local dust (Dietze et al., 2013). The sediment represented by this rEM was probably transported from proximal sources as coarse local dust that was lifted into near-surface, short-term suspension clouds, when shear velocities between 0.2 and 0.8 m s^{-1} prevailed (Tsoar and Pye, 1987). Although these shear velocities are observed in all seasons (Fig. 5, Table 4), sediment needs to be available in the catchments as in periods of unfrozen soil and limited vegetation cover during spring and early summer. Sun et al. (2003) found the same grain-size modes in present-day dust deposited after major spring–summer dust storms on the Chinese Loess Plateau. On the Tibetan Plateau the sandy loess probably has its source in local rivers, sand sheets at the slopes, periglacial and glaciofluvial debris (Stauch et al., 2012). It corresponds to the coarsest-grained loess fraction distinguished by Vandenberghe (2013; i.e. sediment type 1.b.3, also associated with loess deposited during high wind intensities at the Loess Plateau).

5.3   The "background deposition" end members

5.3.1   Coarse regional dust

A rEM present in all studied Tibetan lake sediment records had modes between 10 and 40 μm, with broader maxima at 17 μm (Nam Co and Tangra Yumco) and 21 μm (TT Lake) (Figs. 2 and 4). The mode was very well sorted and symmetrical at Taro Co (TOC) between 23 and 42 μm with a maximum at 33 μm. It contributed almost 50 % of the sediment to the deep position of this westernmost lake (TOC11/6). At Targo Xian, rEM 4 (Fig. 2b) had a slightly finer, but much broader mode with a maximum at 11.8 μm. Donggi Cona core sediments consisted of a very broad mode of rEM 3 between 5 and 40 μm with a maximum at 22 μm (Fig. 2b).

The grain sizes of this rEM are similar to the loess described from Donggi Cona catchment with a suggested source area in the Qaidam Basin, $\sim$ 100 km away (Stauch et al., 2012) – a similar distance as reported from a provenance study of Romanian loess of the same size (Pendea et al., 2009). Particles of these sizes can easily be captured on water surfaces, but also in snow and ice or on (vegetated) slopes (Goossens, 2005; Stauch et al., 2012), which was the suggested depositional process for the robust "remote dust" grain-size end member in Donggi Cona surface and lake high-stand sediments (Dietze et al., 2012, 2013). The typical Tibetan loess was slightly coarser in the Yarlong Zangbo River valley south-east of Nam Co compared to the Donggi Cona loess at the north-eastern Tibetan Plateau (20-44 μm, Sun et al., 2007 compared to 10–30 μm, Stauch et al., 2012).

This rEM represents the size fractions of standard loess that can keep in short-term suspension in the lower atmosphere (Tsoar and Pye, 1987). It could episodically be transported over larger distances and longer time compared to coarser fractions (Pendea et al., 2009; Stauch et al., 2012). Particles < 20 μm can stay in suspension for days during storms with mean shear velocities > 0.3 m s^{-1} (Tsoar and Pye, 1987) that occur throughout the year, but are less likely during autumn (Fig. 5). Hence, this mode should be associated with large-scale, plateau-wide dust storms that deposit coarse regional dust discontinuously during dry periods (e.g. during Tibetan spring storms before the onset of the monsoonal rain season). It can be transported occasionally also in summer times by cyclonic circulations to the northern and western Tibetan Plateau with northeasterly winds (Chen et al., 2013). This rEM represents a combination of the silt sediment types 1.b.2 and 1.b.3 of Vandenberghe (2013).

If transported by wind alone, the mode would be well sorted as in the case of Taro Co sediments (Fig. 4). However, it is also likely that loess was reworked from the onshore slopes (e.g. by unconfined overland flow), as observed in summers 2006 and 2009 at lake Donggi Cona, which might also be likely at the terrestrial site Targo Xian. These sediment-saturated flows enter the lakes as density over- or interflows (Sturm and Matter, 1972) and can be distributed by water currents within the lake also to more distal sites. This would not alter the grain-size mode because the aeolian signature can still be well preserved (Vandenberghe, 2013), especially when the fluvial transport ways were short and no further sorting occurred. A comparison of terrestrial sheetflood and loess surface samples of the Donggi Cona catchments showed similar grain-size distributions supporting this assumption (Dieckmann and Dietze, unpublished data). However, the environmental conditions and implications for using this rEM as a proxy in palaeoclimate studies would be slightly different for direct aeolian deposition or reworking by overland flow.

5.3.2 Remote dust

A rEM in the very fine silt had a mode between 5 and 10 μm, with maxima at 6.2 and 7.6 μm at Taro Co (finer at the deeper location) and 8.2 μm at lakes Tangra Yumco and Nam Co (Fig. 4). At Nam Co it constitutes the main mode of the multimodal rEM sediment (Kasper et al., 2012). It was absent in the small TT lake, but could be integrated in Targo Xian rEM 4 (Figs. 2 and 4), which consisted mainly of the fine short-term suspension end member and seemed to be a reworked loess. The more prominent robust clay end member that appeared all over the Tibetan Plateau had a mode between 2 and 5 μm, with maxima at 2.4 (Nam Co), ~ 3.4 (TT Lake, Donggi Cona, Taro Co), and 4.2 μm in lake Tangra Yumco (Fig. 4). Light laminae of subspherical particles ~ 4 μm represented this mode in lake Kuhai (Fig. 1), although their alteration with darker, coarser laminae could not be related to seasonal or annual cycles (Mischke et al., 2009).

These rEM represent the finest aeolian sediments distinguished by Vandenberghe (2013; type 1.c.2: 2–10 μm). According to Sun et al. (2004) particles of the size fractions represented by both rEM can be transported over large distances (1000–2000 km) in atmospheric levels up to 7 km high. Hence, these rEM grain-size modes represent a continuous background dust deposition all year long (Sun et al., 2004; Vandenberghe, 2013) and can be traced even in the North Pacific (Rea, 1994). On the Tibetan Plateau the remote dust is assumed to originate mainly from the northwestern and northern Central Asian deserts transported by the high-level westerly jet stream in winter and episodic northerly winds in summer time along a main transport route from northwest to southeast (Fig. 5; Zhang et al., 2001; Sun et al., 2002; Stauch et al., 2012; Chen et al., 2013).

In contrast to dry deposition of modern Australian dust (20–40 μm) considerably finer grains (3.5–15 μm) were accumulated by wet deposition (Hesse and McTainsh, 1999). However, Zhang et al. (2001) showed that wet deposition accounted for less than 10 % of the total Tibetan dust deposition. Hence, dry deposition can be assumed as the major depositional process behind these rEMs. Dry remote dust can also be trapped directly on the water or ice surface of the lakes. These fractions need a calm water column to set-tle down to the lake bottom, which is probably only possible when the lakes are partially or completely frozen during winter (Francus et al., 2008).

The two background dust rEM were separated for different reasons: (a) the lake size could play a role – the finest-silt rEM occurred only in the larger lakes, which may be due to different sediment trapping properties than in small lakes and on land (Goossens, 2005), (b) the steepness of the catchment close to the lake could influence local luv-lee and katabatic wind dynamics (Fig. 1), and/or (c) these two rEM indicate two different source regions, high-level wind directions and wind speeds that could change seasonally. The finest-silt rEM can be compared to modes of 5–8 μm of the Chinese Loess Plateau red clays (Vandenberghe, 2013) deposited under intense winter monsoon conditions related to the Siberian High similar to Quaternary loess (Vandenberghe et al., 2004), whereas a finer Red Clay fraction < 5.5 μm representing the clay rEM is assumed to reflect more westerly wind directions from its isotopic signatures (Vandenberghe et al., 2004). Hence, these two background dust end members should be discussed together with other mineralogical or geochemical proxies to distinguish different sources and wind directions in future palaeoenvironmental discussions.

An overprint of the main transport mode similar to other aeolian sediments could be occasional summer fluvial and alluvial flow in the catchment that erode and rework the slope deposits together with soil clay aggregates. Hence, in Donggi Cona surface and lake high-stand sediments this rEM was interpreted more generally as "suspension" load (Dietze et al., 2012) and associated with high lake levels (Dietze et al., 2013). In the context of a multi-proxy approach (Kasper et al., 2013) characteristic grain-size distributions at lake Nam Co were attributed to certain lake levels. Grain-size distributions with the secondary mode (rEM 2, Fig. 2b) were interpreted as representative for a lake level rise (Kasper et al., 2012).

5.4 Fluvial clay

The finest rEM had a modal size of 0.7–2 μm and could only be found at lake Taro Co as an individual rEM with a maximum at ~ 1.2 μm (TOCall and TOC11/4; Fig. 4). It explained 22 % of the grain-size composition in the 68 m water-depth core TOC11/4 (Fig. 2) and was absent in the deeper position. It may have contributed to the broader clay mode that dominated the lowermost lithological unit in the record of lake Nam Co, which was deposited during times of a higher lake level (Kasper et al., 2012). At Targo Xian it was part of a broad clay end member with a mode between 0.7 and 5 μm (maximum at 1.7 μm).

A likely transport mechanism is fluvial input of clay minerals from weathered catchment soils. At Taro Co they could be washed into the ~ 160 km long river course from the south (Fig. 1) and from the nearby slopes during extreme precipitation events. At palaeo-lake Targo Xian and Nam Co the broad and very fine rEM dominated only the lowermost parts of the stratigraphies that were associated with higher lake stands (Kasper et al., 2012). This would indicate wetter times, when also weathering of soils and the amount of runoff were higher. Depending on lake stratification and water currents, finest clay minerals with a size < 1 μm, were transported farthest into lake Donggi Cona compared to other coarser clay minerals (Opitz, unpublished data).

Clays can only reach the lake during summer runoff and remain in the water column as long as settling is possible by ice cover of the lake. Freezing of lakes normally starts in the protected bays, but also depends on wind and wave activity (Kropáček et al., 2013). In comparison with the other large lakes of this study, Taro Co had the lowest duration of complete ice cover in the analysis of modern day ice phenology (Kropáček et al., 2013), although no causes were discussed by the authors. Hence, it is more likely that the shallow areas freeze and allow an earlier settlement of the finest particles. Over the deepest lake parts surficial wind-induced currents would still prevent fine material from settling. Another reason might be the west-east extension of the Taro Co basin (Fig. 1), which allows the greatest fetch along the dominant westerly wind direction (Fig. 5). Hence, surface water currents probably bring particles predominantly to the eastern bay.

Furthermore, this size range (< 2 μm) was also found in high percentages (up to 74 % of total dust deposits) on high Tibetan mountain ranges, where it was trapped in glacier ice (Wu et al., 2010) indicating an aeolian origin of the material from the upper troposphere. The occurrence of grain-size modes between 0.8 and 2 μm in Greenland ice cores (Ruth et al., 2003) suggests a global distribution via long range transport and also that potential source areas could be much farer away from Tibet than for the regional background dust. It could be deposited in the lakes in aggregated forms from direct dust deposition on water and disperse in the water column to later settle down under calm conditions. If the contributions from the finest, long range dust would be comparable to other transport modes, there would be some signal visible in all lake systems and not only at lake Taro Co (especially at the shallow site). Hence, the finest background dust does not seem to explain the sediment source of this rEM.

Therefore, it is much more likely that the robust clay end member results from pedogenic clay that was fluvially transported to the lake. It seems to be of local origin, but may include the finest parts of remote dust. Both fractions settle only when the lake surface is ice-covered.

5.5 The relative contribution of grain-size end members

Assuming the reliability of the rEM interpretation, the relative contributions of the three major sediment transport processes can be estimated in a first approach using the sum of the explained variances of the mean rEM in % (Fig. 2b) grouped into fluvial (sand and clay), local dust and remote

dust for each core (Fig. 4). Accordingly, these processes contribute to the sediment accumulation in the studied Tibetan lakes as follows: (a) fluvial processes induced by precipitation and glacier or snow melt contribute 18.8 ± 4.5 % (Targo Xian rEM 1, TOC11/4 and TOCall rEM 5; mean ± standard deviation from Fig. 2b), (b) short-range local and regional dust 42.3±13.7 % and (c) the remote dust 50.9±10.5 %. The broad and ambiguous rEM 5 (Targo Xian), rEM 3 (Donggi Cona) and Nam Co rEM 2 and 3 were not included in this calculation. Hence, fluvial sands and clays from the catchment could make up less than a quarter of total detrital sediment components in this region, whereas aeolian sedimentation seems to be the dominant detrital sediment input in Tibetan lakes and catchments (the dominant aeolian signature appears to be preserved despite subsequent short-term fluvial/alluvial reworking). In comparison with climatic conditions (Fig. 5), sediment availability and deposition seem to vary seasonally.

Similarly, Wu et al. (2010) found that there was a 5.6 times higher winter dust deposition on high-altitude Dasuopu Glacier compared to summer times, although a differentiation after grain size was not provided. In comparison with more humid regions, aeolian processes are a key factor in the Tibetan sediment cycle similar to other desert regions. However, total annual dust deposition rate on the Tibetan Plateau (local plus remote sources; dry plus wet deposition) was only a third of that in Chinese deserts in a study of dust fluxes during non-dust storm days in 1993 and 1994 (Zhang et al., 2001).

5.6 Limitations

Although robust grain-size end members seem to be comparable across the Tibetan Plateau, some limitations have to be considered. First, end-member modelling of multimodal grain-size distributions still has some numerical problems concerning unexplainable secondary modes that are not yet fully understood. For example, the secondary mode in Nam Co rEM 2 could either be an artifact or reflect a depositional process that integrated different grain-size classes (e.g. glacial processes, Diekmann, 1990). However, the performance of EMMA was never tested using classical diamicts.

Second, a likely contribution of authigenic diatom production has to be taken into account. The preservation, distribution and importance of diatoms in most Tibetan lakes is generally low and Tibetan diatom records are sparse (Gasse et al., 1991). Therefore, it was not considered to be an important factor during pretreatment of the studied samples. However, temporal shifts in the silica limitation can alter diatom abundance and thus the grain-size distributions of the sediment.

Third, the relationship between grain-size distributions (i.e. all end members) and transport processes depend on further factors, for example, shear velocities related to land cover and surface roughness, (already sorted) sediment sources, sediment concentrations, pathways and carrying capacities of the transport medium (Tsoar and Pye, 1987; Vandenberghe, 2013). Sediment sorting considers also further grain characteristics like grain shape and density that are related to the respective grain source material. Grains, especially cohesive fine silts and clays, may enter the transport medium in the form of soil or sediment aggregates or can form floccules in water. Hence, fine material would not necessarily disperse into individual particles in the lake (Schieber et al., 2007) like during pretreatment for laser particle size analysis. However, after a long transport in suspension either in rivers or in lake currents the finest material would settle down individually under very calm conditions and form clay caps as observed in clastic varves after ice cover (Cockburn and Lamoureux, 2008; Francus et al., 2008).

Finally, it is difficult to compare Tibetan lake sedimentation with other studies (e.g. from arctic or alpine lakes), where sedimentation monitoring strategies helped to better assess lacustrine deposition (Cockburn and Lamoureux, 2008; Francus et al., 2008; Kämpf et al., 2012). First of all, seasonal laminations are rare in the Tibetan lakes (Chu et al., 2011), and a considerable effort would be needed to link them to sediment trap and meteorological observations at this remote site. Second, in the high-altitude Tibetan Plateau the amount of snow and snow melt is small due to low winter precipitation (Maussion et al., 2013), high insolation and sublimation compared to other permafrost regions (French, 2007). Greatest runoff and peak precipitation occur during summer and are associated with local convection, cyclonic circulation and thunderstorms (Maussion et al., 2013). They can cause local overland flows that erode slope material complicating the palaeoclimatic interpretation of certain grain-size fractions being deposited at a certain time.

6 Conclusions

Sediment transported and sorted along different pathways is deposited as mixed allochthonous material on the bottom of lakes. Modelling and "unmixing" grain-size end members can be an attractive, quantitative alternative to classic approaches of grain-size data evaluation that help to better understand past environmental changes.

In this study, different lake systems that store sediments from various time spans, depositional environments and catchment configurations were investigated. Grain-size end member modelling of Tibetan lake sediments resulted in statistically robust loadings related to the initial grain-size sorting processes. Robust grain-size end members were interpreted as sedimentation processes using links to local catchment configurations and the modern climatic background (i.e. lake ice phenology, precipitation). To our knowledge, wind-shear data was analysed for the first time on the Tibetan Plateau and helped to better understand aeolian processes. Most sedimentation processes are consistent across the Tibetan Plateau. Assessing the relative contributions

of sediment transport mechanisms local (catchment-wide), regional (plateau-wide) and remote aeolian transport account for most of the clastic sedimentation in Tibetan lakes, whereas the direct fluvial component is of minor importance. Driven by changes in atmospheric circulation, processes of sediment transport and deposition may differ between seasons, i.e. aeolian transport dominates during times of high wind speeds and shear velocities. Future work will link these grain-size–process relationships with the past (using end-member scores, that is, rEM composition in depth/time, Dietze et al., 2012) to assess changes in the process contributions in space and time and facilitate the palaeoenvironmental interpretation of lake sediment archives – a concept that is transferable to other sediment archives and regions of the world.

Since the interpretation of the robust grain-size end members can be seen in general terms of transport energies and agents it is not restricted to certain time periods, lacustrine records and/or to the Tibetan Plateau. For future palaeoenvironmental reconstructions, robust grain-size end members can provide valuable, independent, and quantitative proxies of typical sedimentation processes that are related to past or present environmental and climatic conditions in many other settings. However, grain-size end members should be seen as a continuum and grain-size boundaries that were used to group them are far from being fixed. The background conditions for sediment availability, transport and deposition may vary in time in concert with regional changes in climate (precipitation, wind, ice phenology), or local changes of human activity (affecting land cover; Schlütz and Lehmkuhl, 2009) and tectonics (affecting basin and catchment configuration; Dietze et al., 2013). Hence, sedimentation end members might change their main mode positions and their relative importance, when certain time slices, regions or different types of sediment archives (e.g. loess, dune sequences) are studied individually in a high resolution.

To better assess their meaning robust grain-size end members should be compared with other proxies, for example, micropalaeontological, geochemical and mineralogical data, as well as with all available information from basin and catchment morphologies and the regional climate system. The transferability of the information that robust grain-size end members can provide should be tested in similar studies in other climatic and environmental settings. When considering that end members are not a stand-alone proxy and special local end members might occur, a high transferability of the method and helpful information on past sedimentation processes related to (palaeo-)environmental change are gained.

Acknowledgements. This study integrates several groups of the German Research Foundation (DFG) priority programme "Tibetan Plateau: Formation–Climate–Ecosystems" (SPP 1372). E. Dietze thanks T. Swierczynski, U. Kienel, P. Schulte, and M. Dietze for the fruitful discussions. C. Kirchner is acknowledged for grain-size analyses in Jena. K. Henkel would like to thank the Max

Planck Society and T. Kasper the Carl Zeiss Foundation for the funding of their PhD positions. The HAR was produced within the WET project (code 03G0804A), financed by the German Federal Ministry of Education and Research (BMBF) Program "Central Asia – Monsoon Dynamics and Geo-Ecosystems". We further thank Keely Mills, Jef Vandenberghe and an anonymous reviewer for helpful suggestions that improved the manuscript.

Edited by: K. Mills

References

Bartholdy, J., Christiansen, C., and Pedersen, J. B. T.: Comparing spatial grain-size trends inferred from textural parameters using percentile statistical parameters and those based on the log-hyperbolic method, Sediment. Geol., 202, 436–452, 2007.

Blott, S. J. and Pye, K.: GRADISTAT: a grain size distribution and statistics package for the analysis of unconsolidated sediments, Earth Surf. Proc. Land., 26, 1237–1248, 2001.

Böhner, J.: General climatic controls and topoclimatic variations in Central and High Asia, Boreas, 35, 279–295, 2006.

Chen, S., Huang, J., Zhao, C., Qian, Y., Leung, L. R., and Yang, B.: Modeling the transport and radiative forcing of Taklimakan dust over the Tibetan Plateau: A case study in the summer of 2006, J. Geophys. Res., 118, 797–812, 2013.

Chu, G., Sun, Q., Yang, K., Li, A., Yu, X., Xu, T., Yan, F., Wang, H., Liu, M., Wang, X., Xie, M., Lin, Y., and Liu, Q.: Evidence for decreasing South Asian summer monsoon in the past 160 years from varved sediment in Lake Xinluhai, Tibetan Plateau, J. Geophys. Res., 116, D02116, doi:10.1029/2010JD014454, 2011.

Cockburn, J. H. and Lamoureux, S.: Inflow and lake controls on short-term mass accumulation and sedimentary particle size in a High Arctic lake: implications for interpreting varved lacustrine sedimentary records, J. Paleolimnol., 40, 923–942, 2008.

Cohen, A. S.: Paleolimnology: the history and evolution of lake systems, Oxford University Press, New York, 500 pp., 2003.

Diekmann, B.: Granulometry and sand grain morphoscopy of alpine glacial sediments, Zbl. Geo. Pal., Teil 1, 1989, Heft 9/10, 1407–1421, 1990.

Dietze, E., Wünnemann, B., Diekmann, B., Aichner, B., Hartmann, K., Herzschuh, U., IJmker, J., Jin, H., Kopsch, C., Lehmkuhl, F., Li, S., Mischke, S., Niessen, F., Opitz, S., Stauch, G., and Yang, S.: Basin morphology and seismic stratigraphy of Lake Donggi Cona, north-eastern Tibetan Plateau, China, Quatern. Int., 218, 131–142, 2010.

Dietze, E., Hartmann, K., Diekmann, B., IJmker, J., Lehmkuhl, F., Opitz, S., Stauch, G., Wünnemann, B., and Borchers, A.: An end-member algorithm for deciphering modern detrital processes from lake sediments of Lake Donggi Cona, NE Tibetan Plateau, China, Sediment. Geol., 243–244, 169–180, 2012.

Dietze, E., Wünnemann, B., Hartmann, K., Diekmann, B., Jin, H., Stauch, G., Yang, S., and Lehmkuhl, F.: Early to mid-Holocene lake high-stand sediments at Lake Donggi Cona, northeastern Tibetan Plateau, China, Quaternary Res., 79, 325–336, 2013.

Dietze, M. and Dietze, E.: EMMAgeo: End-member modelling algorithm and supporting functions for grain-size analysis, R package version 0.9.0., available at: http://CRAN.R-project.org/package=EMMAgeo (last access: 10 December 2013), 2013.

Doberschütz, S., Frenzel, P., Haberzettl, T., Kasper, T., Wang, J., Zhu, L., Daut, G., Schwalb, A., and Mäusbacher, R.: Monsoonal forcing of Holocene paleoenvironmental change on the central Tibetan Plateau inferred using a sediment record from Lake Nam Co (Xizang, China), J. Paleolimnol., 1–14, doi:10.1007/s10933-013-9702-1, 2013.

Flemming, B. W.: The influence of grain-size analysis methods and sediment mixing on curve shapes and textural parameters: Implications for sediment trend analysis, Sediment. Geol., 202, 425–435, 2007.

Folk, R. L. and Ward, W. C.: Brazos River bar [Texas]; a study in the significance of grain size parameters, J. Sediment. Res., 27, 3–26, 1957.

Francus, P., Bradley, R., Lewis, T., Abbott, M., Retelle, M., and Stoner, J.: Limnological and sedimentary processes at Sawtooth Lake, Canadian High Arctic, and their influence on varve formation, J. Paleolimnol., 40, 963–985, 2008.

French, H. M.: The Periglacial Environment, John Wiley & Sons, Chichester, UK, 458 pp., 2007.

Gasse, F., Arnold, M., Fontes, J. C., Fort, M., Gibert, E., Huc, A., Bingyan, L., Yuanfang, L., Qing, L., Melieres, F., Campo, E. V., Fubao, W., and Qingsong, Z.: A 13,000-year climate record from western Tibet, Nature, 353, 742–745, 1991.

Goossens, D.: Quantification of the dry aeolian deposition of dust on horizontal surfaces: an experimental comparison of theory and measurements, Sedimentology, 52, 859–873, 2005.

Hartmann, D.: From reality to model: Operationalism and the value chain of particle-size analysis of natural sediments, Sediment. Geol., 202, 383–401, 2007.

Hesse, P. P. and McTainsh, G. H.: Last glacial maximum to early Holocene wind strength in the mid-latitudes of the Southern Hemisphere from aeolian dust in the Tasman Sea, Quaternary Res., 52, 343–349, 1999.

IJmker, J., Stauch, G., Dietze, E., Hartmann, K., Diekmann, B., Lockot, G., Opitz, S., Wünnemann, B., and Lehmkuhl, F.: Characterisation of transport processes and sedimentary deposits by statistical end-member mixing analysis of terrestrial sediments in the Donggi Cona lake catchment, NE Tibetan Plateau, Sediment. Geol., 281, 166–179, 2012.

Immerzeel, W. W., van Beek, L. P. H., and Bierkens, M. F. P.: Climate change will affect the Asian water towers, Science, 328, 1382–1385, 2010.

Kämpf, L., Brauer, A., Dulski, P., Lami, A., Marchetto, A., Gerli, S., Ambrosetti, W., and Guilizzoni, P.: Detrital layers marking flood events in recent sediments of Lago Maggiore (N. Italy) and their comparison with instrumental data, Freshwater Biol., 57, 2076–2090, 2012.

Kasper, T., Haberzettl, T., Doberschütz, S., Daut, G., Wang, J., Zhu, L., Nowaczyk, N., and Mäusbacher, R.: Indian Ocean Summer Monsoon (IOSM)-dynamics within the past 4 ka recorded in the sediments of Lake Nam Co, central Tibetan Plateau (China), Quaternary Sci. Rev., 39, 73–85, 2012.

Kasper, T., Frenzel, P., Haberzettl, T., Schwarz, A., Daut G., Meschner S., Wang, J., Zhu, L., and Mäusbacher, R.: Interplay between redox conditions and hydrological changes in sediments from Lake Nam Co (Tibetan Plateau) during the past 4000 cal BP inferred from geochemical and micropaleontological analyses, Palaeogeogr. Palaeocl., 392, 261–271, 2013.

Keil, A., Berking, J., Mügler, I., Schütt, B., Schwalb, A., and Steeb, P.: Hydrological and geomorphological basin and catchment characteristics of Lake Nam Co, South-Central Tibet, Quatern. Int., 218, 118–130, 2010.

Klovan, J. E. and Imbrie, J.: An algorithm and FORTRAN-IV program for large-scale Q-mode factor analysis and calculation of factor scores, Math. Geol., 3, 61–77, 1971.

Kropáček, J., Maussion, F., Chen, F., Hoerz, S., and Hochschild, V.: Analysis of ice phenology of lakes on the Tibetan Plateau from MODIS data, The Cryosphere, 7, 287–301, doi:10.5194/tc-7-287-2013, 2013.

Kürschner, H., Herzschuh, U., and Wagner, D.: Phytosociological studies in the north-eastern Tibetan Plateau (NW China) A first contribution to the subalpine scrub and alpine meadow vegetation, Botanische Jahrbücher der Systematik, 126, 273–315, doi:10.1127/0006-8152/2005/0126-0273, 2005.

Kutzbach, J. E., Prell, W. L., and Ruddiman, W. F.: Sensitivity of Eurasian Climate to Surface Uplift of the Tibetan Plateau, J. Geol., 101, 177–190, 1993.

Liu, J., Wang, S., Yu, S., Yang, D., and Zhang, L.: Climate warming and growth of high-elevation inland lakes on the Tibetan Plateau, Global Planet. Change, 67, 209–217, 2009.

Long, H., Lai, Z., Frenzel, P., Fuchs, M., and Haberzettl, T.: Holocene moist period recorded by the chronostratigraphy of a lake sedimentary sequence from Lake Tangra Yumco on the south Tibetan Plateau, Quat. Geochronol., 10, 136–142, 2012.

Maussion, F., Scherer, D., Finkelnburg, R., Richters, J., Yang, W., and Yao, T.: WRF simulation of a precipitation event over the Tibetan Plateau, China – an assessment using remote sensing and ground observations, Hydrol. Earth Syst. Sci., 15, 1795–1817, doi:10.5194/hess-15-1795-2011, 2011.

Maussion, F., Scherer, D., Mölg, T., Collier, E., Curio, J., and Finkelnburg, R.: Precipitation seasonality and variability over the Tibetan Plateau as resolved by the High Asia Reanalysis, J. Climate, online first, doi:10.1175/JCLI-D-13-00282.1, 2013.

Miehe, S., Miehe, G., van Leeuwen, J. F. N., Wrozyna, C., van der Knaap, W. O., Duo, L., and Haberzettl, T.: Persistence of Artemisia steppe in the Tangra Yumco Basin, west-central Tibet, China: despite or in consequence of Holocene lake-level changes?, J. Paleolimnol., J. Paleolimnol., 1–19, doi:10.1007/s10933-013-9720-z, 2013.

Miesch, A. T.: Q-mode factor analysis of compositional data, Comput. Geosci., 1, 147–159, 1976.

Mischke, S., Zhang, C., Börner, A., and Herzschuh, U.: Lateglacial and Holocene variation in aeolian sediment flux over the north-eastern Tibetan Plateau recorded by laminated sediments of a saline meromictic lake, J. Quaternary Sci., 25, 162–177, 2009.

Nedell, S. S., Andersen, D. W., Squyres, S. W., and Love, F. G.: Sedimentation in ice-covered Lake Hoare, Antarctica, Sedimentology, 34, 1093–1106, 1987.

North, C. P. and Davidson, S. K.: Unconfined alluvial flow processes: Recognition and interpretation of their deposits, and the significance for palaeogeographic reconstruction, Earth-Sci. Rev., 111, 199–223, 2012.

Opitz, S., Wünnemann, B., Aichner, B., Dietze, E., Hartmann, K., Herzschuh, U., IJmker, J., Lehmkuhl, F., Li, S., Mischke, S., Plotzki, A., Stauch, G., and Diekmann, B.: Late Glacial and Holocene development of Lake Donggi Cona, north-eastern Tibetan Plateau, inferred from sedimentological analysis, Palaeogeogr. Palaeocl., 337–338, 159–176, 2012.

Pendea, I. F., Gray, J. T., Ghaleb, B., Tantau, I., Badarau, A. S., and Nicorici, C.: Episodic build-up of alluvial fan deposits during the Weichselian Pleniglacial in the western Transylvanian Basin, Romania and their paleoenvironmental significance, Quatern. Int., 198, 98–112, 2009.

Rea, D. K.: The paleoclimatic record provided by eolian deposition in the deep sea: The geologic history of wind, Rev. Geophys., 32, 159–195, 1994.

Ruth, U., Wagenbach, D., Steffensen, J. P., and Bigler, M.: Continuous record of microparticle concentration and size distribution in the central Greenland NGRIP ice core during the last glacial period, J. Geophys. Res., 108, 4098, doi:10.1029/2002JD002376, 2003.

Schieber, J., Southard, J., and Thaisen, K.: Accretion of mudstone beds from migrating floccule ripples, Science, 318, 1760–1763, 2007.

Schlütz, F. and Lehmkuhl, F.: Holocene climatic change and the nomadic Anthropocene in Eastern Tibet: palynological and geomorphological results from the Nianbaoyeze Mountains, Quaternary Sci. Rev., 28, 1449–1471, 2009.

Stauch, G. and Lehmkuhl, F.: Reconstruction of a sediment cascade on the north eastern Tibetan Plateau, EGU General Assembly, Vienna, Austria, 22–27 April 2012, EGU2012-11280, 2012.

Stauch, G., IJmker, J., Pötsch, S., Zhao, H., Hilgers, A., Diekmann, B., Dietze, E., Hartmann, K., Opitz, S., Wünnemann, B., and Lehmkuhl, F.: Aeolian sediments on the north-eastern Tibetan Plateau, Quaternary Sci. Rev., 57, 71–84, 2012.

Sturm, M. and Matter, A.: Sedimente und Sedimentationsvorgänge im Thunersee, Eclogae Geol. Helv., 65, 563–590, 1972 (in German).

Stuut, J.-B. W., Prins, M. A., Schneider, R. R., Weltje, G. J., Jansen, J. H. F., and Postma, G.: A 300-kyr record of aridity and wind strength in southwestern Africa: inferences from grain-size distributions of sediments on Walvis Ridge, SE Atlantic, Mar. Geol., 180, 221–233, 2002.

Sun, D., Bloemendal, J., Rea, D. K., Vandenberghe, J., Jiang, F., An, Z., and Su, R.: Grain-size distribution function of polymodal sediments in hydraulic and aeolian environments, and numerical partitioning of the sedimentary components, Sediment. Geol., 152, 263–277, 2002.

Sun, D., Chen, F., Bloemendal, J., and Su, R.: Seasonal variability of modern dust over the Loess Plateau of China, J. Geophys. Res., 108, 4665, doi:10.1029/2003JD003382, 2003.

Sun, D., Bloemendal, J., Rea, D. K., An, Z., Vandenberghe, J., Lu, H., Su, R., and Liu, T.: Bimodal grain-size distribution of Chinese loess, and its palaeoclimatic implications, Catena, 55, 325–340, 2004.

Sun, J., Li, S.-H., Muhs, D. R., and Li, B.: Loess sedimentation in Tibet: provenance, processes, and link with Quaternary glaciations, Quaternary Sci. Rev., 26, 2265–2280, 2007.

Torres, V., Vandenberghe, J., and Hooghiemstra, H.: An environmental reconstruction of the sediment infill of the Bogotá basin
(Colombia) during the last 3 million years from abiotic and biotic
proxies, Palaeogeogr. Palaeocl., 226, 127–148, 2005.

Tsoar, H. and Pye, K.: Dust transport and the question of desert
loess formation, Sedimentology, 34, 139–153, 1987.

Vandenberghe, J.: Grain size of fine-grained windblown sediment:
A powerful proxy for process identification, Earth-Sci. Rev., 121,
18–30, 2013.

Vandenberghe, J., Lu, H., Sun, D., van Huissteden, J., and Konert,
M.: The late Miocene and Pliocene climate in East Asia as
recorded by grain size and magnetic susceptibility of the Red
Clay deposits (Chinese Loess Plateau), Palaeogeogr. Palaeocl.,
204, 239–255, 2004.

Vriend, M. and Prins, M. A.: Calibration of modelled mixing patterns in loess grain-size distributions: an example from the north-
eastern margin of the Tibetan Plateau, China, Sedimentology, 52,
1361–1374, 2005.

Wang, J., Zhu, L., Daut, G., Ju, J., Lin, X., Wang, Y., Zhen, X.:
Investigation of bathymetry and water quality of Lake Nam Co,
the largest lake on the central Tibetan Plateau, China, Limnology,
10, 149–158, 2009.

Wang, J., Peng, P., Ma, Q., and Zhu, L.: Modern limnological features of Tangra Yumco and Zhari Namco, Tibetan Plateau, J.
Lake Sci., 22, 629–632, 2010 (in Chinese).

Weltje, G.: End-member modeling of compositional data:
Numerical-statistical algorithms for solving the explicit mixing
problem, Math. Geol., 29, 503–549, 1997.

Weltje, G. J. and Prins, M. A.: Genetically meaningful decomposition of grain-size distributions, Sediment. Geol., 202, 409–424,
2007.

Wrozyna, C., Frenzel, P., Steeb, P., Zhu, L., van Geldern, R., Mackensen, A., and Schwalb, A.: Stable isotope and ostracode species
assemblage evidence for lake level changes of Nam Co, southern
Tibet, during the past 600 years, Quatern. Int., 212, 2–13, 2010.

Wu, G., Yao, T., Xu, B., Tian, L., Zhang, C., and Zhang, X.: Dust
concentration and flux in ice cores from the Tibetan Plateau over
the past few decades, Tellus B, 62, 197–206, 2010.

Xu, H., Hou, Z. H., Ai, L., and Tan, L. C.: Precipitation at Lake
Qinghai, NE Qinghai-Tibet Plateau, and its relation to Asian
summer monsoons on decadal/interdecadal scales during the past
500 years, Palaeogeogr. Palaeocl., 254, 541–549, 2007.

Zhang, X. Y., Arimoto, R., Cao, J. J., An, Z. S., and Wang, D.:
Atmospheric dust aerosol over the Tibetan Plateau, J. Geophys.
Res., 106, 18471–18476, 2001.

Changes in Mediterranean circulation and water characteristics due to restriction of the Atlantic connection: a high-resolution ocean model

R. P. M. Topper[1,*] and P. Th. Meijer[1]

[1]Department of Earth Sciences, Utrecht University, Budapestlaan 4 3584CD Utrecht, the Netherlands
[*]now at: MARUM – Center for Marine Environmental Sciences and Department of Geosciences, University of Bremen, P.O. Box 330440, 28334, Germany

Correspondence to: R. P. M. Topper (rtopper@marum.de)

Abstract. A high-resolution parallel ocean model is set up to examine how the sill depth of the Atlantic connection affects circulation and water characteristics in the Mediterranean Basin. An analysis of the model performance, comparing model results with observations of the present-day Mediterranean, demonstrates its ability to reproduce observed water characteristics and circulation (including deep water formation). A series of experiments with different sill depths in the Atlantic–Mediterranean connection is used to assess the sensitivity of Mediterranean circulation and water characteristics to sill depth. Basin-averaged water salinity and, to a lesser degree, temperature rise when the sill depth is shallower and exchange with the Atlantic is lower. Lateral and interbasinal differences in the Mediterranean are, however, largely unchanged. The strength of the upper overturning cell in the western basin is proportional to the magnitude of the exchange with the Atlantic, and hence to sill depth. Overturning in the eastern basin and deep water formation in both basins, on the contrary, are little affected by the sill depth.

The model results are used to interpret the sedimentary record of the Late Miocene preceding and during the Messinian Salinity Crisis. In the western basin, a correlation exists between sill depth and rate of refreshment of deep water. On the other hand, because sill depth has little effect on the overturning and deep water formation in the eastern basin, the model results do not support the notion that restriction of the Atlantic–Mediterranean connection may cause lower oxygenation of deep water in the eastern basin. However, this discrepancy may be due to simplifications in the surface forcing and the use of a bathymetry different from that in the Late Miocene. We also tentatively conclude that blocked outflow, as found in experiments with a sill depth ≤ 10 m, is a plausible scenario for the second stage of the Messinian Salinity Crisis during which halite was rapidly accumulated in the Mediterranean.

With the model setup and experiments, a basis has been established for future work on the sensitivity of Mediterranean circulation to changes in (palaeo-)bathymetry and external forcings.

1 Introduction

Ever since the closure of the connection to the Indian Ocean in the Middle Miocene, water exchange between the Mediterranean Sea and the global ocean has been through one or multiple gateways in the Gibraltar arc (e.g. Dercourt et al., 2000). In the Mediterranean, the water loss by evaporation exceeds, on an annual basis, the fresh water input from precipitation and river input. The resulting fresh water deficit, i.e. evaporation – precipitation – river input, is compensated for by a net inflow from the Atlantic. Without a connection to the global ocean, the deficit would not be replenished and Mediterranean sea level would rapidly drop (Meijer and Krijgsman, 2005). At present, the fresh water deficit in combination with the relatively small connection to the Atlantic results in a higher salinity in the Mediterranean than in the Atlantic. This salinity difference, and hence density difference, between the Mediterranean and Atlantic drives a deep outflow of Mediterranean water. The outflow is compensated for with an inflow of less saline Atlantic water at the surface (e.g. Bryden and Kinder, 1991; Astraldi et al., 1999).

This pattern of surface inflow and deep outflow is known as anti-estuarine and has been the dominant mode of exchange since at least the closure of the Indian Ocean connection (Karami et al., 2011; Seidenkrantz et al., 2000; de la Vara et al., 2015). On a geological time scale, the fresh water budget of the Mediterranean has changed in accord with global climate. Superimposed on long-term variations are orbitally driven changes on shorter timescales. Precession and eccentricity have been shown to have a significant impact on Mediterranean river discharge, evaporation and precipitation (e.g. Tuenter et al., 2003). Due to the small exchange with the Atlantic, precessional changes in the fresh water budget have a pronounced effect on the Mediterranean circulation. The Mediterranean sedimentary record contains evidence for reduced bottom water ventilation during periods of precession-induced increased river discharge in the form of sapropels since the Miocene (Hilgen et al., 1995; Cramp and O'Sullivan, 1999).

In the Neogene and Quaternary sedimentary record, the occurrence of sapropels has been related to relatively strong precession minima, clustered in intervals with particularly high eccentricity (100 and 400 kyr cycle). However, at the end of the Miocene (7.16–5.332 Ma), sapropels were formed in nearly every precession cycle (Seidenkrantz et al., 2000), indicative of a heightened sensitivity of Mediterranean circulation to its fresh water budget. Within an interval of increasing bottom water salinity starting at ≈ 8 Ma, a sharp change in the foraminiferal assemblage and a shift in $\delta^{13}C$ at 7.16 Ma indicate reduced bottom water ventilation in several relatively deep marginal basins of the Mediterranean (Seidenkrantz et al., 2000; Kouwenhoven et al., 2003; Kouwenhoven and van der Zwaan, 2006). Low oxygen and high salinity conditions expand from deep marginal basins to marginal basins at shallower depths towards the onset of the Messinian Salinity Crisis (MSC; (Gennari et al., 2013)). The concurrent increase in sensitivity to changes in the fresh water budget, the increase in salinity, and the drop in bottom water ventilation, have been interpreted as indicators of reduced exchange with the Atlantic due to tectonically driven restriction of the Atlantic–Mediterranean connection (Kouwenhoven and van der Zwaan, 2006).

The ultimate expression of a strong tectonically driven restriction in Atlantic–Mediterranean exchange is the MSC, an event which is represented in the sedimentary record by widespread and voluminous evaporites in the Mediterranean. Circum-Mediterranean climate changed little during this event (Fauquette et al., 2006; Bertini, 2006), corroborating the results of model studies that demonstrate that a restriction of the Atlantic–Mediterranean connection can be the sole driver of the MSC (Topper et al., 2011; Topper and Meijer, 2013). A vast halite layer (1–2 km in the deep basins) formed during the short second phase of the MSC (5.61–5.55 Ma, Roveri and Manzi, 2006). A scenario with inflow from the Atlantic but no outflow, the so-called blocked-outflow scenario, has been proposed to explain the fast accumulation of salt in the Mediterranean (Krijgsman et al., 1999a; Meijer, 2006; Krijgsman and Meijer, 2008). According to hydraulic control theory, blocked outflow can be attained in the Atlantic–Mediterranean gateway when the water depth is just a few metres (Meijer, 2012).

In the Late Miocene, Atlantic–Mediterranean exchange occurred through two shallow marine gateways in northern Morocco and southern Spain, respectively the Rifian and Betic corridors (Benson et al., 1991; Betzler et al., 2006). Due to their geometry, i.e. shallow, wide and long, and position at the interface of ongoing Africa–Eurasia convergence, these gateways were vulnerable to changes in width and depth by tectonics (Duggen et al., 2003; Weijermars, 1988; Govers, 2009). Opening and closure of both gateways is recorded in the sedimentary record of the subbasins that make up the corridors. Closure of the Rifian corridor took place somewhere in the interval 6.8–6.0 Ma (Krijgsman et al., 1999b; Ivanovic et al., 2013; van Assen et al., 2006). The Betic corridor has long thought to have been closed by 7.8 Ma (Soria et al., 1999; Betzler et al., 2006; Martín et al., 2009). Recent work has disputed this, stating that the Betic corridor might have been open after 7.8 Ma (Hüsing et al., 2010; Pérez-Asensio et al., 2012). Wherever the connection may have been, the volume of evaporites formed during the MSC can only be explained by a continued inflow of saline water from the Atlantic (Sonnenfeld and Finetti, 1985; Krijgsman and Meijer, 2008; Topper and Meijer, 2013).

The purpose of this study is to examine with a high-resolution ocean circulation model how sill depth of the Atlantic connection, influenced by long-term tectonics, affects circulation and water characteristics in the Mediterranean Basin. The physics-based insight thus gained is used to evaluate the Late Miocene sedimentary record: that of the MSC and, in particular, that of the period just preceding the actual crisis. Notwithstanding our focus on the Late Miocene, the basin bathymetry is – except near the Atlantic connection – that of the present Mediterranean Sea. The reasons for this are that (1) we isolate the changes in circulation and water characteristics that are caused by a different depth of the Atlantic connection, and (2) a validation of model performance is only possible in a present-day model setup because quantitative data on water properties and circulation does not exist for other time periods. Such an assessment of model performance is necessary because the model has not been applied to the Mediterranean before. Thus taking the present-day bathymetry as our reference, the depth of the Strait of Gibraltar is modified to gain insight into the role of depth of the Atlantic connection. Even though basin shape was different in the Late Miocene and the connection to the ocean had a different geometry and location, our model results form a good starting point for this past time as well.

The reason for this is that, in terms of features that govern the overall circulation and water properties, the present-day and Late Miocene basin geometry are largely similar. Both consist of two subbasins, comprise limited shelf areas and deep-basinal portions, are subject to net evaporation and possess parts located at relatively high latitude where cooling of the surface water occurs. Thus, insight gained about the change in circulation and water properties due to variation in sill depth is expected to be largely generic. A minimal model of the Mediterranean with a highly idealized surface forcing has been shown to be able to capture the important characteristics of Mediterranean circulation (Meijer and Dijkstra, 2009). A similar idealized surface forcing is applied in this study, a simplification of the model setup that is justified when the focus is on the influence of the sill depth.

Water exchange in the present-day Strait of Gibraltar has been studied extensively because of its influence on Mediterranean water characteristics and, therewith, circum-Mediterranean climate (Bryden and Stommel, 1984; Sannino et al., 2002; Astraldi et al., 1999; Hopkins, 1999; Candela, 1991; Bryden et al., 1994). The present-day Mediterranean circulation and the influence of temporal variations in the surface forcing, e.g. fresh water budget and heat flux, has also been widely studied, often in the context of sapropels (Meijer and Tuenter, 2007; Meijer and Dijkstra, 2009; Samuel et al., 1999; Myers et al., 1998a, b; Myers, 2002). Data and simple modelling studies have demonstrated the importance of the size of the Atlantic–Mediterranean connection in controlling sedimentation and circulation in the Mediterranean. Recently, ocean circulation models have been used to investigate the influence of gateway geometry on water characteristics and circulation in an idealized marginal basin (Iovino et al., 2008; Pratt and Spall, 2008) and realistic marginal basins, e.g. the Miocene Arctic Ocean (Thompson et al., 2010). Similar studies, regarding the influence of gateway geometry on Mediterranean circulation and water characteristics, are, as yet, few in number. The sensitivity of circulation in the Late Miocene Mediterranean to changes in the palaeogeography has been examined by Meijer et al. (2004). More recently, Alhammoud et al. (2010) have examined how the sill depth of the Strait of Gibraltar affects the thermohaline circulation in a highly idealized coarse resolution ocean circulation model.

The next section will provide a description of the model and boundary conditions used. In Sect. 3, the results of a reference experiment with the present-day sill depth will be compared to water characteristics and circulation as observed or modelled for the present-day Mediterranean to assess model performance. Keeping all other boundary conditions constant, the sill depth of the Atlantic gateway will be both increased, up to 500 m, and reduced, down to 5 m, in a series of experiments. Model results will be discussed in the context of the Late Miocene restriction of the Atlantic–Mediterranean connection and the blocked-outflow scenario for the MSC.

Table 1. Overview of parameter values used to set up the model. SIUS: Smolarkiewicz iterative upstream scheme.

Parameter	Detail	Value
tprni	Inverse Prandtl number	0.2
horcon	Smagorinsky diffusivity coefficient	0.2
nitera	Number of iterations of the SIUS	2
sw	Smoothing parameter of the SIUS	0.8
t_E	External (2-D) time step	30 s
t_I	Internal (3-D) time step	1800 s
t_{RELAX}	Relaxation time scale of surface temperature forcing	1 day

2 Model description

For a good representation of water exchange through a shallow gateway, a model with a high number of vertical layers in shallow water is required. A model with vertical sigma coordinates meets this requirement better than a model with z coordinates. With sigma coordinates, regardless of water depth, the number of vertical grid points in the water column is equal. The distance between vertical grid points is a percentage of the total water depth at each point, i.e. layers are further apart in deep water than in shallow water. In this study, we use sbPOM (Jordi and Wang, 2012), a recently developed parallel, free-surface, sigma-coordinate, primitive equations ocean modelling code based on the Princeton Ocean Model (Blumberg and Mellor, 1987). Differences between sbPOM and the 2008 version of POM only concern the parallelization: the code is rearranged in several files and a message-passing interface using two-dimensional data decomposition of the horizontal domain has been implemented. POM has been applied extensively to the Mediterranean and the Strait of Gibraltar (e.g. Zavatarelli and Mellor, 1995; Ahumada and Cruzado, 2007; Jungclaus and Mellor, 2000; Drakopoulos and Lascaratos, 1999; Alhammoud et al., 2010; Sannino et al., 2002; Beckers et al., 2002). In an application to the Mediterranean, POM is set up with a true evaporation instead of a virtual salt flux. Evaporative water loss in the Mediterranean therefore drives a net flow through the Atlantic gateway which is necessary for a realistic representation of exchange.

Vertical mixing coefficients are calculated in the model with the Mellor and Yamada (1982) level 2.5 turbulence closure scheme. The pressure gradient scheme used is a fourth order scheme using the McCalpin method (Berntsen and Oey, 2010). The inverse Prandtl number, the ratio between horizontal mixing coefficients for diffusivity and viscosity (calculated with a Smagorinsky formulation), has been set to 0.2. This and other model parameters, chosen after an extensive sensitivity study, are listed in Table 1. The model parameter with the largest impact on model results proves to be the smoothing parameter of the Smolarkiewicz iterative upstream scheme.

The bathymetry used in this study (Fig. 1) derives from the 1 min resolution global surface (ETOPO1) provided by the

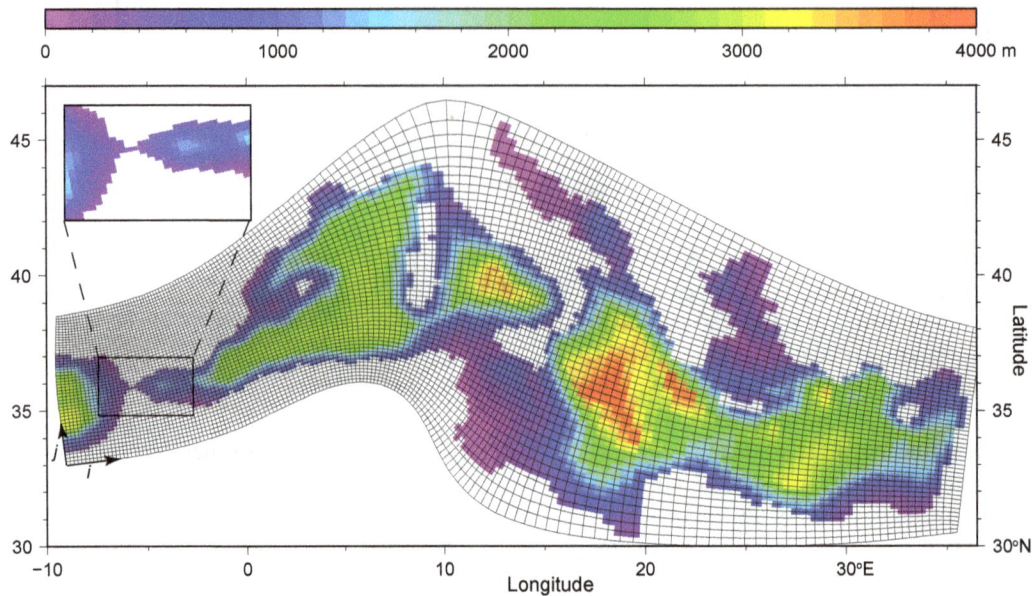

Figure 1. Model grid and bathymetry. Curvilinear grid coordinates i and j increase along curved lines towards, roughly, the east and north. The small inset shows the geometry of the Strait of Gibraltar in more detail.

National Geographical Data Center of the NOAA (Amante and Eakins, 2009). In the model, a curvilinear grid is used in the horizontal, similar to that of Zavatarelli and Mellor (1995). A curvilinear grid has the advantages that it allows for variable horizontal resolution and a high percentage of grid points in the basin, i.e. less grid points are needed compared to a rectangular grid. Before interpolation of the bathymetry to the horizontal grid (162×50 grid points), a Gaussian filter has been applied to remove short wavelength variations from the bathymetry. Furthermore, the parameter used to control the maximum slope between two adjacent grid cells is set to 0.15 in order to minimize pressure gradient errors and ensure model solution stability (Mellor et al., 1994). Horizontal grid resolution is highest near the Strait of Gibraltar and lowest in the eastern Mediterranean Basin. The minimum/maximum size of a grid cell in the i direction (Fig. 1) and j direction are respectively 11.7/62.9 and 10.7/69.6 km. The width of the Strait of Gibraltar is 12.9 km, equal to its smallest width in the present day. The orientation of the curvilinear coordinates (i and j) varies along the grid and deviates from the absolute longitudinal and latitudinal direction. For simplicity, the i and j directions along the curvilinear grid will be referred to as longitudinal and latitudinal, respectively. Likewise, we will take "zonal" to mean "along the curvilinear i coordinate". In the vertical direction, the grid comprises 40 sigma levels. Grid points have the highest density near the surface and bottom in order to resolve the thermocline and bottom currents. In order to optimize the time step, maximum depth in the Atlantic part of the model is set to 3000 m, and in the Mediterranean to 4000 m. The maximum difference with the actual bathymetry is less than

100 m. In the current setup, the model can be integrated for 100 yr in 28 h on 4 processors of an Intel Core i7-950 workstation running at 3.06 GHz with 12 GB memory.

Initial conditions for temperature and salinity derive from Levitus' World Ocean Atlas (Antonov et al., 1998). In the Atlantic part of the model, temperature and salinity are horizontally averaged before interpolation to sigma coordinates to avoid density contrasts near the western boundary of the grid, which is an open boundary, and provides water of constant temperature and salinity at each depth near the Strait of Gibraltar.

The surface forcing is idealized and constant in time, and resembles the forcing used in Alhammoud et al. (2010). The surface water flux is a constant evaporation of 0.5 m yr^{-1} over the whole model domain, a value close to present-day evaporation – precipitation – river input (Mariotti et al., 2002), but also suitable for the Late Miocene (Gladstone et al., 2007; Schneck et al., 2010). Sea surface temperatures are relaxed to a best fit for zonally and yearly averaged air temperatures from ECMWF data (Fig. 2 from Alhammoud et al., 2010). The relaxation time scale is 1 day; sensitivity experiments have shown that the relaxation time scale can be as long as 30 days before model results are significantly affected. Winds over the Mediterranean are highly variable throughout the year; a yearly averaged wind pattern is therefore unusable. Since wind stresses have been shown to affect only the strength of upper ocean circulation without changing the overall circulation pattern (Myers et al., 1998b; Meijer and Dijkstra, 2009), wind stresses have not been included.

The western boundary in the Atlantic part of the model domain is open. For the barotropic mode (2-D) a zero-gradient

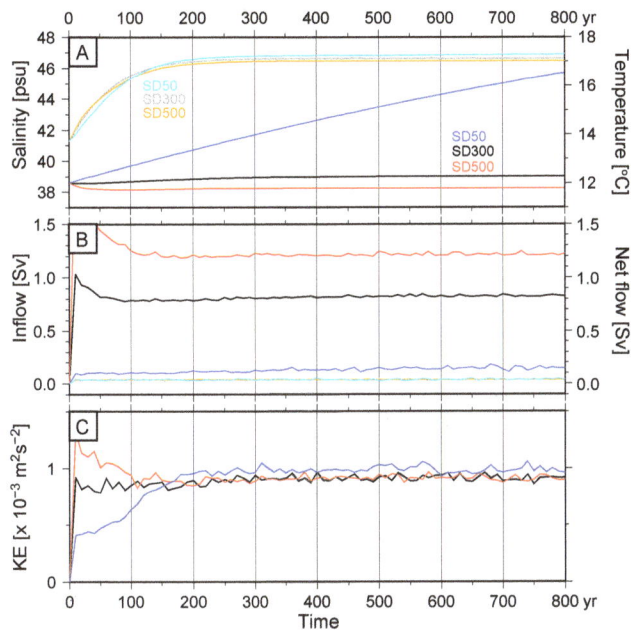

Figure 2. Temporal evolution of diagnostic variables for the reference experiments (SD300, black/grey lines), SD500 (red/orange) and SD50 (blue/light blue). (**a**) Basin-averaged salinity (dark colours, left vertical axis) and basin-averaged temperature (light colours, right vertical axis). (**b**) Inflow (dark colours) and net flow (light colours) through the Strait of Gibraltar. (**c**) Kinetic energy measure.

condition is applied to the free surface elevation, a clamped inflow condition is used for the depth-averaged velocity normal to the open boundary, and a zero velocity is used for the boundary parallel velocity. For the baroclinic mode (3-D), a Sommerfeld radiation condition is applied to the velocity normal to the open boundary and a zero velocity is used, again, for the boundary parallel velocity. The inflow in the external mode compensates for the water volume lost by evaporation in the Mediterranean, keeping water volume constant. To avoid large temperature and/or salinity contrast between the inflow through the open boundary, which has a prescribed salinity and temperature equal to initial conditions, and the Atlantic part of the model, a relaxation back to initial Atlantic conditions is applied in the first 18 columns of the grid (the Strait of Gibraltar starts at the 22nd column). The relaxation timescale in the surface layer varies from 3 months at the westernmost grid cell to 1 month at the easternmost grid cell of the relaxation area, and decreases exponentially in deeper layers.

3   Results and analysis

The results of a reference experiment, with idealized surface forcing and the sill depth of the Atlantic gateway set to the present-day depth of 300 m, will be compared with observations on the present-day Mediterranean. The experiments

used to assess the role of sill depth only differ from the reference experiment in the sill depth and the bathymetry near the sill. The different experiments will be referred to as SD500 to SD5 according to their *sill depth* in metres. Each of the experiments has been run for 800 yr. The time needed to reach steady state conditions in an experiment is inversely proportional to the depth of the sill. As a consequence, experiments with a sill depth shallower than 100 m have not reached a steady state in 800 yr. This will be pointed out where it affects the results.

3.1   Reference experiment

The temporal variation of basin-averaged temperature, salinity, a measure of kinetic energy, and strait transport is illustrated in Fig. 2 for the reference experiment (SD300). Temperature, salinity and kinetic energy are averaged in the Mediterranean part of the model domain, i.e. east of the Strait of Gibraltar at $-5.5°$ E. The average basinal velocity squared is used as a measure of the kinetic energy which, in turn, gives an indication of the intensity of flow in the basin. In the first 300 yr of the reference experiment, salinity and temperature move towards their respective steady state values of 39.02 psu and 17.12°C. At the same time, inflow through the Strait of Gibraltar also stabilizes at 0.83 Sv ($1\,\mathrm{Sv}=10^6\,\mathrm{m}^3\,\mathrm{s}^{-1}$). The difference between inflow and outflow, the net flow, is equal to the fresh water deficit in the Mediterranean, 0.04 Sv, from the start of the experiment. Salinity, temperature and strait transport all reach steady state values at the same time. This is the expected behaviour since salinity and temperature determine the density difference with the Atlantic, which, in turn, drives the exchange at Gibraltar. The kinetic energy measure reaches a value close to its steady state value within the first 100 yr and slowly increases further towards a steady state, which is also reached at 300 yr.

At steady state a balance exists between the transport of heat and salt at the Strait of Gibraltar and the surface forcings in the Mediterranean, which is reflected by quasi-constant properties of the exchange. Velocity, temperature and salinity profiles from the Strait of Gibraltar are shown in Fig. 3. Relatively warm Atlantic water with a close to normal marine salinity flows eastwards through the Strait of Gibraltar in the upper layer, while more saline water flows westward at depth. The anti-estuarine circulation pattern observed in the present-day Strait of Gibraltar is thus reproduced. Moreover, the velocity profile is similar to observed velocity profiles at Camarinal Sill (e.g. Bryden et al., 1994; Tsimplis, 2000; Hopkins, 1999; Candela, 2001), velocity and salinity profiles of other model studies (Sannino et al., 2002; Xu et al., 2007), and theoretical velocity profiles (Hopkins, 1999). Small differences exist in the height of the interface and absolute velocities of inflow and outflow. These are mainly due to the idealized shape of the cross sectional area of the gateway in our model. The curves in Fig. 3 exhibit four features that are

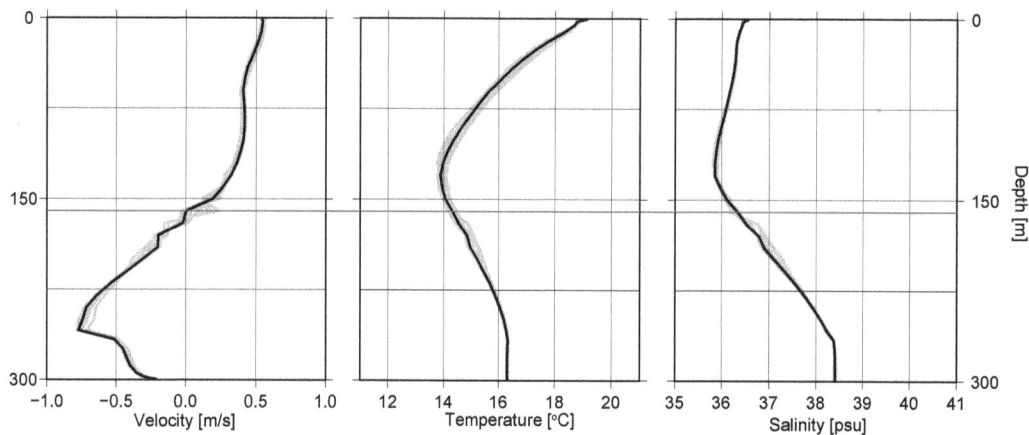

Figure 3. Vertical profiles of zonal velocity (left), temperature (middle), and salinity (right) in the middle of the Strait of Gibraltar in the reference experiment. Grey shading in each frame indicates the range of variation of the variable in the last 100 yr of integration, the solid black line is the average over the same 100 yr. The red line indicates the depth of the interface between inflow and outflow.

noteworthy: (1) the depth of the interface between the inflow and outflow is not exactly halfway but slightly deeper (165 m), (2) velocity changes gradually with depth, i.e. there is no sharp transition across the interface between inflow and outflow, (3) the maximum velocity in the outflow is higher than in the inflow because velocities are strongly reduced near the bottom, and (4) temperature and salinity gradually change with depth, indicating that mixing takes place between inflow and outflow in the strait. All four features are important when exchange in the experiments with different sill depths is compared to the exchange predicted by hydraulic control theory (Sect. 4.2).

Figure 4 illustrates the horizontal and vertical patterns of salinity and temperature in two horizontal slices at 10 m and 300 m and a vertical section crossing the whole domain from west to east. The inflowing Atlantic water starts to change its temperature and salinity in response to the surface forcing as soon as it enters the Mediterranean. At the surface, low saline Atlantic water can be traced along the southern coast into the eastern Mediterranean. The constant evaporation at the surface drives an increase in salinity towards the eastern basin, from 36.5 to 39 psu. Highest salinities are accordingly reached in the easternmost Mediterranean. Due to the strong surface temperature relaxation, surface temperatures are similar to the latitudinally decreasing air temperature in the larger part of the Mediterranean. Only the northern part of the western basin and the area directly east of the Sicily Strait deviate from this pattern. In both areas, advection of heat is faster than the surface forcing.

A difference in water characteristics does not only exist between the Atlantic and Mediterranean, a clear difference is also visible between western and eastern Mediterranean. The Sicily Strait restricts exchange between the basins. In combination with the fresh water deficit of the eastern Mediterranean, this drives an eastward surface flow and westward deep flow, i.e. an anti-estuarine circulation with respect to

the western basin. The western basin has an average salinity of 38.3 psu, compared to 39.5 psu in the eastern basin. In spite of the significantly higher surface temperatures in the eastern basin, differences in basin-averaged temperature are relatively small (16.1 vs. 17.5 °C).

The circulation pattern evidenced by the salinity and temperature distribution in the Mediterranean is confirmed by the "zonal" overturning stream function which is calculated along the curvilinear coordinates (Fig. 5). A positive overturning cell, representing clockwise circulation in an W–E profile, extends from the Atlantic through the whole Mediterranean. The vertical extent and strength of the positive overturning cell is larger in the western than in the eastern Mediterranean. At depth a negative overturning cell is present in both basins with a larger vertical extent and strength in the eastern basin.

On the vertical temperature and salinity profiles, the area east of 25° E stands out due to its significantly higher salinity and temperature at depths up to 500 m. The zonal overturning stream function confirms that this is an area of downwelling. Following the approach from other studies (Dijkstra, 2008; Ezer and Mellor, 1997; Alhammoud et al., 2010), the rate of intermediate water formation can be derived from the stream function. In the area east of 25° E, the rate of intermediate water formation is 0.75 Sv. The maximum rate of Levantine Intermediate Water formation in this area is estimated at 1.2 Sv (Lascaratos and Nittis, 1998), slightly higher than our estimate. The tilted isotherm in the deep eastern basin is consistent with intermediate and deep water formation in the eastern part of the basin. The inclined isohalines in the upper 300 m of the western basin are not caused by vertical water movements but by the flow of more saline water from the eastern basin to the west.

For further analysis of the intermediate and deep water formation, Fig. 6a shows the distribution of the average mixed layer depth. The average mixed layer depth is relatively shal-

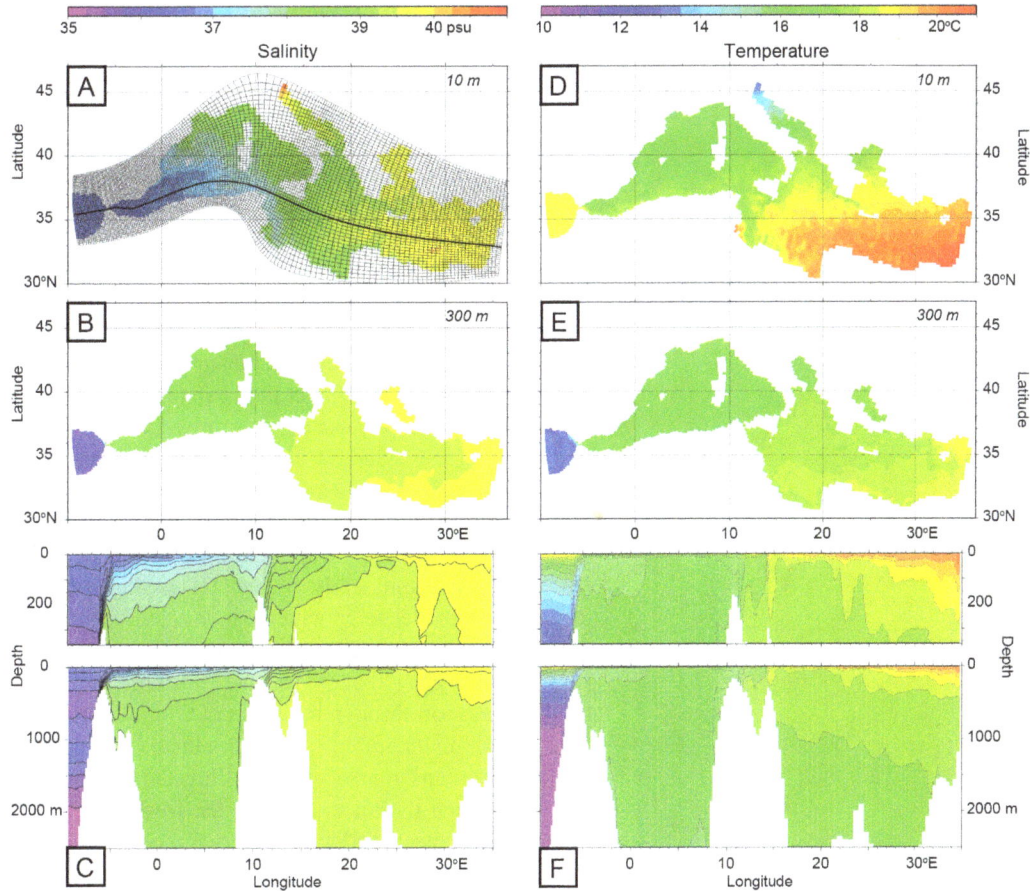

Figure 4. Two horizontal and a vertical cross section through the three-dimensional salinity (**a, c, e**) and temperature (**b, d, f**) fields averaged over the last 10 yr of the reference experiment. Horizontal slices are shown in the surface layer (10 m) and at the sill depth of the Strait of Gibraltar (300 m). The vertical cross section is along the curved path indicated in (**a**) that crosses both the Gibraltar and Sicily straits.

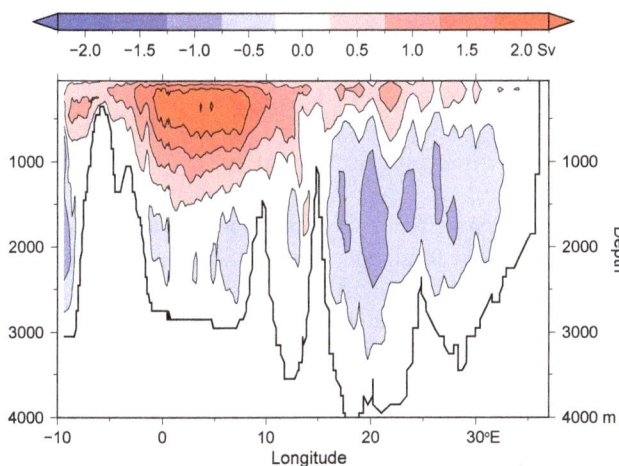

Figure 5. "Zonal" overturning stream function of the reference experiment. The contour interval is 0.5 Sv. Positive values (red colours) indicate clockwise circulation in this view, negative values (blue) anti-clockwise circulation.

low because of a high spatial and temporal variability in the mixed layer depth, i.e. mixed layer depths often vary between almost zero and the maximum shown in Fig. 6b. Hence, the maximum mixed layer depth can be used to illustrate the variation in depth and location of the intermediate and deep water formation sites. Deep water formation occurs where the mixed layer is anomalously deep, e.g. the northern part of the western basin where mixing occurs up to 3000 m, and where the mixed layer depth (nearly) equals the water depth, as in the Adriatic Sea, the Aegean Sea and the north-east corner of the Levantine Basin. Figure 6 shows the path of water particles that have been tracked for 30 days in the lowest sigma layer, i.e. the bottom currents. The idea of visualizing bottom velocities to track dense water currents is taken from Zavatarelli and Mellor (1995). Where dense water is formed in shallow areas, it flows downslope to the deep basin until it reaches a depth where its density equals that of the water at that depth. While dense water from the Adriatic clearly reaches large depths, dense water from the Aegean only reaches intermediate depths. In addition to deep water formation and dense water currents flowing into the

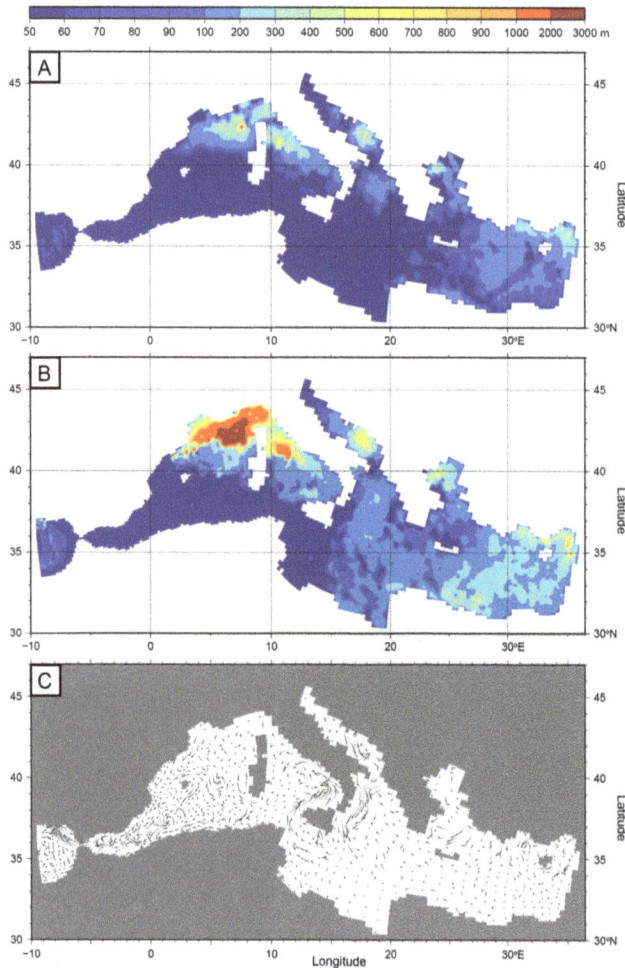

Figure 6. The average (**a**) and maximum (**b**) mixed layer depth, calculated from the last 50 yr of integration, and bottom currents (**c**) in the reference experiment. The mixed layer depth is defined as the depth, measured from the surface, of the minimum vertical mixing parameter. Bottom currents are visualized by tracking water particles for 30 days in the average velocity field of the last 10 yr of integration. Trajectories start at light grey and proceed to black at the 30th day.

deep basin, Fig. 6 also shows the trajectories of dense water flowing from the eastern to the western basin and from the Mediterranean to the Atlantic. The northward deflection of both westward directed flows is due to the Coriolis effect.

3.1.1 Model results compared to present-day observations

Initial conditions in the model are based on recent observations of salinity and temperature (Levitus fields). Therefore, the difference between initial and steady state salinity and temperature is equal to the difference between modelled and observed values. Due to the simple constant surface forcings, we cannot expect to capture annual variability in circulation

and deep water formation. Basin-averaged temperatures are 3.4 °C higher than observed (17.12 vs. 13.71 °C). The bulk of this difference can be ascribed to significantly higher deep water temperatures. Temperature differences near the surface are comparatively small and, because only 10 % of the basinal volume is contained in the surface layer, do not significantly affect basin averages.

In the present-day Mediterranean, deep water formation is strongest during the winter months (e.g. Lascaratos et al., 1999). Hence, most deep water is formed when sea surface temperatures are below the annual average. The use of mean annual air temperatures in our model gives rise to deep water formation throughout the year and an associated overestimation of deep water temperatures. Basin-averaged salinity, on the other hand, is close to the value from observations (39.0 vs. 38.6 psu). Annual variability of evaporation in the Mediterranean is small. The use of a constant evaporation rate is therefore a simplification of the surface forcing that does not significantly impact the basinal salinity. Furthermore, a model setup with a constant surface forcing provides a convenient starting point in unravelling a possible correlation between sill depth and the overturning.

Transport through the Strait of Gibraltar has been the subject of innumerable studies. Estimates have been based on observations, numerical modelling and hydraulic control theory. Given the uncertainty in the fresh water budget of the Mediterranean and the wide range of approaches, the volume transport at the Strait of Gibraltar has been estimated at 0.8–1.8 Sv, with most recent estimates at the lower end of this range (e.g. Astraldi et al., 1999; Tsimplis and Bryden, 2000). The inflow of 0.83 Sv in the reference experiment is thus within this range. In a model with annual variation in the surface forcing, the low deep water temperatures observed in the Mediterranean are reproduced (unpublished results). Compared to the results of the reference experiment, lower deep water temperatures will lead to a small increase of exchange due to the resultant larger density contrast with the Atlantic. However, for the objective of examining the role of strait depth, water exchange with the Atlantic is sufficiently reproduced.

Often, basin-scale circulation models with realistic and idealized atmospheric forcing have had difficulties reproducing deep water formation and, in particular, deep overturning in the western basin (e.g. Meijer et al., 2004; Meijer and Dijkstra, 2009). The reference experiment, set up with idealized forcing, is able to produce both. It should, however, be noted that the locations of deep water formation in the model differ, especially in the western Mediterranean, from locations inferred from observations.

In a series of sensitivity experiments, deep water formation has been compared in models with different latitudinal gradients in the air temperature. With a reduced latitudinal air temperature gradient, the mixed layer depth is still largest in the northern part of the western basin, the Adriatic and the Aegean Sea. However, rates of deep water formation and the strength of the deep overturning cell are lower while the formation of intermediate water in the eastern basin and the strength of the upper overturning cell in both basins are higher. These results are in agreement with Alhammoud et al. (2010) and Somot et al. (2006) who found that deep water formation is mainly controlled by the surface temperature forcing. These authors also found that intermediate water formation is mainly controlled by evaporation. A series of sensitivity experiments with different rates of evaporation confirms this.

The strength of the surface overturning cell in the reference experiment is slightly larger than in most other high-resolution model studies (Meijer et al., 2004; Meijer and Dijkstra, 2009; Somot et al., 2006; Stratford et al., 2000; Adloff et al., 2011). The vertical and horizontal extent, however, is similar for both the western and eastern basin. Also, the ratio between minimum and maximum zonal overturning is equal to other studies at ≈ 0.33. The results of a series of sensitivity experiments demonstrate that the degree of bathymetry smoothing, the maximum slope in the bathymetry and the smoothing parameter of the Smolarkiewicz scheme have a significant and predictable impact on the strength of circulation. In general, less smoothing, steeper slopes and a lower smoothing parameter result in a stronger overturning with more lateral variation. The bathymetry used here has relatively low maximum slopes and minimal smoothing while the smoothing parameter is relatively low.

The overall good agreement between observed and modelled strait transports, water characteristics and circulation, quantitatively as well as qualitatively, shows that the model setup of the reference experiment captures all important processes of the Mediterranean thermohaline circulation. Having validated the model setup of the reference experiment, we will describe in the next section the changes in Mediterranean water characteristics and circulation due to changes in the sill depth of the Atlantic connection.

3.2 The role of sill depth

The temporal variations of water characteristics and transport in two experiments with a shallow (50 m, SD50) and deep (500 m, SD500) gateway and the reference experiment can be compared in Fig. 2. The increase in basin-averaged temperature with run time in all three experiments is comparable in both magnitude and duration. A steady state is reached after 300 yr with a slightly lower value for SD500 (17.01 °C) and a higher value for SD50 (17.30 °C). The temporal variation of basin-averaged salinities, on the other hand, is significantly different. SD500 reaches a steady state within 100 yr at 38.6 psu, SD50 does not reach a steady value within 800 yr. The average rate of salinity increase in SD50 is 8.9 psu kyr^{-1}. However, the rate drops slowly with time; in the first 100 yr it is 10.7 psu kyr^{-1}, in the last 100 yr "only" 6.8 psu kyr^{-1}. When the density difference between Atlantic and Mediterranean increases, inflow and outflow also in-

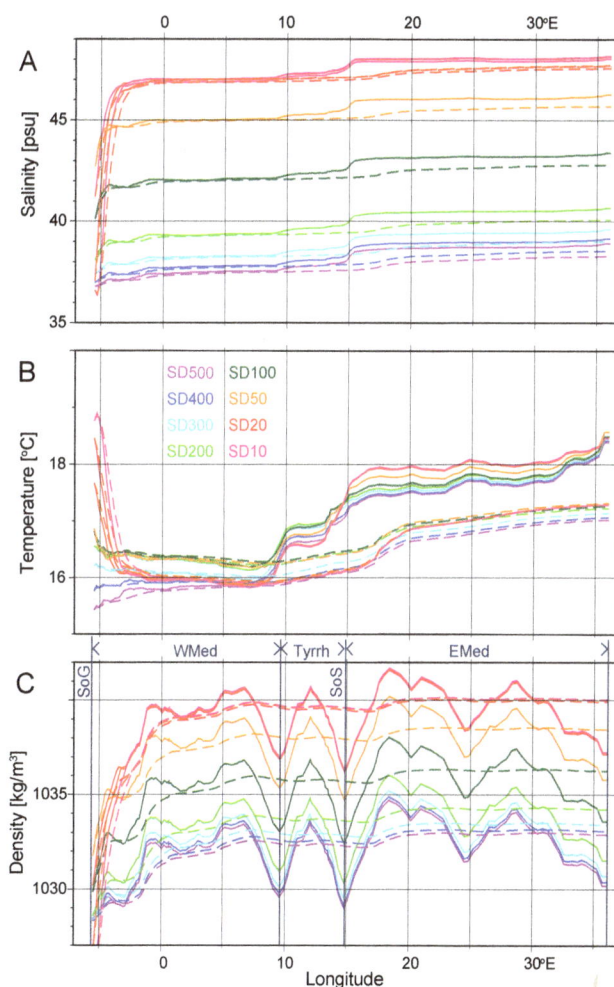

Figure 7. Latitude-depth-averaged profiles of salinity (**a**), temperature (**b**) and density (**c**) for SD500–SD10. Dashed lines indicate the volume-averaged salinity/temperature/density of the volume between the Strait of Gibraltar and the indicated longitude. Therefore, the values at 36° E are the averages of the whole basin. Indicated in (**c**) are the longitudinal ranges corresponding to the western basin (WMed), the Tyrrhenian Sea (Tyrrh), and the eastern basin (EMed) and location of the Strait of Gibraltar (SoG) and Strait of Sicily (SoS).

crease. When a steady state is reached, the inflow volume multiplied with its salinity equals the outflow volume multiplied with its salinity. Before steady state more salt flows in than out, Mediterranean salinity increases, and, consequently, transport increases. The increasing transport reduces the difference between salt volume in inflow and outflow and, hence, the salinity rise slows down towards steady state. After 800 yr, all experiments with a sill depth ≥ 100 m (SD100–SD500) are in equilibrium, i.e. salinity and strait transport have reached a steady state.

The basin averages in Fig. 2 do not show where the differences in salinity and temperature arise between the different experiments. For this purpose, Fig. 7a shows the salinity averaged over depth and along the curvilinear j coordinate ("latitude") for every i coordinate. Due to the curvilinear grid, these depth-j slices are not exactly north–south. Latitude-depth-averaged temperature and density are calculated in a similar fashion (Fig. 7b, c). Also shown, in dashed lines, is the average temperature/salinity/density of the basin from the Gibraltar Strait up to each longitude. The temperature/salinity/density at the eastern end (36° E) is thus the basin-averaged value.

Only where the bathymetry was modified to accommodate a different sill depth, i.e. between −6 to −2° E, does the shape of salinity curves differ between SD5–SD500. For a shallower sill depth the whole curve, apart from this westernmost segment, shifts to higher salinities. In each curve, salinity increases in two distinct steps at longitudes corresponding to the western basin–Tyrrhenian Sea connection (9° E) and the Sicily Strait (15° E) (Fig. 7c). Although salinities in each basin – western, Tyrrhenian and eastern basin – are higher at shallower sill depths, the difference between them is nearly constant: ≈ 0.4 between the western basin and the Tyrrhenian, ≈ 0.8 between the Tyrrhenian and eastern basin. These steps are visible in this representation due to fast changes in the water properties at the gateways, the cumulative average salinity increases more gradually.

Latitude-depth-averaged temperatures in Fig. 7b have a large spread near the gateway. Water depths near the gateway are shallower when the sill depth is shallower, hence the influence of warm surface water on the average temperature in this area is larger and the temperature higher. In contrast to the salinity curves, the relative position of the temperature curves is not the same in all subbasins; in the western basin temperatures are relatively low in SD500–SD300 and SD20–SD5, in the eastern basin temperatures increase from SD500 to SD5. Given that surface forcing is constant, these differences must be caused by differences in the circulation which we will elaborate on below. Compared to the steps in the salinity curves, the changes in temperature occur over a broad range of longitudes. Due to the shallow depths near the connections between the basins, the warm surface layer becomes more important for the local average. Notwithstanding local differences, the trend in the basinal averages is towards higher temperatures at shallower sill depths.

The density curves in Fig. 7c illustrate the combined effect of salinity and temperature. Because salinity is fairly stable in each subbasin, density changes in subbasins are caused by temperature differences. Because these temperature differences are mainly caused by different latitudinally averaged water depths, the density curve reflects the average water depth. Shallow low density areas are the connections between the subbasins, the Aegean Sea and the easternmost eastern basin.

The main features of the basinal circulation can be captured in the minimum and maximum strengths of the overturning cells and the depths where these occur in the western and eastern basin. These parameters are shown as a function of sill depth in Fig. 8. The upper cell is consistently deeper and stronger in the western than in the eastern basin (as visible in Fig. 5 for SD300). For the deep cell this pattern is reversed: it is consistently deeper and stronger in the eastern basin than in the western basin. In the experiments that have reached a steady state, i.e. SD100–SD500, the strength and depth of all cells except the upper cell in the western basin are strikingly similar. In SD5–SD50, where a steady state has not yet been reached, the strength of the overturning cells changes significantly between 400 and 800 yr (light and dark symbols, respectively). Over this period, the strengths of the western deep cell and the eastern deep and upper cell in SD5–SD200 move towards the steady state value that is reached in SD100–SD500. This suggests that a similar overturning circulation will eventually be reached regardless of the sill depth. A difference not captured in Fig. 8 is the depth of the interface between the upper and deep overturning cells. Even though the depth of the minimum and maximum strength remains the same, the interface shifts to shallower depths when the upper overturning cell loses strength.

Compared to the overturning in steady state, deep overturning in the eastern basin is more vigorous before a steady state is reached. In the western basin deep overturning is weaker before a steady state, as are the surface cells in the western and eastern basin. Towards a steady state, deep water formation in the eastern basin slows down when the vertical density gradient in the basin stabilizes. At the same time, intermediate water formation in both basins and deep water formation in the western basin pick up when vertical temperature differences decrease.

The only cell that stabilizes at a different strength depending on sill depth, is the upper cell in the western basin. The increase in overturning strength from SD100 to SD500 (+0.84 Sv) is similar to the increase in exchange with the Atlantic (+0.91 Sv, Fig. 9). From this similarity it can be inferred that the upper overturning cell in the western basin is mainly controlled by the exchange with the Atlantic. The temperatures in the western basin are consistent with this: temperatures are lower, i.e. closer to Atlantic values, when the exchange with the Atlantic is larger (SD100 to SD500). SD5–SD20 deviate from this trend (Fig. 7b). However, not being in steady state, the deep water temperature in the western basin is still rising after 800 yr in these experiments.

The temperature in the eastern basin is only affected by the magnitude of the Atlantic-derived inflow in the surface layer near the Strait of Sicily. Therefore, a reduced Atlantic inflow cannot explain the small temperature increase from SD500 to SD100 in the eastern basin (Fig. 7b). This temperature increase is most pronounced below 250 m and associated with a small change in overturning circulation. The upper and lower cell in the eastern basin are minimally stronger in SD100 compared to SD500. In the reference experiment, deep water was formed where sea surface temperatures were low, and intermediate water where maximum salinities were reached near Cyprus. At higher salinities, the change in den-

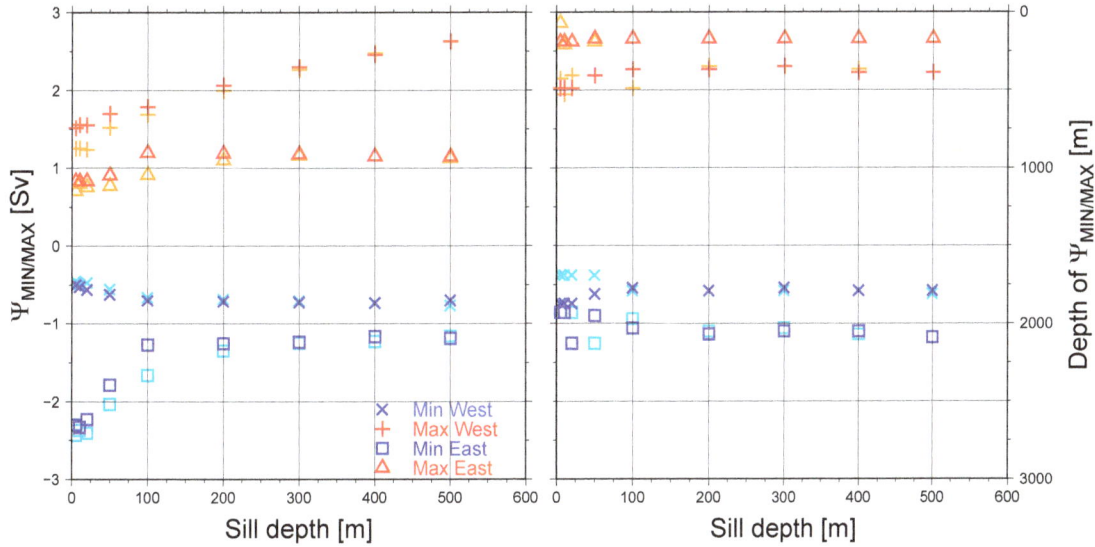

Figure 8. Strength (left) and depth of overturning extrema (right) of the upper (red) and deep (blue) zonal overturning cells in the western and eastern basin. To illustrate temporal changes, light colours indicate the value after 400 yr, dark colours the value after 800 yr.

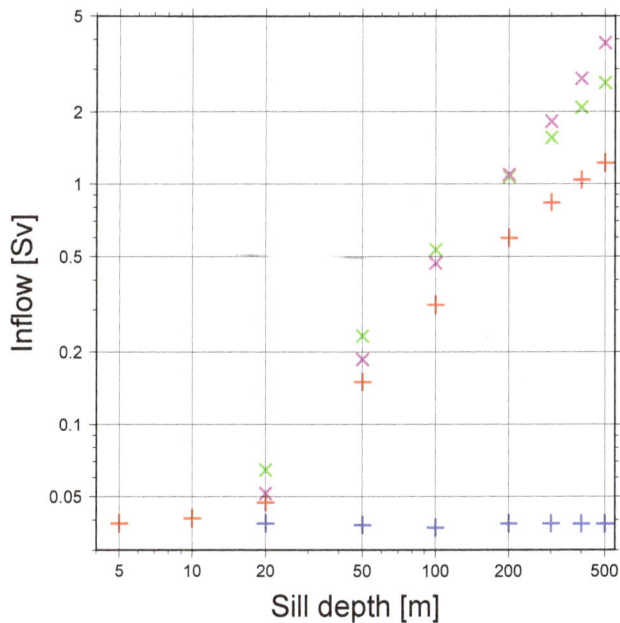

Figure 9. Modelled inflow (red plus signs) and net flow (blue plus signs) in SD500–SD5. Green crosses show the inflow calculated with hydraulic control theory from the basin-averaged densities of the Atlantic and Mediterranean. Purple crosses indicate the inflow calculated with hydraulic control theory using the average density of inflow and outflow in the gateway instead of basin averages.

sity caused by cooling of surface water is comparatively larger. As a consequence, the density of the surface waters increases more due to the same cooling and the mixed layer depth increases significantly throughout the eastern basin. Furthermore, due to the slightly higher sea surface temper-

atures at shallower sill depths (Fig. 7b), the relaxation to the annual mean air temperature results in a stronger heat loss, a larger increase in density, and again a deeper mixed layer. Both mechanisms drive a stronger intermediate water formation and slightly enhance the strength of the upper overturning cell. Noteworthy is the appearance of a new branch of intermediate water formation in the northern Ionian Basin in lower sill depth experiments. This area receives less relatively low saline water from the western basin at shallower Atlantic sill depths due to a lower density difference between the basins. Consequently, water becomes dense enough to sink in this area. Compared to the dense water formation sites in the Adriatic and Aegean, surface water is warmer in the Ionian Basin. Due to the reduced vertical density gradient at shallower sill depths, dense water formed in the Adriatic flows downslope to greater depths, thereby enhancing the strength of the deep overturning cell. In summary, lower density gradients at shallow sill depths enhance upper and deep overturning circulation in the eastern basin which leads to the small increase in intermediate and deep water temperatures observed in Fig. 7b.

3.2.1 Strait transport

The magnitude of modelled strait transport as a function of sill depth is illustrated in Fig. 9 (red plus signs). Down to a sill depth of 20 m, inflow decreases steadily towards the value of net flow (blue plus signs) which is essentially constant in all runs. Inflow is almost linearly proportional to sill depth; only towards deeper sill depths does the increase in inflow flatten slightly. Outflow is not shown because it shows the same trend as the inflow (outflow being inflow minus net flow).

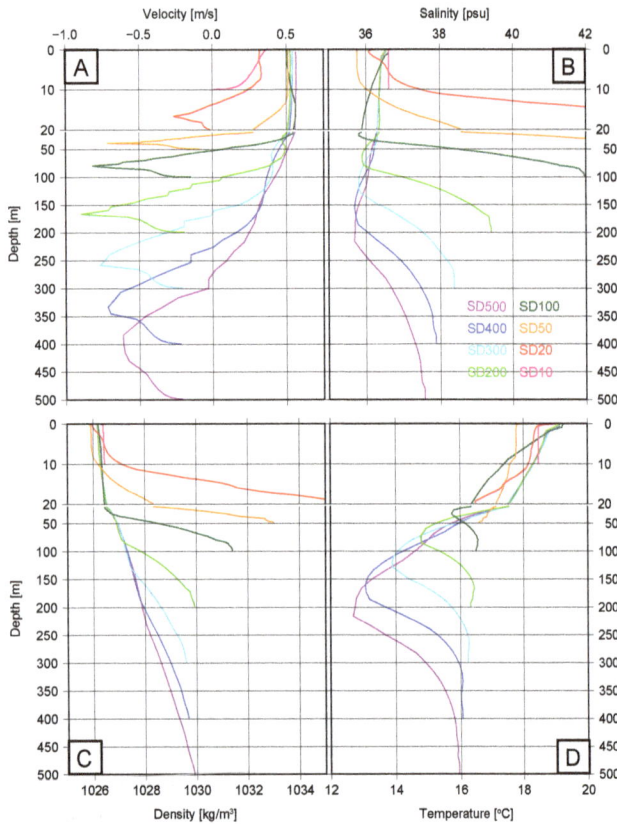

Figure 10. Vertical profiles of zonal velocity (**a**), density (**c**) and temperature (**d**) at the sill in the Strait of Gibraltar in experiments SD500–SD10. Note the change in vertical scale at 20 m. Temperature and salinity profiles of SD50 stand out with incongruent values near the surface. This is caused by a numerical overshoot due to steep slopes near the gateway in this experiment.

Even though the Atlantic–Mediterranean exchange appears to be a simple two-way flow from Fig. 9, velocity, salinity, temperature and density profiles in the gateway show that this is not true (Fig. 10). Velocity (Fig. 10a) is positive in the top layer and decreases gradually towards the interface between inflow and outflow, at a depth lower than half the sill depth, and decreases again towards the bottom due to bottom friction. Salinity (Fig. 10b) in the top layer is near constant at all sill depths, being supplied with Atlantic water with relatively constant salinity. The lower layer becomes increasingly more saline towards shallower sill depths in accordance with the average basinal salinity. Temperatures (Fig. 10d) decrease in the top layer from the surface to the interface depth, and increase again towards the bottom. The high temperatures at the surface are caused by the heat flux at the surface. In the Atlantic, temperatures drop significantly with depth (Fig. 4). In the surface inflow this results in a lower temperature near the interface depth at deeper sill depths. In the Mediterranean, temperatures decrease less with depth. Therefore, temperatures in the outflow move back to higher values away from the interface. The gradual change of ve-

locity, temperature and salinity between inflow and outflow indicates mixing between the two layers. The combined effect of salinity and temperature is expressed in the density profiles (Fig. 10). Towards shallower sill depths the density difference between the Atlantic surface flow and Mediterranean outflow becomes larger. i.e. density changes faster over a shorter vertical distance.

In the present day, the Mediterranean is a heat sink for the Atlantic, i.e. the Mediterranean has a net surface heat loss. The present-day net surface heat loss is estimated to be $5\,\mathrm{W\,m^{-2}}$ based on heat transport measurements in the Strait of Gibraltar (Macdonald et al., 1994). Numerical models and reanalyses, however, show a much larger range of estimates, e.g. -21–$40\,\mathrm{W\,m^{-2}}$ (Sanchez-Gomez et al., 2011). Table 2 gives an overview of heat transport through the Strait of Gibraltar for different sill depths. By spreading the net flow at the Strait over the whole Mediterranean surface area, we get the equivalent net surface heat loss (ENSHL) presented in the last column. Even though the absolute value is not exactly reproduced, the sign of the Mediterranean heat budget is correct in the reference experiment. Heat inflow and outflow both decrease almost linearly towards shallower sill depths, until at 10 m the outflow stops. The net flow is small compared to the inflow and outflow and increases to an almost constant value of 3 TW at sills shallower than 200 m. Only in SD500 is the inflow from the Atlantic cold enough at depth, in combination with a warm outflow, to cause a net heat flow from the Mediterranean to the Atlantic. In all other experiments the Atlantic inflow is dominated by surface waters significantly warmer than the outflow.

As noted in Fig. 10, the basin-averaged temperature is higher when the sill is shallower. This is counterintuitive since one would expect the surface heat loss to dominate over the smaller Atlantic inflow and make Mediterranean temperatures drop. However, the Atlantic inflow becomes increasingly warmer at shallower sill depths because a larger portion of the inflow is drawn from the warm Atlantic surface layer. Surface temperatures in the Mediterranean are only notably affected by the Atlantic inflow temperature in SD500 and SD400, in all other experiments surface temperatures and hence the net surface heat loss are similar. Because the warmer Atlantic inflow is not compensated for by a larger surface heat loss, the Mediterranean warms up until the outflow is warm enough to compensate for the warmer inflow and the Mediterranean heat budget is back to zero.

In SD5 and SD10, the inflow is constant and equal to the net flow, i.e. there is no outflow. Although outflow is blocked in these two experiments, neither experiment has reached a steady state yet. Salinity rises constantly at $11.6\,\mathrm{psu\,kyr^{-1}}$ as long as outflow is blocked because all salt that enters the Mediterranean through inflow ($4.3 \times 10^{13}\,\mathrm{kg\,yr^{-1}}$) is retained in the Mediterranean. In SD20 and SD50, like SD5 and SD10 not yet in steady state, the ongoing salinity rise in the Mediterranean is accompanied by a steady increase in the outflow and, hence, inflow. It is possible that the increas-

Table 2. Overview of heat transport through the Strait of Gibraltar. ENSHL: equivalent net surface heat loss.

Sill depth (m)	Inflow (TW)	Outflow (TW)	Net flow (TW)	ENSHL ($W\,m^{-2}$)
500	73.604	74.108	−0.504	−0.212
400	63.121	62.326	0.795	0.334
300	52.984	50.583	2.401	1.010
200	39.231	36.028	3.204	1.348
100	21.241	18.142	3.099	1.304
50	10.542	7.563	2.980	1.253
20	3.487	0.607	2.880	1.212
10	3.035	0.000	3.035	1.277

ing density difference between Mediterranean and Atlantic in SD10 and SD5 will eventually incite two-way flow. However, the anticipated salinities for the steady states in these experiments are >100 psu, which will take at least another 5000 yr to be reached. Therefore, due to the additional model run time required, reaching steady states in SD5 and SD10 is unfeasible in the current model setup. Moreover, the equation of state – which relates temperature, salinity and pressure to density – is only valid up to 42 psu. At higher salinities, a linear extrapolation is used to calculate the density increase due to salinity. This will give an increasingly larger error in the density determination towards higher salinities. Also, the viscosity of water will change significantly at such high salinities, a process not included in the model.

In summary, exchange between the Mediterranean and Atlantic is proportional to the sill depth until at 10 m sill depth outflow is blocked and only inflow remains. The interface between inflow and outflow is consistently deeper than half the sill depth. The highest velocities are reached in the saline, warm Mediterranean outflow which is overlain by a less saline, colder Atlantic inflow.

4   Discussion

4.1   Constant forcing

Even though the usage of a minimal constant forcing is a deliberate choice in our model setup, it may cause changes in the Mediterranean circulation and water characteristics compared to a model with a non-constant (seasonal) forcing. Whereas most deep water formation would be episodic, i.e. concentrated in the winter months, with a seasonal cycle, a constant forcing drives a constant deep water formation. Deep water forms year-round in the northern part of the western and eastern basins due to cooling at the surface, i.e. a net surface heat loss of 1.3 $W\,m^{-2}$ (Table 2). This cooling is not the strong winter cooling (5 $W\,m^{-2}$) that would occur with a seasonal forcing, hence deep water is formed at relatively high temperatures. The deep basin is filled with warm saline water which is, on average, 3.4 °C warmer than observed in the present-day Mediterranean (Fig. 4). The depth of deep

water formation (Fig. 6) and also the rate are, however, not significantly affected by the constant forcing because deep basinal water and dense water formed at deep water formation sites are both warmer and their respective densities both change with a similar amount. Furthermore, the deep water formation sites are at or close to the actual observed locations because the overall circulation, including the transport of preconditioned water to the deep water formation sites, and the buoyancy forcing are reproduced to an acceptable degree. Mediterranean salinity is relatively constant throughout the year in the present day and is not significantly affected by the use of a constant forcing.

Strait transport is not significantly affected by the high deep water temperatures and associated warm Mediterranean outflow. A change in salinity affects density, and therefore Mediterranean outflow, four times more than temperature, i.e. a change of 1 $g\,L^{-1}$ has the same impact on density as 4 °C. Mediterranean outflow in a model forced with a seasonal cycle would be slightly stronger due to lower intermediate/deep water temperatures in the Mediterranean and the associated increase in the density contrast with the Atlantic. Also, the amount of heat lost through Mediterranean outflow would be lower because a larger part of the heat from the Atlantic inflow is lost at the surface instead.

In summary, although the constant forcing results in year-round formation of relatively warm deep water, the overall circulation, deep water formation, strait transport, and salinity are sufficiently reproduced to study the impact of sill depth on them.

4.2   Hydraulic control

A comparison of exchange in the model and that predicted by hydraulic control theory is called for since hydraulic control theory has been extensively used to describe the Atlantic–Mediterranean exchange. Hydraulic control theory can be used to calculate the flow through a narrow strait if the geometry and density difference along the strait are known. A recent overview of the basic principles underlying hydraulic control theory and an application to the Mediterranean can be found in Meijer (2012). Differences between exchange calculated with hydraulic control theory and that from the model are to be expected due to the more complex physics in the model. For comparison with the modelled exchange, an expression for hydraulic control for a strait with a rectangular cross section and zero net flow is used (Farmer and Armi, 1986; Bryden and Kinder, 1991):

$$Q_A = Q_M = 0.208 \cdot \sqrt{g'H}\,HW, \qquad (1)$$

where Q_A is the inflow from the Atlantic which is equal to the outflow from the Mediterranean, Q_M; H is the sill depth, W the strait width, and the reduced gravity, g', is defined by

$$g' = g\,(\rho_M - \rho_A)/\rho_M, \qquad (2)$$

where g is the gravitational acceleration, ρ_M the Mediterranean outflow water density and ρ_A Atlantic water density.

Besides the modelled strait transport, Fig. 9 also shows the inflow calculated with Eq. (1) for a strait with the same dimensions (width and depth) as in the model; Atlantic and Mediterranean basin-averaged densities are used as ρ_A and ρ_M respectively. Compared to the strait transport predicted by hydraulic control, the modelled transports are consistently lower. The absolute difference as well as the ratio between inflow predicted by hydraulic control and inflow from the model decrease towards shallower sill depths, i.e. modelled transport is closer to that predicted by hydraulic control in shallow gateways.

In hydraulic control theory, the velocity and density profiles at the gateway are envisaged to be a step function with a constant positive velocity and low density in the upper layer and a constant negative velocity and high density in the bottom layer. In contrast, profiles derived from the model (Fig. 10) show a gradual change in water properties near the interface between inflow and outflow. It must be noted, however, that density and salinity profiles at shallow sill depths are closer to a step function than those at deep sill depths; a larger change in salinity/density occurs in a smaller depth interval. This may partly explain why hydraulic control theory better matches modelled transports at shallower sill depths.

When transport is calculated by inserting the average density of inflow and outflow in Eq. (1), instead of averages of Atlantic and Mediterranean basins, it is closer to modelled values at shallow sill depths and further from modelled values at deep sill depths. The density difference between inflow and outflow is larger than between the basin averages at deep sill depths, while it is smaller for shallower sill depths. At sill depths > 100 m, mixing between the inflow and outflow reduces the temperature of the outflow, which is already more saline than the inflow, and lowers the density. Hence, the density contrast between inflow and outflow is increased compared to the difference between the basin averages. Consequently, the inflow predicted by hydraulic control with the densities in the gateway is higher than that with basin average densities. At sill depths < 100 m, the density difference between the inflow and outflow is smaller than the difference between basin averages because the largest difference

in temperature between the basins occurs at greater depths than those involved in the exchange. Accordingly, the inflow predicted by hydraulic control with densities in the gateway is lower than that from basin-averaged densities and closer to modelled transports.

Regardless of the densities used, exchange calculated with hydraulic control theory is always larger than modelled transport. The aforementioned vertical mixing between inflow and outflow is one obvious cause of the difference. Another important factor is friction; at the bottom it slows down the outflow and friction between the inflow and outflow slows down both. The Coriolis force does not play a role here due to the narrow width of the gateway.

In summary, transport at shallow sill depths is closer to that predicted by hydraulic exchange theory than at deep sill depths because mixing between inflow and outflow is not as effective in reducing the difference between them. Furthermore, using basin-averaged density differences for calculation of the exchange with hydraulic control gives an overestimation of the exchange because water characteristics at the depths involved in the exchange are not representative of the whole basin.

4.3   The role of sill depth

In experiments with sill depths in the range of 500–5 m, basin-averaged salinities and temperatures are consistently higher at shallower sill depths. Spatial differences, e.g. between the western and eastern basin, are largely independent of sill depth. The upper overturning cell in the western basin is controlled by the exchange with the Atlantic and is weaker at shallower sill depths. The upper overturning cell in the eastern basin and the deep overturning cells in both basins, however, are practically constant in depth and strength at all sill depths as soon as a steady state has been reached. Dense water formation is slightly more temperature-driven when basin averaged salinities are higher; at high salinities surface cooling causes a larger increase of density than at low salinities. However, overall, the locations and rate of dense water formation change little.

The influence of sill depth on circulation and water characteristics found in this study is, mainly due to differences in model setup, different from that found by Alhammoud et al. (2010) (AMD10). In AMD10, a highly idealized representation of the Mediterranean was used, which consists of a single large basin with a depth of 1500 m gradually shallowing towards the margins. In accordance with our findings for the western basin they found a steadily increasing inflow and stronger upper zonal overturning cell in experiments with increasingly larger sill depths. Their deep overturning cell, however, almost disappears at large sill depths whereas it is here found to be constant in strength regardless of sill depth. The shift of the interface between upper and deep overturning cells in their experiments does resemble the shift found here. Because their basin extends to only 1500 m, the deep overturning cell is suppressed by stronger surface cells when these extend to greater depths, while it persists below 1500 m in our experiments. As in our experiments, deep water circulation, although strongly reduced, never entirely stopped in the experiments of AMD10.

The low resolution used in AMD10 resulted in a 222 km wide Strait of Gibraltar. Due to this width, exchange with the Atlantic was found to be consistent with rotational control on the flow instead of hydraulic control. The width of the Strait of Gibraltar in our model (13 km) is well below the Rossby radius and, hence, rules out rotational control. The larger strait transports in AMD10 kept Mediterranean salinities closer to Atlantic values, but the overall trend towards higher salinities at lower sill depths is consistent with our findings.

4.4 The interpretation of the Late Miocene sedimentary record

As argued in the introduction, we expect that the change in circulation and water properties due to variation in sill depth that we calculated for a basin with the shape of the present Mediterranean forms a starting point for understanding the past as well. In this section we relate our findings to the Late Miocene sedimentary record.

Stable isotopes and faunal changes in the pre-MSC interval of the Late Miocene suggest that in an interval with increasing salinity, the average deep water oxygenation decreased steadily (Kouwenhoven and van der Zwaan, 2006). As already noted by Kouwenhoven and van der Zwaan (2006) and Krijgsman et al. (2000), the occurrence of sapropels in this interval indicates precessional variation of the oxygenation of Mediterranean deep water on top of the long-term trend. The estimated depth of the marginal basins in which the Monte del Casino, Metochia, Faneromeni and Gibliscemi sections accumulated is 300–1200 m. In these basins the decreasing oxygenation and sapropelic sedimentation are observed. All are located in the eastern Mediterranean basin in which, in our model, the upper overturning cell is less than 500 m deep. If rates of deep water formation and the strength of deep overturning are lower at shallower sill depths, oxygen conditions in the deep water layer would decrease concomitant with a decreasing gateway depth. Model results, however, do not show a significant change in either deep water formation or deep overturning in the eastern basin.

In the western basin, depths up to 1200 m are in the upper overturning cell in SD500–SD100. Oxygenation of the marginal basins in this setting would not be by deep water formation, but by intermediate water formation in the upper overturning cell. The simultaneous decrease of exchange with the Atlantic and strength of the upper overturning cell in the western basin towards shallower sill depths leads to a longer residence time of the water in the upper cell of the western basin. Consequently, water in the marginal basins will be replenished more slowly with oxygenated water from

the surface. Therefore, model results from the western basin could explain the suggested correlation between sill depth and oxygenation. Because AMD10 represents the Mediterranean with only one basin, their correlation between observations and model results is similar to what is suggested here for the western basin.

Model results thus seem to contradict the notion that restriction of the Atlantic–Mediterranean gateway induces lower oxygenation of deep water in the eastern basin. A lower oxygenation may, however, be inferred for the upper overturning cell in the western basin. Due to the relatively shallow connection between the western and eastern basin, only the western overturning is affected by sill depth in a way that seems consistent with the data. A possible cause of the discrepancy between model results and observations is the present-day bathymetry used in the model. If the connection between both basins was deeper in the Late Miocene, the upper overturning cell could have extended further and deeper into the eastern basin. Without the Sicily Strait, model results are expected to be similar to those in AMD10 who indeed had a single surface overturning cell in the whole Mediterranean.

In the present day, deep water formation is mainly driven by cooling of surface water during the winter. Despite the fact that deep water formation in the model is continuous, the forcing that drives deep water formation is the same. We therefore argue that deep water formation is sufficiently represented in our setup to reproduce possible changes in deep water oxygenation that may occur in a less idealized case.

Another simplification in the surface forcing, on the other hand, may influence the correlation of model results and observations: the inclusion of river discharge in a uniform surface flux equal to E-P-R. Increased river discharge during precession minima is generally accepted to be an important factor in the establishment of low-oxygen conditions during sapropel formation. River discharge is thought to form a fresh-water-lid at the surface, hindering deep water formation by reducing surface layer densities. Due to mixing with more saline water and evaporative concentration, water originating from rivers will lose its fresh water signature when it moves away from the outlet. If the receiving basin is at higher salinity, the density difference between river water and the basin is larger and stratification will be stronger. Although depending on the volume and location of river input, basin circulation, evaporation – precipitation, and the rate of mixing with surrounding water, we can assume that stratification due to river discharge is stronger when the basin-averaged salinity is higher. Hence, deep water formation may be more effectively reduced at lower sill depth. If a river drains into a marginal basin with restricted exchange with the deep basin, stratification is presumably more severe and deep water oxygenation even more strongly reduced.

A spatially heterogeneous distribution of precipitation may have the same effect as river discharge. Precipitation, however, does not give a continuous fresh water input at the same location like river input. Hence, its influence with respect to river discharge is expected to be lower.

4.5   Blocked outflow

Notwithstanding the fact that SD10 and SD5 are not yet in steady state, it is the first time that a blocked outflow has been observed in an ocean circulation model in this context. From hydraulic control theory, one layer flow is predicted to occur only when the sill depth is a few metres for a gateway with a width of 13 km (Meijer, 2012). If bottom friction is taken into account, this depth is expected to be somewhat deeper. Because the density difference between the Atlantic and Mediterranean is still rising at the time model results are shown for, it may be that outflow will commence if the experiment is run to steady state.

Our model results suggest that if a sill depth of ≈ 10 m existed during the MSC, either due to a global sea level drop or local uplift due to flexure or tectonics, the salinity rise in the Mediterranean would be maximal. Consequently, the salt gain of the Mediterranean is highest in this scenario, allowing for the fast accumulation of evaporites. During blocked outflow, 15 km^3 of halite is transported to the Mediterranean every year. At this rate it would take 33–133 kyr to form the 0.5–2 million km^3 of halite observed on seismics (Ryan, 2008). Halite formation in the deep Mediterranean basins took place during an 60 kyr interval which encompasses two glacials: 5.61–5.55 Ma with glacials TG12 and TG14. The growth of icecaps on the poles during this interval would reduce global sea level, lowering the relative sill depth. Furthermore, glacials are characterized by a relatively high fresh water deficit in the Mediterranean due to reduced river discharge and precipitation. In this situation, salinity rise and salt gain will be even higher than found in our model. During blocked outflow, inflow could thus bring in the observed volume of salt in the 60 kyr MSC interval. The blocked-outflow scenario is therefore plausible. To examine whether blocked-outflow endures at higher density contrasts, a model should be set up to represent only the Strait of Gibraltar, or another gateway thought to be open during the Messinian. This, however, is impossible with the current model because the equation of state implemented in POM is not valid at salinities larger than 42 psu and anticipated viscosity changes at high salinities cannot be dealt with.

5   Conclusions

In this study, a parallel version of POM (sbPOM) has been used to examine changes in Mediterranean circulation and water characteristics due to restriction of the Atlantic connection. Model results have implications for the interpretation of the Late Miocene sedimentary record in the Mediterranean. Compared to earlier models of the (Miocene) Mediterranean, the use of a curvilinear grid and parallel code allows for the use of a higher resolution and more realistic bathymetry even in long model runs (800 yr). A comparison of the results from our model with observations of the present day shows that, despite an idealized and constant surface forcing, Mediterranean circulation and water properties are generally well reproduced.

The model setup presented in this study would seem to provide a valuable basis and reference for examination of additional aspects of Mediterranean palaeoconfigurations. This may relate to other aspects of the Miocene evolution, for example, but our setup is also applicable to the Last Glacial Maximum when lower sea level was responsible for a reduction in sill depth.

The main results and implications for the Late Miocene Mediterranean are the following:

– Basin-averaged salinity, temperature and density increase when the sill is shallower. However, spatial distribution and inter-basinal differences in water properties in the Mediterranean are largely unaffected by sill depth.

– The strength of the upper overturning cell in the western Mediterranean is proportional to the magnitude of water exchange with the Atlantic. Overturning in the eastern basin is not significantly affected by the depth of the sill.

– Temperature-driven dense water formation operates regardless of the basin-averaged salinity. At shallower sill depths, the higher salinity in the Mediterranean results in a stronger salinity-driven dense water formation in the eastern basin.

– Modelled strait transport is always smaller than that predicted by hydraulic control theory. This difference is due to friction, vertical mixing and a difference between basin-averaged density and the density of the water involved in the exchange with the Atlantic.

– Outflow is blocked in (at least the first 800 yr of) experiments with sill depths ≤ 10 m. Future work is needed to establish whether blocked-outflow is a viable scenario for the interval with halite deposition in the MSC.

– With the present-day bathymetry, restriction of the Atlantic–Mediterranean connection does not significantly alter Mediterranean deep water circulation and refreshening. Hence, model results do not affirm the hypothesis that deep water ventilation decreases at shallower sill depths.

Acknowledgements. The authors would like to thank Rinus Wortel for valuable input on drafts of this article. The manuscript benefited from reviews by two anonymous reviewers and Mike Rogerson. R. P. M. Topper was supported by the Netherlands Research Center for Integrated Solid Earth Science. Computational resources for this work were also provided by ISES (ISES 3.2.5 High End Scientific Computation Resources). Figures in this paper were created using GMT version 4.5.1 (Wessel and Smith, 1991).

Edited by: U. Mikolajewicz

References

Adloff, F., Mikolajewicz, U., Kuc(era, M., Grimm, R., Maier-Reimer, E., Schmiedl, G., and Emeis, K.-C.: Upper ocean climate of the Eastern Mediterranean Sea during the Holocene Insolation Maximum – a model study, Clim. Past, 7, 1103–1122, doi:10.5194/cp-7-1103-2011, 2011.

Ahumada, M. A. and Cruzado, A.: Modeling of the circulation in the Northwestern Mediterranean Sea with the Princeton Ocean Model, Ocean Sci., 3, 77–89, doi:10.5194/os-3-77-2007, 2007.

Alhammoud, B., Meijer, P. Th., and Dijkstra, H. A.: Sensitivity of Mediterranean thermohaline circulation to gateway depth: A model investigation, Paleoceanography, 25, PA2220, doi:10.1029/2009PA001823, 2010.

Amante, C. and Eakins, B. W.: ETOPO1 1 arc-minute global relief model: procedures, data sources and analysis, NOAA Technical Memorandum NESDIS NGDC-24, 19 pp., 2009.

Antonov, J. I., Levitus, S., Boyer, T. P., Conkright, M. E., and O'Brien and C. Stephens, T. D.: World Ocean Atlas Data 1998, NOAA Atlas NESDIS, 27, 166 pp., 1998.

Astraldi, M., Balopoulos, S., Candela, J., Font, J., Gacic, M., Gasparini, G. P., Manca, B., Theocharis, A., and Tintoré, J.: The role of straits and channels in understanding the characteristics of Mediterranean circulation, Prog. Oceanogr., 44, 65–108, 1999.

Beckers, J.-M., Rixen, M., Brasseur, P., Brankart, J.-M., Elmoussaoui, A., Crépon, M., Herbaut, Ch., Martel, F., Van den Berghe, F., Mortier, L., Lascaratos, A., Drakopoulos, P., Korres, G., Nittis, K., Pinardi, N., Masetti, E., Castellari, S., Carini, P., Tintore, J., Alvarez, A., Monserrat, S., Parrilla, D., Vautard, R., and Speich, S.: Model intercomparison in the Mediterranean: MEDMEX simulations of the seasonal cycle, J. Marine Syst., 33, 215–251, 2002.

Benson, R. H., Bied, K. R.-E., and Bonaduce, G.: An important current reversal (influx) in the Rifian corridor (Morocco) at the Tortonian-Messinian boundary: the end of the Tethys ocean, Paleoceanography, 6, 164–192, 1991.

Berntsen, J. and Oey, L.-Y.: Estimation of the internal pressure gradient in σ-coordinate ocean models: comparison of second-, fourth-, and sixth-order schemes, Oc. Dynam., 60, 317–330, 2010.

Bertini, A.: The Northern Apennines palynological record as a contribute for the reconstruction of the Messinian palaeoenvironments, Sediment. Geol., 188–189, 235–258, 2006.

Betzler, C., Braga, J. C., Martín, J. M., Sánchez-Almazo, I. M., and Lindhorst, S.: Closure of a seaway: stratigraphic record and facies (Guadix basin, Southern Spain), Internat. J. Earth Sci., 95, 903–910, 2006.

Blumberg, A. F. and Mellor, G. L.: A description of a three-dimensional coastal ocean circulation model, Coast. Estuar. Sci., 4, 1–16, 1987.

Bryden, H. L. and Kinder, T. H.: Steady two-layer exchange through the Strait of Gibraltar, Deep-Sea Res., 38, S445–S463, 1991.

Bryden, H. L. and Stommel, H. M.: Limiting processes that determine basic features of the circulation in the Mediterranean Sea, Oceanol. Ac., 7, 289–296, 1984.

Bryden, H. L., Candela, J., and Kinder, T. H.: Exchange through the Strait of Gibraltar, Prog. Oceanogr., 33, 201–248, 1994.

Candela, J.: The Gibraltar Strait and its role in the dynamics of the Mediterranean Sea, Dynam. Atmos. Oc., 15, 267–299, 1991.

Candela, J.: Mediterranean water and global circulation, vol. 77 of International Geophysics, chap. 5.7, Academic Press, San Diego, 419–429, 2001.

Cramp, A. and O'Sullivan, G.: Neogene sapropels in the Mediterranean: a review, Mar. Geol., 153, 11–28, 1999.

de la Vara, A., Meijer, P. Th., and Wortel, M. J. R.: Model study of the circulation in the Miocene Mediterranean Sea and Paratethys: closure of the Indian gateway, Clim. Past Discuss., 9, 4385–4424, doi:10.5194/cpd-9-4385-2013, 2013.

Dercourt, J., Gaetani, M., Vrielynck, B., Barrier, E., Biju-Duval, B., Brunet, M. F., Cadet, J. P., Crasquin, S., and Sandulescu, M. (Eds.): Peri-Tethys palaeogeographical atlas, CCGM/CGMW, 2000.

Dijkstra, H. A.: Scaling of the Atlantic meridional overturning circulation in a global ocean model, Tellus, 60A, 749–760, 2008.

Drakopoulos, P. G. and Lascaratos, A.: Modelling the Mediterranean Sea: climatological forcing, J. Mar. Syst., 20, 157–173, 1999.

Duggen, S., Hoernle, K., van den Bogaard, P., Rüpke, L., and Phipps Morgan, J.: Deep roots of the Messinian salinity crisis, Nature, 422, 602–606, 2003.

Ezer, T. and Mellor, G. L.: Simulations of the Atlantic Ocean with a free surface sigma coordinate ocean model, J. Geophys. Res., 102, 15647–15657, 1997.

Farmer, D. M. and Armi, L.: Maximal two-layer exchange over a sill and through the combination of a sill and contraction with barotropic flow, J. Fluid Mechan., 164, 53–76, 1986.

Fauquette, S., Suc, J.-P., Bertini, A., Popescu, S.-M., Warny, S., Bachiri Taoufiq, N., Perez Villa, M.-J., Chikhi, H., Feddi, N., Subally, D., Clauzon, G., and Ferrier, J.: How much did climate force the Messinian salinity crisis? Quantified climatic conditions from pollen records in the Mediterranean region, Palaeogeogr. Palaeoclimatol. Palaeocol., 238, 281–301, 2006.

Gennari, R., Manzi, V., Angeletti, L., Bertini, A., Biffi, U., Ceregato, A., Costanza, C., Gliozzi, E., Lugli, S., Menichetti, E., Rosso, A., Roveri, M., and Taviani, M.: A shallow water record of the onset of the Messinian salinity crisis in the Adriatic foredeep (Legnagnone section, Northern Apennines), Palaeogeogr. Palaeoclimatol. Palaeoecol., 386, 145–164, 2013.

Gladstone, R., Flecker, R., Valdes, P., Lunt, D., and Markwick, P.: The Mediterranean hydrologic budget from a Late Miocene global climate simulation, Palaeogeogr. Palaeoclimatol. Palaeocol., 251, 254–267, 2007.

Govers, R.: Choking the Mediterranean to dehydration: The Messinian salinity crisis, Geology, 37, 167–170, 2009.

Hilgen, F. J., Krijgsman, W., Langereis, C. G., Lourens, L. J., Santarelli, A., and Zachariasse, W. J.: Extending the astronomical (polarity) time scale into the Miocene, Earth Planet. Sci. Lett., 136, 495–510, 1995.

Hopkins, T. S.: The thermohaline forcing of the Gibraltar exchange, J. Mar. Syst., 20, 1–31, 1999.

Hüsing, S. K., Oms, O., Agustí, J., Garcés, M., Kouwenhoven, T. J., Krijgsman, W., and Zachariasse, W.-J.: On the late Miocene closure of the Mediterranean–Atlantic gateway through the Guadix basin (southern Spain), Palaeogeogr. Palaeoclimatol. Palaeocol., 291, 167–179, 2010.

Iovino, D., Straneo, F., and Spall, M. A.: On the effect of a sill on dense water formation in a marginal sea, J. Mar. Res., 66, 325–345, 2008.

Ivanovic, R. F., Flecker, R., Gutjahr, M., and Valdes, P. J.: First Nd isotope record of Mediterranean–Atlantic water exchange through the Moroccan Rifian Corridor during the Messinian Salinity Crisis, Earth Planet. Sci. Lett., 368, 163–174, 2013.

Jordi, A. and Wang, D.-P.: sbPOM: A parallel implementation of Princeton Ocean Model, Environ. Modell. Software, 38, 59–61, 2012.

Jungclaus, J. H. and Mellor, G. L.: A three-dimensional model study of the Mediterranean outflow, J. Marine Syst., 24, 41–66, 2000.

Karami, M. P., De Leeuw, A., Krijgsman, W., Meijer, P. Th., and Wortel, M. J. R.: The role of gateways in the evolution of temperature and salinity of semi-enclosed basins: An oceanic box model for the Miocene Mediterranean Sea and Paratethys, Glob. Planet. Change, 79, 73–88, 2011.

Kouwenhoven, T. J. and van der Zwaan, G. J.: A reconstruction of late Miocene Mediterranean circulation patterns using benthic foraminifera, Palaeogeogr. Palaeoclimatol. Palaeocol., 238, 373–385, 2006.

Kouwenhoven, T. J., Hilgen, F. J., and van der Zwaan, G. J.: Late Tortonian–early Messinian stepwise disruption of the Mediterranean–Atlantic connections: constraints from benthic foraminiferal and geochemical data, Palaeogeogr. Palaeoclimatol. Palaeocol., 198, 303–319, 2003.

Krijgsman, W. and Meijer, P. Th.: Depositional environments of the Mediterranenan "Lower Evaporites" of the Messinian salinity crisis: Constraints from quantitative analysis, Mar. Geol., 253, 73–81, 2008.

Krijgsman, W., Hilgen, F. J., Raffi, I., Sierro, F. J., and Wilson, D. S.: Chronology, causes and progression of the Messinian Salinity Crisis, Nature, 400, 652–655, 1999a.

Krijgsman, W., Langereis, C. G., Zachariasse, W. J., Boccaletti, M., Moratti, G., Gelati, R., Iaccarino, S., Papani, G., and Villa, G.: Late Neogene evolution of the Taza-Guercif Basin (Rifian Corridor, Morocco) and implications for the Messinian salinity crisis, Mar. Geol., 153, 147–160, 1999b.

Krijgsman, W., Garcés, M., Agustí, J., Raffi, I., Taberner, C., and Zachariasse, W. J.: The "Tortonian salinity crisis" of the eastern Betics (Spain), Earth Planet. Sci. Lett., 181, 497–511, 2000.

Lascaratos, A. and Nittis, K.: A high-resolution three-dimensional numerical study of intermediate water formation in the Levantine Sea, J. Geophys. Res., 103, 18497–18511, 1998.

Lascaratos, A., Roether, W., Nittis, K., and Klein, B.: Recent changes in deep water formation and spreading in the eastern Mediterranean Sea: a review, Progress in oceanography, 44, 5–36, 1999.

Macdonald, A. M., Candela, J., and Bryden, H. L.: An estimate of the net heat transport through the Strait of Gibraltar, in: Seasonal and interannual variability of the Western Mediterranean Sea, edited by: Viollette, P. E. L., 13–32, Am. Geophys. Union, Washington DC, 13–32, 1994.

Mariotti, A., Struglia, M. V., Zeng, N., and Lau, K.-M.: The hydrological cycle in the Mediterranean region and implications for the water budget of the Mediterranean Sea, J. Climate, 15, 1674–1690, 2002.

Martín, J. M., Braga, J. C., Aguirre, J., and Puga-Bernabéu, A.: History and evolution of the North-Betic Strait (Prebetic Zone, Betic Cordillera): A narrow, early Tortonian, tidal–dominated, Atlantic–Mediterranean marine passage, Sediment. Geol., 216, 80–90, 2009.

Meijer, P. Th.: A box model of the blocked-outflow scenario for the Messinian Salinity Crisis, Earth Planet. Sci. Lett., 248, 486–494, 2006.

Meijer, P. Th.: Hydraulic theory of sea straits applied to the onset of the Messinian Salinity Crisis, Mar. Geol., 326–328, 131–139, 2012.

Meijer, P. Th. and Dijkstra, H. A.: The response of Mediterranean thermohaline circulation to climate change: a minimal model, Clim. Past, 5, 713–720, doi:10.5194/cp-5-713-2009, 2009.

Meijer, P. Th. and Krijgsman, W.: A quantitative analysis of the desiccation and re-filling of the Mediterranean during the Messinian Salinity Crisis, Earth Planet. Sci. Lett., 240, 510–520, 2005.

Meijer, P. Th. and Tuenter, E.: The effect of precession-induced changes in the Mediterranean freshwater budget on circulation at shallow and intermediate depth, J. Marine Syst., 68, 349–365, 2007.

Meijer, P. Th., Slingerland, R., and Wortel, M. J. R.: Tectonic control on past circulation of the Mediterranean Sea: a model study of the Late Miocen, Paleoceanography, 19, 1–19, 2004.

Mellor, G. L. and Yamada, T.: Development of a turbulence closure model for geophysical fluid problems, Rev. Geophys., 20, 851–875, 1982.

Mellor, G. L., Ezer, T., and Oey, L. Y.: The pressure gradient conundrum of sigma coordinate ocean models, J. Atmos. Oc. Technol., 11, 1126–1134, 1994.

Myers, P. G.: Flux-forced simulations of the paleocirculation of the Mediterranean, Paleoceanography, 17, 9–1, 2002.

Myers, P. G., Haines, K., and Josey, S.: On the importance of the choice of wind stress forcing to the modeling of the Mediterranean Sea circulation, J. Geophys. Res. (1978–2012), 103, 15729–15749, 1998a.

Myers, P. G., Haines, K., and Rohling, E. J.: Modeling the paleocirculation of the Mediterranean: The Last Glacial Maximum and the Holocene with emphasis on the formation of sapropel S1, Paleoceanography, 13, 586–606, 1998b.

Pérez-Asensio, J. N., Aguirre, J., Schmiedl, G., and Civis, J.: Impact of restriction of the Atlantic-Mediterranean gateway on the Mediterranean Outflow Water and eastern Atlantic circulation during the Messinian, Paleoceanography, 27, PA3222, doi:10.1029/2012PA002309, 2012.

Pratt, L. J. and Spall, M. A.: Circulation and exchange in choked marginal seas, J. Phys. Oceanogr., 38, 2639–2661, 2008.

Roveri, M. and Manzi, V.: The Messinian salinity crisis: Looking for a new paradigm?, Palaeogeogr. Palaeoclimatol. Palaeoecol., 238, 386–398, 2006.

Ryan, W. B. F.: Modeling the magnitude and timing of evaporative drawdown during the Messinian salinity crisis, Stratigraphy, 5, 227–243, 2008.

Samuel, S., Haines, K., Josey, S., and Myers, P. G.: Response of the Mediterranean Sea thermohaline circulation to observed changes in the winter wind stress field in the period 1980–1993, J. Geophys. Res., 104, 7771–7784, 1999.

Sanchez-Gomez, E., Somot, S., Josey, S. A., Dubois, C., Elguindi, N., and Déqué, M.: Evaluation of Mediterranean Sea water and heat budgets simulated by an ensemble of high resolution regional climate models, Clim. Dynam., 37, 2067–2086, doi:10.1007/s00382-011-1012-6, 2011.

Sannino, G., Bargagli, A., and Artale, V.: Numerical modeling of the mean exchange through the Strait of Gibraltar, J. Geophys. Res., 107, 9-1–9-24, 2002.

Schneck, R., Micheels, A., and Mosbrugger, V.: Climate modelling sensitivity experiments for the Messinian Salinity Crisis, Palaeogeogr. Palaeoclimatol. Palaeocol., 286, 149–163, 2010.

Seidenkrantz, M.-S., Kouwenhoven, T. J., Jorissen, F. J., Shackleton, N. J., and van der Zwaan, G. J.: Benthic foraminifera as indicators of changing Mediterranean–Atlantic water exchange in the late Miocene, Mar. Geol., 163, 387–407, 2000.

Somot, S., Sevault, F., and Déqué, M.: Transient climate change scenario simulation of the Mediterranean Sea for the twenty-first century using a high-resolution ocean circulation model, Clim. Dynam., 27, 851–879, 2006.

Sonnenfeld, P. and Finetti, I.: Messinian evaporites in the Mediterranean: a model of continuous inflow and outflow, in: Geological Evolution of the Mediterranean Basin, edited by: Stanley, D. J. and Wezel, F.-C., Springer, 347–353, 1985.

Soria, J. M., Fernández, J., and Viseras, C.: Late Miocene stratigraphy and palaeogeographic evolution of the intramontane Guadix Basin (Central Betic Cordillera, Spain): implications for an Atlantic–Mediterranean connection, Palaeogeogr. Palaeoclimatol. Palaeoecol., 151, 255–266, 1999.

Stratford, K., Williams, R. G., and Myers, P. G.: Impact of the circulation on sapropel formation in the eastern Mediterranean, Global Biogeochem. Cy., 14, 683–695, 2000.

Thompson, B., Nilsson, J., Nycander, J., Jakobsson, M., and Döös, K.: Ventilation of the Miocene Arctic Ocean: An idealized model study, Paleoceanography, 25, PA4216, doi:10.1029/2009PA001883, 2010.

Topper, R. P. M. and Meijer, P. Th.: A modelling perspective on spatial and temporal variations in Messinian evaporite deposits, Mar. Geol., 336, 44–60, 2013.

Topper, R. P. M., Flecker, R., Meijer, P. Th., and Wortel, M. J. R.: A box model of the Late Miocene Mediterranean Sea: Implications from combined $^{87}Sr/^{86}Sr$ and salinity data, Paleoceanography, 26, PA3223, doi:10.1029/2010PA002063, 2011.

Tsimplis, M. N.: Vertical structure of tidal currents over the Camarinal Sill at the Strait of Gibraltar, J. Geophys. Res., 105, 19709–19728, 2000.

Tsimplis, M. N. and Bryden, H. L.: Estimation of the transports through the Strait of Gibraltar, Deep-Sea Res. Pt. I, 47, 2219–2242, 2000.

Tuenter, E., Weber, S. L., and Lourens, L. J.: The response of the African summer monsoon to remote and local forcing due to precession and obliquity, Glob. Planet. Change, 36, 219–235, 2003.

van Assen, E., Kuiper, K. F., Barhoun, N., Krijgsman, W., and Sierro, F. J.: Messinian astrochronology of the Melilla Basin: stepwise restriction of the Mediterranean–Atlantic connection through Morocco, Palaeogeogr. Palaeoclimatol. Palaeoecol., 238, 15–31, 2006.

Weijermars, R.: Neogene tectonics in the Western Mediterranean may have caused the Messinian Salinity Crisis and an associated glacial event, Tectonophysics, 148, 211–219, 1988.

Wessel, P. and Smith, W. H. F.: Free software helps map and display data, EOS Transitions AGU, 72, 441–446, 1991.

Xu, X., Chassignet, E. P., Price, J. F., Özgökmen, T. M., and Peters, H.: A regional modeling study of the entraining Mediterranean outflow, J. Geophys. Res., 112, C12005, doi:10.1029/2007JC004145, 2007.

Zavatarelli, M. and Mellor, G. L.: A numerical study of the Mediterranean Sea circulation, J. Phys. Oceanogr., 25, 1384–1414, 1995.

6

Early deglacial Atlantic overturning decline and its role in atmospheric CO_2 rise inferred from carbon isotopes ($\delta^{13}C$)

A. Schmittner[1] and D. C. Lund[2]

[1]College of Earth, Ocean, and Atmospheric Sciences, Oregon State University, Corvallis, Oregon, USA
[2]Department of Marine Sciences, University of Connecticut, USA

Correspondence to: A. Schmittner (aschmitt@coas.oregonstate.edu)

Abstract. The reason for the initial rise in atmospheric CO_2 during the last deglaciation remains unknown. Most recent hypotheses invoke Southern Hemisphere processes such as shifts in midlatitude westerly winds. Coeval changes in the Atlantic meridional overturning circulation (AMOC) are poorly quantified, and their relation to the CO_2 increase is not understood. Here we compare simulations from a global, coupled climate–biogeochemistry model that includes a detailed representation of stable carbon isotopes ($\delta^{13}C$) with a synthesis of high-resolution $\delta^{13}C$ reconstructions from deep-sea sediments and ice core data. In response to a prolonged AMOC shutdown initialized from a preindustrial state, modeled $\delta^{13}C$ of dissolved inorganic carbon ($\delta^{13}C_{DIC}$) decreases in most of the surface ocean and the subsurface Atlantic, with largest amplitudes (more than $1.5\permil$) in the intermediate-depth North Atlantic. It increases in the intermediate and abyssal South Atlantic, as well as in the subsurface Southern, Indian, and Pacific oceans. The modeled pattern is similar and highly correlated with the available foraminiferal $\delta^{13}C$ reconstructions spanning from the late Last Glacial Maximum (LGM, ~ 19.5–$18.5\,ka\,BP$) to the late Heinrich stadial event 1 (HS1, ~ 16.5–$15.5\,ka\,BP$), but the model overestimates $\delta^{13}C_{DIC}$ reductions in the North Atlantic. Possible reasons for the model–sediment-data differences are discussed. Changes in remineralized $\delta^{13}C_{DIC}$ dominate the total $\delta^{13}C_{DIC}$ variations in the model but preformed contributions are not negligible. Simulated changes in atmospheric CO_2 and its isotopic composition ($\delta^{13}C_{CO_2}$) agree well with ice core data. Modeled effects of AMOC-induced wind changes on the carbon and isotope cycles are small, suggesting that Southern Hemisphere westerly wind effects may have been less important for the global carbon cycle response during HS1 than previously thought. Our results indicate that during the early deglaciation the AMOC decreased for several thousand years. We propose that the observed early deglacial rise in atmospheric CO_2 and the decrease in $\delta^{13}C_{CO_2}$ may have been dominated by an AMOC-induced decline of the ocean's biologically sequestered carbon storage.

1 Introduction

Earth's transition from the LGM (Last Glacial Maximum) (23–19 ka BP), into the modern warm period of the Holocene (10–0 ka BP) remains enigmatic (Denton et al., 2006). Evidence of an early warming of the Southern Hemisphere and atmospheric CO_2 increase (Petit et al., 1999; Denton et al., 2010) has prompted hypotheses of a Southern Hemisphere trigger for the deglaciation (Stott et al., 2007; Timmermann et al., 2009). But the early rise in atmospheric CO_2, although an important forcing for deglacial global warming (Shakun et al., 2012), remains unexplained. Various mechanisms have been proposed. Prominent recent studies suggest wind (Anderson et al., 2009; Denton et al., 2010; Toggweiler et al., 2006) and/or stratification (Watson and Naveira Garrabato 2006; Schmittner et al. 2007; Sigman et al. 2007; Tschumi et al., 2011) changes in the Southern Ocean and/or changes in the North Pacific circulation (Menviel et al., 2014).

Others have suggested that the deglaciation was initiated by a collapse of the AMOC (Atlantic meridional overturning circulation) caused by the melting of Northern Hemisphere ice sheets (Clark et al., 2004; Sigman et al. 2007; Anderson et al., 2009; Denton et al., 2010; Shakun et al., 2012; He et al., 2013) and abrupt North Atlantic climate changes (Broecker et al., 1985). This idea is appealing since the AMOC is known from theory to exhibit multiple steady states with the possibility of rapid transitions between them (Stommel, 1961). Moreover, AMOC variations are consistent with the observed antiphasing of surface temperatures between the hemispheres (Crowley 1992; Blunier et al., 1998; Schmittner et al., 2003; Stocker and Johnson, 2003; EPICA community members 2006; Shakun et al., 2012) and evidence for ITCZ (intertropical convergence zone) migration (Menviel et al., 2008). However, surface temperatures and tropical rainfall patterns alone do not allow robust inferences about the AMOC (Kurahashi-Nakamura et al., 2014) and evidence from the deep ocean for circulation variations remains sparse. One widely cited record of protactinium : thorium ratios (^{231}Pa / ^{230}Th) from the subtropical North Atlantic has been interpreted as an AMOC collapse around 19–18 ka BP followed by a rapid resumption $\sim$ 15 ka BP in the warm Bølling–Allerød period (McManus et al., 2004). However, this interpretation has been questioned (Keigwin and Boyle, 2008) and a subsequent set of ^{231}Pa/^{320}Th records (Gherardi et al., 2009) suggested that a complete AMOC cessation during HS1 (Heinrich stadial event 1) was unlikely. Moreover, our understanding of ^{231}Pa / ^{230}Th in the modern ocean continues to evolve (Anderson and Hayes, 2013) and inferences on the basin or global scale circulation from a single site require validation with multiple proxies from a range of oceanographic locations. A quantitative deglacial AMOC reconstruction constrained by distributed interior ocean observations continues to be lacking. Here we attempt a first step towards such a reconstruction by combining model simulations with δ^{13}C measurements of sediment samples.

Deep-sea reconstructions based on δ^{13}C are more common than ^{231}Pa / ^{230}Th, the processes governing δ^{13}C are better understood, and realistic three-dimensional models exist (e.g., Schmittner et al., 2013), providing necessary ingredients for quantitative hypothesis testing. Here we compile deep-ocean δ^{13}C reconstructions at a high temporal resolution from the early deglaciation and compare them with model simulations of δ^{13}C changes caused by AMOC variations in order to test the hypothesis that the AMOC was reduced during HS1. We also compare our model results to observations of atmospheric CO_2 concentrations and the δ^{13}C of atmospheric CO_2 in order to assess mechanisms of the early deglacial CO_2 rise. Here we do not address the full deglaciation but restrict our investigation to its initial phase from the late LGM ($\sim$19.5–18.5 ka BP) to the late HS1 ($\sim$16.5–15.5 ka BP).

Various modeling studies have previously examined the effect of AMOC changes on atmospheric CO_2, with sometimes conflicting results (Marchal et al., 1998; Marchal et al., 1999; Scholze et al., 2003; Köhler et al., 2005; Schmittner et al., 2007a; Obata, 2007; Schmittner and Galbraith, 2008; Menviel et al., 2008, 2012, 2014; Bozbiyik et al., 2011). Schmittner and Galbraith (2008) found that a large AMOC reduction decreases the efficiency of the ocean's biological

pump if North Atlantic Deep Water (NADW) is more depleted in preformed nutrients than water masses sourced in the south (Antarctic Bottom Water (AABW) and Antarctic Intermediate Water (AAIW)), and it thus leads to the outgassing of CO_2 into the atmosphere and gradually increasing atmospheric CO_2 over several thousand years, consistent with theory (Ito and Follows 2005; Marinov et al., 2008a, b) and ice core CO_2 reconstructions (Ahn and Brook, 2007). Some of the differences in model responses may thus be due to the simulations of preformed nutrients. Whereas Schmittner and Galbraith (2008) have demonstrated consistency of their model with modern preformed nutrient observations, such a validation is not published, to our knowledge, for other models (e.g., the LOVECLIM model used by Menviel et al., 2008, 2014). Several studies found a dependency of the results on the initial state (Köhler et al., 2005; Schmittner et al., 2007a; Menviel et al., 2008), suggesting that starting from glacial conditions may give a different answer than starting from modern conditions. However, none of these studies have validated their initial deep-ocean LGM states with reconstructions. Thus inferences from these studies regarding the sensitivity of the real ocean carbon cycle to initial conditions remain subject to considerable uncertainty.

Effects of Southern Hemisphere westerly winds on atmospheric CO_2 have also been quantified with models before (Winguth et al., 1999; Menviel et al., 2008; Tschumi et al., 2008, 2011; d'Orgeville et al., 2010; Lee et al., 2011; Völker and Köhler, 2013). Most of these studies conclude that reasonably expected changes in the strength and/or latitude of westerly winds cannot explain a large fraction of the observed 100 ppm glacial–interglacial CO_2 amplitude (Winguth et al., 1999, Menviel et al., 2008, Tschumi et al., 2008, 2011, d'Orgeville et al., 2010, Völker and Köhler, 2013). An exception is the work by Lee et al. (2011), who find a 20–60 ppm CO_2 increase for a 25 % strengthening of the westerly winds. However, their wind stress forcing was calculated from an atmosphere only model, which was forced with a very large heat flux anomaly and leads to large areas of the subtropical North Atlantic experiencing extreme cooling of more than 10 °C, much more than reconstructed (Bard et al., 2000). We will show below that more realistic coupled ocean–atmosphere model simulations of an AMOC collapse result in much smaller wind stress changes. For comparison, a complete removal of the Antarctic Ice Sheet leads only to a 50 % reduction in Southern Hemisphere westerly winds (Schmittner et al., 2011). Tschumi et al. (2011), whose simulations include δ^{13}C and are forced with wind stress changes over the Southern Ocean, conclude that stratification changes there can explain the observed rise in atmospheric CO_2 and the decrease in δ^{13}C$_{CO2}$ during HS1. Here we propose a different mechanism, namely changes in the AMOC and its effect on the efficiency of the biological pump, as an alternative hypothesis for the ice core observations. Another difference from previous studies is that we directly and quantitatively test Anderson et al.'s (2009) hypothesis that AMOC changes

Table 1. Sediment cores used in this study. Note that cores 1 and 2 have been averaged for the comparison with the model simulations shown in Figs. 7–10 and Table 2.

	Core	Longitude	Latitude	Depth (m)	References	Age model (if different)
1	ODP984	61° N	24° W	1649	Praetorius et al. (2008)	Lund et al. (2014)
2	NEAP4K	61° N	24° W	1627	Rickaby and Elderfield (2005)	
3	RAPiD-10-1P	62° N	17° W	1237	Thornalley et al. (2010)	Lund et al. (2014)
4	ODP980	55° N	15° W	2179	Benway et al. (2010)	
5	NA87-22	55° N	14° W	2161	Vidal et al. (1997)	Waelbroeck et al. (2011)
6	KN166-14-JPC-13	53° N	33° W	3082	Hodell et al. (2010)	
7	MD01-2461	52° N	13° W	1153	Peck et al. (2007)	See text (Section 2)
8	SO75-26KL	37° N	10° W	1099	Zahn et al. (1997)	Lund et al. (2014)
9	MD99-2334K	37° N	10° W	3146	Skinner and Shackleton (2004)	
10	MD95-2037	37° N	32° W	2159	Labeyrie et al. (2005)	Waelbroeck et al. (2011)
11	KNR166-2-26JPC	24° N	83° W	546	Lynch-Stieglitz et al. (2014)	
12	M35003-4	12° N	61° W	1299	Zahn and Stuber (2002)	Lund et al. (2014)
13	KNR159-5 90GGC	28° S	46° W	1105	Lund et al. (2014); Curry and Oppo (2005)	
14	KNR159-5 36GGC	28° S	46° W	1268	Curry and Oppo (2005); Sortor and Lund (2011)	
15	KNR159-5 17JPC	28° S	46° W	1627	Tessin and Lund (2013)	
16	KNR159-5 78GGC	28° S	46° W	1820	Tessin and Lund (2013)	
17	KNR159-5 33GGC	28° S	46° W	2082	Tessin and Lund (2013)	
18	KNR159-5 42JPC	28° S	46° W	2296	Curry and Oppo (2005); Tessin and Lund (2013)	
19	KNR159-5 30GGC	28° S	46° W	2500	Tessin and Lund (2013)	
20	KNR159-5 125GGC	30° S	45° W	3589	Tessin and Lund (2013); Hoffman and Lund (2012)	
21	RC11-83	41° S	9° E	4718	Charles et al. (1996)	
22	MD01-2588	41° S	25° E	2907	Ziegler et al. (2013)	
23	74KL	14° N	57° E	3212	Sirocko et al. (1993)	
24	NIOP905	10° N	52° E	1580	Jung et al. (2009)	
25	MD97-2120	45° S	174° E	1210	Pahnke and Zahn (2005)	
26	W8709A-13PC	42° N	126° W	2710	Lund et al. (2011); Mix et al. (1999)	

affect atmospheric circulation and wind stress in the Southern Ocean to such a degree that the outgassing of CO_2 contributes importantly to the total CO_2 rise during HS1. We do this by applying realistic wind stress changes from a coupled ocean–atmosphere model simulation of an AMOC shutdown to a global carbon cycle model including isotopes.

2 Methods

We have compiled 25 published deep-ocean records covering the early deglaciation at a high temporal resolution (Table 1). Mostly published age models are used, except in some cases where the radiocarbon calibration was updated as described in Lund et al. (2014). In order to be consistent with the treatment of the other cores in Lund et al. (2014), we have updated the age model of MD01-2461 by recalibrating the radiocarbon ages using INTCAL13 and reservoir ages estimated by Stern and Lisiecki (2013). The ages may have considerable (O(1 ka)) uncertainties. However, we believe that the records are of sufficient resolution and their age models are well enough constrained to evaluate multimillennial changes. The purpose of this paper is to present an initial comparison to model results focusing on model analysis. The quantification of age uncertainties and their effects on the results are beyond the scope of this paper. The sediment data compilation is available in the supplement to this paper.

We employ the Model of Ocean Biogeochemistry and Isotopes (MOBI 1.4), a coupled climate–biogeochemical system that includes $\delta^{13}C$ cycling in the three-dimensional ocean, land, and atmosphere to explore the effect of AMOC variations on carbon isotopes (see the Appendix for a more detailed model description). MOBI's large-scale ocean distribution of $\delta^{13}C_{DIC}$ in dissolved inorganic carbon (DIC) is consistent with modern water column observations (Schmittner et al., 2013). It is embedded in the University of Victoria climate model of intermediate complexity version 2.9 and run to a *preindustrial* equilibrium with prognostic atmospheric CO_2 and $\delta^{13}C_{CO_2}$. Subsequently four numerical experiments, each ~ 3500 years long, were conducted with varying amplitudes (0.05, 0.1, 0.15, and 0.2 Sv; referred to as FW0.05, FW0.1, FW0.15, and FW0.2, respectively; Sv stands for Sverdrup, i.e. 10^6 m^3 s^{-1}) of a stepwise, 400-year-long freshwater input to the North Atlantic between 45–65° N and 60–0° W (Fig. 1a). The added freshwater is not compensated for elsewhere, but it affects surface tracer concentrations though dilution. Lower salinity and increased buoyancy of surface waters causes the AMOC to slow down. Note that these are idealized experiments, designed to examine only how AMOC variations impact the global $\delta^{13}C$ distribution. Realistic initial conditions for the LGM are currently not available. Thus, we do not attempt realistic deglacial simulations. However, it is well known that the $\delta^{13}C$ distribution of the LGM ocean (Curry and Oppo, 2005; Gebbie, 2014)

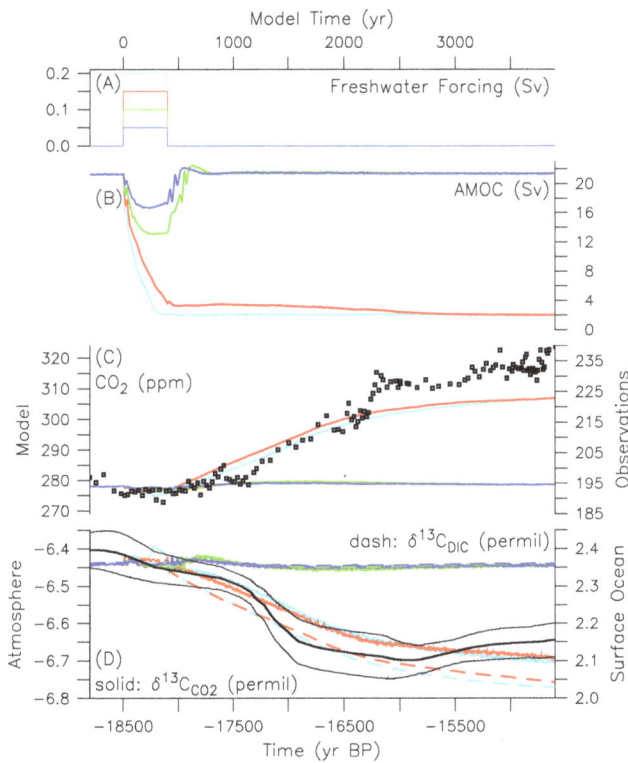

Figure 1. Time series of (**A**) North Atlantic freshwater forcing, (**B**) AMOC response, (**C**) atmospheric CO_2 concentrations, and (**D**) $\delta^{13}C$ of atmospheric CO_2 (solid, left axis) and global mean surface ocean $\delta^{13}C_{DIC}$ (dashed, right axis) for four model simulations (color lines). Symbols in (**C**) and thick black curve (error estimates are indicated by thin lines) in (**D**) show ice core measurements (Marcott et al., 2014, Schmitt et al., 2012) (bottom and right axes in (**C**)).

and atmospheric CO_2 concentrations (Monnin et al., 2001; Lourantou et al., 2010; Parrenin et al., 2013; Marcott et al., 2014) were different from the preindustrial. In order to account for these differences in initial conditions, our data–model comparison focuses on anomalies rather than absolute values. Possible sensitivity of the results to initial conditions is further discussed in the Discussion section below. Selected model data are available in the supplement to this paper.

3   Results

3.1   Simulated circulation changes

The AMOC reduces in all experiments (Fig. 1b). However, in FW0.05 and FW0.1 the reduction is reversible, and, after hosing is stopped, the AMOC quickly returns to its initial state. In experiments FW0.15 and FW0.2, on the other hand, the AMOC collapses permanently (Fig. 2). Reduced salt input to the deep ocean by North Atlantic Deep Water (NADW) leads to a freshening of the deep ocean and salin-

ification of the surface (not shown here; see Figs. 5 and 6 and Plate 2 in Schmittner et al., 2007a), deeper mixed layers, and decreased stratification in the Southern Ocean (Schmittner et al., 2007a) and North Pacific (Saenko et al., 2005), as illustrated in Fig. 2 by the thickening of isopycnal layers between $26.8 \leq \sigma_\Theta \leq 27.63$.

3.2   Simulated carbon cycle changes

The effect on atmospheric CO_2 in model simulations with partial and short AMOC reductions (FW0.05 and FW0.1) is negligible. In contrast, in the simulations with a large and prolonged AMOC decline (FW0.15 and FW0.2) CO_2 starts to rise about 500 years after the beginning of the hosing. It continues to increase gradually by ~ 25 ppm until year 2000, after which its rate of change slows. The amplitude and rate of change of the simulated CO_2 increase agrees well with the long-term trend of measurements of HS1 air recovered from Antarctic ice (Monin et al., 2011; Parrenin et al., 2013; Marcott et al., 2014), but the model does not reproduce the rapid increase around 16 250 BP.

The simulated atmospheric CO_2 increase in FW0.15 and FW0.2 is due to a loss of ocean carbon to the atmosphere. Initially net primary production (NPP) declines within a few hundred years from 64 to $54 \, \mathrm{Pg \, C \, yr^{-1}}$, consistent with Schmittner (2005; not shown here), which reduces the production of dissolved organic carbon (DOC), whereas dissolved inorganic carbon increases initially until around year 600, after which it starts to decline. By model year 3500 the ocean has lost $\sim 120 \, \mathrm{Pg \, C}$ (Fig. 3d) in FW0.15, most of which ($\sim 90 \, \mathrm{Pg \, C}$) is due to DIC and less ($\sim 30 \, \mathrm{Pg \, C}$) due to DOC. The ocean's DIC loss is caused by a reduced efficiency of the biological pump as indicated by the large loss of remineralized DIC ($\sim 400 \, \mathrm{Pg \, C}$; Fig. 3g), most of which is due to less organic matter oxidation (DIC_{org}; Fig. 4), whereas it is buffered by the increase in preformed DIC due to rising surface ocean DIC and atmospheric CO_2, consistent with previous results and theory (Schmittner and Galbraith, 2008; Ito and Follows, 2005; Marinov et al., 2008a, b).

3.3   Simulated carbon isotope changes

The loss of biologically sequestered, isotopically light ($\delta^{13}C_{org} = -20\permil$) organic carbon increases deep-ocean $\delta^{13}C_{DIC}$ by $\sim 0.06\permil$ (Fig. 3f). Accumulation of this isotopically light carbon in the surface ocean and atmosphere decreases their $\delta^{13}C_{DIC}$ (by $\sim -0.3\permil$) and $\delta^{13}C_{CO_2}$ by $\sim -0.25\permil$, respectively (Fig. 1d). Modeled land carbon storage increases (Fig. 3a) and its average $\delta^{13}C_L$ decreases (Fig. 3c), implying that land cannot be the cause of the atmospheric changes. The simulated atmospheric $\delta^{13}C_{CO_2}$ decline in models FW0.15 and FW0.2 is consistent, both in amplitude and the rate of change, with ice core measurements (Fig. 1d; Schmitt et al., 2012), but the model response may

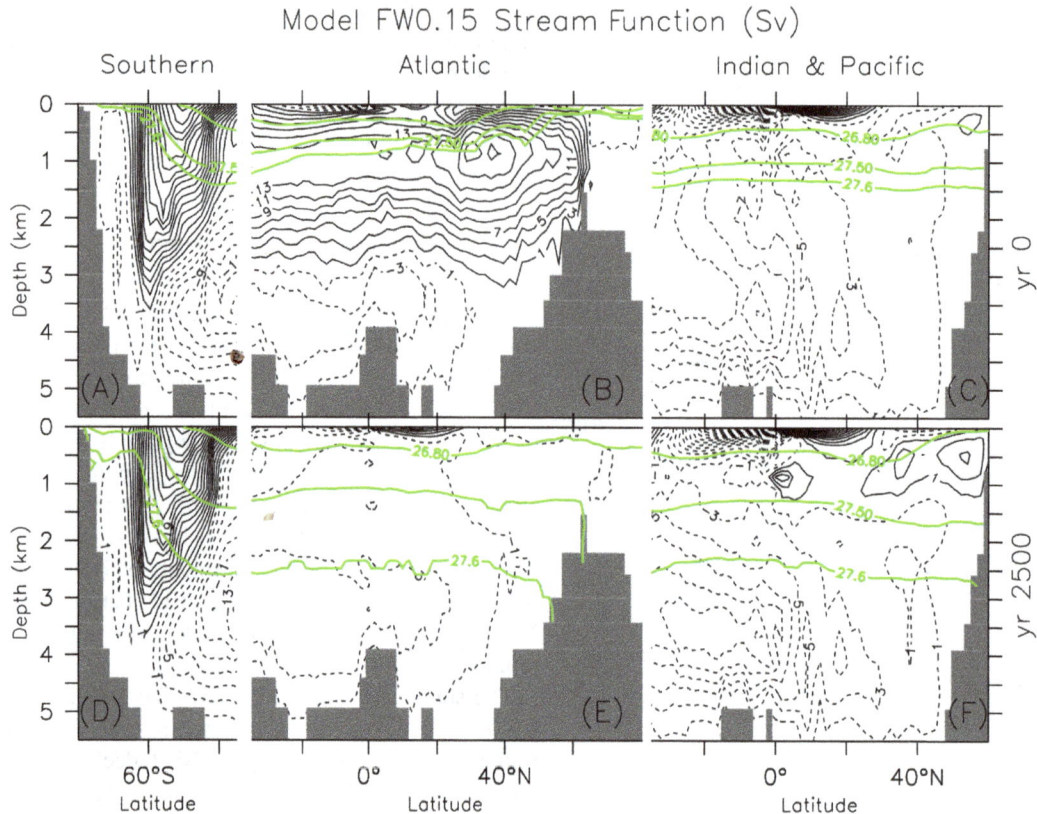

Figure 2. Meridional overturning stream function (black contour lines; solid and dashed lines denote negative and positive values, respectively, and clockwise and counterclockwise flow) in the Southern (**A, D**), Atlantic (**B,E**), and Indian and Pacific (**C, F**) oceans at year 0 (averaged from year −1000 to 0) (**A, B, C**) and 2500 (averaged from year 2000 to 3000) (**D, E, F**). Three zonally averaged potential density (σ_Θ) isolines (27.63, 27.5, 26.8) are shown as green lines.

depend on boundary and/or initial conditions and this agreement may be fortuitous.

The simulated preindustrial (model year 0; Fig. 5a–c) distribution of $\delta^{13}C_{DIC}$ in the ocean is characterized by high values in the surface and deep North Atlantic and low values in the deep North Pacific, consistent with a previous model version and observations (Schmittner et al., 2013). Sinking of isotopically well-equilibrated surface waters with low nutrient and respired carbon content causes high $\delta^{13}C_{DIC}$ values in the deep North Atlantic, whereas aging and accumulation of isotopically light respired organic matter progressively decreases $\delta^{13}C_{DIC}$ as deep waters flow into the South Atlantic and further into the Indian and Pacific oceans. Thus, the modern interbasin difference in deep water $\delta^{13}C_{DIC}$ is caused by the interbasin meridional overturning circulation (MOC) (Boyle and Keigwin, 1982). Hence, as the AMOC collapses, the $\delta^{13}C_{DIC}$ difference between North Atlantic and North Pacific deep waters is reduced (Fig. 5d–f).

Differences between years 2500 and 0 (Fig. 5g–i) show the largest $\delta^{13}C_{DIC}$ decreases at intermediate depths (1–2.5 km) in the northern North Atlantic. Anomalies decrease further south, but a pronounced minimum emerges at the depth of NADW (2–3 km, Fig. 2) in the South Atlantic with positive anomalies below, at the depth of Antarctic Bottom Water, and above, at the depth of Antarctic Intermediate Water. South of 40° S in the Atlantic as well as in the Indian and Pacific oceans, $\delta^{13}C_{DIC}$ increases everywhere below ∼ 500 m due to reduced export of ^{13}C-depleted carbon from the photic zone. The weakening of the biological pump causes surface ocean $\delta^{13}C$ to decrease by 0.2–0.4‰ in the Indian and Pacific basins, possibly explaining the onset of planktonic $\delta^{13}C$ minima on glacial terminations (Spero and Lea, 2002). The deep-ocean signal dominates the global mean $\delta^{13}C_{DIC}$ increase of 0.04‰ by year 2500 (Fig. 3f). In the North Pacific $\delta^{13}C_{DIC}$ shows the largest increase at a depth of around 1 km, owing to reduced stratification and intensified intermediate water formation (Saenko et al., 2005), which decreases the amount of respired carbon and increases remineralized $\delta^{13}C$ ($\delta^{13}C_{rem}$; Fig. 6) there. $\delta^{13}C_{rem}$ increases in most of the deep Pacific, Indian and Southern oceans due to the loss of respired carbon, whereas, in the Atlantic, respired carbon accumulates, leading to a decrease in $\delta^{13}C_{rem}$. Although changes in $\delta^{13}C_{rem}$ dominate the spatial variations of the total $\delta^{13}C_{DIC}$ changes, preformed $\delta^{13}C$ ($\delta^{13}C_{pre}$) variations are not negligible, particularly in the Atlantic, where they decrease by more than 0.3 ‰.

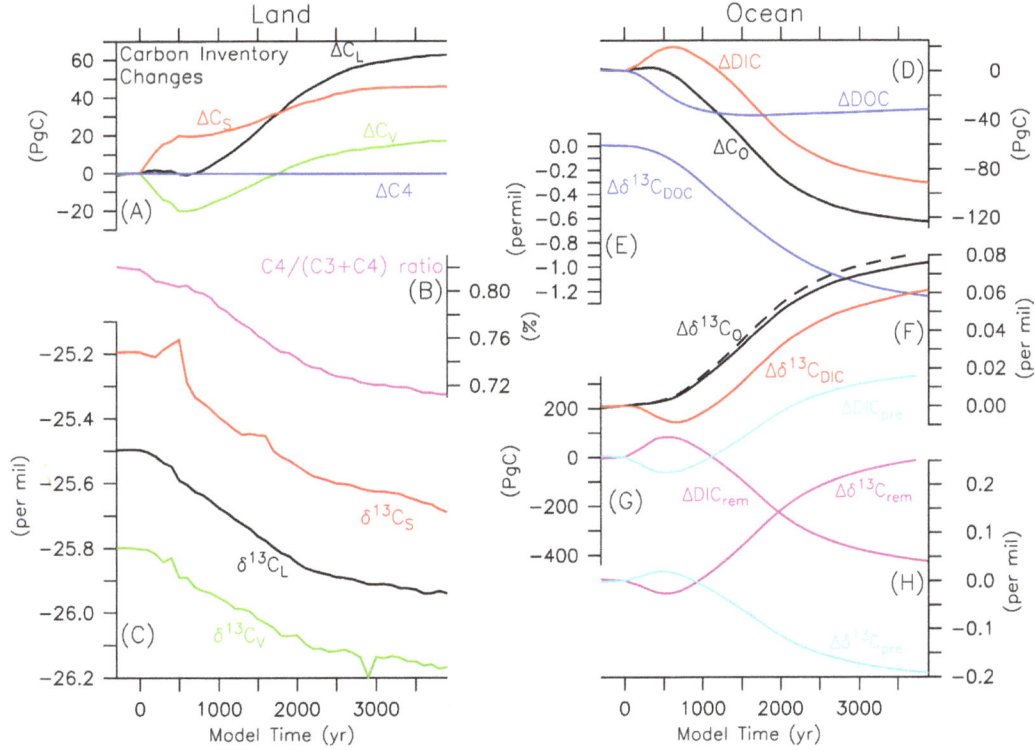

Figure 3. Simulated changes in global land (left) and ocean (right) carbon inventories (in Pg C; 1 ppm = 2.1 Pg C) and their averaged δ^{13}C (in per mil) in model FW0.15. Changes in **(A)** land carbon $\Delta C_L = \Delta C_V + \Delta C_S$ (black) are due to vegetation ΔC_V (green) and soil ΔC_S (red) changes. Vegetation is composed of C_3 and C_4 plants ($C_V = C_3 + C_4$), but C_4 plants contribute only a small fraction **(B)** to the total. Changes in C_4 plant biomass (blue line in **A**) are negligible compared to those in C_3 plants (difference between green and blue lines). Panel **(C)** shows biomass weighted mean δ^{13}C of the land $\delta^{13}C_L = (\sum_i C_{v,i} \times \delta^{13}C_i + C_s \times \delta^{13}C_s)/(\sum_i C_{v,i} + C_s)$ (black), vegetation $\delta^{13}C_V = (\sum_i C_{v,i} \times \delta^{13}C_i)/(\sum_i C_{v,i})$ (green), and soil $\delta^{13}C_s$ (red). Ocean carbon changes $\Delta C_O = \Delta DIC + \Delta DOC + \underbrace{\Delta POC}_{\sim 0}$ (**D**, black) are due to dissolved organic (DOC, blue) and inorganic (DIC, red) carbon and negligible changes in particulate organic carbon (POC, not shown). Total ocean $\delta^{13}C_O = (DIC \cdot \delta^{13}C_{DIC} + DOC \cdot \delta^{13}C_{DOC})/(DIC + DOC)$ (**F**, black) is dominated by changes in $\delta^{13}C_{DIC}$ (red). $\delta^{13}C_{DOC}$ (blue line in panel **E**) changes play only a minor role for $\Delta\delta^{13}C_O$, as illustrated by the dashed black line in **(F)**, which was calculated assuming a constant $\delta^{13}C_{DOC} = -21.5$ ‰. However, the relative contribution of DOC to C_O decreases by about 10 %, which explains the difference between the solid red and dashed black lines in **(F)**. DIC changes are further separated into remineralized (DIC$_{rem}$, $\Delta\delta^{13}C_{rem}$, purple) and preformed (DIC$_{pre}$, $\Delta\delta^{13}C_{pre}$, light blue) components in **(G)** and **(H)**, following Schmittner et al. (2014). All anomalies are shown relative to model year 0, at which absolute numbers are $C_L = 1785$ Pg C, $C_O = 37\,390$ Pg C, $C_{DIC} = 37\,191$ Pg C, $C_{DOC} = 297$ Pg C, $C_{POC} = 2$ Pg C, $\delta^{13}C_{DIC} = 0.72$ ‰, and $\delta^{13}C_{DOC} = -21.5$ ‰.

Table 2. Statistical indices of comparison for the reconstructed HS1 (15.5–16.5 ka BP) minus LGM (18.5–19.5 ka BP) ocean δ^{13}C changes with those from the model simulations (model years 2000–3000 mean minus years −1000 to 0 mean): correlation coefficients (r), root-mean-squared errors (rms), bias (model mean minus observed mean), and the ratio of model over observed standard deviations (rstd). The number of data points is $n = 25$.

Model	r	rms	bias	rstd
FW0.05	0.76	0.45	0.26	0.02
FW0.1	0.76	0.45	0.26	0.03
FW0.15	0.85	0.49	−0.29	1.75
FW0.2	0.85	0.64	−0.38	2.11

3.4 Observed carbon isotope changes during HS1

Observations from the North Atlantic show large $\delta^{13}C_{DIC}$ decreases early in the deglaciation (Fig. 7a–e; Fig. 8a). The largest amplitudes (~ 1 ‰) are found in high-resolution records from the northern North Atlantic (61° N) at intermediate (1.3–1.6 km) depths (Praetorius et al., 2008; Rickaby and Elderfield, 2005; Thornalley et al., 2010). Further south and in deeper water, the $\delta^{13}C_{DIC}$ decrease is smaller (0.4 to 0.7 ‰) (Vidal et al., 1997; Hodell et al., 2010; Zahn et al., 1997; Skinner and Shackleton, 2004; Labeyrie et al., 2005; Zahn and Stuber, 2002). Changes simulated at the same locations by model experiments FW0.15 and FW0.2, which exhibit multimillennial AMOC collapses, are generally similar in amplitude, albeit somewhat larger and in some cases

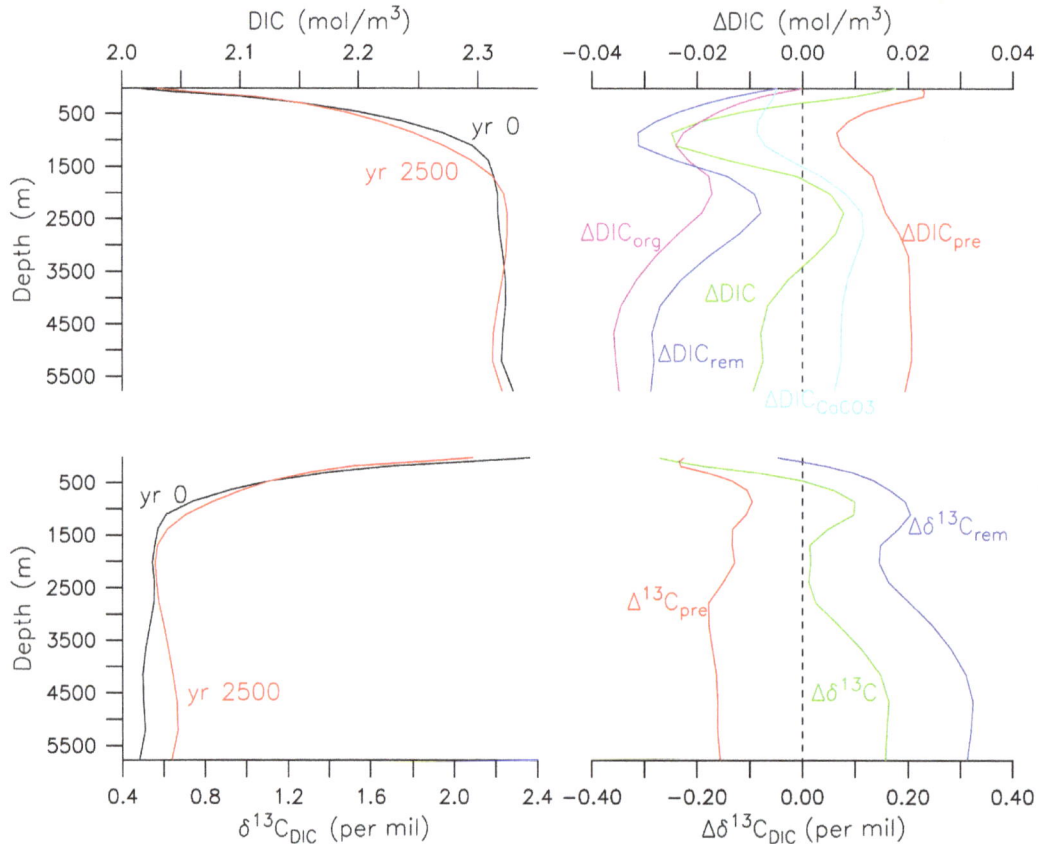

Figure 4. Vertical profiles of globally horizontally averaged ocean DIC (top left) and δ^{13}C (bottom left) at years 0 (black) and 2500 (red) of experiment FW0.15. Right panels show changes (year 2500 minus year 0) in DIC = DIC_{pre} + DIC_{rem} and δ^{13}C = $\delta^{13}C_{pre}$ + $\delta^{13}C_{rem}$ (green) as well as preformed DIC_{pre} and remineralized DIC_{rem} = DIC_{org} + DIC_{CaCO_3}. See Schmittner et al. (2013) for the calculation of the individual terms. The differences between the blue and green lines are due to changes in preformed DIC and δ^{13}C.

earlier. Note that the timing of the rapid $\delta^{13}C_{DIC}$ decrease in core NEAP4K is later than in the two nearby cores ODP984 and RAPID-10-1P, presumably because the age model of NEAP4K is not as well constrained as those from the other two cores.

Despite similar AMOC evolutions, model FW0.15 shows smaller amplitudes than model FW0.2, in better agreement with the reconstructions, due to local effects of the different freshwater forcing on stratification and $\delta^{13}C_{DIC}$. The overall spatial distribution of the observed $\delta^{13}C_{DIC}$ changes, with largest amplitudes at intermediate depths in the northern North Atlantic and decreases further south and in deeper waters, is in best agreement with the results from model FW0.15 (Figs. 7 and 8; Table 2).

A new data set from the Brazil Margin in the South Atlantic (Figs. 7f–k, 9) (Tessin and Lund, 2013; Lund et al., 2015) shows increasing $\delta^{13}C_{DIC}$ by 0.2–0.4‰ between 1.1 and 1.3 km depth and decreasing $\delta^{13}C_{DIC}$ by ~ 0.5‰ between 1.6 and 2.1 km depth, whereas deeper in the water column the data are noisier without a clear trend. Model FW0.15's initial $\delta^{13}C_{DIC}$ values at the Brazil Margin are higher than the observations mainly for two reasons (Fig. 9).

First, the model does not consider the whole ocean lowering of $\delta^{13}C_{DIC}$ due to the reduction in land carbon during the LGM, and, second, it does not include the shoaling of NADW and very low $\delta^{13}C_{DIC}$ values in South Atlantic bottom waters (Curry and Oppo, 2005; Gebbie, 2014). Thus the simulated $\delta^{13}C_{DIC}$ decrease extends deeper than in the observations and shows a substantial reduction below 2.2 km. However, the reconstructed pattern of opposing δ^{13}C signal between shallow–intermediate and mid-depths agrees well with the simulated changes due to large AMOC reductions (Fig. 9). The rapid initial increase at intermediate depths appears to be influenced by two factors. First, there is a reduced return flow of low $\delta^{13}C_{DIC}$ from the Indian ocean (not shown). Second, less upwelling (Schmittner et al., 2005) of low $\delta^{13}C_{DIC}$ deep water into the upper and surface Southern Ocean leads to a deepening of the high $\delta^{13}C_{DIC}$ Antarctic Intermediate and Subantarctic Mode waters, which, together with decreased stratification and deeper mixed layers (Schmittner et al., 2007a; Fig. 2), increases $\delta^{13}C_{DIC}$ by ~ 0.3‰ at 1.2 km depth in all ocean basins at mid-southern latitudes (Fig. 5g–i).

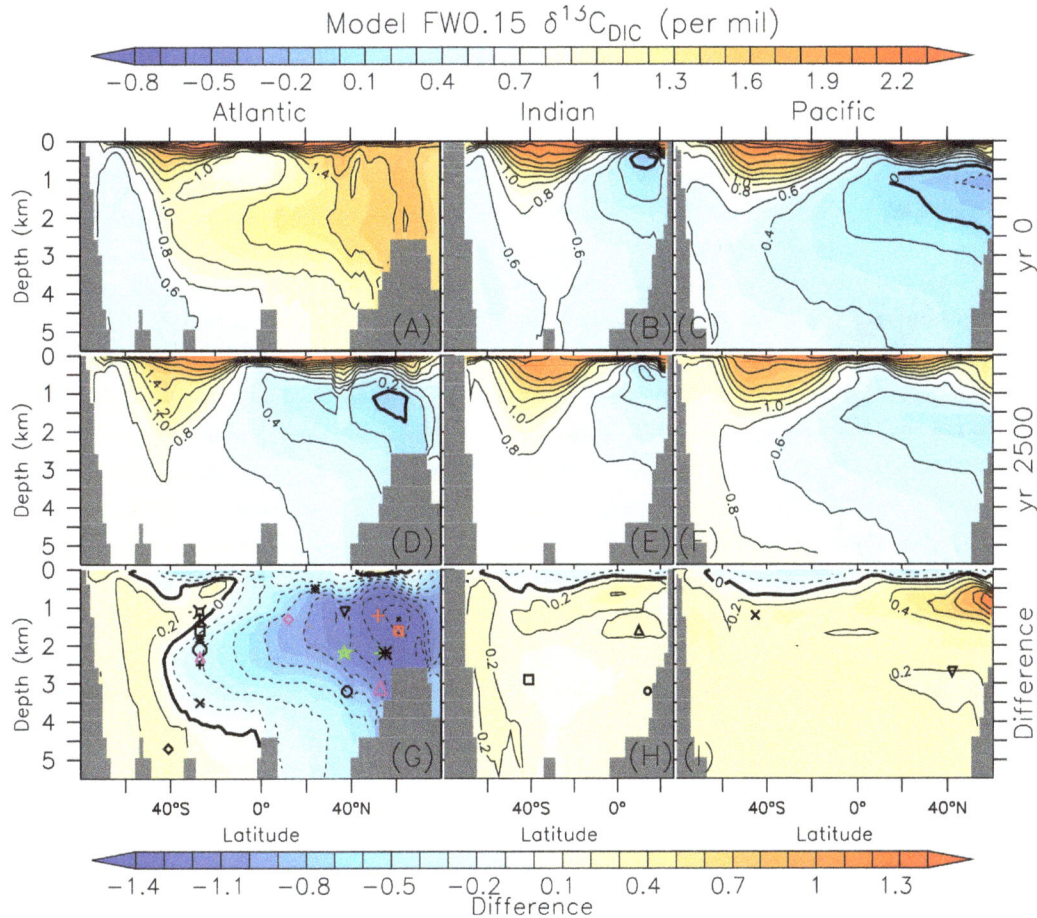

Figure 5. Zonally averaged distributions of $\delta^{13}C_{DIC}$ (color shading and black isolines) as a function of latitude and depth simulated by model FW0.15 in the Atlantic (left), Indian (center) and Pacific (right) ocean basins at model years 0 (**A–C**) and 2500 (**D–F**; **A–F** use top color scale), and the difference (**G–I**; bottom color scale). Symbols in bottom panels denote locations of observations shown in Fig. 7.

The simulated $\delta^{13}C_{DIC}$ increase at 1.2 km depth in the southwest Pacific ($\sim 0.5\,‰$) and at 1.6 km depth in the tropical Indian Ocean ($\sim 0.3\,‰$) agrees well with local reconstructions (Fig. 7p and o), but the simulated changes happen 1.5 ka earlier than in the sediment data. The lack of age model error estimates for the sediment data currently prevents a more detailed assessment of the simulated temporal evolution. In deep waters of the Southern and Indian oceans, the reconstructions are noisy and no clear trend can be identified (Fig. 7l–n). Core W8709A-13PC from the deep eastern North Pacific (Lund et al., 2011) shows no trend in contrast to models FW0.15 and FW0.2. More data from the deep North Pacific are needed in order to better assess model simulations there.

3.5 Sensitivity to wind changes

The model results discussed above did not include the effects of wind changes. Winds enter the UVic (University of Victoria) model in three ways:

1. moisture advection velocities u_q determine convergence of specific humidity and thus precipitation;

2. wind stress τ supplies momentum to the surface ocean and sea ice;

3. wind speed u modulates the air–sea exchange of heat, water, and gases (CO_2, O_2).

Figure 11 shows the annual mean fields derived from the National Center for Environmental Prediction (NCEP) reanalysis (Kalnay et al., 1996) used in the above runs.

In order to test the sensitivity of our results to these variables, we performed three additional simulations, in which we use anomalies calculated from hosing experiments with the Oregon State University/University of Victorial (OSU-Vic) model (bottom panels in Fig. 11). The OSUVic model includes a fully coupled dynamical atmosphere at T42 resolution (Schmittner et al., 2011), whereas the other components are identical to the UVic model version 2.8 without dynamic vegetation. In response to an AMOC shutdown, OS-UVic simulates a large anticyclonic anomaly over the North

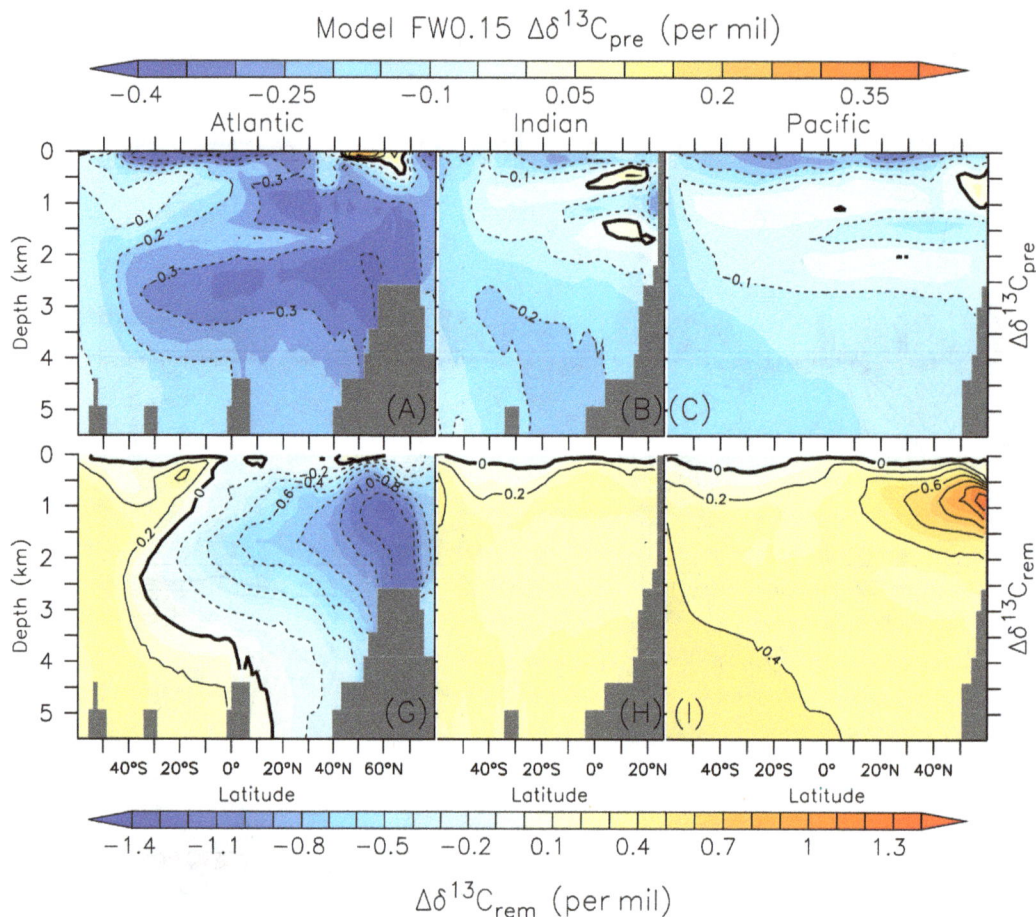

Figure 6. Impact of AMOC collapse on $\delta^{13}C_{pre}$ (top) and $\Delta\delta^{13}C_{rem}$ (bottom). Zonally averaged changes between year 2500 and year 0 of model run FW0.15 in the Atlantic (left), Indian Ocean (center), and Pacific (right). Note the different color scales and isoline differences used. For absolute values of $\delta^{13}C_{pre}$ and $\Delta\delta^{13}C_{rem}$, see Figs. 6 and 12 in Schmittner et al. (2013).

Atlantic, a cyclonic anomaly over the North Pacific, a southward shift of the ITCZ particularly over the Atlantic, and a southward shift of Southern Hemisphere westerlies consistent with previous studies (Timmermann et al., 2007; Zhang and Delworth, 2005; Schmittner et al., 2007b). Note that the changes in Southern Hemisphere westerlies are generally less than 10 % of the absolute values of the control simulation and thus much smaller than those used by Lee et al. (2011).

The OSUVic wind anomalies are applied at model year 400 of the FW0.15 simulation (blue dashed line in panel A of Fig. 12). The wind changes have only a modest impact on simulated carbon cycle and isotope distributions (Fig. 12). The largest effect is due to changes in moisture advection velocities, which lead to a rapid decrease in vegetation and soil carbon around year 400. This causes a rapid CO_2 increase by a few parts per million and a rapid decrease of $\delta^{13}C_{CO_2}$ by a few hundredths of a per mil. It also delays the oceanic carbon loss by a few hundred years. However, the multimillennial response and our conclusions are not impacted much by the wind changes.

4 Discussion

Taken together, the changes in the sedimentary deep ocean $\delta^{13}C_{DIC}$ reconstructions from the LGM to the late HS1 are most consistent with simulations of a severe and prolonged, multimillennial AMOC reduction. Model FW0.15 fits the reconstructions best, as indicated by a high correlation coefficient ($r_{FW0.15} = 0.85$; Fig. 10; Table 2), a root-mean-squared error ($rms_{FW0.15} = 0.49$) just slightly larger than for models FW0.05 and FW0.1, and a standard deviation closest to the observations ($rstd_{FW0.15} = 1.75$). However, $\delta^{13}C_{DIC}$ changes in the North Atlantic are about twice as large in the model than in the reconstructions, which causes the standard deviation in model FW0.15 to be 75 % larger than that of the observations. One reason for this discrepancy may be that AMOC changes during HS1 were smaller than those simulated here (Gherardi et al., 2009; Lund et al., 2015). A second reason could be the mismatch in initial conditions. If the LGM AMOC was weaker and shallower than in the model's preindustrial simulation, as indicated by a number of

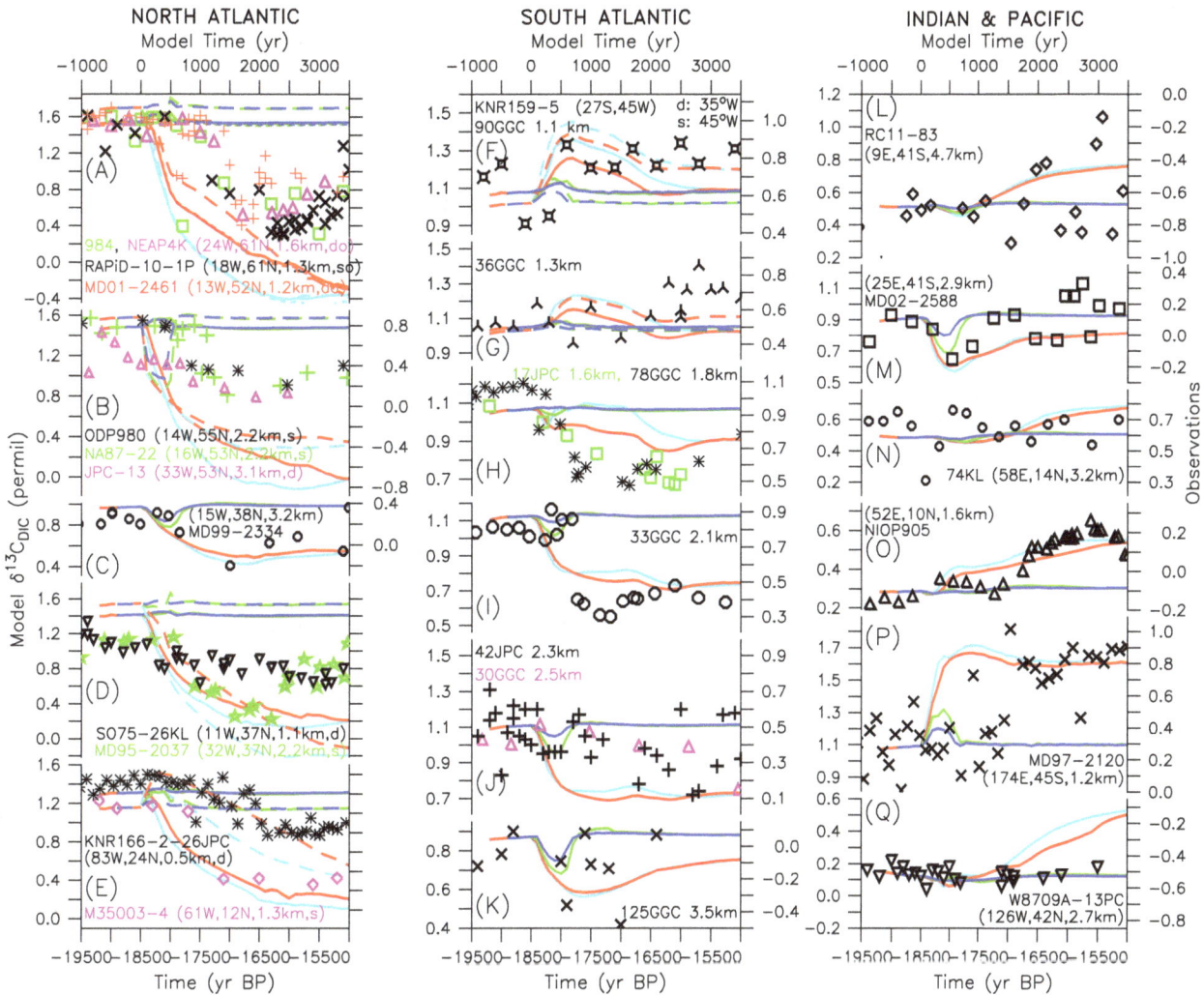

Figure 7. Comparison of simulated (lines as in Fig. 1; left and top axes) and observed (symbols as in Fig. 5; right and bottom axes) $\delta^{13}C_{DIC}$ time series in the North Atlantic (**A–E**), South Atlantic (**F–L**), Indian (**M–O**), and Pacific (**P**) oceans. If no numbers are given on the right axis, the scale is identical to the left axis. If numbers are given on the right axis, the scale is different but the range (max–min) is identical to that of the left axis. Note that different ranges of the vertical axis are used for the different columns, whereas, within each column they are the same.

reconstructions (Lynch-Stieglitz et al., 2007; Gebbie, 2014), the model would overestimate changes in volume fluxes and perhaps carbon isotopes even if a complete AMOC collapse did occur during HS1. A third reason may be biases in the foraminifera-based $\delta^{13}C_{DIC}$ reconstructions, e.g., due to a dependency on carbonate ion concentrations (Spero et al., 1997) or dampened records of the actual $\delta^{13}C_{DIC}$ changes by smoothing and averaging due to bioturbation and/or age model errors. The former would affect particularly regions with large changes in carbonate ion concentrations such as the North Atlantic in the case of an AMOC collapse and the latter may affect particularly low-resolution sediment cores as indicated by reduced agreement with lower-resolution data from a previous study (Sarnthein et al., 1994) ($r_{FW0.15} = 0.80$; $rms_{FW0.15} = 0.60$; Fig. 10). Resolving the likelihood

of these different possibilities will be an important task for future research.

Due to these issues our results can only be regarded as semiquantitative. Qualitatively, they support the interpretation of McManus et al.(2004) of the $^{231}Pa / ^{230}Th$ record, but they cannot rule out the possibility of a reduced but not necessarily collapsed HS1 AMOC based on analyses of Atlantic carbon and oxygen isotope data (Lund et al., 2015; Oppo and Curry, 2015). More work is needed for a truly quantitative reconstruction of early deglacial AMOC changes.

Our simulations suggest that an AMOC decline during HS1 could have caused the observed rise in atmospheric CO_2 and the decrease in $\delta^{13}C_{CO_2}$ by modulating the global efficiency of the ocean's biological pump. This is in contrast with ideas that invoke Southern Ocean (Toggweiler et al.,

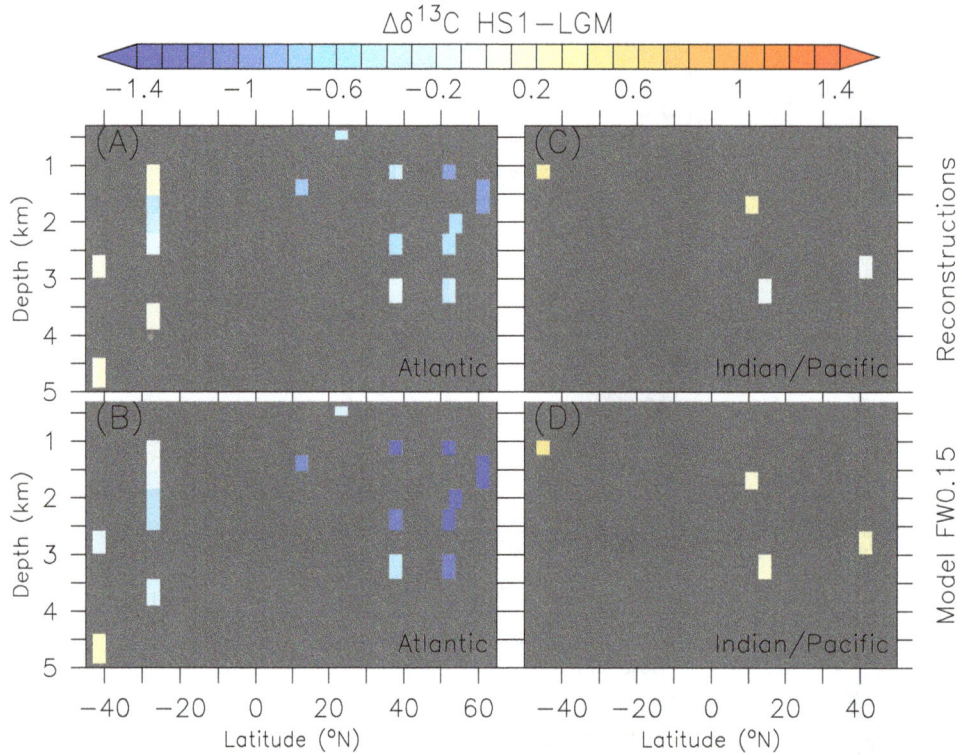

Figure 8. Heinrich Stadial 1 (16.5–15.5 ka BP) minus LGM (19.5–18.5 ka BP) difference in $\delta^{13}C_{DIC}$ in the Atlantic (left) and Indian and Pacific (right) basins from our high-resolution synthesis of reconstructions averaged on the model grid (top) compared to model FW0.15 results (bottom; averages of model years 2000 to 3000 minus averages of model years −1000 to 0.).

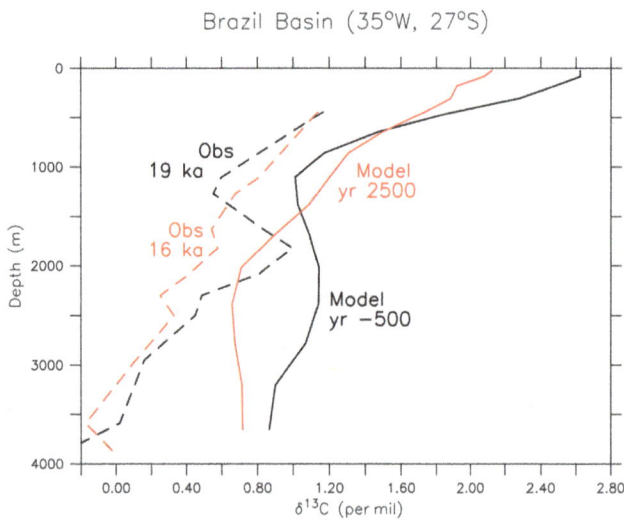

Figure 9. Simulated (solid; model FW0.15) and observed (dashed) vertical profiles of $\delta^{13}C_{DIC}$ at the Brazil Margin in the South Atlantic before (black) and after (red) the AMOC collapse. Observations show 1 ka averages of smoothed (2 ka) data. Results for model FW0.2 are very similar to FW0.15 (not shown). However, models FW0.05 and FW0.1 show almost no changes from their initial (year −500) distribution (not shown).

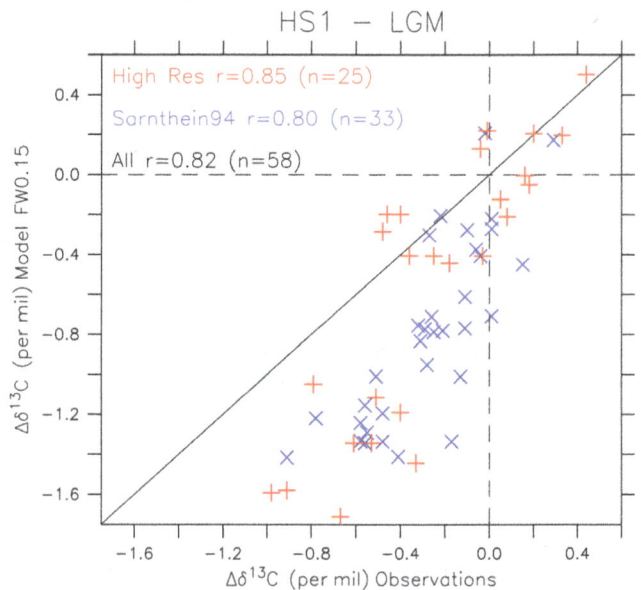

Figure 10. HS1 minus LGM change in $\delta^{13}C$ from Sarnthein et al. (1994; blue) our high-resolution compilation (red) vs. changes between years −500 and 2500 (1000-year centered averages) predicted by model experiment FW0.15 at the same locations. The diagonal 1 : 1 line corresponds to a perfect match.

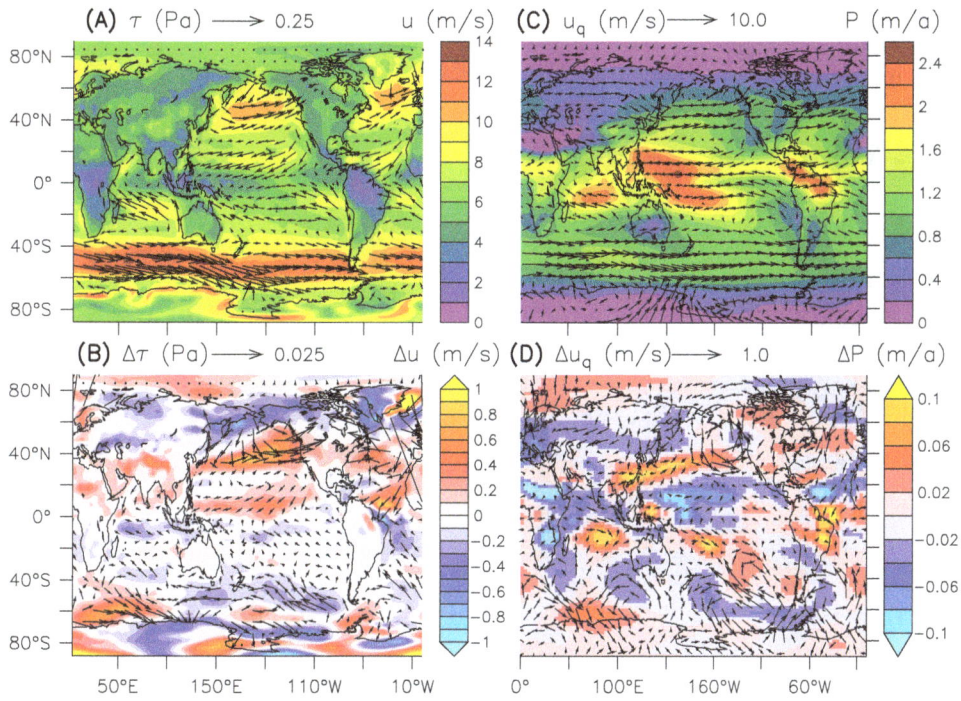

Figure 11. (A) Annual mean wind stress τ (arrows) and wind speed u (color) fields used in the control run of the UVic model. **(B)** Changes in annual mean wind stress $\Delta\tau$ and wind speed Δu derived from a hosing simulation with the OSUVic model. **(C)** and **(D)** as **(A)** and **(B)** but for moisture advection velocities u_q (arrows) and precipitation (color). Note the differences in scales between the top and bottom panels.

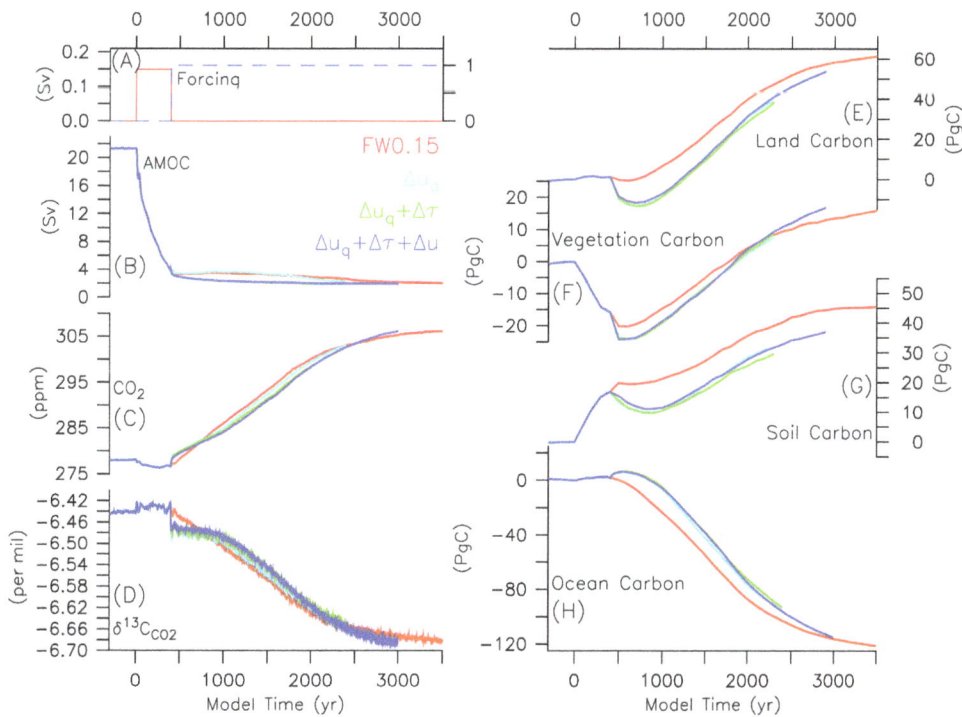

Figure 12. Sensitivity to changes in winds. Experiment FW0.15 (red) is repeated with changes in moisture advection velocities u_q (light blue), u_q plus wind stress τ (green), and $u_q + \tau$ plus wind speed u (dark blue) calculated from the OSUVic model. See Fig. 11 and text for more details.

δ^{13}C Land (per mil)

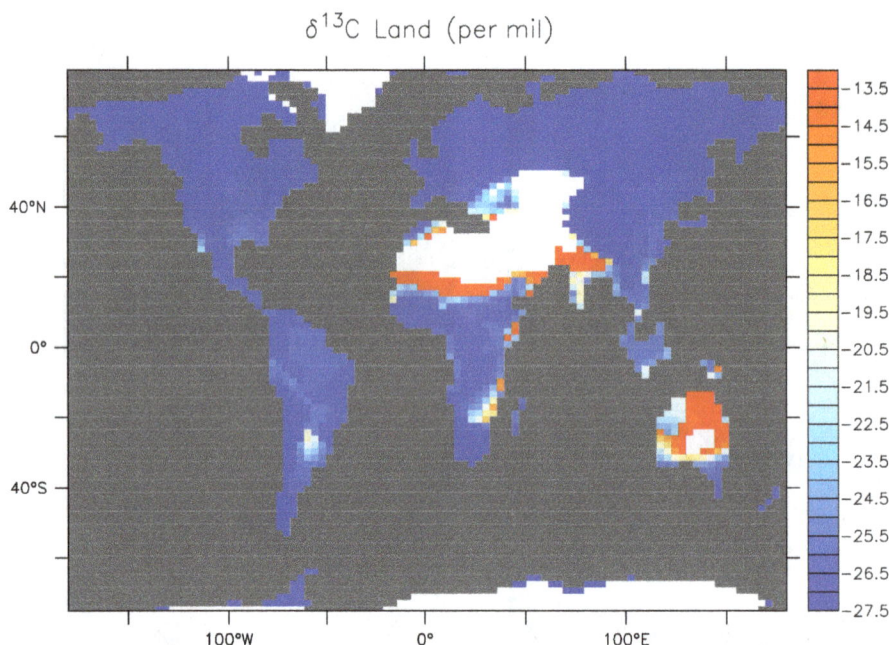

Figure 13. Simulated average preindustrial land δ^{13}C distribution (model year 0). Each pool's (five vegetation plant functional types, PFTs, and one soil, S, carbon compartment) δ^{13}C value was weighted by its mass in calculating the average as explained in the caption for Fig. 2. Desert regions with negligible vegetation carbon ($< 10\,\mathrm{g\,m^{-2}}$) are shown in white.

2006; Anderson et al., 2009; Tschumi et al., 2011; Lee et al., 2011) and/or North Pacific (Menviel et al., 2014) mechanisms for the early deglacial CO_2 rise. However, as discussed above, one possible explanation for the overestimated North Atlantic $\delta^{13}C_{DIC}$ changes in the model is that the early deglacial AMOC changes were smaller. If this was the case, the model could possibly also overestimate the effects on atmospheric CO_2 and $\delta^{13}C_{CO_2}$ and the agreement with the ice core observations could be fortuitous.

A critical test of our hypothesis that the AMOC reduction caused a decrease in the efficiency of the biological pump may come from additional $\delta^{13}C_{DIC}$ reconstructions from the deep Pacific and Indian oceans, which hold most of the ocean's carbon and where the model predicts $\delta^{13}C_{DIC}$ to increase but where few sedimentary data are currently available (Figs. 4 and 7). Indeed, our mechanism relies on changes in the inflow of (low preformed nutrient) Atlantic deep water into the Southern, Indian, and Pacific oceans. Currently it is not known if this inflow was weaker during the LGM. Gebbie (2014) suggests a similar AMOC to the modern one. Kwon et al. (2012) suggest an even stronger influence of North Atlantic water in the global deep ocean at the LGM. These findings may indicate that our simulations with modern initial conditions may also be applicable to the early deglacial, but a solid quantitative assessment remains to be performed. Such an assessment requires simulations with more realistic initial conditions, which will be an important task for future work.

Our wind stress experiments show much smaller impacts on the carbon cycle than those caused by the buoyancy forcing and suggest only a minor effect on the overall rise in atmospheric CO_2 during HS1, but they are subject to the same caveats as discussed above with respect to initial conditions. However, changes in tropical winds associated with ITCZ shifts impact the hydrological cycle and terrestrial carbon and cause a jump of CO_2 by a few parts per million (Fig. 12). Although this is much smaller than the 12 ppm jump recently observed around 16 250 years BP by Marcott et al. (2014), it suggests a mechanism that could explain rapid increases in atmospheric CO_2.

We have not discussed the later parts of the deglacial period such as the Bølling–Allerød (15–13 ka BP), during which the AMOC was presumably reinvigorated (McManus et al. 2004). In our model this would lead to a decrease in atmospheric CO_2, whereas ice core data show stable concentrations (Monin et al., 2011; Parrenin et al., 2013; Marcott et al. 2014), suggesting that an additional process counteracted the AMOC effect. We speculate that this process may be related to a deepening of the AMOC beyond LGM depths and the erosion of the deep South Atlantic reservoir of respired carbon, consistent with recent reconstructions that show that $\delta^{13}C_{DIC}$ decreases there later in the deglaciation (Lund et al., 2015).

5 Conclusions

A comparison of distributed deep-ocean $\delta^{13}C_{DIC}$ reconstructions with our model simulations suggests that, during HS1, the AMOC was substantially reduced for several thousand years. However, due to remaining model–sediment-data differences and a mismatch in initial conditions, we cannot assess the likelihood of a partial AMOC reduction versus a complete shutdown.

Agreement of simulated atmospheric CO_2 and $\delta^{13}C_{CO_2}$ with ice core data, if not fortuitous, supports our hypothesis of an AMOC-induced reduction of the oceans' biological pump during HS1. However, this idea needs further testing with more realistic simulations in the future, improving initial conditions and transient forcing.

AMOC-induced wind changes simulated in a coupled ocean–atmosphere model only have a small impact on the carbon cycle and isotope distributions in our carbon cycle model, suggesting that wind changes were less important than previously thought in controlling atmospheric CO_2 and $\delta^{13}C_{CO_2}$ during HS1. However, effects of wind shifts on the hydrological cycle and terrestrial carbon could explain some of the recently observed rapid CO_2 increases (Marcott et al. 2014).

Appendix A: Model description

The University of Victoria Earth System Climate Model (UVic ESCM) (Weaver et al., 2001), is used in version 2.9 (Eby et al., 2009). It consists of a coarse-resolution ($1.8° \times 3.6°$, 19 vertical layers) ocean general circulation model coupled to a one-layer atmospheric energy–moisture balance model and a dynamic thermodynamic sea ice model, both at the same horizontal resolution. The model is forced with seasonally varying solar irradiance at the top of the atmosphere, cloud albedo, wind stress, wind speed, and moisture advection velocities. This seasonal forcing does not change between different years. Atmospheric CO_2 and $\delta^{13}C$ are calculated in a single box assuming rapid mixing.

A1 Description of land carbon isotopes ($\delta^{13}C$ and $\delta^{14}C$) model

The land carbon isotopes model has not been published before. Therefore, we provide a description and evaluation here. It is based on TRIFFID, the Top-down Representation of Interactive Foliage and Flora Including Dymamics dynamic vegetation model by Cox (2001), as modified by Meissner et al. (2003) and Matthews et al. (2004), which solves prognostic equations for total vegetation carbon density $C_v = {}^{12}C_v + {}^{13}C_v$ and fractional coverage $v_i \in (0, 1)$ of five plant functional types (PFTs; $i = 1, ..., 5$):

$$\frac{\partial}{\partial t}\left(C_{v,i} v_i\right) = v_i \Pi_i - v_i \Lambda_i, \qquad (A1)$$

where Π_i is net primary production (NPP) and Λ_i is litter production, which enters the soil carbon pool. Total soil carbon density is calculated according to

$$\frac{\partial}{\partial t}C_s = \sum_i \Lambda_i - R_s. \qquad (A2)$$

We added prognostic equations for the heavy carbon isotopes ^{13}C and ^{14}C to both vegetation,

$$\frac{\partial}{\partial t}\left({}^{13}C_{v,i} v_i\right) = \gamma_{\Pi,i}^{13} v_i \Pi_i - \gamma_{\Lambda,i}^{13} v_i \Lambda_i, \qquad (A3)$$

and soil,

$$\frac{\partial}{\partial t}{}^{13}C_s = \sum_i \gamma_{\Lambda,i}^{13} \Lambda_i - \gamma_R^{13} R_s, \qquad (A4)$$

$$\frac{\partial}{\partial t}\left({}^{14}C_{v,i} v_i\right) = \gamma_{\Pi,i}^{14} v_i \Pi_i - \gamma_{\Lambda,i}^{14} v_i \Lambda_i - \kappa v_i {}^{14}C_{v,i}, \qquad (A5)$$

and

$$\frac{\partial}{\partial t}{}^{14}C_s = \sum_i \lambda_{\Lambda,i} \Lambda_i - \gamma_R^{14} R_s - \kappa {}^{14}C_{s,i}, \qquad (A6)$$

where fractionation during photosynthesis is indicated by factors

$$\gamma_{\Pi}^{13} = \frac{\beta_{\Pi}^{13}}{1 + \beta_{\Pi}^{13}} \qquad (A7)$$

and

$$\beta_{\Pi}^{13} = \alpha_{\Pi}^{13} R_A^{13}, \qquad (A8)$$

where

$$R_A^{13} = \frac{{}^{13}C_{CO_2}}{{}^{12}C_{CO_2}} \qquad (A9)$$

is the heavy to light isotope ratio of atmospheric CO_2.

Fractionation factors are different for C_3 and C_4 plants:

$$\alpha_{NPP,i}^{13} = \begin{cases} 0.979, & \text{for } C_3 \\ 0.993, & \text{for } C_4 \end{cases}, \qquad (A10)$$

which corresponds to a fractionation of $\varepsilon^{13} = (1 - \alpha^{13}) = -7‰$ for C_4 plants and $\varepsilon^{13} = -21‰$ for C_3 plants (O'Leary, 1988).

No fractionation occurs during litter production or respiration:

$$\gamma_{\Lambda}^{13} = \frac{\beta_{\Lambda}^{13}}{1 + \beta_{\Lambda}^{13}}, \qquad (A11)$$

$$\beta_{\Lambda}^{13} = R_{v,i}^{13} = \frac{{}^{13}C_v}{C_v - {}^{13}C_v}, \qquad (A12)$$

$$\gamma_R^{13} = \frac{\beta_R^{13}}{1 + \beta_R^{13}}, \qquad (A13)$$

$$\beta_R^{13} = R_s^{13} = \frac{{}^{13}C_s}{C_s - {}^{13}C_s}. \qquad (A14)$$

For radiocarbon Eqs. (A5) and (A6), radioactive decay is considered though $\kappa = 1.210 \times 10^{-4} a^{-1}$, which corresponds to a half life of 5730 years, and twice the fractionation during NPP is assumed with respect to $\delta^{13}C$ $\varepsilon^{14} = 2\varepsilon^{13}$, such that

$$\alpha_{NPP,i}^{14} = \begin{Bmatrix} 0.958, \text{for } C_3 \\ 0.986, \text{for } C_4 \end{Bmatrix}. \tag{A15}$$

The simulated spatial distribution of average $\delta^{13}C$ (Fig. 13) varies from $-13\%_o$ in regions dominated by C_4 grasses such as North Africa and Australia to $-27\%_o$ in most other regions, which are dominated by C_3 plants, due to the differences in fractionation factors for C_3 and C_4 plants used in the model. This distribution is broadly consistent with previous independent estimates (Still and Powell, 2010; Powell et al., 2012).

A2   Description of ocean carbon isotope model

We use the Model of Ocean Biogeochemistry and Isotopes (MOBI) version 1.4. The ocean carbon isotope component is described in detail in Schmittner et al. (2013). Here we only describe differences with respect to that publication. The physical UVic model version was updated to version 2.9 (Schmittner et al. (2013) used 2.8). The ocean ecosystem model has been modified by changing zooplankton grazing, using a slightly different approach to consider iron limitation of phytoplankton growth as described in detail in Keller et al. (2012). This model gives very similar results to model FeL (iron limitation), which uses a simple mask to consider iron limitation of phytoplankton growth, in Schmittner et al. (2013).

The implementation of the carbon isotope equations has been changed from the "alpha" formulation to the "beta" formulation, courtesy of Christopher Somes. In the alpha formulation, the change in the heavy (rare) isotope carbon density ^{13}C (in $mol\,C\,m^{-3}$) of the product (e.g., phytoplankton) of a given process (e.g., photosynthesis),

$$\frac{\partial}{\partial t}{}^{13}C = \alpha R^{13}\frac{\partial}{\partial t}{}^{12}C = \alpha R^{13}\frac{\partial}{\partial t}C, \tag{A16}$$

is calculated as the product of the total carbon change $\partial C/\partial t$ times the fractionation factor α for that process times the heavy to light isotope ratio of the source (e.g., sea water DIC) $R^{13} = {}^{13}C/{}^{12}C$. This formulation assumes total carbon,

$$C = {}^{12}C + {}^{13}C \approx {}^{12}C, \tag{A17}$$

is equal to ^{12}C, which is a good approximation since ^{13}C is usually 2 orders of magnitude smaller than ^{12}C.

However, the assumption (Eq. 17) can be avoided by using the beta formulation, in which the heavy isotope change is calculated according to

$$\frac{\partial}{\partial t}{}^{13}C = \frac{\beta^{13}}{1+\beta^{13}}\frac{\partial}{\partial t}C, \tag{A18}$$

where $\beta^{13} = \alpha^{13}R^{13}$.

In order to convert isotope ratios to delta values,

$$\delta^{13}C = (R/R_{std} - 1), \tag{A19}$$

we now use the conventional standard ratio $R_{std}^{13} = 0.0112372$ instead of $R_{std}^{13} = 1$, which was used in Schmittner et al. (2013). For radiocarbon $R_{std}^{14} = 1.17 \times 10^{-12}$ is used.

MOBI 1.4 includes dissolved organic carbon (DOC) cycling described in Somes et al. (2014). The close agreement of the preindustrial $\delta^{13}C_{DIC}$ distributions with model FeL of Schmittner et al. (2013) suggests that none of the changes described above have a major impact on the simulated $\delta^{13}C_{DIC}$.

Acknowledgements. A. Schmittner has been supported by the National Science Foundation's Marine Geology and Geophysics program grant OCE-1131834. Most reconstructions listed in Table 2 were extracted from the National Oceanic and Atmospheric Administration's (NOAA) National Climatic Data Center (NCDC) Paleoclimatology Program Database (http://www.ncdc.noaa.gov/data-access/paleoclimatology-data). We are grateful to all data generators who make their data available on public databases. We are also thankful to Claire Waelbroeck, Rainer Zahn, and Ros Rickaby for generously sharing data used in this study. Thanks to Christopher Somes for rewriting the carbon isotope model in the beta formulation.

Edited by: H. Fischer

References

Anderson, R. and Hayes, C.: New insights into gochemical proxies from GEOTRACES, 11th International Conference on Paleoceanography, Sitges, Spain, 2013.

Anderson, R. F., Ali, S., Bradtmiller, L. I., Nielsen, S. H. H., Fleisher, M. Q., Anderson, B. E., and Burckle, L. H.: Wind-driven upwelling in the Southern Ocean and the deglacial rise in atmospheric CO_2, Science, 323, 1443–1448, 2009.

Bard, E., Rostek, F., Turon, J-L., and Gendreau, S.: Hydrological impact of Heinrich Events in the subtropical northeast Atlantic, Science, 289, 1321–1324, 2010.

Benway, H. M., McManus, J. F., Oppo, D. W., and Cullen, J. L.: Hydrographic changes in the eastern sub polar North Atlantic during the last deglaciation, Quaternary Sci. Rev., 29, 3336–3345, 2010.

Blunier, T., Chappellaz, J., Schwander, J., Dällenbach, A., Staffer, B., Stocker, T. F., Raynaud, D., Jouzel, J., Clausen, H. B., Hammer, C. U., and Johnsen, S. J.: Asynchrony of Antarctic and Greenland climate change during the last glacial period, Nature, 394, 739–743, 1998.

Boyle, E. A. and Keigwin, L. D.: Deep circulation of the North Atlantic over the last 200 000 years: Geochemical evidence, Science, 218, 784–787, 1982.

Bozbiyik, A., Steinacher, M., Joos, F., Stocker, T. F., and Menviel, L.: Fingerprints of changes in the terrestrial carbon cycle in response to large reorganizations in ocean circulation, Clim. Past, 7, 319–338, doi:10.5194/cp-7-319-2011, 2011.

Broecker, W. S., Peteet, D. M., and Rind, D.: Does the ocean–atmosphere system have more than one stable mode of operation?, Nature, 315, 21–26, 1985.

Charles, C. D., Lynch-Stieglitz, J., Ninnemann, U. S., and Fairbanks, R. G.: Climate connections between the hemisphere revealed by deep sea sediment core ice core correlations, Earth Planet. Sc. Lett., 142, 19–27, 1996.

Clark, P. U., McCabe, A. M., Mix, A. C., and Weaver, A. J.: Rapid rise of sea level 19 000 years ago and its global implications, Science, 304, 1141–1144, doi:10.1126/Science.1094449, 2004.

Cox, P.: Description of the "TRIFFID" Dynamic Global Vegetation Model, Hadley Center, Technical Note 24, 1–16, 2001.

Crowley, T. J.: North Atlantic Deep Water cools the southern hemisphere, Paleoceanography, 7, 489–497, 1992.

Curry, W. B. and Oppo, D. W.: Glacial water mass geometry and the distribution of delta C-13 of Sigma CO_2 in the western Atlantic Ocean, Paleoceanography, 20, PA1017, doi:10.1029/2004PA001021, 2005.

Denton, G., Broecker, W. S., and Alley, R. B.: The mystery invervall 17.5 to 14.5 kyr ago, PAGES News, 14, 14–16, 2006.

Denton, G. H., Anderson, R. F., Toggweiler, J. R., Edwards, R. L., Schaefer, J. M., and Putnam, A. E.: The last glacial termination, Science, 328, 1652–1656, 2010.

Eby, M., Zickfeld, K., Montenegro, A., Archer, D., Meissner, K. J., and Weaver, A. J.: Lifetime of anthropogenic climate change: millennial time scales of potential CO_2 and surface temperature perturbations, J. Climate, 22, 2501–2511, 2009.

EPICA community members: One-to-one coupling of glacial climate variability in Greenland and Antarctica, Nature, 444, 195–198, 2006.

Gebbie, G.: How much did glacial North Atlantic water shoal?, Paleoceanography, 29, 190–209, 2014.

Gherardi, J., Labeyrie, L., Nave, S., Francois, R., McManus, J., and Cortijo, E.: Glacial-interglacial circulation changes inferred from $^{231}Pa/^{230}Th$, Paleoceanogr., 24, PA2204, doi:10.1029/2008PA001696, 2009.

He, F., Shakun, J. D., Clark, P. U., Carlson, A. E., Liu, Z., Otto-Bliesner, B. L., and Kutzbach, J. E.: Northern Hemisphere forcing of Southern Hemisphere climate during the last deglaciation, Nature, 494, 81–85, 2013.

Hodell, D. A., Evans, H. F., Channell, J. E. T., and Curtis, J. H.: Phase relationships of North Atlantic ice-rafted debris and surface-deep climate proxies during the last glacial period, Quaternary Sci. Rev., 29, 3875–3886, 2010.

Hoffman, J., and Lund, D. C.: Refining the stable isotope budget for Antarctic Bottom Water: New results from the abyssal southwestern Atlantic, Paleoceanogr., 27, PA1213, doi:10.1029/2011PA002216, 2012.

Ito, T. and Follows, M. J.: Preformed phosphate, soft tissue pump and atmospheric CO_2, J. Mar. Res., 63, 813–839, 2005.

Jung, S. J. A., Kroon, D., Ganssen, G., Peeters, F., and Ganeshram, R.: Enhanced Arabian Sea intermediate water flow during glacial North Atlantic cold phases, Earth Planet. Sc. Lett., 280, 220–228, 2009.

Kalnay, E., Kanamitsu, M., Kistler, R., Collins, W., Deaven, D., Gandin, L., Iredell, M., Saha, S., White, G., Woollen, J., Zhu, Y., Chelliah, M., Ebisuzaki, W., Higgins, W., Janowiak, J.,

Mo, K. C., Ropelewski, C., Wang, J., Leetmaa, A., Reynolds, R., Jenne, R., and Joseph, D.: The NCEP/NCAR 40 year reanalysis project, B. Am. Meteorol. Soc., 77, 437–471, 1996.

Keigwin, L. D. and Boyle, E. A.: Did North Atlantic overturning halt 17,000 years ago?, Paleoceanography, 23, PA1101, doi:10.1029/2007PA001500, 2008.

Keller, D. P., Oschlies, A., and Eby, M.: A new marine ecosystem model for the University of Victoria Earth System Climate Model, Geosci. Model Dev., 5, 1195–1220, doi:10.5194/gmd-5-1195-2012, 2012.

Köhler, P., Joos, F., Gerber, S., and Knutti, R.: Simulated changes in vegetation distribution, land carbon storage, and atmospheric CO_2 in response to a collapse of the North Atlantic thermohaline circulation, Clim. Dyn., 25, 689–708, 2005.

Kurahashi-Nakamura, T., Losch, M., and Paul, A.: Can sparse proxy data constrain the strength of the Atlantic meridional overturning circulation?, Geosci. Model Dev., 7, 419–432, doi:10.5194/gmd-7-419-2014, 2014.

Kwon, E. Y., Hein, M. P., Sigman, D. M., Galbraith, E. D., Sarmiento, J. L., and Toggweiler, J. R.: North Atlantic ventilation of "southern-sourced" deep water, Paleoceanography, 27, PA2208, doi:10.1029/2011PA002211, 2012.

Labeyrie, L., Waelbroeck, C., Cortijo, E., Michel, E., and Duplessy, J. C.: Changes in deep water hydrology during the Last Deglaciation, CR Geosci., 337, 919–927, 2005.

Lourantou, A., Lavric, J. V., Kohler, P., Barnola, J. M., Paillard, D., Michel, E., Raynaud, D., and Chappellaz, J.: Constraint of the CO2 rise by new atmospheric carbon isotopic measurements during the last deglaciation, Global Biogeochem Cy, 24, GB2015, doi:10.1029/2009gb003545, 2010.

Lund, D. C., Tessin, A. C., Hoffman, J. L., and Schmittner, A.: Southwest Atlantic watermass evolution during the last deglaciation, Paleoceanography, in review, 2015.

Lund, D. C., Mix, A. C., and Southon, J.: Increased ventilation age of the deep northeast Pacific Ocean during the last deglaciation, Nat. Geosci., 4, 771–774, 2011.

Lynch-Stieglitz, J., Adkins, J. F., Curry, W. B., Dokken, T., Hall, I. R., Herguera, J. C., Hirschi, J. J. M., Ivanova, E. V., Kissel, C., Marchal, O., Marchitto, T. M., McCave, I. N., McManus, J. F., Mulitza, S., Ninnemann, U., Peeters, F., Yu, E. F., and Zahn, R.: Atlantic meridional overturning circulation during the Last Glacial Maximum, Science, 316, 66–69, 2007.

Lynch-Stieglitz, J., Schmidt, M. W., Henry, L. G. Curry, W. B., Skinner, L. C., Mulitza, S., Zhang, R., and Chang, P.: Muted change in Atlantic overturning circulation over some glacial-aged Heinrich events, Nature Geosci., 7, 144–150, 2014.

Marchal, O., Stocker, T. F., and Joos, F.: Impact of oceanic reorganizations on the ocean carbon cycle and atmospheric carbon dioxide content, Paleoceanography, 13, 225–244, 1998.

Marchal, O., Stocker, T. F., Joos, F., Indermühle, A., Blunier, T., and Tschumi, J.: Modeling the concentration of atmospheric CO_2 during the Younger Dryas climate event, Climate Dynamics, 15, 341–354, 1999.

Marcott, S., A., Bauska, T., K., Buizert, C., Steig, E., J., Rosen, J., L., Cuffey, K., M., Fudge, T., J., Severinghaus, J., P., Ahn, J., Kalk, M., L., McConnell, J., R., Sowers, T., Taylor, K., C., White, J., W., C., Brook, E., J.: Centennial-scale changes in the global carbon cycle during the last deglaciation, Nature, 514, 616–619, 2014.

Marinov, I., Follows, M., Gnanadesikan, A., Sarmiento, J. L., and Slater, R. D.: How does ocean biology affect atmospheric pCO_2?, Theory and models, J. Geophys. Res., 113, C07032, doi:10.1029/2007jc004598, 2008a.

Marinov, I., Gnanadesikan, A., Sarmiento, J. L., Toggweiler, J. R., Follows, M., and Mignone, B. K.: Impact of oceanic circulation on biological carbon storage in the ocean and atmospheric pCO_2, Global Biogeochem. Cy., 22, GB3007, doi:10.1029/2007GB002958, 2008b.

Matthews, H. D., Weaver, A. J., Meissner, K. J., Gillett, N. P., and Eby, M.: Natural and anthropogenic climate change: incorporating historical land cover change, vegetation dynamics and the global carbon cycle, Clim. Dynam., 22, 461–479, doi:10.1007/s00382-004-0392-2, 2004.

McManus, J. F., Francois, R., Gherardi, J. M., Keigwin, L. D., and Brown-Leger, S.: Collapse and rapid resumption of Atlantic meridional circulation linked to deglacial climate changes, Nature, 428, 834–837, 2004.

Meissner, K. J., Weaver, A. J., Matthews, H. D., and Cox, P. M.: The role of land surface dynamics in glacial inception: a study with the UVic Earth System Model, Clim. Dynam., 21, 515–537, doi:10.1007/S00382-003-0352-2, 2003.

Menviel, L., Timmermann, A., Mouchet, A., and Timm, O.: Meridional reorganizations of marine and terrestrial productivity during Heinrich events, Paleoceanography, 23, PA1203, doi:10.1029/2007pa001445, 2008.

Menviel, L., Joos, F., and Ritz, S. P.: Simulating atmospheric CO_2, ^{13}C and the marine carbon cycle during the Last GlacialeInterglacial cycle: possible role for a deepening of the mean remineralization depth and an increase in the oceanic nutrient inventory, Quaternary Sci. Rev., 56, 46–68, 2012.

Menviel, L., England, M. H., Meissner, K. J., Mouchet, A., and Yu, J.: Atlantic-Pacific seesaw and its role in outgassing CO2 during Heinrich events, Paleoceanography, 29, 58–70, 2014.

Mix, A. C., Lund, D. C., Pisias, N. G., Boden, P., Bornmalm, L., Lyle, M., and Pike, J.: Rapid climate oscillations in the Northeast Pacific during the last deglaciation reflect northern and southern sources, in: Mechanisms of Global Climate Change at Millennial Time Scales, edited by: Clark, P., Webb, R. S., and Keigwin, L. D., Geophysical Monograph Series, 112, American Geophysical Union, Washington DC, 127–148, 1999.

Monnin, E., Indermühle, A., Dällenbach, A., Flückiger, J., Stauffer, B., Stocker, T. F., Raynaud, D., and Barnola, J.-M.: Atmospheric CO_2 concentrations over the Last Glacial Termination, Science, 291, 112–114, 2001.

Obata, A.: Climate-carbon cycle model response to Freshwater discharge into the North Atlantic, J. Climate, 20, 5962–5976, 2007.

O'Leary, M. H.: Carbon isotopes in photosynthesis, Bioscience, 38, 328–336, 1988.

Oppo, D. W. and Curry, W. B.: What can benthic $\delta^{13}C$ and $\delta^{18}O$ data tell us about the Atlantic circulation during Heinrich Stadial 1?, Paleoceanography, in review, 2015.

Pahnke, K. and Zahn, R.: Southern Hemisphere water mass conversion linked with North Atlantic climate variability, Science, 307, 1741–1746, 2005.

Parrenin, F., Masson-Delmotte, V., Köhler, P., Raynaud, D., Paillard, D., Schwander, J., Barbante, C., Landais, A., Wegner, A., and Jouzel, J.: Synchronous change of atmospheric CO_2 and Antarctic temperature during the last deglacial warming, Science, 339, 1060–1063, 2013.

Peck, V. L., Hall, I. R., Zahn, R., and Scourse, J. D.: Progressive reduction in NE Atlantic intermediate water ventilation prior to Heinrich events: Response to NW European ice sheet instabilities?, Geochem. Geophys. Geosyst., 8, Q01N10, doi:10.1029/2006GC001321, 2005.

Petit, J. R., Jouzel, J., Raynaud, D., Barkov, N. I., Barnola, J. M., Basile, I., Bender, M., Chappellaz, J., Davis, M., Delaygue, G., Delmotte, M., Kotlyakov, V. M., Legrand, M., Lipenkov, V. Y., Lorius, C., Pepin, L., Ritz, C., Saltzman, E., and Stievenard, M.: Climate and atmospheric history of the past 420 000 years from the Vostok ice core, Antarctica, Nature, 399, 429–436, 1999.

Powell, R. L., Yoo, E.-H., and Still, C. J.: Vegetation and soil carbon-13 isoscapes for South America: integrating remote sensing and ecosystem isotope measurements, Ecosphere, 3, art109, doi:10.1890/ES12-00162.1, 2012.

Praetorius, S. K., McManus, J. F., Oppo, D. W., and Curry, W. B.: Episodic reductions in bottom-water currents since the last ice age, Nat. Geosci., 1, 449–452, 2008.

Rickaby, R. E. M., and Elderfield, H.: Evidence from the high-latitude North Atlantic for variations in Antarctic intermediate water flow during the last deglaciation, Geochem. Geophy. Geosy., 6, Q05001, doi:10.1029/2004GC000858, 2005.

Saenko, O., Schmittner, A., and Weaver, A. J.: The Atlantic–Pacific Seesaw, J. Climate, 17, 2033–2038, 2004.

Sarnthein, M., Winn, K., Jung, S. J., Duplessy, J. C., Labeyrie, L., Erlenkeuser, H., and Ganssen, G.: Changes in east Atlantic deep-water circulation over the last 30 000 years: eight time slice reconstructions, Paleoceanography, 9, 209–267, 1994.

Schmitt, J., Schneider, R., Elsig, J., Leuenberger, D., Lourantou, A., Chappellaz, J., Köhler, P., Joos, F., Stocker, T. F., Leuenberger, M., and Fischer, H.: Carbon isotope constraints on the deglacial CO_2 rise from ice cores, Science, 336, 711–714, 2012.

Schmittner, A., Saenko, O. A., and Weaver, A. J.: Coupling of the hemispheres in observations and simulations of glacial climate change, Quaternary Sci. Rev., 22, 659–671, 2003.

Schmittner, A.: Decline of the marine ecosystem caused by a reduction in the Atlantic overturning circulation, Nature, 434, 628–633, 2005.

Schmittner, A. and Galbraith, E. D.: Glacial greenhouse-gas fluctuations controlled by ocean circulation changes, Nature, 456, 373–376, 2008.

Schmittner, A., Brook, E. J., and Ahn, J.: Impact of the ocean's overturning circulation on atmospheric CO_2, in: Ocean Circulation: Mechanisms and Impacts, edited by: Schmittner, A., Chiang, J. C. H., and Hemming, S. R., Geophysical Monograph Series, 173, American Geophysical Union, Washington DC, 315–334, 2007a.

Schmittner, A., Galbraith, E. D., Hostetler, S. W., Pedersen, T. F., and Zhang, R.: Large fluctuations of dissolved oxygen in the Indian and Pacific oceans during Dansgaard-Oeschger oscillations caused by variations of North Atlantic Deep Water subduction, Paleoceanography, 22, PA3207, doi:10.1029/2006PA001384, 2007b.

Schmittner, A., Silva, T. A., Fraedrich, K., Kirk, E., and Lunkeit, F.: Effects of mountains and ice sheets on global ocean circulation, J. Climate, 24, 2814–2829, 2011.

Schmittner, A., Gruber, N., Mix, A. C., Key, R. M., Tagliabue, A., and Westberry, T. K.: Biology and air–sea gas exchange controls on the distribution of carbon isotope ratios (δ^{13}C) in the ocean, Biogeosciences, 10, 5793–5816, doi:10.5194/bg-10-5793-2013, 2013.

Scholze, M., Knorr, W., and Heimann, M.: Modelling terrestrial vegetation dynamics and carbon cycling for an abrupt climatic change event, The Holocene, 13, 327–333, 2003.

Shakun, J. D., Clark, P. U., He, F., Marcott, S. A., Mix, A. C., Liu, Z. Y., Otto-Bliesner, B., Schmittner, A., and Bard, E.: Global warming preceded by increasing carbon dioxide concentrations during the last deglaciation, Nature, 484, 49–54, 2012.

Sigman, D. M., de Boer, A. M., and Haug, G. H.: Antarctic stratification, atmospheric water vapor, and Heinrich Events: A hypothesis for late Pleistocene Deglaciations, in: Ocean Circulation: Mechanisms and Impacts, edited by: Schmittner, A., Chiang, J. C. H., and Hemming, S. R., Geophysical Monograph Series, 173, American Geophysical Union, Washington DC, 315–334, 2007a.

Sirocko, F., Sarnthein, M., Erlenkeuser, H., Lange, H., Arnold, M., and Duplessy, J. C.: Century-scale events in monsoonal climate over the past 24,000 years, Nature, 364, 322–324, 1993.

Skinner, L. C. and Shackleton, N.: Rapid transient changes in Northeast Atlantic deep water ventilation age across termination I, Paleoceanography, 19, doi:10.1029/2003pa000983, 2004.

Sortor, R. N. and Lund, D. C.: No evidence for a deglacial intermediate water Δ^{14}C anomaly in the SW Atlantic, Earth Planet. Sci. Lett., 310, 65–72, 2011.

Spero, H. J. and Lea, D. W.: The cause of carbon isotope minimum events on glacial terminations, Science, 296, 522–525, 2002.

Stern, J. V. and Lisiecki, L. E.: North Atlantic circulation and reservoir age changes over the past 41 000 years, Geophys. Res. Lett., 40, 3693–3697, 2013.

Still, C. J. and Powell, R. L.: Continental-Scale Distributions of Vegetation Stable Carbon Isotope Ratios, edited by: West, J. B., Bowen, G. J., Dawson, T. E., and Tu, K. P., Springer Netherlands, Dordrecht, 179–193, 2010.

Stommel, H.: Thermohaline convection with two stable regimes of flow, Tellus, 13, 224–230, 1961.

Stott, L., Timmermann, A., and Thunell, R.: Southern Hemisphere and deep-sea warming led deglacial atmospheric CO_2 rise and tropical warming, Science, 318, 435–438, 2007.

Tessin, A. C. and Lund, D. C.: Isotopically depleted carbon in the mid-depth South Atlantic during the last deglaciation, Paleoceanography, 28, 296–306, 2013.

Thornalley, D. J. R., Elderfield, H., and McCave, I. N.: Intermediate and deep water paleoceanography of the northern North Atlantic over the past 21 000 years, Paleoceanography, 25, PA1211, doi:10.1029/2009pa001833, 2010.

Timmermann, A., Okumura, Y., An, S. I., Clement, A., Dong, B., Guilyardi, E., Hu, A., Jungclaus, J. H., Renold, M., Stocker, T. F., Stouffer, R. J., Sutton, R., Xie, S. P., and Yin, J.: The influence of a weakening of the Atlantic meridional overturning circulation on ENSO, J. Climate, 20, 4899–4919, 2007.

Timmermann, A., Timm, O., Stott, L., and Menviel, L.: The roles of CO_2 and orbital forcing in driving Southern Hemispheric temperature variations during the Last 21 000 yr, J. Climate, 22, 1626–1640, 2009.

Toggweiler, J. R., Russell, J. L., and Carson, S. R.: Mid-latitude westerlies, atmospheric CO_2, and climate change during the ice ages, Paleoceanography, 21, Pa2005, doi:10.1029/2005pa001154, 2006.

Tschumi, T., Joos, F., Gehlen, M., and Heinze, C.: Deep ocean ventilation, carbon isotopes, marine sedimentation and the deglacial CO_2 rise, Clim. Past, 7, 771–800, doi:10.5194/cp-7-771-2011, 2011.

Vidal, L., Labeyrie, L., Cortijo, E., Arnold, M., Duplessy, J. C., Michel, E., Becque, S., and vanWeering, T. C. E.: Evidence for changes in the North Atlantic Deep Water linked to meltwater surges during the Heinrich events, Earth Planet. Sc. Lett., 146, 13–27, 1997.

Völker, L. and Köhler, P.: Responses of ocean circulation and carbon cycle to changes in the position of the Southern Hemisphere westerlies during the Last Glacial Maximum, Palaeoceaogr., 28, 726–739, 2013.

Waelbroeck, C., Skinner, L. C., Labeyrie, L., Duplessy, J. C., Michel, E., Vazquez Riveiros, N., Gherardi, J. M., and Dewilde, F.: The timing of deglacial circulation changes in the Atlantic, Paleoceanography, 26, PA3213, doi:10.1029/2010pa002007, 2011.

Watson, A. J., and Naveira Garabato, A.: The role of Southern Ocean mixing and upwelling in glacial-interglacial atmospheric CO_2, Tellus B, 58, 73–87, 2011.

Weaver, A. J., Eby, M., Wiebe, E. C., Bitz, C. M., Duffy, P. B., Ewen, T. L., Fanning, A. F., Holland, M. M., MacFadyen, A., Matthews, H. D., Meissner, K. J., Saenko, O., Schmittner, A., Wang, H. X., and Yoshimori, M.: The UVic Earth System Climate Model: model description, climatology, and applications to past, present and future climates, Atmos. Ocean, 39, 361–428, 2001.

Zahn, R. and Stuber, A.: Suborbital intermediate water variability inferred from paired benthic foraminiferal Cd/Ca and delta C-13 in the tropical West Atlantic and linking with North Atlantic climates, Earth Planet. Sci. Lett., 200, 191–205, 2002.

Zahn, R., Schonfeld, J., Kudrass, H. R., Park, M. H., Erlenkeuser, H., and Grootes, P.: Thermohaline instability in the North Atlantic during meltwater events: stable isotope and ice-rafted detritus records from core SO75-26KL, Portuguese margin, Paleoceanography, 12, 696–710, 1997.

Zhang, R. and Delworth, T. L.: Simulated tropical response to a substantial weakening of the Atlantic thermohaline circulation, J. Climate, 18, 1853–1860, 2005.

Ziegler, M., Diz, P., Hall, I. R., and Zahn, R.: Millennial-scale changes in atmospheric CO_2 levels linked to the Southern Ocean carbon isotope gradient and dust flux, Nat. Geosci., 6, 457–461, 2013.

7

Global sensitivity analysis of the Indian monsoon during the Pleistocene

P. A. Araya-Melo, M. Crucifix, and N. Bounceur

Université catholique de Louvain, Earth and Life Institute, Georges Lemaître Centre for Earth and Climate Research, 1348 Louvain-la-Neuve, Belgium

Correspondence to: M. Crucifix (michel.crucifix@uclouvain.be)

Abstract. The sensitivity of the Indian monsoon to the full spectrum of climatic conditions experienced during the Pleistocene is estimated using the climate model HadCM3. The methodology follows a global sensitivity analysis based on the emulator approach of Oakley and O'Hagan (2004) implemented following a three-step strategy: (1) development of an experiment plan, designed to efficiently sample a five-dimensional input space spanning Pleistocene astronomical configurations (three parameters), CO_2 concentration and a Northern Hemisphere glaciation index; (2) development, calibration and validation of an emulator of HadCM3 in order to estimate the response of the Indian monsoon over the full input space spanned by the experiment design; and (3) estimation and interpreting of sensitivity diagnostics, including sensitivity measures, in order to synthesise the relative importance of input factors on monsoon dynamics, estimate the phase of the monsoon intensity response with respect to that of insolation, and detect potential non-linear phenomena.

By focusing on surface temperature, precipitation, mixed-layer depth and sea-surface temperature over the monsoon region during the summer season (June-July-August-September), we show that precession controls the response of four variables: continental temperature in phase with June to July insolation, high glaciation favouring a late-phase response, sea-surface temperature in phase with May insolation, continental precipitation in phase with July insolation, and mixed-layer depth in antiphase with the latter. CO_2 variations control temperature variance with an amplitude similar to that of precession. The effect of glaciation is dominated by the albedo forcing, and its effect on precipitation competes with that of precession. Obliquity is a secondary effect, negligible on most variables except sea-surface temperature.

It is also shown that orography forcing reduces the glacial cooling, and even has a positive effect on precipitation.

As regards the general methodology, it is shown that the emulator provides a powerful approach, not only to express model sensitivity but also to estimate internal variability and detect anomalous simulations.

1 Introduction

Since the pioneering studies of Kutzbach and Street-Perrott (1985), modelling efforts with general circulation models have routinely been used to understand, quantify and identify the causes of past changes in monsoon dynamics.

One general approach to this end has been to perform snapshot experiments for specific time slices in the past. The general circulation model is run with a particular set of initial conditions for a perpetual year for a long computational time until equilibrium is reached. The epoch used for defining the astronomical forcing and boundary conditions is one for which specific efforts are being undertaken to collect observations. This is the general spirit of projects such as COHMAP (Anderson et al., 1988) and PMIP (Braconnot et al., 2007). Specifically, the COHMAP project focused on a series of time slices spaced every 3000 years throughout the deglaciation (Kutzbach and Guetter, 1986; Anderson et al., 1988), while PMIP historically focused on the mid-Holocene and the Last Glacial Maximum, though on this basis an increasing number of periods are being considered, including the Eemian (Braconnot et al., 2008) and the last interglacials (Yin and Berger, 2012).

Based on these experiments, it is now well understood that glacial boundary conditions, typical, for example, of the Last

Glacial Maximum, induce a weakening of moisture transport over the Indian subcontinent and a reduction of precipitation in East Asia (see also Kutzbach and Guetter, 1986; Felzer et al., 1998; Yanase and Abe-Ouchi, 2007; Braconnot et al., 2007). On the other hand, an increase in northern summer insolation compared to a reference state strengthens monsoon dynamics, in agreement with general considerations on the dynamics of heat transport and on the location of the Intertropical Convergence Zone. These effects may be combined. For example, Masson et al. (2000) showed the possibility of intense Indian monsoon under glacial conditions, more specifically stage 6.5, when the astronomical configuration is favourable.

These past climate simulations are often complemented with additional sensitivity experiments. One classical experimental setup consists in considering two end-member states, often the pre-industrial and one well-defined past period, and intermediate configurations for which one or several forcing components are "activated" while the others are left as the pre-industrial configuration (e.g. Felzer et al., 1998; Masson et al., 2000; Yin et al., 2009). Such sensitivity studies will be referred here as "local" approaches, in the sense that only a small set of forcing conditions are explicitly considered out of the space of possible forcings.

Palaeoclimate modelers are also concerned with the phase relationship between forcing and climate. In particular, climatic precession may be seen as a quasi-periodic rotation of the point of smallest Earth–Sun distance (it will be referred to here as the perigee because we work in geocentric coordinates) and the vernal equinox. By considering specific periods in the past for GCM experiments one can only already develop a partial understanding of the phase relationships. Specifically, Braconnot and Marti (2003) showed that an "early-phase" configuration (perigee reached in April) produces a stronger monsoon, which occurs earlier in the year than a "late-phase" configuration (perigee reached in September). Alternatively, Kutzbach et al. (2008) (see also Chen et al., 2011) proposed the use of long transient simulations to study the evolutionary response to orbital forcing of global summer monsoon over the past 280 000 years. They showed that north tropical sea-surface temperature leads June insolation by about 40°. This particular work did not consider CO_2 and ice boundary condition effects. At the time of writing, such experimental setups can only be afforded with fairly low-resolution models (these authors used FOAM) with an acceleration technique: one model year actually represents 100 years of simulation time.

Here, we will experiment with an alternative approach that will enable us to simultaneously document the *sensitivity* of a general circulation model (HadCM3); the independent and combined effects of different forcing components on monsoon dynamics, namely astronomical forcing, CO_2 and ice boundary conditions; and, finally, estimate the phase relationship between monsoon response and insolation forcing.

The starting point of this approach consists in performing an ensemble of snapshot simulations. The ensemble is designed such that experiments span the space of possible forcing configurations that the Earth encountered during the late Pleistocene (ca. the last 800 000 years). For this reason the approach will be qualified as "global"; more specifically, this is a *global sensitivity analysis* because we do not explicitly consider a reference state. Thus, a statistical model is used to estimate the state of the system at any input point within the space spanned by the experiment ensemble. To this end, we consider a statistical model that is commonly referred to as an "emulator" in the statistical literature (O'Hagan, 2006). In particular, the term emulator refers to the following properties (O'Hagan, 2006; Petropoulos et al., 2009):

– it is derived from a small number of model runs filling the entire multidimensional input space;

– once the emulator is built, it is not necessary to perform any additional runs with the model.

The emulator is then used to generate visual diagnostics and numerical indices summarising the sensitivity of the model to the different elements of the forcing.

This technique of emulation is beginning to be commonly used to estimate uncertainties on climate model outputs, given probability distributions on uncertain quantities such as model parameters (Lee et al., 2011) or elements of the forcing (Carslaw et al., 2013). Such approaches may also integrate information from observations following a Bayesian formalism in order to construct posterior distributions of model parameters and update current knowledge on predictive quantities such as climate sensitivity (Holden et al., 2010; Schmittner et al., 2011). The inference model may in particular include a statistical quantity called model discrepancy, used to express the distance between the model and the real world (Sexton et al., 2012).

Compared to this series of works the present objective is a bit different. As stated, we are interested in input quantities which we know varied in the past, though we will assume that they varied sufficiently slowly to justify a hypothesis of quasi-stationarity of the ocean–atmosphere system with the forcing. Our purpose is to estimate the contribution of input factors to the temporal climate variance that can be observed in palaeoclimate records. To this end we refer to the statistical theory of global sensitivity analysis with emulation formalised by Oakley and O'Hagan (2002) based on general principles of global sensitivity analysis (Homma and Saltelli, 1996) and experiment design (Sacks et al., 1989), but adapted to our particular objective.

The paper is structured as follows. Section 2 provides a description of the emulator and the simulations used. The section is admittedly technical and contains material that has been published before in the statistical literature. However, following the practice of recent articles of climate literature (e.g. Lee et al., 2013), we choose to walk the reader through

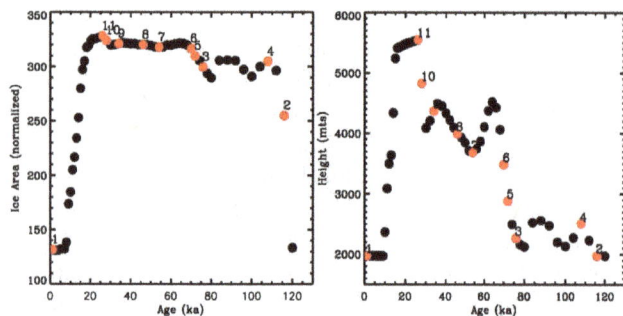

Figure 1. Left panel: ice area, in normalized units, and maximum height (in meters) in the region 45–75° N and 240–275° W (Laurentide Ice Sheet), as a function of time in the boundary conditions used in the Singarayer and Valdes (2010) experiment. Right panel: Height (in meters) of the ice sheets. This shows that, although the volume of the ice masses is quite different, their area is not. Red circles indicate the boundary conditions used for this specific study.

the details of emulation design (see also video in the supplementary material). This also gives us the opportunity to document in detail technical statistical modelling choices. The hasty reader may, however, jump to the Sect. 3, where the results of applying the emulator on the Indian monsoon region are discussed. We focus, on the one hand, on the performance of the emulator as such and, on the other hand, on the climatic lessons emerging from this experiment. In particular, the specific influence of ice sheet topographic forcing is quantified. Conclusions follow in Sect. 4.

2 Methodology

2.1 Experiment design

The first task is to define the space of input configurations to be explored with an ensemble of experiments. We consider five input factors: the three elements of astronomical forcing (eccentricity e, longitude of perigee ϖ, where $\varpi = 0$ when perigee is in March, and obliquity ε), the concentration in carbon dioxide (CO_2), and a variable called the ice or glaciation level, which combines ice and orography forcings associated with the presence of continental ice in the Northern Hemisphere.

The three elements of astronomical forcing are combined under the form of $e\sin\varpi$, $e\cos\varpi$ and obliquity ε. This choice is justified by the fact that these combinations produce orthogonal patterns in the season–latitude space, and generally insolation at any point and time in year is well approximated as a linear combination of those terms (Loutre, 1993). The factors $e\sin\varpi$ and $e\cos\varpi$ are sampled in the range $[-0.05, 0.05]$, while ε is varied in the range 22–25°. Atmospheric CO_2 concentration is sampled in the range 180–280 ppm.

The glaciation level is determined as follows. Our purpose is to select 11 realistic boundary conditions representative of glacial–interglacial dynamics. Pragmatically, we sampled these boundary conditions among the series prepared by Singarayer and Valdes (2010), and kindly supplied to us by Prof. Paul Valdes, University of Bristol. Level 1 corresponds to present-day conditions, and levels 2 to 11 are chosen as such to represent approximately 10 equally spaced top altitudes of the North American Ice Sheet, within the glaciation phase. One limitation of this design for the present purpose is that levels 3 to 11 effectively represent similar ice sheet areas – thus similar albedo forcing – even though they sample very different ice sheet volume (see Fig. 1).

The next step is to define an ensemble of experiments to run with the climate model in order to efficiently span the input space. The choice of the number of experiments and, for each experiment, the choice of input parameters is called the design. A *design point* refers in this context to a specific experiment. The construction of the design should conform to rules of good practice explained, for example, in Santner et al. (2003). In particular, we want the design to be *space filling*, and theoretical considerations and experience point to the Latin hypercube design (McKay et al., 1979; Morris and Mitchell, 1995; Sacks et al., 1989; Urban and Fricker, 2010) as a good starting point. The principle for a Latin hypercube design of n elements is to divide the ranges covered by each input factor into n distinct categories, each experiment sampling one of the n categories without replacement. However, many Latin hypercubes could be constructed in this way, and the design most appropriate for emulation should satisfy additional constraints. Following Santner et al. (2003, p. 167) and Joseph and Hung (2008) we combine two criteria. First, we select, among the possible Latin hypercube designs, those maximising the minimum Euclidean distance found between any two design points. This is called the maxi–min criteria. Among those designs, we chose those maximising the determinant of $X'X$, so that the resulting design is also near-orthogonal.

For this application, two additional constraints need to be accounted for in order to avoid sampling unrealistic inputs that would be uninformative for the sensitivity analysis of climate over the Pleistocene: exclude forcings with $e > 0.05$ and exclude combinations of high CO_2 and high glaciation levels (and conversely), delineated by an ellipse with large and small axes as shown in Fig. 2. To satisfy these constraints, the design points generated by the Latin hypercube sampling procedure lying in the exclusion zone are geometrically projected on the allowed region. This procedure may break some of the original properties of the design (maxi–min and orthogonality), but it offers the practical advantage of enhancing the coverage of the input space near its boundary.

Note that this design is in principle suitable for continuous factor ranges only. The glaciation level used for experiments is an integer obtained by rounding the value obtained by this

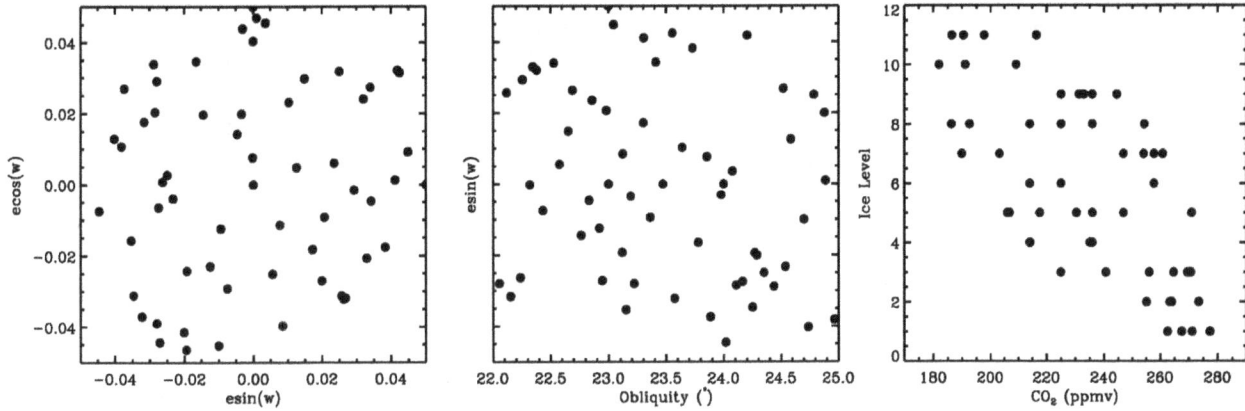

Figure 2. Experiment plan design, optimised to maximise the minimum distance between points and to achieve orthogonality (maximise the determinant of the covariance of input factors). Right: $e\cos\varpi$–$e\sin\varpi$ space distribution; middle: $e\sin\varpi$–obliquity space distribution; right: glaciation level–CO_2 space distribution.

Table 1. Experiment setup: simulation name and number, astronomical parameters (eccentricity, longitude of the perigee and obliquity), CO_2 concentration and glaciation level.

No.	Name	e –	ϖ (°)	ε (°)	CO_2 (ppm)	Ice level –	No.	Name	e –	ϖ (°)	ε (°)	CO_2 (ppm)	Ice level –
1	xadba	0.0527	53.52	23.6	277.3	1	32	xadfa	0.0383	334.53	23.8	257.8	6
2	xadbb	0.0520	211.44	22.9	267.5	1	33	xadfb	0.0417	139.99	24.5	214.1	6
3	xadbc	0.0309	218.44	23.1	262.6	1	34	xadfc	0.0480	215.67	23.2	225.0	6
4	xadbd	0.0201	350.24	23.2	271.2	1	35	xadfd	0.0404	140.60	22.1	225.0	6
5	xadka	0.0282	256.84	24.2	264.1	2	36	xadga	0.0301	194.43	22.4	254.1	7
6	xadkb	0.0466	228.06	24.2	263.4	2	37	xadgb	0.0261	208.55	22.9	189.8	7
7	xadkc	0.0411	88.21	23.3	273.5	2	38	xadgc	0.0503	202.65	24.3	260.8	7
8	xadkd	0.0077	358.66	22.3	255.1	2	39	xadgd	0.0389	122.16	22.3	257.8	7
9	xadaa	0.0403	316.14	22.1	270.6	3	40	xadge	0.0345	97.90	23.4	246.8	7
10	xadab	0.0263	271.85	22.2	270.7	3	41	xadgf	0.0362	299.18	22.2	246.8	7
11	xadac	0.0416	140.71	22.7	269.6	3	42	xadgg	0.0440	355.96	24.0	260.9	7
12	xadad	0.0257	167.54	22.6	256.1	3	43	xadgh	0.0422	287.83	24.7	203.2	7
13	xadae	0.0406	167.95	23.1	240.7	3	44	xadha	0.0436	51.20	22.5	192.6	8
14	xadaf	0.0460	305.89	23.9	224.9	3	45	xadhb	0.0333	26.49	22.7	254.3	8
15	xadag	0.0293	93.07	22.3	264.7	3	46	xadhc	0.0461	205.77	24.3	186.2	8
16	xadda	0.0244	323.78	22.8	214.1	4	47	xadhd	0.0386	246.02	23.1	214.1	8
17	xaddb	0.0421	114.71	23.7	214.2	4	48	xadhe	0.0405	38.22	24.8	225.0	8
18	xaddc	0.0253	23.96	23.6	235.9	4	49	xadhf	0.0491	221.00	23.6	235.9	8
19	xaddd	0.0469	1.20	24.9	235.1	4	50	xadia	0.0150	341.91	22.8	244.4	9
20	xadei	0.0000	0.00	23.0	230.4	5	51	xadib	0.0457	78.40	23.0	235.9	9
21	xadej	0.0500	90.00	23.0	230.4	5	52	xadic	0.0226	113.92	23.0	225.0	9
22	xadek	0.0500	0.00	23.0	230.4	5	53	xadid	0.0400	53.05	22.4	232.9	9
23	xadel	0.0000	0.00	24.0	230.4	5	54	xadie	0.0336	143.57	24.9	231.3	9
24	xadea	0.0155	217.23	23.4	205.9	5	55	xadja	0.0452	260.43	24.0	182.0	10
25	xadeb	0.0527	52.54	24.2	235.9	5	56	xadjb	0.0444	319.59	24.4	209.2	10
26	xadec	0.0456	4.52	24.1	206.6	5	57	xadjc	0.0463	192.48	24.7	191.0	10
27	xaded	0.0135	68.81	24.6	246.8	5	58	xadca	0.0350	305.63	24.1	190.5	11
28	xadee	0.0236	260.39	24.5	217.6	5	59	xadcb	0.0137	145.99	23.9	216.4	11
29	xadef	0.0396	285.78	25.0	246.8	5	60	xadcc	0.0250	136.64	23.3	186.4	11
30	xadeg	0.0251	276.28	24.3	271.0	5	61	xadcd	0.0243	75.55	22.9	197.7	11
31	xadeh	0.0404	359.97	23.5	206.9	5							

process to the closest integer. Designs specifically adapted for input spaces mixing categorical and continuous variables could best be implemented in the future (see, for example, MacCalman, 2013, for an up-to-date review).

Table 1 lists the simulations with their input parameters. The choice of 61 members is a conservative implementation of the recommendation of 10 experiments per input factors (Loeppky et al., 2009). In fact, a first 57-member design was produced using the method above, to which 4 members were added (experiments 20–23). These experiments are idealised orbital changes that were performed during the first phase of this project in order to locally explore the model sensitivity to astronomical forcing.

2.2 Climate simulator

The climate model – referred to in this context as the simulator – is the general circulation model HadCM3 (Gordon et al., 2000), using the MOSES2 dynamic land surface scheme (Essery et al., 2003). The atmospheric component dynamics and physics are resolved on a $3.75° \times 2.5°$ longitude–latitude grid. The oceanic component has a horizontal resolution of $1.25° \times 1.25°$.

Initial conditions are the final state of the PMIP2 0K experiment featured in Braconnot et al. (2007). Each simulation is run for 400 years, except for the $xadk\#$ set. Accidentally, the first 200 years did not account for ice sheet topography. This was corrected for the following 200 years. In the case of the $xadk\#$ simulations, they were run for 300 years, accounting for ice sheet topography from the beginning. Typical residual deep-ocean temperature trends are of the order of $10^{-4}\,°C\,year^{-1}$.

The last 100 years of all simulations with orographic forcing were retained for analysis. Over this interval, the top-of-the-atmosphere imbalance ranges between -0.2 and $-0.1\,W\,m^{-2}$. The last 100 years of the experiment section without orographic forcing are also used for an investigation of the specific effect of the orographic forcing (cf. Sect. 3.6).

2.3 Emulator

At this stage we suppose that the simulator HadCM3 has been run for all design points. We now show that it is possible to *estimate*, with quantified uncertainty, the output that one would have obtained by running HadCM3 at any input lying within the parameter space spanned by the design.

To this end, we need to develop a statistical model that can interpolate the outputs obtained with the simulator at the design points. The procedure is akin to geospatial interpolation, except that the input field is here five-dimensional, instead of two- or three-dimensional as in most geospatial applications (cf. video in the supplementary material).

In particular, we follow Oakley and O'Hagan (2002) and use a Gaussian process model, with a Bayesian formalism.

Although there is no strict practice, the term *emulator* is often reserved to such Bayesian meta-models.

The calibration of the emulator is mathematically described as follows. Let x_j be the set of input values of the jth member of the design (here: a vector of which the components are the astronomical forcing, ice level and CO2). The output of the climate model is modelled as a stochastic process combining a global response function (the regressors) with a local component. It is fully specified by the mean $\widetilde{m}$ and a covariance $\widetilde{V}$ function, which have the following priors:

$$\widetilde{m}(x) = h(x)'\beta \tag{1}$$

$$\widetilde{V}(x, x^\star) = \sigma^2 c(x, x^\star) \tag{2}$$

where $c(x, x^*)$ is the Gaussian process correlation function, and σ^2 its variance; $h(x)$ is a $(q \times 1)$ vector of a priori known regression functions; and β is the vector of corresponding regression coefficients. Note that the $()'$ is used to denote a horizontal vector. The definition of the correlation function is given below.

Let $f(x)$ denote the climate model output when run at input vector x. In Bayesian language, we say that the fact of actually running the model at the design n points allows us to *update* our knowledge of $f(x)$ at any input point.

We also need to make a choice regarding the values of β and σ^2. Given that we do not know their true value, we proceed, in the Bayesian way, by defining prior probabilities for these quantities. We would like not to introduce specific information on β and σ^2. Given that σ^2 is a scale factor, theoretical considerations show that the prior $(\beta, \sigma^2) \propto \sigma^{-2}$ is appropriate as a vague prior, i.e., all values of β are a priori equally plausible and the probability density of σ^2 decays in a way that preserves independence on unit choices (Berger et al., 2001).

In these conditions, the *posterior* estimate of $f(x)$ is a Student t distribution with $n - q$ degrees of freedom, with the following mean and variance (Oakley and O'Hagan, 2002; Bastos and O'Hagan, 2009):

$$
\begin{aligned}
m(x) &= h(x)'\hat{\beta} + T(x^\star)'A^{-1}(y - H\hat{\beta}), &(3)\\
V(x, x^\star) &= \hat{\sigma}^2[c(x, x^\star) - T(x)A^{-1}T(x)' &(4)\\
&\quad + P(x)(H'A^{-1}H)^{-1}P(x^\star)'],
\end{aligned}
$$

respectively, with

$$
\begin{aligned}
\hat{\sigma}^2 &= \frac{1}{n-q-2}(y - H\hat{\beta})'A^{-1}(y - H\hat{\beta}) \text{ and} &(5)\\
\hat{\beta} &= (H'A^{-1}H)^{-1}H'A^{-1}y,
\end{aligned}
$$

where y is a matrix of n lines, of which each line gathers the input of the respective experiments; $T(x)_j = c(x, x_j)$; and $P(x) = h(x)' - T(x)A^{-1}H$. In the following, we conveniently approximate the Student t distribution by a normal distribution. Although in principle is true only as $n \to \infty$, is accurate enough in practice for values of $n-q$ larger than 20.

Remember that x_j are the input parameters (astronomical configuration, etc.) of experiment j of the design. Hence, for example, $T(x)_j$ is a scalar, obtained by applying the so-called correlation function defined below between the input vector x – at which one wants to predict the simulator output – and the input x_j of the design. Consequently, the quantity $T(x)$ is treated as an n-component vector, of which the respective components are associated with the different elements of the design. With this framework, the choices of the regression functions $h(x)$ and the Gaussian process correlation function $c(x, x^*)$ are application-dependent. This is where the user has the opportunity to inject knowledge on the expected response of the simulator.

For this application, linear regression is an adequate choice because the seasonal and annual forcings are almost linear with the input factors, except possibly for glaciation level. Hence, $h(x)' = (1, x')$.

The correlation function $c(x, x^*)$ is a linear measure of how informative the simulator output at x is about the simulator output at x^*. It is thus a key component of the emulator. We use here the classical exponential decay (Oakley and O'Hagan, 2002):

$$c(x, x^*) = \exp[-(x' \Lambda^{-2} x^*)]. \tag{6}$$

The scaling matrix Λ is diagonal, with components λ_i called the length scales. The interpretation is thus that the correlation between the outputs of two experiments decreases exponentially as the normalised distance between two input factors decreases. The normalisation factors are the length scales. Intuitively, the length scale may thus be interpreted as a measure of the roughness of the surface response: the larger the length scale, the smoother the response surface (see video animation in the Supplement).

There is a further correction to be accounted for before using this function. The quantity we are interested in emulating is the hypothetic mean of an infinitely long experiment that has perfectly reached the stationary state. In practice, we have to be content with the mean of a finite-length experiment, obtained for a specific set of initial conditions and which may not have perfectly reached the stationary state. The difference between the output of an experiment and the ideal experiment average is expected to be small yet impossible to predict exactly because it may chaotically depend on initial conditions. It may effectively be accounted for in the emulator as follows. Observe that the function $c(x, x^*)$ always appears as filling the elements of a matrix (Eqs. 2 and 4). This matrix is further modified by adding a small element along the diagonal called the nugget ν, which will absorb the effects mentioned about the experiment sample being only an estimate of the stationary state. The error tolerance will be of the order of $\hat{\sigma}^2 \nu$.

The nugget has another benefit: it regularises the problem for large length scales, and it may in particular be shown that posterior means converge to the solution of a linear regression problem for $\lambda_i \to \infty$ (Andrianakis and Challenor, 2012).

The remaining problem is to estimate the hyperparameters λ_i and ν completely. Following Kennedy and O'Hagan (2000), we maximise the emulator likelihood (the expression used here is from Andrianakis and Challenor, 2012):

$$\log L(\nu, \Lambda) = -\frac{1}{2} \left(\log \left(|A| |H^T A^{-1} H| \right) + (n - q) \log(\hat{\sigma}^2) \right).$$

In order to guarantee that the emulator is at least no less informative than would be linear regression, Andrianakis and Challenor (2012) recommend the use of a *penalised likelihood* as follows:

$$\log L^P(\nu, \Lambda) = \log L(\nu, \Lambda) - 2 \frac{\overline{M}(\nu, \Lambda)}{\epsilon \overline{M}(\infty)}, \tag{7}$$

where $\overline{M}(\nu, \Lambda)$ is the mean squared error between the training points and the emulator's posterior mean at the design points, and $\overline{M}(\infty)$ is its asymptotic value at $\lambda_i \to \infty$. We use $\epsilon = 1$.

It is worth noting that, in our case, using the normal likelihood or the penalised one has practically no effect on the results.

2.4 Sensitivity measures

We are now in a position to estimate the simulator output at potentially any input point spanned by the design. It is now possible to develop indices, of which the purpose is to summarise the sensitivity of the simulator to individual or combined factor throughout the whole input space. This is the general idea of global sensitivity analysis.

In particular, one of the early applications of Bayesian emulators (as we use here) was to estimate sensitivity measures to quantify the uncertainty on a simulator output arising from the fact that the inputs are themselves uncertain (Oakley and O'Hagan, 2004). In this context, the uncertain inputs may be quantified by means of a multivariate probability density function $\rho(x)$. The problem of interest here is slightly different because we know how the inputs varied in the past. The theory of global sensitivity analysis may, however, be recycled by giving $\rho(x)$ a frequentist interpretation. In other words, we use $\rho(x)$ to describe the time-wise occupation density of the input space estimated by considering the history of the late Pleistocene.

In particular, the occupation density along the components of the astronomical forcing can be estimated with histograms of long time series generated with known astronomical solutions, such as those presented by Berger (1978). We then consider the following empirical distribution to broadly capture the observed covariance between CO_2 and glaciation level (see Fig. 3):

$$\rho(c^*, i^*) \propto \begin{cases} \mathcal{N}\left(0.5, \frac{3}{8} \begin{pmatrix} 1 & \frac{1}{3} \\ -1 & \frac{1}{3} \end{pmatrix}^2 \right) & \text{where } 0 < c^* < 1, \quad 0 < i^* < 1 \\ 0 & \text{elsewhere,} \end{cases} \tag{8}$$

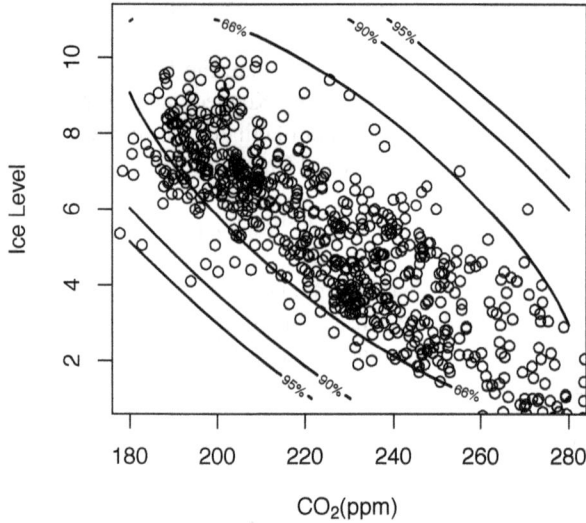

Figure 3. Lines: 66, 90 and 95 % percentiles of the empirical distribution used to describe the probability distribution in the CO_2–ice space (Eq. 8). Dots: observations of CO_2 (Luethi et al., 2008; Siegenthaler et al., 2005; Petit et al., 1999) and estimates of ice level assuming a linear relationship with the LR04 stack of benthic foraminifera $\delta^{18}O$ (Lisiecki and Raymo, 2005) over the last 800 000 years. Based on these observations, the empirical distribution appears to be slightly biased towards high ice level at low CO_2.

where c^* and i^* are inputs standardised as follows:

$$c^* = (CO_2 - 180\,\text{ppm})/(100\,\text{ppm}), \tag{9}$$

$$i^* = (\text{glaciation level} - 1)/10. \tag{10}$$

In order to relate output variances with input variances, we first define what is known in the global sensitivity literature as the *main effect* associated with an input p (e.g. Saltelli et al., 2004, Chapter 1):

$$\eta(\boldsymbol{x}_p) = \int_{\mathcal{X}_{\overline{p}}} f(\boldsymbol{x})\rho(\boldsymbol{x}_{\overline{p}}|\boldsymbol{x}_p)\,\mathrm{d}\boldsymbol{x}_{\overline{p}}, \tag{11}$$

where we have denoted $\mathcal{X}_{\overline{p}}$ as the space spanned by all the components of $\boldsymbol{x}$ but p, and $\rho(\boldsymbol{x}_{\overline{p}}|\boldsymbol{x}_p)$ is the density of occupation of the space $\mathcal{X}_{\overline{p}}$ given the vector $\boldsymbol{p}$. The main effect is thus the expected mean of the simulator output, given a known value of $\boldsymbol{x}_p$ but no more information than the prior on the other components of $\boldsymbol{x}$.

Given that we cannot run the model at every point of the space $\mathcal{X}_p$, this quantity is uncertain, but its mean and variance may be estimated with the emulator:

$$m_p(\boldsymbol{x}_p) = \mathbb{E}_f(\eta(\boldsymbol{x}_p)) = \int_{\mathcal{X}_{\overline{p}}} \rho(\boldsymbol{x}_{\overline{p}}|\boldsymbol{x}_p)\,\mathrm{d}\boldsymbol{x}_{\overline{p}}, \tag{12}$$

$$V_{pp}(\boldsymbol{x}_p, \boldsymbol{x}_p^\star) = \mathbb{V}ar_f(\eta(\boldsymbol{x}_p)) = \tag{13}$$

$$\iint_{\mathcal{X}_{\overline{p}} \times \mathcal{X}_{\overline{p}}} V(\boldsymbol{x}, \boldsymbol{x}^\star)\rho(\boldsymbol{x}_{\overline{p}}|\boldsymbol{x}_p)\rho(\boldsymbol{x}_{\overline{p}}^\star|\boldsymbol{x}_p)\,\mathrm{d}\boldsymbol{x}_{\overline{p}}\mathrm{d}\boldsymbol{x}_{\overline{p}}^\star,$$

where $\mathbb{E}_f$ and $\mathbb{V}ar_f$ denote mean and variance due to using the emulator instead of actually running the simulator at all points. On this basis, it is possible to define two measures of sensitivity of the outputs to input $\boldsymbol{x}_p$:

$$S_p = \mathbb{E}_f \mathbb{V}ar(\eta(\boldsymbol{x}_p)) \quad \text{and} \tag{14}$$

$$\bar{S}_p = \mathbb{E}_f \left[\mathbb{V}ar(\eta(\boldsymbol{x})) - \mathbb{V}ar(\eta(\boldsymbol{x}_{\overline{p}})) \right]. \tag{15}$$

The quantity S_p, called the main effect index[1] is the loss in output variance that would occur assuming that $\boldsymbol{x}_p$ is known and constant, compared to a situation where all factors vary. More precisely, this is the *expected* loss, averaged over all possible values of x_p (e.g. Saltelli et al., 2004, Chapter 1). On the other hand, $\bar{S}_p$ is the output variance that occurs when factor p is variable; all other factors assumed to be known and constant. This is the total effect index[2]. The distinction between main and total effect is particularly important when there is a covariance between input factors. This is the case here: CO_2 and ice volume co-vary. More precisely, the main effect index associated with, for example, ice volume, includes an implicit contribution associated with the fact that CO_2 co-varies with ice level. The total effect index does not include this contribution. Therefore, we use the total effect index.

In order to compute S_p and $\bar{S}_p$, we define the auxiliary quantities:

$$\Sigma_p = \int_{\mathcal{X}_p} \left[m_p(\boldsymbol{x}_p)^2 + V_{pp}(\boldsymbol{x}_p, \boldsymbol{x}_p) \right] \mathrm{d}\rho(\boldsymbol{x}_p), \tag{16}$$

$$\Sigma_0 = \left[m_0(\boldsymbol{x})^2 + V_{00}(\boldsymbol{x}, \boldsymbol{x}) \right], \tag{17}$$

$$\Sigma = \int_{\mathcal{X}} \left[m(\boldsymbol{x})^2 + V(\boldsymbol{x}, \boldsymbol{x}) \right] \mathrm{d}\rho(\boldsymbol{x}), \tag{18}$$

where the subscripts 0 and 00 imply that the space $\mathcal{X}_{\overline{p}}$ referred to in the intergrals (12) and (13) is the full input space. It may then be shown that (Oakley and O'Hagan, 2004)

$$S_p = \Sigma_p - \Sigma_0, \tag{19}$$

$$\overline{S}_p = \Sigma - \Sigma_{\overline{p}}. \tag{20}$$

---

[1]Strictly speaking, the word *index* applies when this quantity is divided by the total output variance.

[2]As above.

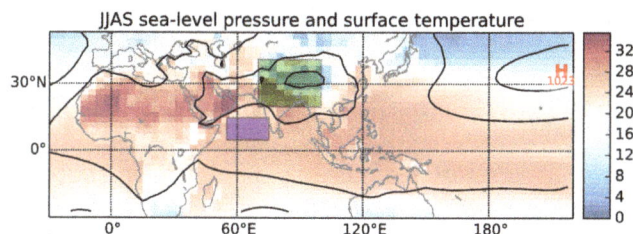

Figure 4. JJAS sea-level pressure and surface temperature of the two regions depicted: NI and IO. Units are in °C.

3 Results

In order to study the Indian monsoon, we define two regions: northern India (NI), with coordinates 70–100° E, 20–40° N, and the northwestern Indian Ocean (IO), with coordinates 55–75° E, 5–15° N (see Zhao et al., 2005). The chosen regions are depicted in Fig. 4, in which the sea-level pressure and surface temperature of one of the simulations are shown. The NI region covers the Indian subcontinent and part of the Tibetan Plateau (which is dry today), while IO covers the northwestern part of the Indian Ocean. In the supplementary material we explore another continental region which does not include the Tibetan Plateau (Chen et al., 2011).

We focus specifically on four physical variables representative of the summer Indian monsoon process: June-July-August-September (JJAS) temperature and precipitation on the continental box, and JJAS sea-surface temperature (SST) and mixed-layer depth on the Indian Ocean box. Over the experiment design, continental temperature varies between 15 and 21 °C. Precipitation varies between 72 and 230 mm month^{-1}, SST between 25 and 31 °C, and mixed-layer depth between 29 and 59 m. For emulation, the logarithms of precipitation and mixed-layer depth are used, because these distributions are more Gaussian than those of the absolute values.

3.1 Emulation validation

An emulator using all 61 experiments is calibrated using the procedure given in Sect. 2.3, with scales λ_i (with $i = 1, \ldots, 5$) and nugget determined by maximisation of the penalised likelihood. The performance of the emulator is then assessed following a leave-one-out cross-validation approach, that is, we construct 60 emulators to predict the experiment being left out. Figure 5 shows the result of this leave-one-out cross-validation procedure for SST and mixed-layer depth only, the other variables being discussed later.

This leads us to the following observations:

1. For $e \sin \varpi$, $e \cos \varpi$ and ice volume, the length scales λ are of the same order of magnitude as the range covered by the input factors. This is the ideal scenario: the space between two experiments is consistent with the decorrelation length of the simulator.

2. There are some instances where length scales are much greater than the scale of the variables: this is observed on all output variables for the response to CO_2 and, to a lesser extent, for obliquity. A large covariance scale implies that response is linear with respect to the factor, which is indeed a realistic outcome for CO_2, in the range considered. This is not a problem on its own. It simply informs the user that a sparser sampling of this factor would have worked as well.

3. The leave-one-out cross-validation plot shows that two experiments are not well captured by the Gaussian process model for SST (experiments 11 and 40), and one for mixed-layer depth (experiment 40). The emulator fails to predict the outputs within an error of less than 3 standard deviations when they are left out of the calibration procedure. The effects of these experiments on the emulator output are well visible in Fig. 6 (top panels). These plots, which will be commented on in more detail in Sect. 3.5, represent the mean model response (Eq. 11) as a function of glaciation level and $e \sin \varpi$, and assuming CO_2 fixed. The figure reveals departure from smooth gradients contours, most notably the 26.25 and 26.5 °C isotherms on the SST plot and the 38.5 m iso-depth that conflict with our expectation of a smooth response structure.

At this stage one could consider an alternative emulator, calibrated on a 59-member experiment design in which the two problematic simulations are omitted.

This new emulator with new scales λ_i and nugget (see Table 2) presents a much more satisfactory performance (Fig. 8):

1. All ancillary emulators constructed for the leave-one-out diagnostic capture between 38 (mixed-layer depth) and 43 (continental temperature) of the leave-one-out experiments within 1 standard deviation, and between 56 and 58 within 2 standard deviations, which roughly correspond to the 66 and 95 % ratios expected for a normal distribution.

2. The normalised errors are compatible with a normal distribution based on the Shapiro–Wilk normality test, except for continental temperature (normality rejected with 97 % confidence).

3. There is no error exceeding 3 standard deviations.

4. Finally, the suspicious anomalies generated on the glaciation/precession plots are cleared (Fig. 6, bottom panels).

Based on our experience with HadCM3 we are inclined to give more credit to this new emulator as a predictor of HadCM3 outputs, rather than the one obtained with simulations 11 and 40. Of course, this choice leaves us with the task

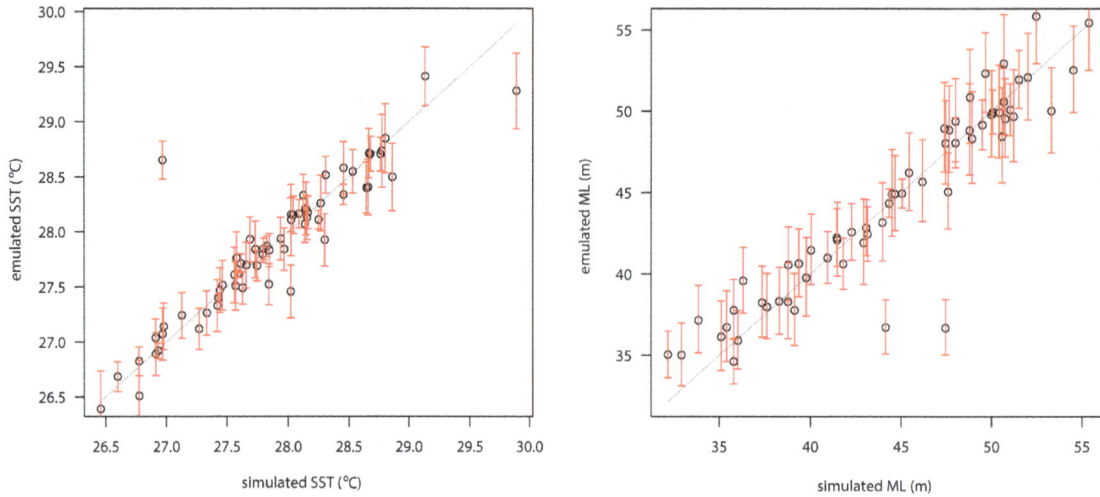

Figure 5. Diagnostic of emulator performance considering experiments 11 and 40. Shown are the mean and standard deviations of sea-surface temperature (left panel) and mixed-layer depth (right panel). Clearly seen are the two bad predictions, especially in the case of sea-surface temperature.

Table 2. Emulator scales for the different fields under study. In general, scales are commensurate with the range covered by the input factors. However, for CO_2 and sometimes obliquity, the scales are much larger than the fields' scale. This simply indicates that the response is linear with respect to the factor.

	Length scales					Nugget
	$\lambda_{e\cos\varpi}$ –	$\lambda_{e\sin\varpi}$ –	λ_{ε} (°)	λ_{CO_2} (ppm)	λ_{ice} –	
Land temperature	0.0704	0.0914	3.191	940	3.348	0.0047
Land precipitation	0.1153	0.3037	20.221	12 588	2.2807	0.0188
Sea surface temperature	0.1118	0.1142	600.	9786	7.307	0.0035
Mixed-layer depth	0.0767	0.0308	3.7724	411	10.6960	0.0439

of explaining what went wrong with these two simulations. It seems that we have to leave it as an open case. Further inspection of these particular experiments reveals a clear warm–cold–warm pattern in the North Atlantic, and cooling over the rest of the ocean, exemplified here by comparing experiments 11 and 15 (Fig. 7). This pattern has been seen before in HadCM3, most notably in early experiments of the Last Glacial Maximum (Hewitt, 2003). It was associated with an enhancement of the North Atlantic Overturning Circulation cell, and can be annealed by addition of freshwater in the North Atlantic (Hewitt et al., 2006). Experiments 11 and 40 have, however, low to moderate glaciation levels, and reasons why their behaviour should differ from the other experiments are far from clear. Based on further inspection of time series as well as that of longer experiments, we are left with the speculation that the particular 100 years used to construct climatic averages correspond to some meta-stable state of the ocean circulation, possibly excited by the spin-up procedure.

Although we appreciate the difficulty, from a statistical inference prospective, of rejecting problematic experiments for the calibration of the emulator, we find it in fact positive that the emulator is effective in identifying experiments that behave unexpectedly compared to the bulk of the design.

Let us now consider the nugget.

As explained, this quantity quantifies the uncertainty of the simulation, i.e. how representative of the mean model state are the 100-year simulations.

The residual error in the emulator is of the order of $\hat{\sigma}^2 \nu$, but it can be estimated precisely by looking at the posterior variance at design points. Here, the obtained nuggets induce residual errors with standard deviations of 0.04 °C on continental temperature, 2.3 % on precipitation, 0.05 °C on SST, and 0.7 % on mixed-layer depth. All these values are consistent with the 100-year variances of the corresponding quantities in HadCM3.

Thus, remarkably, the emulator calibration has successfully estimated model internal variability using only 100-year means, which we take as one more argument to use the recalibrated emulator.

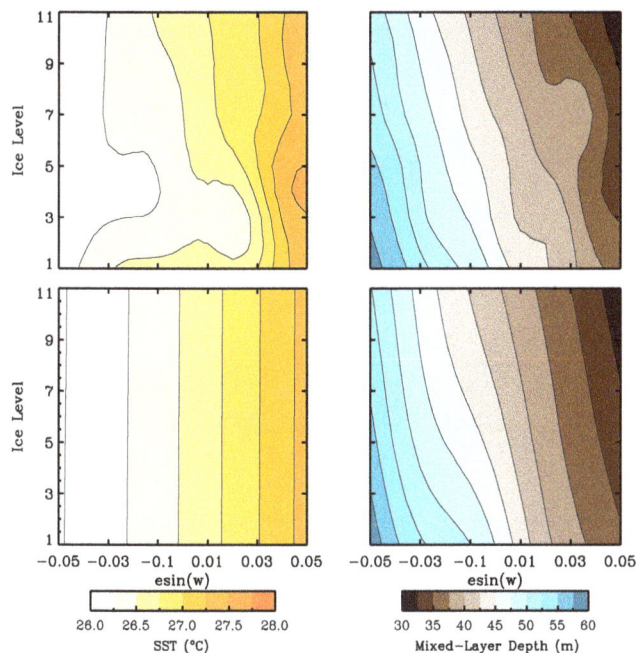

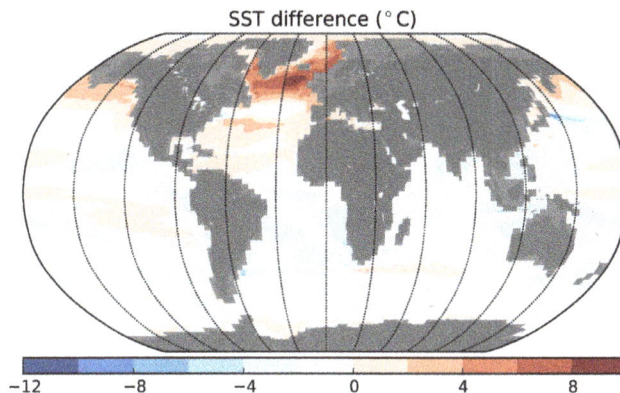

Figure 7. Sea surface temperature difference between simulations 11 and 15 (see Table 1). There is a clear warming pattern in the North Atlantic, which affects the mean sea-surface temperature.

Figure 6. Sensitivity to glaciation level and $e \sin \varpi$ for sea-surface temperature and mixed-layer depth. Top panels: the contour plots include the experiments 11 and 40. The effect of these experiments are clearly visible in both cases, ice level 3 in the case of sea-surface temperature and glaciation level 7 for mixed-layer depth. Bottom panels: the removal of these experiments a smooth response of the emulator, as clearly seen in the contour plots.

3.2 Sensitivity measures

Figure 9 summarises the sensitivities of the four different variables to the external factors. $e \cos \varpi$ and $e \sin \varpi$ are grouped together under the term "precess", for climatic precession.

The figure shows that continental summer temperature is primarily determined by precession, CO_2 and, to a lesser extent, ice volume. It shows no significant sensitivity to obliquity. Continental precipitation is also mainly driven by precession and less to ice volume. In contrast to temperature, it exhibits no sensitivity to CO_2.

Similar to continental temperature, SST is primarily driven by precession and CO_2 and, to a lesser extent, ice volume. It also shows a larger response to obliquity. Finally, mixed-layer depth shows a pattern similar to precipitation, except that the response to obliquity is not significant compared to the sources of uncertainty induced by the emulation and sampling variance.

3.3 Sensitivity to precession

Figure 10 displays the effects of precession on the four variables retained for analysis. The choice here is to show the effects by fixing ice and CO_2 concentration at three distinct levels representative of the course of glaciation (from top to bottom): glaciation level $1/CO_2 = 280$ ppm, glaciation level $5/CO_2 = 230$ ppm and glaciation level $11/CO_2 = 180$ ppm. Quantities are further averaged over obliquity. In order to ease the interpretation, the months representing the time at which perigee is reached are written on the plots: June for $\varpi = 90°$, September for $\varpi = 180°$, etc. That is, neglecting slow transient effects that could be associated with the deep ocean response, this graphical representation provides an indication of the phase lag between the climate response and the precession forcing of insolation.

We see that the temperature response is in phase with June insolation at low glaciation levels, and in phase with July insolation at mid- and high-glaciation stages.

This feature may physically be understood by considering the summer precipitation response. Precipitation enhances latent heat cooling when perigee is around July. This effect gradually weakens as glaciation takes place and the total amount of precipitation declines, hence the drift towards a more linear response. At higher glaciation levels the JJAS temperature response phase also aligns with July insolation.

The maximum precipitation is obtained when perigee is reached in early July. Among the series of experiments shown by Braconnot et al. (2008), it is indeed the 126 000 year BP experiment (i.e. July perigee) experiment that shows the strongest precipitation response over India.

Furthermore, continental precipitation and mixed-layer depth show opposite response phases to precession. This result is consistent with the earlier findings of Zhao et al. (2005), who identified a shoaling of the mixed-layer depth in this region by about 6 m, consistent across different models, in 6000-year experiments (September perigee). Braconnot and Marti (2003) examined also two nearly opposite precession configurations with the IPSL model, corresponding to perigee in April and October, respectively, and they found a shoaling of the mixed-layer depth compared to the present-day (perigee in January) in both cases.

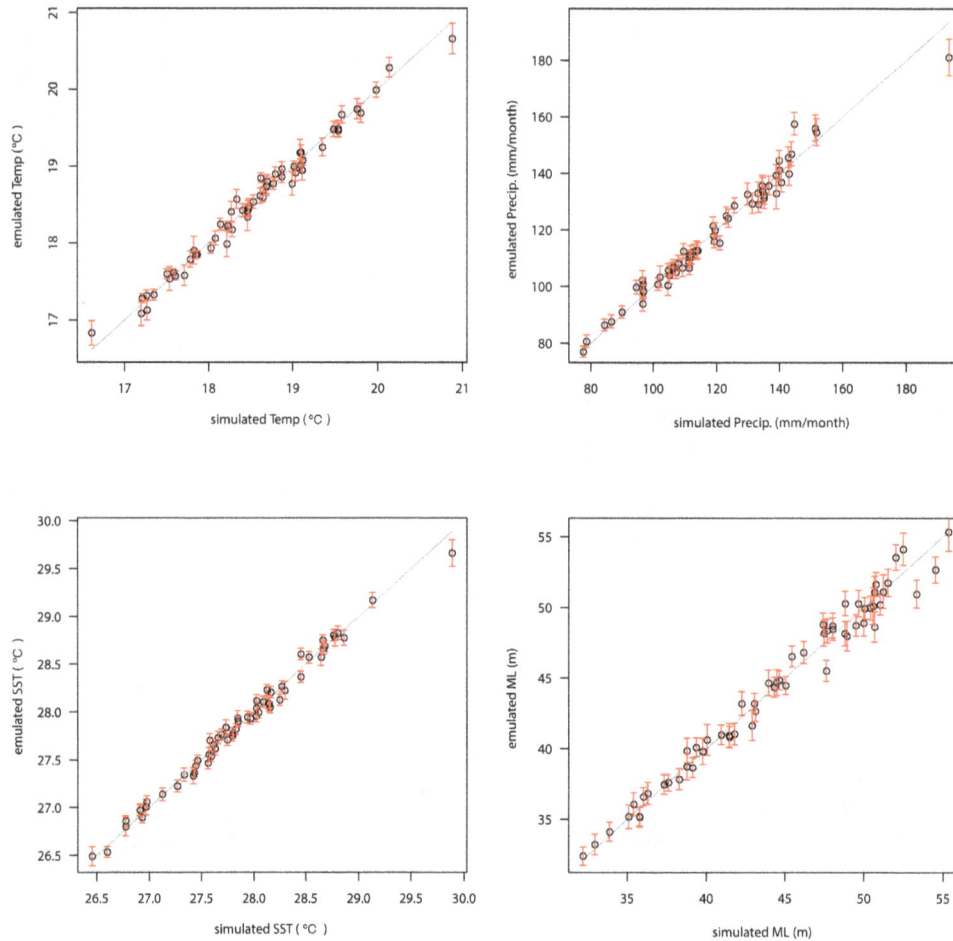

Figure 8. Diagnostic of emulator performance. Shown are the mean and standard deviation of the simulated and the emulated data points for the all the simulations with the exception of simulation number 11 and 40. Top left panel: continental temperature; top right panel: continental precipitation; bottom left panel: sea-surface temperature; bottom right panel: mixed-layer depth.

Zhao et al. (2005) attributed the mixed-layer depth shoaling to a stratification effect involving the response of SST. On this point, our analysis reveals that the maximum SST response occurs when perigee is reached in May. This is not so surprising given that the ocean thermal inertia generally imposes a lag of a few months between the forcing and the response. This response, however, induces an asymmetry between perigee in April and perigee in October, the first one only showing anomalously high SSTs. This is consistent with the analysis of seasonal cycle response provided by Braconnot and Marti (2003).

3.4   Sensitivity to obliquity

The response of obliquity is mostly linear, as we can infer from the high values of the length scales (see Table 2).

The range of obliquity covered during the Pleistocene induces negligible continental temperature response over the west Indian box. It also induces a slight increase in precipitation. Regarding the Indian Ocean box, there is a somewhat

larger effect on SST compared to continental temperature, but not significant. As for the mixed-layer depth, the response to obliquity is negligible.

In order to better understand the effect of obliquity, we considered the four *idealised* experiments (simulations 20–23; see Table 1). In particular, we discuss here experiments 22 and 23, termed OBL23 and OBL24. They use zero eccentricity, the same CO_2 concentration and glaciation level, and differ by the configuration of obliquity (24 and 23°, respectively). The temperature difference map for JJAS reveals the signature of obliquity-induced insolation changes, with a warming of Northern Hemisphere continents, and slight cooling of significant areas of the tropical oceans (see Fig. 11).

3.5   Sensitivity to CO_2 and glaciation level

The response of all variables to CO_2 is best captured by linear processes (optimal λ_i largely exceeds the range covered by the experiment design). Hence, the contribution of CO_2

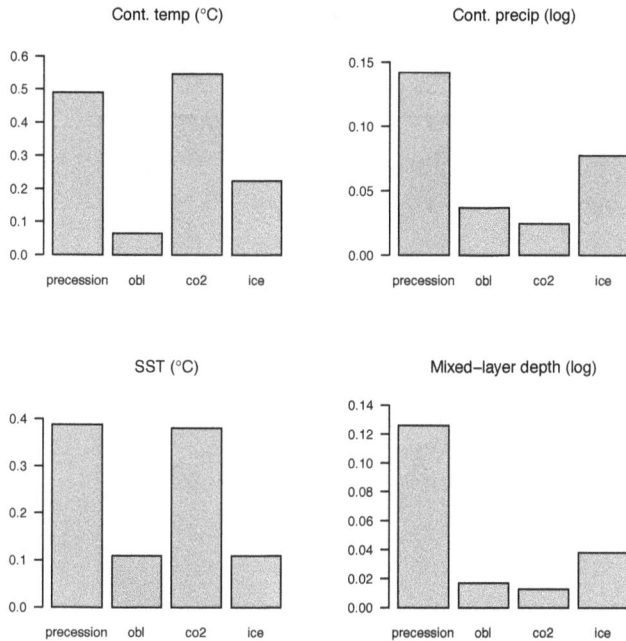

Figure 9. Sensitivity analysis: shown is the standard deviation of model outputs ($\sqrt{S}$) of each variable, induced by variations in input factors during the Pleistocene. From left to right, top to bottom: continental precipitation, continental temperature, sea-surface temperature and mixed-layer depth.

to the climate response may be estimated straightforwardly from the coefficients $\hat{\beta}$, given by Eq. (5). Specifically, the continental temperature and SST responses to the 100 ppm range covered by the experiment design are 2.03 and 1.40 °C, respectively. This corresponds to CO_2 doubling sensitivities of 3.20 and 2.21 °C, in line with the reported HadCM3 sensitivity in CO_2 doubling experiments (see, for example, Fig. 5 of Williams et al., 2001) The responses of precipitation and mixed-layer depth are, again, opposite and very moderate: +6 % of precipitation over 100 ppm and −0.5 % of mixed-layer depth.

Figure 12 shows the response of continental temperature (left panel), sea-surface temperature (middle panel) and mixed-layer depth (right panel) to the variations of CO_2 concentration and glaciation level. The temperature ranges covered by CO_2 and glaciation levels are of the order of 1 and 2 °C for the continent and ocean surface, respectively. The continental ice effect is mainly present between glaciation levels 1 and 3. With the ice sheet reconstructions used here, the ice area extent which is responsible for the shortwave forcing almost reaches its maximum value at glaciation level 3. Further increasing the glaciation levels affects climate predominantly through the orography forcing (cf. Sect. 3.6).

3.6 Orographic effect

Finally, we consider the differences between the simulations with and without orography forcing of the ice sheets. The latter is potentially important given that mountains and elevated land masses affect the atmospheric circulation and precipitation patterns, and then the whole climate system. To this end, an emulator was calibrated on the available present-day orography experiments.

The net effect orography can then be seen in Fig. 13, where all four variables are plotted as a function of the glaciation level. Black solid lines show the respective variables obtained with the standard experiment design, while red solid lines show the response obtained with the experiment design assuming pre-industrial orography, regardless of the presence of ice sheets. The value plotted is obtained from Eq. (11). Note that by construction this value is also implicitly a function of CO_2 concentration, which enters Eq. (11) via the factor $\rho(x|x_{\text{ice}})$. Dotted lines indicate a 1σ deviation, in both cases, based on Eq. (13), using $x_p = x_p^*$.

A clear deviation is seen around glaciation level 3. This effect is due to the fact that, as explained in Sect. 2.1, levels 3–11 represent effectively similar ice sheet area, but significantly higher orography (see Fig. 1). Hence, the albedo forcing dominates over the lower range of glaciation levels (1–3), with decreasing temperatures, precipitation and mixed-layer depth shoaling. The orography–no-orography differences appear more markedly above index 3: orography reduces the cooling trends by as much as 1 °C on the continent at glaciation level 11, and even reverses the precipitation trend. As stated in the Introduction, it is known that ice orography forcing may impact monsoon precipitation regimes, but to our knowledge the specific effect of Northern Hemisphere ice sheet orography on the Indian monsoon is yet to be documented. The warming signal caused by orography may be understood by considering the increase in surface potential temperature over elevated regions, similar to what is seen today over the Tibetan Plateau. Because of these high potential temperatures, down-sloping air is effectively warmer than it would be in the absence of orography forcing, and contributes here to increasing the Northern Hemisphere continental surface temperatures. Orographic forcing generally induces atmospheric circulation anomalies and effects on ocean circulation and stratification. For example, Fig. 13 suggests a weak positive effect on mixed-layer depth, quite small compared to the astronomical forcing effects. An in-depth analysis of these effects falls beyond the scope of the present contribution.

4 Conclusions

We present a first application of a global sensitivity analysis theory to study the climate response of the Indian monsoon to the climate factors which evolved during the Pleis-

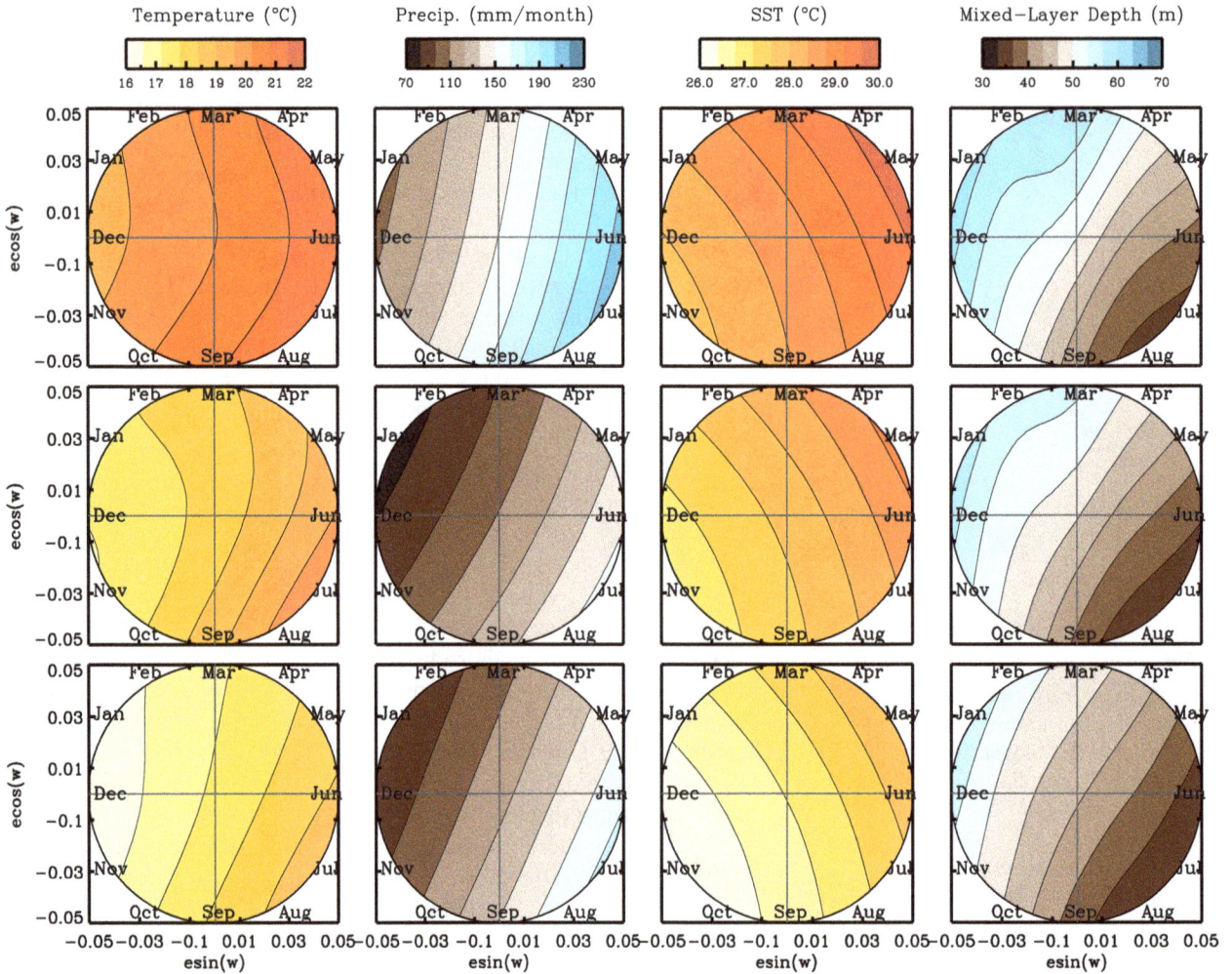

Figure 10. Sensitivity to $e\cos(\varpi)$ and $e\sin(\varpi)$ for all fields. Each panel, from top to bottom, shows the four fields with a different configuration of glaciation level – CO_2 concentration. Top panels: glaciation level = 1 and CO_2 = 280 ppmv. Middle panels: glaciation level = 5 and CO_2 = 230. Bottom panels: glaciation level = 11 and CO_2 = 180. All fields were integrated over obliquity.

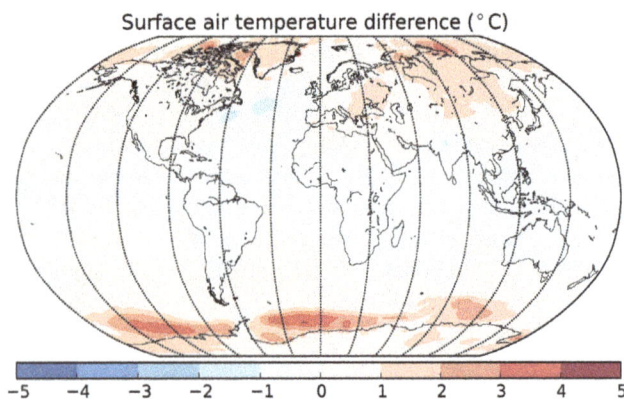

Figure 11. Sea surface temperature difference for two *idealised* simulations. CO_2 concentration, glaciation level and precession remained fixed, the only difference being obliquity (23 and 24°).

tocene, namely the astronomical forcing ($e\sin(\varpi)$, $e\cos(\varpi)$, ε), CO_2 concentration and glaciation level.

We focus, in particular, on four variables: continental temperature, continental precipitation, sea-surface temperature and mixed-layer depth. These variables were averaged for the JJAS season over northern India and northwestern Indian Ocean.

Similar to a number of recent studies based on statistical modelling for global sensitivity analysis of computationally expensive simulators, the technical implementation follows a three-step methodology:

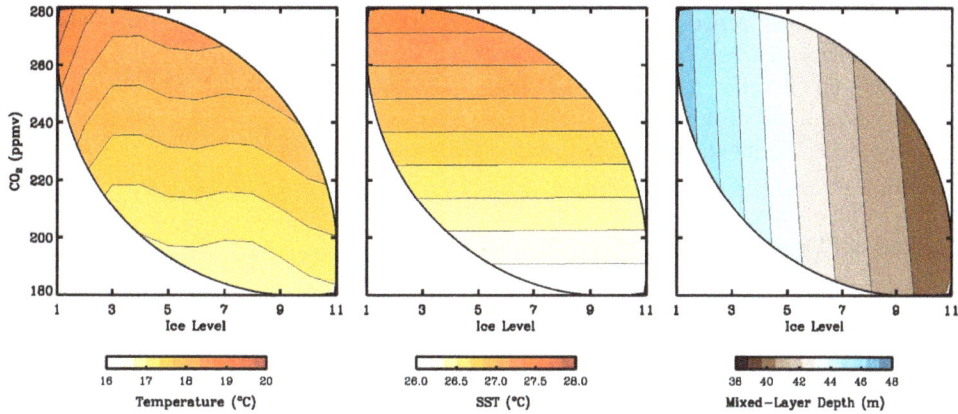

Figure 12. Sensitivity to CO_2 and glaciation level. From left to right: continental temperature, sea-surface temperature and mixed-layer depth. Fields were integrated over $e\sin(\varpi)$, $e\cos(\varpi)$ and obliquity.

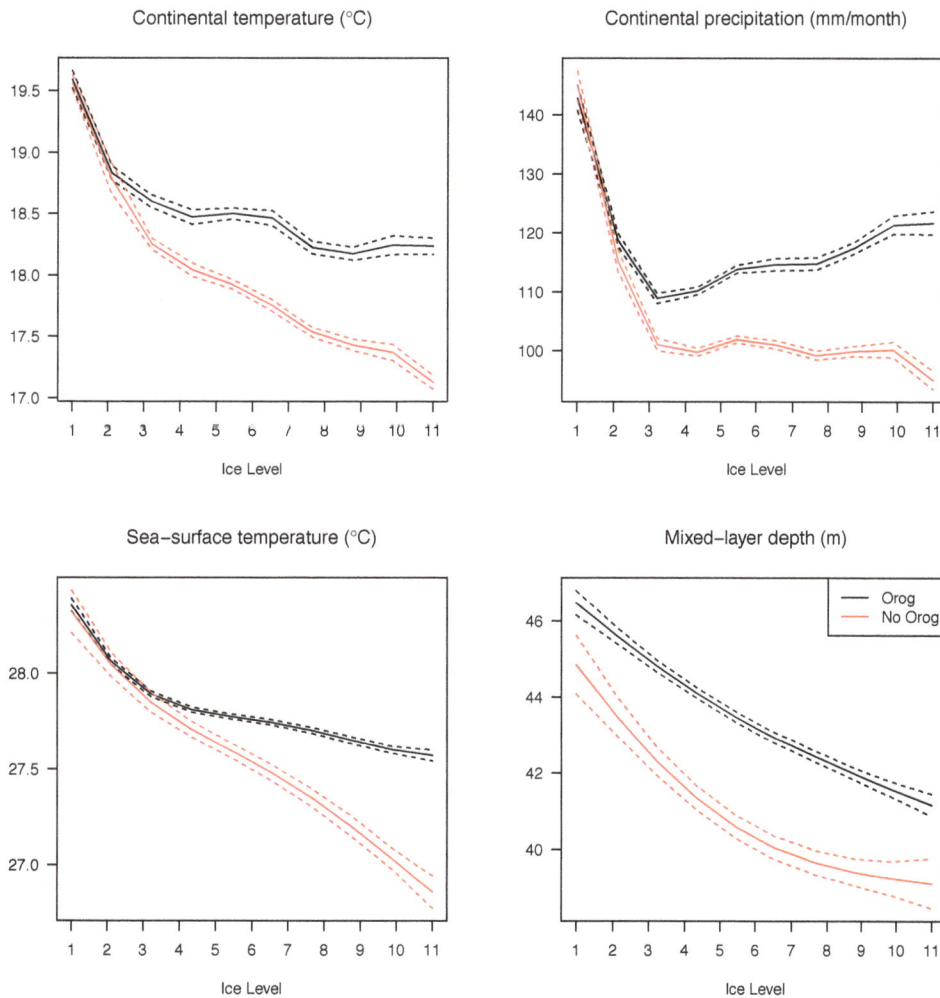

Figure 13. Orography–no-orography difference. From top to bottom, left to right: effect on continental temperature, precipitation, sea-surface temperature, and mixed-layer depth, with orography forcing (black) and without (red). The dotted lines show one standard deviation of the emulator prediction. One may see a departure point from glaciation level 3 in all four fields, as this is the point at which orography forcing becomes the most significant.

– *Designing an experiment plan*. We adopted a Latin hypercube design, optimised following two constraints: maximisation of the minimum distance between two points in the input space – this is called the maxi–min property – and maximisation of the determinant of the matrix of covariance between the input factors – this is a constraint of orthogonality. In addition, the design excludes configurations with excessive eccentricity and unrealistic combinations of CO_2 and glaciation level.

– *Calibration and validation of the emulator*. The validation was performed following a leave-one-out cross-validation approach. Two experiments were excluded of the design as presenting an anomalous North-Atlantic SST patterns. The emulator calibrated on the remaining 59 experiments overall validates the present statistical modelling choices.

– *Quantifying and visualising the individual and combined effects of the different factors on the summer Indian monsoon*, based on sensitivity measures and cross-section plots.

This analysis yielded the following conclusions:

– precession controls the response of four variables: continental temperature in phase with June–July insolation; high glaciation favouring a late-phase response; sea-surface temperature in phase with May insolation; and continental precipitation in phase with July insolation, and mixed-layer depth in antiphase with the latter.

– The effect of CO_2 on continental temperature and SST is of similar size to that of precession on summer continental temperature and SST.

– Obliquity is a secondary effect, negligible on most variables except sea-surface temperature.

– The effect of glaciation is dominated by the albedo forcing, and its effect on precipitation competes with that of precession.

– The orographic forcing reduces the glacial cooling induced by the albedo forcing, and even has a positive effect on precipitation.

The present study confirms the high potential of emulation for exploring and understanding the response of climate models. One originality of the present work was to consider, as inputs, several elements of the climate forcing that (have) varied in the past, and the emulator was used as a method to help us quantify the link between forcing variability and climate variability. The methodology may naturally be applied to other regions of focus and other climate models.

Acknowledgements. Paul Valdes and Gethin Williams (University of Bristol) are thanked for their assistance in setting up HadCM3 on our systems, and for providing us with ice topography boundary conditions. Richard Wilkinson (University of Nottingham) provided important clarifications to statistical notations. Constructive comments on the discussion version of this paper provided by Tamsin Edwards, Guangshan Chen, Yongqiang Yu and Alexander Kislov are warmly acknowledged. This research is a contribution to the ITOP project, ERC-StG grant 239604. M. Crucifix is funded by the Belgian National Fund of Scientific Research (FRS-FNRS) and P. A. Araya-Melo and N. Bounceur are funded through the ERC ITOP project. Computational resources have been provided by the supercomputing facilities of the Université catholique de Louvain (CISM/UCL) and the Consortium des Equipements de Calcul Intensif en Fédération Wallonie Bruxelles (CECI) funded by the FRS-FNRS. The data used in this work, together with the emulator code, are available in the supplementary material.

Edited by: P. Braconnot

References

Anderson, P. M., Barnosky, C. W., Bartlein, P. J., Behling, P. J., Brubaker, L., Cushing, E. J., Dodson, J., Dworetsky, B., Guetter, P. J., Harrison, S. P.,Huntley, B., Kutzbach, J. E., Markgraf, V., Marvel, R., McGlone, M. S., Mix, A., Moar, N. T., Morley, J., Perrott, R. A., Peterson, G. M., Prell, W. L., Prentice, I. C., Ritchie, J. C., Roberts, N., Ruddiman, W. F., Salinger, M. J., Spaulding, W. G., Street-Perrott, F. A., Thompson, R. S., Wang, P. K., Webb III, T., Winkler, M. G., and Wright Jr., H. E.: Climatic changes of the last 18 000 years – observations and model simulations, Science, 241, 1043–1052, 1988.

Andrianakis, I. and Challenor, P. G.: The effect of the nugget on Gaussian process emulators of computer models, Comput. Stat. Data An., 56, 4215–4228, 2012.

Bastos, L. S. and O'Hagan, A.: Diagnostics for Gaussian process emulators, Technometrics, 51, 425–438, 2009.

Berger, A. L.: Long-term variations of daily insolation and Quaternary climatic changes, J. Atmos. Sci., 35, 2362–2367, 1978.

Berger, J. O., De Oliveira, V., and Sansó, B.: Objective Bayesian analysis of spatially correlated data, J. Am. Stat. Assoc., 96, 1361–1374, 2001.

Braconnot, P. and Marti, O.: Impact of precession on monsoon characteristics from coupled ocean atmosphere experiments: changes in Indian monsoon and Indian ocean climatology, Mar. Geol., 201, 23–34, asian Monsoons and Global Linkages on Milankovitch and Sub-Milankovitch Time Scales, 2003.

Braconnot, P., Otto-Bliesner, B., Harrison, S., Joussaume, S., Peterchmitt, J.-Y., Abe-Ouchi, A., Crucifix, M., Driesschaert, E., Fichefet, Th., Hewitt, C. D., Kageyama, M., Kitoh, A., Laîné, A., Loutre, M.-F., Marti, O., Merkel, U., Ramstein, G., Valdes, P., Weber, S. L., Yu, Y., and Zhao, Y.: Results of PMIP2 coupled simulations of the Mid-Holocene and Last Glacial Maximum – Part 1: experiments and large-scale features, Clim. Past, 3, 261–277, doi:10.5194/cp-3-261-2007, 2007.

Braconnot, P., Marzin, C., Grégoire, L., Mosquet, E., and Marti, O.: Monsoon response to changes in Earth's orbital parameters: comparisons between simulations of the Eemian and of the Holocene, Clim. Past, 4, 281–294, doi:10.5194/cp-4-281-2008, 2008.

Carslaw, K. S., Lee, L. A., Reddington, C. L., Pringle, K. J., Rap, A., Forster, P. M., Mann, G. W., Spracklen, D. V., Woodhouse, M. T., Regayre, L. A., and Pierce, J. R.: Large contribution of natural aerosols to uncertainty in indirect forcing, Nature, 503, 67–71, 2013.

Chen, G.-S., Liu, Z., Clemens, S., Prell, W., and Liu, X.: Modeling the time-dependent response of the Asian summer monsoon to obliquity forcing in a coupled GCM: a PHASEMAP sensitivity experiment, Clim. Dynam., 36, 695–710, 2011.

Essery, R. L. H., Best, M. J., Betts, R. A., Cox, P. M., and Taylor, C. M.: Explicit representation of subgrid heterogeneity in a GCM land-surface scheme, J. Hydrometeorol., 4, 530–543, 2003.

Felzer, B., Webb III, T., and Oglesby, R. J.: The impact of ice sheets, CO_2, and orbital insolation on late quaternary climates: sensitivity experiments with a general circulation model, Quaternary Sci. Rev., 17, 507–534, 1998.

Gordon, C., Cooper, C., Senior, C. A., Banks, H., Gregory, J. M., Johns, T. C., Mitchell, J. F. B., and Wood, R. A.: The simulation of SST, sea ice extents and ocean heat transports in a version of the Hadley Centre coupled model without flux adjustments, Clim. Dynam., 16, 147–168, 2000.

Gramacy, R. and Lee, H. H.: Cases for the nugget in modeling computer experiments, Stat. Comput., 22, 713–722, 2012.

Hewitt, C. D.: The effects of ocean dynamics in a coupled GCM simulation of the Last Glacial Maximum, Clim. Dynam., 20, 203–218, 2003.

Hewitt, C. D., Broccoli, A. J., Crucifix, M., Gregory, J. M., Mitchell, J. F. B., and Stouffer, R. J.: The effect of a large freshwater perturbation on the Glacial Atlantic Ocean using a coupled general circulation model, J. Climate, 19, 4436–4447, 2006.

Holden, P., Edwards, N., Oliver, K., Lenton, T., and Wilkinson, R.: A probabilistic calibration of climate sensitivity and terrestrial carbon change in GENIE-1, Climate Dynamics, 35, 785–806, 2010.

Homma, T. and Saltelli, A.: Importance measures in global sensitivity analysis of nonlinear models, Reliab. Eng. Syst. Safe., 52, 1–17, 1996.

Joseph, V. R. and Hung, Y.: Orthogonal-maximin latin hypercube designs, Stat. Sinica, 18, 171–186, 2008.

Kennedy, M. C. and O'Hagan, A.: Predicting the output from a complex computer code when fast approximations are available, Biometrika, 87, 1–13, 2000.

Kutzbach, J. E. and Guetter, P. J.: The influence of changing orbital parameters and surface boundary conditions on climate simulations for the past 18 000 years, J. Atmos. Sci., 43, 1726–1759, 1986.

Kutzbach, J. E. and Ruddiman, W. F.: Model description, external forcing, and surface boundary conditions, in: Global climates since the last glacial maximum, 12–23, 1993.

Kutzbach, J. E. and Street-Perrott, F. A.: Milankovitch forcing of fluctuations in the level of tropical lakes from 18 to 0 kyr BP, Nature, 317, 130–134, 1985.

Kutzbach, J., Liu, X., Liu, Z., and Chen, G.: Simulation of the evolutionary response of global summer monsoons to orbital forcing over the past 280,000 years, Clim. Dynam., 30, 567–579, 2008.

Lee, L. A., Carslaw, K. S., Pringle, K. J., Mann, G. W., and Spracklen, D. V.: Emulation of a complex global aerosol model to quantify sensitivity to uncertain parameters, Atmos. Chem. Phys., 11, 12253–12273, doi:10.5194/acp-11-12253-2011, 2011.

Lee, L. A., Pringle, K. J., Reddington, C. L., Mann, G. W., Stier, P., Spracklen, D. V., Pierce, J. R., and Carslaw, K. S.: The magnitude and causes of uncertainty in global model simulations of cloud condensation nuclei, Atmos. Chem. Phys., 13, 8879–8914, doi:10.5194/acp-13-8879-2013, 2013.

Lisiecki, L. E. and Raymo, M. E.: A Pliocene-Pleistocene stack of 57 globally distributed benthic $\delta^{18}O$ records, Paleoceanography, 20, PA1003, doi:10.1029/2004PA001071, 2005.

Loeppky, J. L., Sacks, J., and Welch, W. J.: Choosing the sample size of a computer experiment: a practical guide, Technometrics, 51, 366–376, doi:10.1198/TECH.2009.08040, 2009.

Loutre, M. F.: Paramètres orbitaux et cycles diurne et saisonnier des insolations, Ph.D. thesis, Université catholique de Louvain, Louvain-la-Neuve, Belgium, 1993.

Luethi, D., Le Floch, M., Bereiter, B., Blunier, T., Barnola, J.-M., Siegenthaler, U., Raynaud, D., Jouzel, J., Fischer, H., Kawamura, K., and Stocker, T. F.: High-resolution carbon dioxide concentration record 650 000–800 000 years before present, Nature, 453, 379–382, 2008.

MacCalman, A. D.: Flexible space-filling designs for complex system simulations, Ph.D. thesis, Naval Postgraduate School, Monterey, California, US, available at: http://hdl.handle.net/10945/34701 (last access: 29 March 2014), 2013.

Masson, V., Braconnot, P., Jouzel, J., de Noblet, N., Cheddadi, R., and Marchal, O.: Simulation of intense monsoons under glacial conditions, Geophys. Res. Lett., 27, 1747–1750, 2000.

McKay, M. D., Beckman, R. J., and Conover, W. J.: A comparison of three methods for selecting values of input variables in the analysis of output from a computer code, Technometrics, 21, 239–245, 1979.

Morris, M. D. and Mitchell, T. J.: Exploratory designs for computational experiments, J. Stat. Plan. Infer., 43, 381–402, 1995.

Oakley, J. and O'Hagan, A.: Bayesian inference for the uncertainty distribution of computer model outputs, Biometrika, 89, 769–784, 2002.

Oakley, J. E. and O'Hagan, A.: Probabilistic sensitivity analysis of complex models: a Bayesian approach, J. Roy. Stat. Soc. B, 66, 751–769, 2004.

O'Hagan, A.: Bayesian analysis of computer code outputs: a tutorial, Reliab. Eng. Syst. Safe., 91, 1290–1300, 2006.

Pepelyshev, A.: The role of the nugget term in the Gaussian process method, in: mODa 9 – Advances in Model-Oriented Design and Analysis, Springer, 149–156, 2010.

Petit, J. R., Jouzel, J., Raynaud, D., Barkov, N. I., Barnola, J.-M., Basile, I., Bender, M., Chappellaz, J., Davis, M., Delaygue, G., Delmotte, M., Kotlyakov, V. M., Legrand, M., Lipenkov, V. Y., Lorius, C., Pepin, L., Ritz, C., Saltzman, E., and Stievenard, M.: Climate and atmospheric history of the past 420 000 years from the Vostok ice core, Antarctica, Nature, 399, 429–436, 1999.

Petropoulos, G., Wooster, M. J., Carlson, T. N., Kennedy, M. C., and Scholze, M.: A global Bayesian sensitivity analysis of the 1d SimSphere soil–vegetation–atmospheric transfer (SVAT) model using Gaussian model emulation, Ecol. Model., 220, 2427–2440, 2009.

Sacks, J., Welch, W. J., Mitchell, T. J., and Wynn, H. P.: Design and analysis of computer experiments, Stat. Sci., 4, 409–423, 1989.

Santner, T., Williams, B., and Notz, W.: The Design and Analysis of Computer Experiments, Springer, New York, 2003.

Saltelli, A., Tarantola, S., Campolongo, F., and Ratto, M.: Sensitivity analysis in practice, John Wiley and Sons, Ltd, Thichester, W. Sussex, England, 2004.

Schmittner, A., Urban, N. M., Shakun, J. D., Mahowald, N. M., Clark, P. U., Bartlein, P. J., Mix, A. C., and Rosell-Melé, A.: Climate Sensitivity Estimated from Temperature Reconstructions of the Last Glacial Maximum, Science, 334, 1385–1388, 2011.

Sexton, D. M., Murphy, J. M., Collins, M., and Webb, M. J.: Multivariate probabilistic projections using imperfect climate models part I: outline of methodology, Clim. Dynam., 38, 2513–2542, 2012.

Siegenthaler, U., Stocker, T. F., Lüthi, D., Schwander, J., Stauffer, B., Raynaud, D., Barnola, J.-M., Fisher, H., Masson-Delmotte, V., and Jouzel, J.: Stable carbon cycle-climate relationship during the Late Pleistocene, Science, 310, 1313–1317, 2005.

Singarayer, J. S. and Valdes, P. J.: High-latitude climate sensitivity to ice-sheet forcing over the last 120 kyr, Quaternary Sci. Rev., 29, 43–55, 2010.

Urban, N. M. and Fricker, T. E.: A comparison of Latin hypercube and grid ensemble designs for the multivariate emulation of an Earth system model, Comput. Geosci., 36, 746–755, 2010.

Williams, K. D., Senior, C. A., and Mitchell, J. F. B.: Transient climate change in the Hadley Centre Models: the role of physical processes, J. Climate, 14, 2659–2674, 2001.

Yanase, W. and Abe-Ouchi, A.: The LGM surface climate and atmospheric circulation over East Asia and the North Pacific in the PMIP2 coupled model simulations, Clim. Past, 3, 439–451, doi:10.5194/cp-3-439-2007, 2007.

Yin, Q. and Berger, A.: Individual contribution of insolation and CO_2 to the interglacial climates of the past 800 000 years, Clim. Dynam., 38, 709–724, 2012.

Yin, Q. Z., Berger, A., and Crucifix, M.: Individual and combined effects of ice sheets and precession on MIS-13 climate, Clim. Past, 5, 229–243, doi:10.5194/cp-5-229-2009, 2009.

Zhao, Y., Braconnot, P., Marti, O., Harrison, S. P., Hewitt, C. D., Kitoh, A., Liu, Z., Mikolajewicz, U., Otto-Bliesner, B., and Weber, S. L.: A multi-model analysis of the role of the ocean on the African and Indian monsoon during the mid-Holocene, Clim. Dynam., 25, 777–800, 2005.

8

Interdependence of the growth of the Northern Hemisphere ice sheets during the last glaciation: the role of atmospheric circulation

P. Beghin[1], S. Charbit[1], C. Dumas[1], M. Kageyama[1], D. M. Roche[1,2], and C. Ritz[3]

[1]Laboratoire des Sciences du Climat et de l'Environnement, CEA-CNRS – UMR8212, Gif-sur-Yvette, France
[2]Earth and Climate Cluster, Faculty of Earth and Life Sciences, Vrije Universiteit Amsterdam, Amsterdam, the Netherlands
[3]Laboratoire de Glaciologie et de Géophysique de l'Environnement, CNRS, Saint Martin d'Hérès, France

Correspondence to: P. Beghin (pauline.beghin@lsce.ipsl.fr)

Abstract. The development of large continental-scale ice sheets over Canada and northern Europe during the last glacial cycle likely modified the track of stationary waves and influenced the location of growing ice sheets through changes in accumulation and temperature patterns. Although they are often mentioned in the literature, these feedback mechanisms are poorly constrained and have never been studied throughout an entire glacial–interglacial cycle. Using the climate model of intermediate complexity CLIMBER-2 coupled with the 3-D ice-sheet model GRISLI (GRenoble Ice Shelf and Land Ice model), we investigate the impact of stationary waves on the construction of past Northern Hemisphere ice sheets during the past glaciation. The stationary waves are not explicitly computed in the model but their effect on sea-level pressure is parameterized. We tested different parameterizations to study separately the effect of surface temperature (thermal forcing) and topography (orographic forcing) on sea-level pressure, and therefore on atmospheric circulation and ice-sheet surface mass balance. Our model results suggest that the response of ice sheets to thermal and/or orographic forcings is rather different. At the beginning of the glaciation, the orographic effect favors the growth of the Laurentide ice sheet, whereas Fennoscandia appears rather sensitive to the thermal effect. Using the ablation parameterization as a trigger to artificially modify the size of one ice sheet, the remote influence of one ice sheet on the other is also studied as a function of the stationary wave parameterizations. The sensitivity of remote ice sheets is shown to be highly sensitive to the choice of these parameterizations with a larger response when orographic effect is accounted for.

Results presented in this study suggest that the various spatial distributions of ice sheets could be partly explained by the feedback mechanisms occurring between ice sheets and atmospheric circulation.

1 Introduction

The Quaternary era is characterized by a succession of glacial and interglacial phases. During the last ice age, large land-ice masses covered Canada and northwestern Eurasia (Peltier, 2004; Lambeck et al., 2006; Tarasov, 2012; Clark et al., 1993; Dyke and Prest, 1987; Svendsen, 2004). These large ice sheets represent a crucial element of the climate system (Clark, 1999). Because of their highly reflective surface and their high altitude, they induce zonal anomalies in surface temperature and topography. These anomalies modify large-scale atmospheric circulation by generating zonal asymmetries often referred to as stationary waves (Cook and Held, 1988).

The relations between atmosphere and ice sheets have been previously investigated with several modeling studies. Based on the analysis of simulations carried out with general circulation models (GCM) under Last Glacial Maximum (LGM) conditions, Broccoli and Manabe (1987) found that ice sheets are the main cause of change of stationary waves during ice age climate. In line with these previous findings, Pausata et al. (2011) have shown that the topography of ice sheets is the dominant factor altering the northern extratropical atmospheric large-scale circulation. Moreover, several authors put forward that one of the main effects of an ice sheet on the atmospheric circulation is to change the strength and the position of the subtropical jet, which has also an influence on storm tracks (Kageyama and Valdes, 2000; Hall et al., 1996; Rivière et al., 2010; Laîné et al., 2008). This leads to changes in the pattern of precipitation and consequently to changes in the accumulation over ice sheets.

Changes in stationary waves also induce changes in surface temperature. Using a simple ice-sheet model based on an idealized geometry coupled with a stationary-wave model, Roe and Lindzen (2001a, b) highlighted the importance of accounting for the feedbacks between ice sheets and the temperatures induced by stationary waves to properly simulate the evolution of an ice sheet. In the same way, with a three-dimensional stationary wave model, Liakka et al. (2011) showed that the southern margin of ice sheets strongly depends on the temperature anomalies due to stationary waves. All these studies illustrate the existence of feedbacks between ice sheets and atmospheric circulation. This suggests that the construction of a given ice sheet (e.g. the North American ice sheet) may influence the growth or the decay of the other one (e.g. the Eurasian ice sheet) through changes in atmospheric circulation that induce modifications in both temperature and precipitation patterns. In turn, these modifications directly influence the surface mass balance of the ice sheets.

However, up to now, no study has been undertaken to understand the role atmospheric circulation may play on ice-sheet evolution throughout a glacial–interglacial cycle. The aim of this study is twofold. First, we examine how the planetary waves may influence the evolution of ice sheets over the last glacial period. Secondly, we investigate how past Northern Hemisphere ice sheets (i.e. Laurentide ice sheet (LIS) and Fennoscandia ice sheet (FIS)) interact together through induced changes in planetary waves. To achieve this goal, we use the climate model of intermediate complexity CLIMBER 2.4 (Petoukhov et al., 2000) fully coupled with the 3-D thermo-mechanical ice-sheet model GRISLI (GRenoble Ice Shelf and Land Ice model) (Peyaud et al., 2007) to carry out a series of numerical experiments covering the last glacial–interglacial cycle.

2   Model description

2.1   The climate model CLIMBER 2.4

The CLIMBER 2.4 model is a revised version of the CLIMBER 2.3 model extensively described in Petoukhov et al. (2000). It is based on simplified representations of the ocean with three zonally averaged ocean basins for the Atlantic, Indian and Pacific oceans ($2.5° \times 20$ uneven layers), of the atmosphere (with resolution $10°$ in latitude, $51°$ in longitude), of the vegetation and of the mutual interactions between these three components. The main advantage of this

model is its fast computational time allowing long-term simulations at glacial–interglacial timescale.

The atmospheric module is designed to only resolve large-scale processes, the effect of synoptic weather systems on heat and moisture transports being parameterized. The atmospheric variables such as humidity and temperature are calculated in 2-D, and their 3-D profiles are then computed from hypotheses on the vertical structure of the atmosphere. Stationary waves are not explicitly resolved but their effect on sea-level pressure (hereafter SLP) is parameterized. In the standard version of the model, it is described by relationship between long-term large-scale azonal temperature and pressure fields in quasi-stationary planetary-scale waves adapted for the CLIMBER model: the longitudinal shift between the pressure and the temperature is neglected because of the coarse spatial resolution of CLIMBER. The azonal sea-level pressure (p_0') is expressed as a function of the azonal component of sea-level temperature T_0' (Petoukhov et al., 2000):

$$p_0 = \overline{p_0} + p_0' \tag{1}$$

with

$$p_0' = p_{\text{th}}' = \frac{-g \, P_0 \, H_{\text{T}}}{2 \, R} \frac{T_0'}{\overline{T_0}^2} \tag{2}$$

with

$$T_0 = \overline{T_0} + T_0', \tag{3}$$

where p_0 is the sea-level pressure, $P_0 = 1012 \, \text{hPa}$ the mean of sea-level pressure. $R = 287.058 \, \text{J kg}^{-1} \, \text{K}^{-1}$ is the specific gas constant for dry air, $g = 9.81 \, \text{m s}^{-1}$ is the gravity constant and H_{T} is the computed tropopause height.

Equation (2) shows that the influence of topography on the azonal SLP component is not explicitly accounted for. Here, the impact of orographic changes on stationary waves is studied through a new parameterization (described in Sect. 3). The use of parameterizations to account for the effect of stationary waves on sea-level pressure offers the opportunity to test the influence of thermal and orographic effects separately.

2.2   The ice-sheet model GRISLI

GRISLI is a three-dimensional thermo-mechanical model which simulates the evolution of ice-sheet geometry (extension and thickness) and the coupled temperature–velocity fields in response to climate forcing. A comprehensive description of the model can be found in Ritz et al. (2001) and Peyaud et al. (2007). Here, we only summarize the main characteristics of this model. The equations are solved on a cartesian grid ($40 \, \text{km} \times 40 \, \text{km}$). Over the grounded part of the ice sheet, the ice flow resulting from internal deformation is governed by the shallow-ice approximation (Morland, 1984; Hutter, 1983). The model also deals with ice flow through ice shelves using the shallow-shelf approximation

(MacAyeal, 1989) and predict the large-scale characteristics of ice streams using criteria based on the effective pressure and hydraulic load. At each time step, the velocity and vertical profiles of temperature in the ice are computed as well as the new geometry of the ice sheet. The temperature field is computed both in the ice and in the bedrock by solving a time-dependent heat equation. The surface mass balance is defined as the sum between accumulation and ablation computed by the positive degree day (PDD) method (e.g. Reeh, 1991; Fausto et al., 2009). This method assumes that melt rates of snow and ice are linearly related to the number of PDD through constant degree-day factors (Braithwaite, 1984; Braithwaite, 1995). The PDD method relies on the assumption that the annual surface melting is proportional to the sum over one year of the excess of temperature above the melting point. The number of positive degree days is calculated from the normal probability distribution around the mean daily temperatures and the variations around the daily mean (Reeh, 1991; Braithwaite, 1984). This approach has been widely used in ice-sheet models and is expressed as

$$\text{PDD} = \frac{1}{\sigma\sqrt{2\pi}} \int_{1\text{year}} dt \int_0^{T_d+2.5\sigma} T \cdot \exp\left(\frac{-(T-T_a(t))^2}{2\sigma^2}\right) dT, \quad (4)$$

where T is the near-surface temperature, $T_a(t)$ is the mean daily near-surface temperature, and σ the standard deviation around the daily mean. The mean daily temperature $T_a(t)$ follows a sine cycle during the year. This formulation allows positive temperatures in a given day even if the mean daily temperature is negative: the higher the σ value, the larger the probability of having positive temperatures is. Therefore, high σ values favor ablation. In the standard PDD formulation (Reeh, 1991), σ is fixed to 5 °C. However, the daily temperature variability is strongly dependent on the altitude. Based on measurements from automatic weather stations in Greenland, Fausto et al. (2009) derived a parameterization of σ expressed as

$$\sigma = \sigma_0 + \alpha z_s, \quad (5)$$

where z_s is the altitude of the ice-sheet surface and σ_0 the sea-level value; in this parameterization, $\sigma_0 = 1.574$ °C and $\alpha = 1.22 \times 10^{-3}$ m^{-1}, corresponding to $\sigma_{3000m} = 5.2$ °C. In the present study we used the same type of relationship between σ and the altitude but with different numerical values. We used different couples of σ_0 and α values (see Sect. 3) in order to modulate the amount of ablation and, thus, to simulate ice sheets of different sizes.

2.3 Coupling procedure between CLIMBER-2.4 and GRISLI

The mean annual and summer surface temperatures computed by CLIMBER, as well as snowfall, are used as inputs to GRISLI to compute surface mass balance. To account for the resolution difference between both models, we apply a specific downscaling procedure: for each CLIMBER grid box and each surface type, the temperature is computed on five vertical levels using the free atmospheric lapse rate to account for the dependency with the altitude. This vertical temperature profile is used to compute the vertical humidity and the resulting vertical precipitation profile. For temperature fields, these calculations are performed for each CLIMBER surface type and then averaged over the GRISLI surface types (land ice, ice-free land and ocean). The three climatic fields used as forcing are then tri-linearly interpolated on the ice-sheet model grid. GRISLI returns to CLIMBER new boundary conditions for each CLIMBER grid box: new surface type, altitude of each grid point and fresh water fluxes coming from the potential melting of ice sheets. A more detailed explanation of the coupling procedure is described in Charbit et al. (2013).

3 Experimental set-up

3.1 Parameterization of stationary waves in CLIMBER

As mentioned in Sect. 2.1, in the standard version of CLIMBER, the azonal component of sea-level pressure only depends on sea-level temperature, so the topography does not exert any direct influence on sea-level pressure. The orographic impact on stationary waves depends on the topography and on the strength of the jet zonal wind (Held et al., 2002; Vallis, 2006; Holton, 1979). We parameterize the impact of the orography on stationary waves following the same basic principle as the one described by Eq. (2). Since the thermal contribution is expressed as a linear relation between the azonal component of sea-level pressure (p_{th}') and the azonal component of temperature (T_0'), we expressed the orographic contribution (p_{oro}') as a function of the deviation of orography from its zonal mean (h_0'). This accounts for the deviation of the zonal wind by an anomaly of topography. This new parameterization is expressed as follows:

$$p_{oro}' \sim P_0 \frac{h_{topo}'}{\overline{H_T}} \max\left(\frac{\Delta T_{E/P}}{\Delta T_{limit}} - 1; 0\right) \quad (6)$$

with

$$h_{topo}' = h_{topo} - \overline{h_{topo}}, \quad (7)$$

where $P_0 = 1012$ hPa is a constant, h_{topo} is the altitude of the surface and $\overline{h_{topo}}$ the mean zonal altitude. Since the zonal wind is driven by the equator-to-pole temperature gradient, we added a dependency to this term ($\Delta T_{E/P}$). Adding this dependency allows to account for seasonal changes of the influence of the orographic effect on the sea-level pressure. The larger this gradient is, the stronger the zonal wind and the influence of topography are. On the other hand, if $\Delta T_{E/P}$ is too small (i.e. below the ΔT_{limit}), the zonal wind and its deviation due to orography will be negligible. Therefore, the ΔT_{limit} represents the limit below which the orographic contribution is negligible.

The numerical value of this parameter has been determined by comparing the present-day sea-level pressure obtained with the new parameterization ($p'_0 = p'_{\text{th}} + p'_{\text{oro}}$) with the NCEP reanalysis (Kalnay et al., 1996). The best correlation has been obtained for $\Delta T_{\text{limit}} = 25\,°\text{C}$.

Using this approach, we can study separately orographic and thermal effects, as well as a combination of both. We tested four SLP parameterizations:

  a. $p_0 = \overline{p_0}$ (without any waves),

  b. $p_0 = \overline{p_0} + p'_{\text{th}}$ (thermal forcing only),

  c. $p_0 = \overline{p_0} + p'_{\text{oro}}$ (orographic forcing only),

  d. $p_0 = \overline{p_0} + p'_{\text{oro}} + p'_{\text{th}}$ (thermal and orographic effect).

Figure 1 shows the azonal sea-level pressure for the northern winter for NCEP reanalysis (Fig. 1a) and for the three parameterizations B, C and D (Fig. 1b–d). The most striking feature is that the patterns of azonal sea-level pressure simulated by CLIMBER are smoother than the NCEP ones due to the spatial resolution. The thermal effect (Fig. 1b) allows to account for the main structures linked to the land–sea temperature difference. Low pressure over North Pacific and North Atlantic and high pressure over the continental regions are clearly represented. However, their meridional extent is too large. This leads to a negative SLP anomaly over Greenland and a positive one over the Scandinavian and the Barents–Kara regions, in contradiction with NCEP reanalysis. Moreover, the amplitudes of the anomalies over the Northern Hemisphere are weaker than those of the NCEP database. This implies that the anti-cyclonic structure over the North American continent is almost absent. The amplitudes of sea-level pressure anomalies are smaller in the Southern Hemisphere (Fig. 1a) than in the Northern Hemisphere and occur over a smaller spatial scale. Due to the coarse horizontal resolution of CLIMBER, these structures are poorly resolved whatever the azonal sea-level parameterization is. With the thermal effect (Fig. 1b) this translates into a large anticyclone centered over South Atlantic that expands over the Indian Ocean, western Pacific and the Antarctic ice sheet. As a result, the bipolar structure over Antarctica observed in NCEP is not represented in the simulations carried out with the thermal forcing. Nevertheless, the Southern Hemisphere regions are beyond the regions of interest for the purpose of the present study.

With the orographic parameterization (Fig. 1c), a high pressure appears over Greenland, in agreement with Pausata et al. (2011). Subsequently this leads to a negative anomaly over the other regions of the same latitudinal band, especially over the Barents–Kara seas and over Scandinavia. The spatial structures are in a better agreement with NCEP but the amplitude of the negative anomaly over the Barents–Kara sector

is still too weak. A high pressure is also simulated centered over the Caspian sea and over Central and North America. The high pressure over North America is weaker in our simulations than in NCEP. This is due to the Rocky Mountains poorly resolved in CLIMBER because of the zonal structure of the model. Finally, the bipolar structure over Antarctica is represented, although the amplitudes of low and high pressures over western and eastern parts respectively are weaker than the NCEP ones. This is also likely due to the spatial resolution.

Combining both effects (thermal and orographic, Fig. 1d) leads to amplitudes of SLP anomalies in better agreement with NCEP, especially over North Atlantic, North Pacific and Eurasia. Over the Barents–Kara seas, the negative anomaly due to the orographic effect clearly appears, but its amplitude is too small and overtaken by the influence of the thermal effect. As expected the high pressure over North America has a larger extent than that simulated under the thermal forcing alone, but its amplitude remains too weak with respect to the NCEP data set.

3.2 Modification of the ablation in GRISLI

Since the early PDD formulation (Reeh, 1991), a number of experimental campaigns over Greenland (Oerlemans and Vugts, 1993; Ambach, 1988) and other glacier locations have revealed strong spatial dependency of degree day factors (see Hock, 2003 and Braithwaite and Zhang, 2000 for a compilation of degree day values). Moreover, owing to the fact that the daily temperature variability is larger in continental climate regions (e.g. Siberia) than in regions with oceanic climate (e.g. northern Europe), σ is unlikely to be spatially constant. In this study, σ is used as a tuning parameter which modulates the amount of ablation and allows to simulate more or less large ice sheets. To achieve that goal, we defined four different areas characterized by different σ values: North America (Alaska excluded) (σ_0^{LIS}), Greenland (σ_0^{GIS}), Fennoscandia (including British Islands) (σ_0^{FIS}) and the rest of the grid (σ_0^{R}). Each region is characterized by a specific σ_0 value, but the slope α is constant over the entire model grid. The amplitude of the daily temperature variability directly affects the number of positive degree days and, thus, the amount of ablation. Various values of σ_0 and α lead therefore to different shapes and sizes of simulated ice sheets. In this study, we choose σ_0 and α values in order to obtain a significant ice volume over North American and Eurasian ice sheets. In doing so, this method leads to differences in the size of both ice sheets which are large enough to investigate their mutual influence.

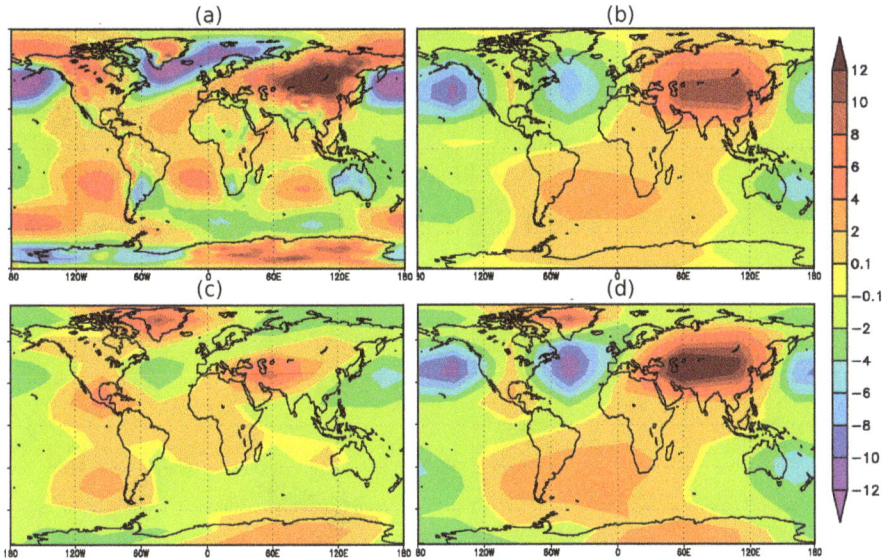

Fig. 1. Difference between sea-level pressure and its zonal mean in winter (DJF) for (**a**) NCEP reanalysis, (**b**) with only thermal parameterization (TH), (**c**) with only orographic parameterization (ORO) and (**d**) for the sum of the two parameterizations (OTH) (hPa).

3.3 Description of the experiments

To explore the relationship between the Laurentide and the Fennoscandian ice sheets and to investigate how it is modulated by planetary waves, we used different values of σ_0^{FIS} and σ_0^{LIS} to obtain larger or smaller Fennoscandian and Laurentide ice sheets, respectively. This allows us to study the impact of the FIS geometry on the LIS and vice versa. We carried out three different experiments for the four parameterizations of sea-level pressure (see Sect. 3.1). These twelve experiments are summarized in Table 1. The initial climatic state is given by a time-slice CLIMBER experiment carried out for 126 ka conditions and the initial ice-sheet topography is set to that of the present-day Greenland. Transient simulations are forced by variations of insolation (Berger, 1978), atmospheric CO_2 concentration (Petit et al., 1999) and sea level (Waelbroeck et al., 2002).

This experimental set-up allows to study the evolution of the coupled Laurentide–Fennoscandia system when the growth of one or the other ice sheet is favored. This evolution is studied in response to each stationary wave parameterization described above. Ice volume changes are analyzed in terms of changes in temperature and accumulation fields. Over the last glacial period, changes in Greenland ice-sheet geometry are not large enough to induce changes in stationary waves, especially at our model resolution. Therefore, the impact of Greenland on the Laurentide and Fennoscandian is likely to be negligible and is not discussed in this paper. Moreover, with the model version used in the present study, we fail simulating the complete deglaciation from the Last Glacial Maximum (LGM) until 0 ka. Actually, previous works suggest that an increase of climatic variability

Table 1. Experimental set-up. For each experiment, the 1st part of the name (NONE, TH, ORO, OTH) corresponds to the stationary wave parameterization while the second part corresponds to the ablation formulation: REF indicates the couple $\sigma_0^{LIS} - \sigma_0^{FIS}$ for the baseline experiment, FIS correspond to a PDD parameterization favoring the growth of Fennoscandian ice sheet, and LIS to a PDD parameterization favoring the growth of Laurentide ice sheet.

Wave parameterization	σ_0^{LIS}	σ_0^{FIS}	Simulation name
Without waves (A)	3.25	0.50	NONE REF
Without waves (A)	3.25	0.25	NONE-FIS
Without waves (A)	3.00	0.50	NONE-LIS
Thermal forcing (B)	3.25	0.50	TH-REF
Thermal forcing (B)	3.25	0.25	TH-FIS
Thermal forcing (B)	3.00	0.50	TH-LIS
Orographic forcing (C)	3.25	0.50	ORO-REF
Orographic forcing (C)	3.25	0.25	ORO-FIS
Orographic forcing (C)	3.00	0.50	ORO-LIS
Therm. and oro. forcing (D)	3.25	0.50	OTH-REF
Therm. and oro. forcing (D)	3.25	0.25	OTH-FIS
Therm. and oro. forcing (D)	3.00	0.50	OTH-LIS

(Bonelli et al., 2009) or the impact of dust deposition on snow albedo (Ganopolski et al., 2010; Bonelli et al., 2009) may have played a critical role in the deglaciation process. Here, we did not account for the influence of these factors to better constrain the mutual influence of past Northern Hemisphere ice sheets driven only by changes in atmospheric circulation. We therefore kept σ_0^{FIS} and σ_0^{LIS} constant through time and we did not include a representation of the impact of dust on snow albedo. Moreover, in the present version of GRISLI,

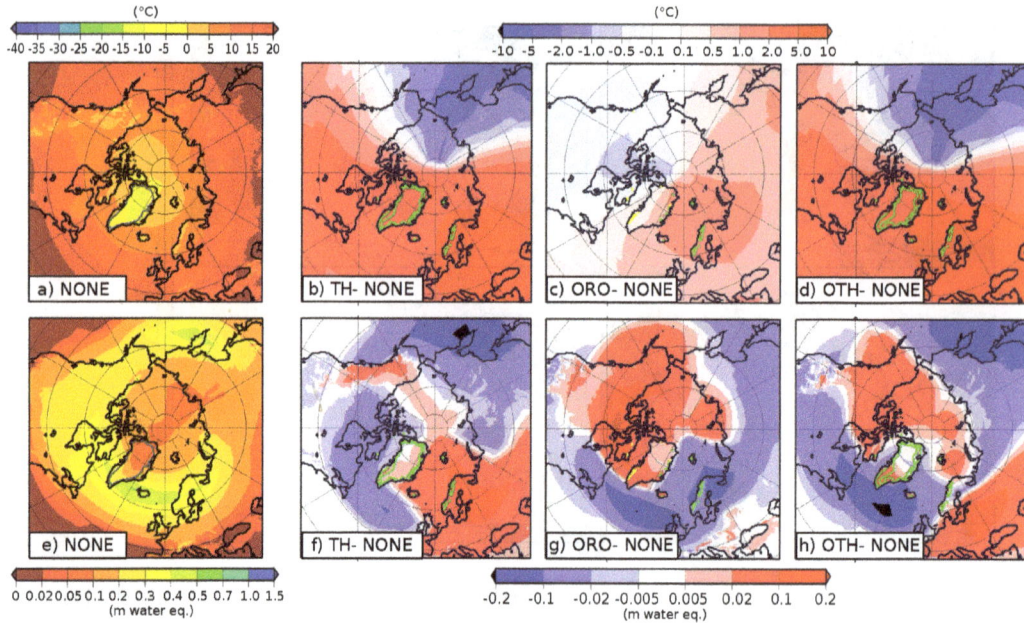

Fig. 2. Summer surface temperature at 125 ka in NONE-REF experiment (parameterization of SLP are removed) (**a**), difference of summer surface air temperature between TH-REF (thermal parameterization) and NONE-REF (**b**); ORO-REF (orographic parameterization) and NONE-REF (**c**) and OTH-REF (combination of both thermal and orographic parameterization) and NONE-REF (**c**) at 125 ka. Same for the snow accumulation (**e–h**). The green line represents the limit where the ice thickness difference exceeds −500 m.

the impact of water-saturated sediments on basal sliding that favor the ice retreat has not been taken into account. These missing mechanisms may explain why the last glacial termination is not simulated satisfactorily in our experiments. Our following analysis therefore focuses on the 126–15 ka period. Our aim here is not to simulate a perfect picture of the last glacial–interglacial cycle in full agreement with all available paleo-data, but rather to investigate the impact of one or the other stationary wave parameterization on the evolution of the ice sheets and their interdependence.

4 Results

4.1 Impact of sea-level pressure on temperature and accumulation patterns

The calculation of sea-level pressure leads to modifications in winds which in turn influence humidity and heat transports, altering cloud formation, precipitation and air temperature. The relation between precipitation, temperature and sea-level pressure is complex due to numerous feedbacks involved in the climate system. Therefore, it is difficult to predict from SLP patterns what the temperature and accumulation patterns will be. However, we compared the CLIMBER response of the sea-level pressure to a change in topography to the response of the IPSL-CM5 model. We used two experiments carried out with both models (with the OTH SLP parameterization in CLIMBER) with exactly the

same experimental set-up. The first one was run under pre-industrial conditions (PI) and the second one under LGM ones (except that the only ice sheet considered as boundary condition was the Laurentide ice sheet). Both model responses exhibit similar SLP patterns (not shown), despite slight disagreements in some places as for the amplitude of high and low pressures. In the present section, we examine the impact of different SLP parameterizations on snow accumulation and surface air temperature and thus on the ice-sheet surface mass balance. This analysis is made at 125 ka because at this time period there is no large ice sheet in the Northern Hemisphere except Greenland. Moreover, the 125 ka climate state has likely influenced the early phase of glacial inception and thus the construction and the further development of both Laurentide and Fennoscandia.

Figure 2 displays the simulated summer temperature (Fig. 2a) and accumulation (Fig. 2e) when the parameterization of sea-level pressure is removed (i.e. NONE experiment) as well as the difference of summer temperature and annual accumulation between each SLP parameterization (i.e. TH, ORO and OTH experiments, Figs. 2b–d and 2f–h). The azonal component in the SLP field produces a warming over a great part of the GRISLI domain (TH and OTH experiments) and over eastern longitudes and North Atlantic Ocean (ORO experiment), especially over Eurasia. Since the effect of the orography is weak in summer (see Sect. 3.1), its impact on summer temperature is less pronounced compared to the impact of the thermal effect (Fig. 2b and d). Over North America, where the orographic effects are accounted for, temperature patterns are dominated by the thermal effect (Fig. 2d). An excess of accumulation with respect to the NONE experiment is produced over western Eurasia with the thermal forcing whereas accumulation is reduced over Canada, Beringia and eastern Eurasia (Fig. 2f). The orographic parameterization has almost the opposite effect (Fig. 2g). The response of accumulation in OTH experiment is a combination of the two previous responses in which the regions of large accumulation rates are northwestern Canada, Barents and Kara seas and southern Europe.

These accumulation and temperature patterns have a direct impact on the construction of ice sheets during glacial inception. The spatial distributions of the simulated ice sheets are displayed in Fig. 3 for 115 ka. At this period, the glacial inception has already started as suggested by the sea-level reconstructions (Waelbroeck et al., 2002). This figure shows that the ice sheet response strongly depends on the SLP parameterization.

Compared to the NONE experiment, results from the ORO simulation show that the orographic effect favors the growth of the Laurentide ice sheet (Fig. 3c). This is due to the large accumulation rate over Canada (Fig. 2g) combined with a cooling effect (Fig. 2c). The Fennoscandian ice sheet is rather sensitive to the thermal forcing (Fig. 3b) which produces a large accumulation rate over Scandinavia (Fig. 2f). However, the simulated Fennoscandian ice sheet is smaller than that simulated by the NONE experiment because the large accumulation rate is widely counterbalanced by the temperature warming effect (Fig. 2b). Combining both SLP perturbations (thermal and orographic) (Fig. 3d) leads to a much smaller ice volume over Scandinavia (with respect to TH experiment) due to less accumulation, and more ice over the Kara Sea (increased accumulation). Likewise, the Laurentide ice volume simulated on the OTH experiment is in between the TH and the ORO ones.

These first results highlight how the representation of stationary waves in CLIMBER (through SLP parameterizations) may influence accumulation and temperature patterns and therefore the construction of the ice sheets. They show that the simulated ice sheets are heavily different depending on the forcing effect that we take into account (i.e. thermal and/or orographic). In the following sections, we examine the mutual interactions of ice sheets under different SLP parameterizations.

4.2 Glacial inception: from 126 to 110 ka

Figure 4 displays the evolution of the simulated LIS and FIS ice volumes from 126 to 110 ka for the twelve experiments described in Sect. 3. For a given parameterization, the comparison between solid and dotted lines (Fig. 4a–b) and between solid and dashed lines (Fig. 4c–d) illustrates the difference between the standard simulations (referred to as X-REF in the following) and the simulations where σ_0^{FIS} (Fig. 4a–b) or σ_0^{LIS} (Fig. 4c–d) is reduced.

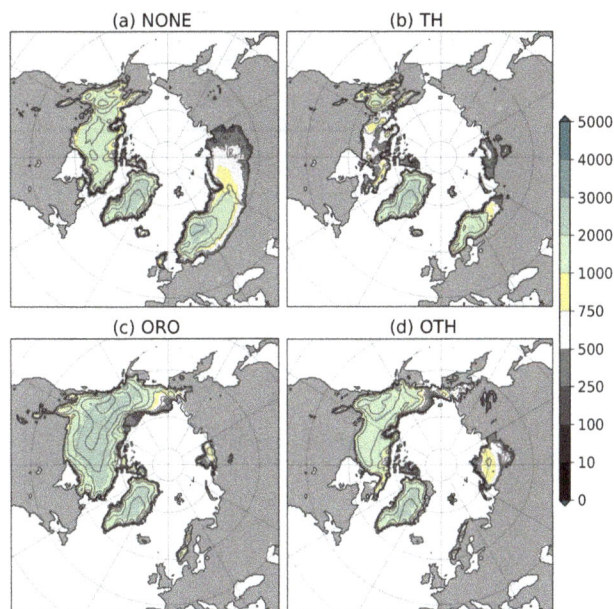

Fig. 3. Ice thickness (colors) and ice-sheet height (contours, isolines every 500 m) simulated at 115 ka with the four parameterizations (m).

4.2.1 Effect of smaller σ_0^{FIS}

As expected, for all SLP parameterizations, decreasing σ_0^{FIS} favors the growth of the FIS (Fig. 4a). However, the amplitude of this growth is more or less pronounced depending on the parameterization used. When stationary waves are off (i.e. azonal SLP component is set to zero), the simulated NONE-FIS ice volume (i.e. σ_0^{FIS} reduced) is around 30 % larger at 110 ka than the ice volume simulated in the baseline experiment. This increase is even more pronounced in TH and OTH experiments (270 % of ice volume difference between TH-FIS and TH-REF and 400 % between ORO-FIS and ORO-REF). In the ORO experiment, the ice volume is increased by 100 % compared to the reference case, but due to the small amount of ice, this increase is not really significant. Figure 4b illustrates how a change in the Fennoscandian ice-sheet geometry influences the construction of the Laurentide ice sheet. Compared to the Laurentide ice volume simulated under orographic forcing (15×10^{15} m^3), the difference in Fennoscandian ice volumes between ORO-FIS and ORO-REF is not large enough (0.5×10^{15} m^3 at 110 ka) to substantially modify the construction of the Laurentide ice sheet. By contrast, when stationary waves are off or when thermal effect is accounted for, the increase of the Fennoscandian ice volume leads to an increase of the volume of the Laurentide ice sheet. This effect may be directly related to the cold temperature anomaly occurring in response to the development of the Fennoscandian ice sheet. The comparison of summer temperature between NONE-FIS and NONE-REF (Fig. 5) at 125 and 120 ka shows that this cold temperature anomaly

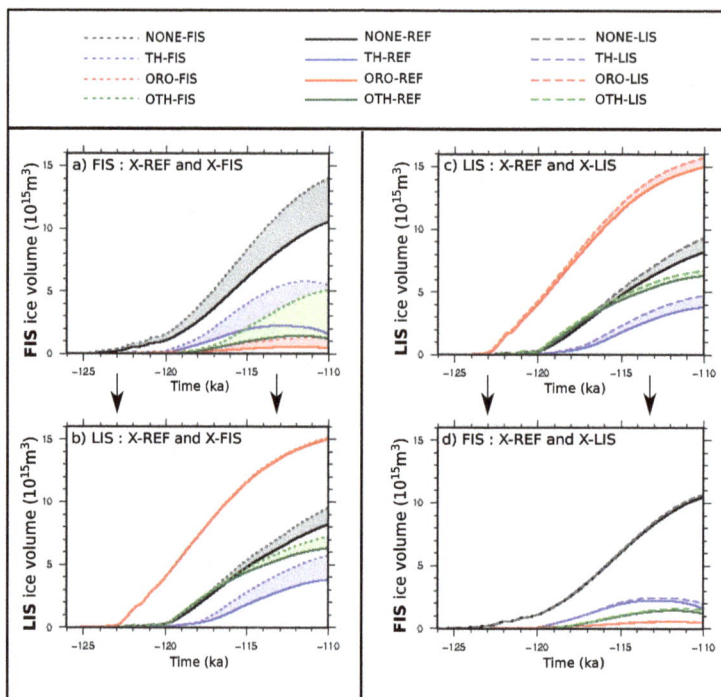

Fig. 4. The left part of the figure shows the evolution of Fennoscandian ice volume (**a**) for the reference experiments (solid lines, black when the waves are off, blue for the thermal parameterization, red for the orographic parameterization, and green when the two effects are on) and for the experiments with a smaller σ_0^{FIS} (plotted line, same color code). (**b**) shows the Laurentide ice volume for the same experiments. The right part of the figure shows the evolution of the Laurentide ice volume (**c**) for the reference experiments (solid lines) and for the ones with smaller σ_0^{LIS} (dotted lines), then the Fennoscandian ice volume (**d**) for the same experiments. For all panels, the evolution of ice volume is shown for the inception.

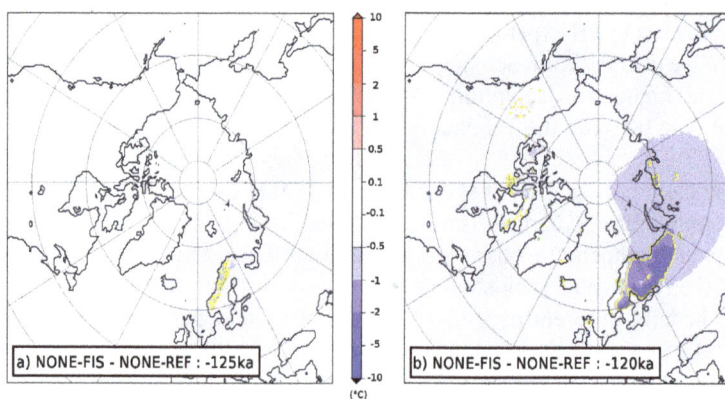

Fig. 5. Summer surface temperature differences between NONE-FIS and NONE-REF at 125 ka (**a**) and 120 ka (**b**). The yellow line represents the limit where the ice thickness difference exceeds 500 m.

progressively spreads all over the GRISLI domain, favoring thereby the growth of the Laurentide ice sheet. The spreading of the cold temperature anomaly can also be observed in OTH and TH experiments. This suggests that the larger the Fennoscandia ice sheet, the greater the effect on the Laurentide ice sheet. However, the differences of Fennoscandian ice volumes between TH-REF and TH-FIS on one hand and between OTH-REF and OTH-FIS on the other hand are similar.

However, the effect on the Laurentide ice sheet is twice as large under thermal forcing alone than the effect produced with the OTH experiment. This means that, in addition to the cooling effect (Fig. 5), another mechanism comes into play. Actually, in OTH experiments, the growth of Laurentide ice sheet leads to a high pressure over Canada (Fig. 6a–b). This causes a westward shift of the accumulation area (Fig. 6c–d) which in turn slows down the growth of the

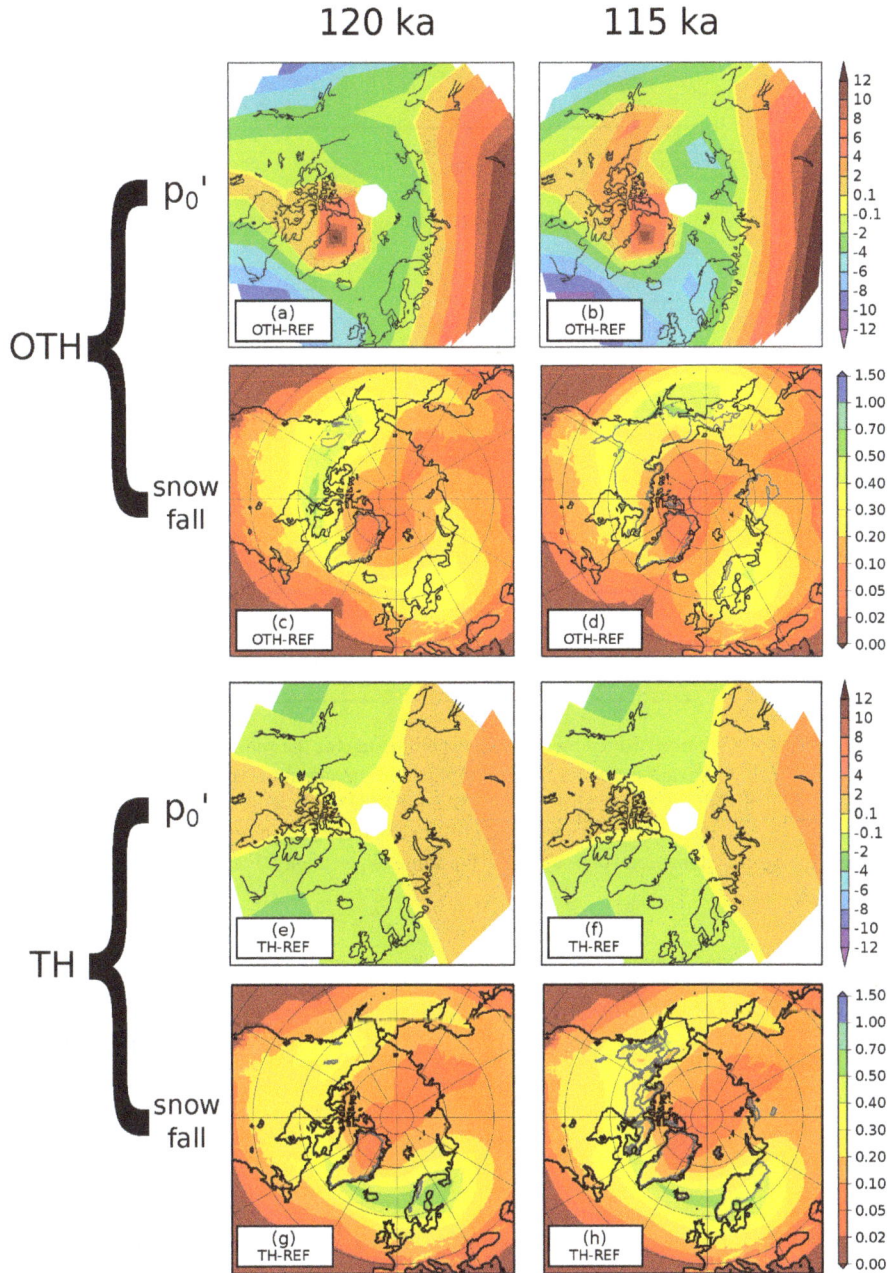

Fig. 6. Azonal sea-level pressure (in hPa) at 120 and 115 ka for OTH (**a–b**) and TH experiments (**e–f**), respectively, and accumulation (in meter water equivalent) at 120 and 115 ka for OTH (**c–d**) and TH experiments (**g-h**). Grey lines indicate the limits of the ice sheets.

Laurentide ice sheet. Note that this westward shift is also observed in the ORO experiments (not shown). Under thermal forcing, a slight extent of the high pressure area is also simulated with the growth of the Laurentide ice sheet. However, this extent remains insufficient to act on the displacement of the accumulation area. This makes the TH-FIS Laurentide ice sheet more sensitive to the cooling effect induced by a larger Fennoscandia than the ice sheet simulated in OTH-FIS. This also explains why the differences in simulated LIS

ice volumes between TH-REF and TH-FIS are larger than those between OTH-REF and OTH-FIS.

4.2.2 Effect of a smaller σ_0^{LIS}

The reduction of σ_0^{LIS} has a smaller effect on the LIS than the effect of a lowered σ_0^{FIS} value on the FIS (Fig. 4c). Although, σ_0^{LIS} and σ_0^{FIS} are both reduced by 0.25 °C, the resulting changes in ablation are not equivalent due to the non-linearity of the PDD formulation. However, the lowering

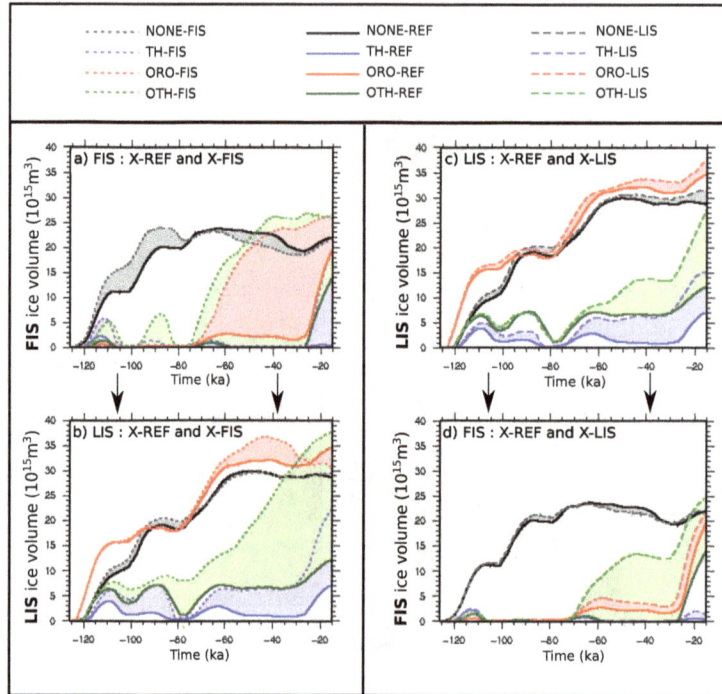

Fig. 7. The left part of the figure shows the evolution of Fennoscandian ice volume (**a**) for the reference experiments (solid lines, black when the waves are off, blue for the thermal parameterization, red for the orographic parameterization, and green when the two effects are on) and for the experiments with a smaller σ_0^{FIS} (plotted line, same color code). (**b**) shows the Laurentide ice volume for the same experiments. The right part of the figure shows the evolution of the Laurentide ice volume (**c**) for the reference experiments (solid lines) and for the ones with smaller σ_0^{LIS} (dotted lines), then the Fennoscandian ice volume (**d**) for the same experiments. For all panels, the evolution of ice volume is shown for the all glaciation.

of σ_0^{LIS} is large enough to produce a larger LIS (Fig. 4c). The ice-sheet growth is smaller in ORO-LIS and OTH-LIS experiments due to the westward shift of the accumulation mentioned above. In all experiments, the differences in the ice-sheet size (with respect to the standard simulations) are too small to significantly influence the construction of FIS (Fig. 4d), at least during glacial inception.

4.3   Full glacial state: from 80 to 30 ka

The evolution of the simulated FIS and LIS ice volumes over the entire last glaciation are displayed in Fig. 7. Around 70 kyr BP, there is a minimum of insolation in northern latitudes combined to a decrease of the atmospheric CO_2 concentration which allows a rapid decrease of summer temperature that accelerates the glaciation process.

4.3.1   Effect of a smaller σ_0^{FIS}

A large difference is observed in Fennoscandian ice volume (Fig. 7a) between ORO-REF (red solid line) and ORO-FIS (dotted-red line) from 75 ka to the end of the simulation. Actually, slight differences in simulated ice volumes between ORO-REF and ORO-FIS appear before 80 ka. They induce small differences in accumulation and temperature patterns

as soon as 80 ka (Fig. 8). A small excess of accumulation over the Kara Sea is simulated in ORO-FIS (with respect to ORO-REF) as well as slightly colder temperatures over the Eurasian region. These tiny differences are sufficient to trigger a massive glaciation of the FIS in the ORO-FIS experiment, whereas in ORO-REF the ice volume remains at a low level ($< 3.0 \times 10^{15}\,\mathrm{m}^3$) until 30 ka. These results show how the decrease of insolation and atmospheric CO_2 concentration may be amplified by a small change in the ice-sheet surface mass balance. The development of both LIS and FIS in the ORO-FIS experiment leads to a cooling all over the GRISLI domain. This shifts southward the snow–rain limit (with respect to ORO-REF), increasing the accumulation rate in the North Atlantic and eastern Canada. This effect combined with the previous cooling effect leads to an acceleration of the LIS growth (Fig. 7b, red solid line with respect to red dashed line).

A similar behavior is observed for the FIS ice volume when both forcings are accounted for (OTH experiment) with an acceleration of the FIS growth in OTH-FIS after 75 ka (Fig. 7a, green lines). The underlying mechanism is similar to the one explained in the ORO case. However, it is more efficient because the differences at 80 ka between the LIS ice volumes simulated in the OTH-REF and OTH-FIS

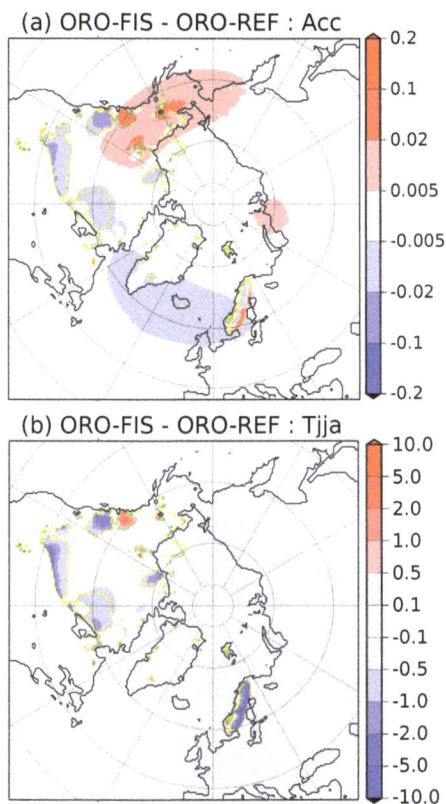

Fig. 8. (a) Accumulation difference (in meter water equivalent) between ORO-FIS and ORO-REF at 80 ka. The yellow line is the limit where the thickness difference exceeds 500 m, the green one where the difference is under −500 m. (b) Summer temperatures differences between ORO-FIS and ORO-REF at 80 ka (in °C).

experiments are larger than the differences between ORO-REF and ORO-FIS (Fig. 7b). Similarly to the ORO experiments, the acceleration of the FIS growth implies an acceleration of the LIS growth (with respect to OTH-REF).

Under the thermal parameterization, there is no difference in the FIS ice volumes between TH-REF and TH-FIS throughout the period spanning from 80 to 30 ka (Fig. 7a, blue lines), whereas the LIS is larger (TH-FIS experiment, Fig. 7b). As explained in Sect. 4.2.1, at the beginning of the glaciation, the decrease of σ_0^{FIS} leads to a larger FIS and in turn to a larger LIS. Actually, the LIS has remained larger in TH-FIS (with respect to TH-REF), even during periods where FIS was entirely melted (e.g. 100 ka). This means that the LIS ice volume differences between TH-FIS and TH-REF after 75 ka does not result from a direct effect of a change in FIS geometry, but rather comes from the fact that the LIS has remained glaciated throughout this period even around 80 ka when the FIS ice volume was very small but not zero.

Finally, we obtain a surprising result when the stationary waves are off: after 75 ka, the NONE-FIS experiment simulates a smaller FIS than the NONE-REF experiment

(Fig. 7a, grey and black lines), despite a smaller σ_0^{FIS}. The comparison of accumulation and temperature patterns between both experiments (not shown) does not explain the decrease of the ice volume after 90 ka in the NONE-FIS experiment (Fig. 7a). Instead, a strong decrease of the simulated ice thickness in the southwestern part of the ice sheet (NONE-FIS) is associated with a decrease of ice flow velocities (not shown). This suggests that the mechanisms responsible for this behavior are closely linked to ice dynamical effects. However, an in-depth analysis of the ice dynamics is beyond the scope of this study.

4.3.2 Effect of a smaller σ_0^{LIS}

At the beginning of the glaciation, the decrease of σ_0^{LIS} does not have a significant influence on the growth of the ice sheets, at least until 75 ka (Fig. 7c and d). After 75 ka, when the LIS is large (NONE and ORO experiments), the decrease of σ_0^{LIS} has only a poor impact on the ice volume. With the TH and OTH parameterizations, the geometry of simulated ice sheets in both reference runs and runs carried out with a smaller σ_0^{LIS} value is similar around 80 ka (TH and OTH experiments). However, after 75 ka, the differences of ice volumes between TH-REF and TH-LIS and between OTH-REF and OTH-LIS become significant, in contrast to results obtained in the ORO and NONE experiments. The impact of a larger LIS on the evolution of FIS is clearly visible in the OTH runs but remains very weak with the TH forcing. To understand the origin of this different behavior, we examined the differences between OTH-LIS and OTH-REF and between TH-LIS and TH-REF in terms of summer temperature and accumulation patterns. To be relevant, this comparison has to be made at a time period when the differences of ice volumes between reference runs and small σ_0^{LIS} runs (hereafter called Δice) are roughly of similar magnitude in both OTH and TH experiments. Similar Δice values are obtained at 75 (OTH) and 71 ka (TH). Figure 9 displays the differences in SLP, accumulation and summer temperature patterns at these periods. Since we take into account the topography effect on sea-level pressure, the response of a similar LIS Δice on sea-level pressure is stronger in the OTH than in the TH experiment (Fig. 9a and b). Changes in accumulation are larger in the OTH case (Fig. 9e and f). A slight excess of accumulation is observed over the Kara Sea region in OTH-LIS experiment (with respect to OTH-REF), which seems to be sufficient to trigger the growth of FIS. Nevertheless, the impact on temperature in the TH experiment is more important than in the OTH experiment, due to a larger LIS Δice (Fig. 9c and d).

These results clearly show that the sensitivity of the atmospheric circulation to the presence of an ice sheet becomes significant when the topography effect is accounted for. Moreover, a change in sea-level pressure due to the growth of an ice sheet seems to mainly alter the accumulation with only a little influence on summer temperature.

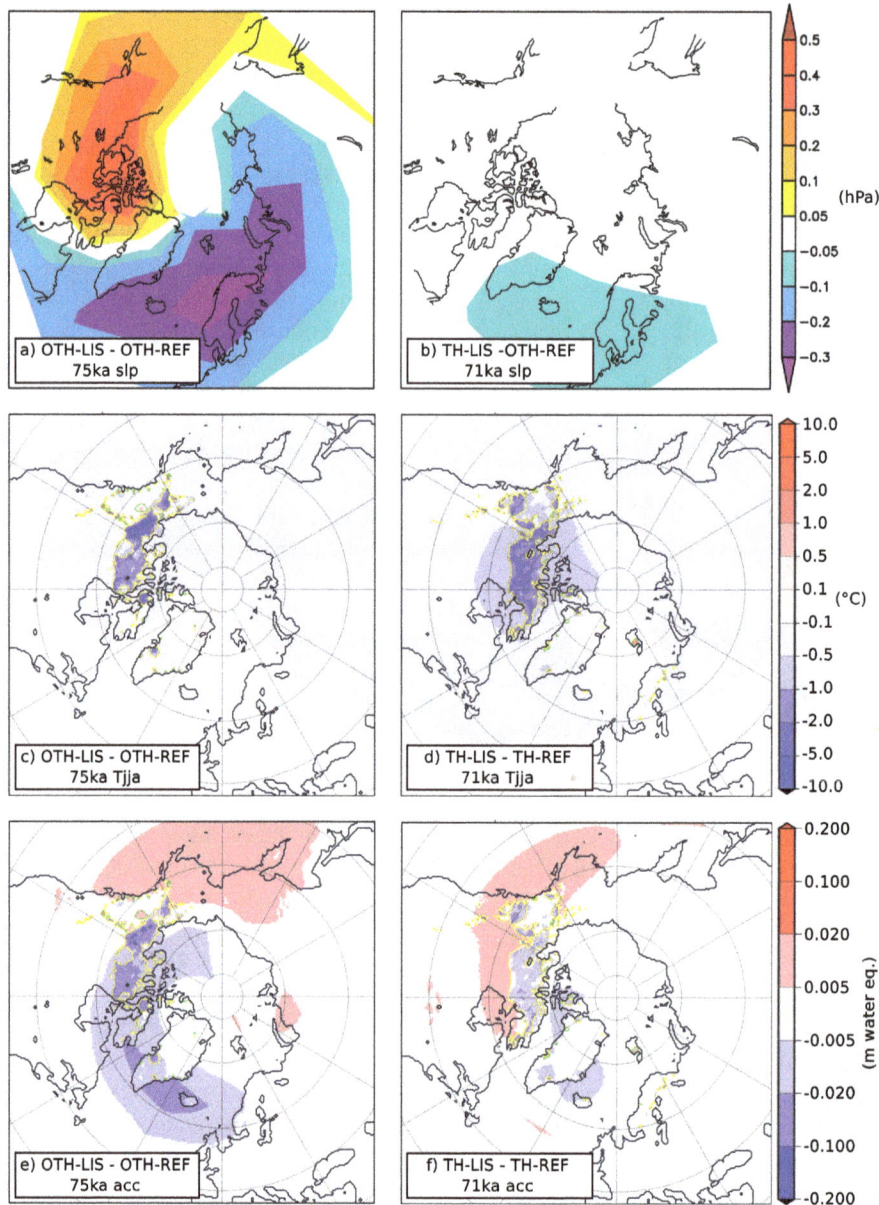

Fig. 9. Sea-level pressure differences (in hPa) between OTH-LIS and OTH-REF at 75 ka (**a**) and between TH-LIS and TH-REF at 71 ka (**b**). Summer temperature difference between OTH-LIS and OTH-REF at 75 ka (**c**); and TH-LIS and TH-REF at 71 ka (**d**); accumulation difference between OTH-LIS and OTH-REF at 75 ka (**e**); and TH-LIS and TH-REF at 71 ka (**f**).

4.4   30 until 15 ka BP

After 30 ka, a new decrease in insolation concomitant with a decrease in atmospheric CO_2 concentration occurs. The response of the ice volume is more or less pronounced depending on the experiment (Fig. 7). As an example, a pronounced response of both ice sheets (Fig. 7a and b, blue dotted line) is simulated in TH-FIS, unlike NONE-REF, NONE-FIS and NONE-LIS. More generally, it should be noted that the response is absent or weak in simulations where both ice sheets have already reached a significant size at 35 ka (more

than $18 \times 10^{15} \, m^3$ for the FIS and more than $28 \times 10^{15} \, m^3$ for the LIS). This can be explained by the dry cold air above continental-scale ice sheet. In contrast, when one ice sheet is large but the other is small (e.g. ORO-REF) at 35 ka, colder temperatures due to the decreasing of CO_2 atmospheric concentration and insolation favor the growth of the smaller ice sheet (e.g. FIS start to grow at 30 ka). In turn, this leads to even much colder temperatures and a further growth of the larger ice sheet too (LIS starting to grow at 35 ka). Therefore, the response of one ice sheet to an insolation/CO_2 decrease also depends on the size of both ice sheets.

5   Conclusions

In the present study, we investigated the atmospheric-based processes that relate the two main Northern Hemisphere ice sheets during the last glaciation. In particular, through appropriate parameterizations, we examined the effect of topography and surface temperature on the sea-level pressure, and we studied the impact of these effects in the relationship between ice sheets. The parameterization of ablation in the ice-sheet model is used as a trigger to change the size of one ice sheet and to investigate the mutual influence of the ice sheets. Our aim was to investigate the effect of a change in one ice-sheet geometry on the other one, depending on the sea-level pressure parameterization. First, at the beginning of the glaciation, we showed that the growth of the Laurentide ice sheet is favored by the topographic effect on sea-level pressure, due to a large accumulation rate in this area, whereas there is more accumulation over Fennoscandia with the thermal parameterization. Secondly, we showed that a larger ice sheet (e.g. FIS) leads locally to a cooling that progressively spreads over the northern latitudes and promotes the growth of the other ice sheet (e.g. LIS). This former mechanism is true whatever the sea-level pressure parameterization is. The second process is the shift of snowfall pattern when an ice sheet grows, altering thereby the accumulation over the other ice sheet. This mechanism is dominant when the orographic effect is on, because it increases the sensitivity of sea-level pressure to the presence of an ice sheet. In that case a change in sea-level pressure mainly affects the accumulation patterns. Owing to the fact that stationary waves are not explicitly computed in CLIMBER and due to the low resolution of the model, the importance of various feedbacks discussed in the present study as well as the amplitude of remote effects of ice sheets may be under- or overestimated. However, this study highlights the key role of topography on sea-level pressure and accumulation, and thus on the evolution of the ice sheets themselves. This means that it is necessary to take into account the mutual influence of past Northern Hemisphere ice sheets to properly understand the mechanisms underlying their own evolution. In the same way, our results suggest that feedbacks between ice sheets and stationary waves could be of key importance to understand the various configurations of ice-sheet shapes during different ice ages of the Quaternary era (e.g. Svendsen, 2004).

Acknowledgements. This work has benefited from fruitful discussions with Daniel Lunt. It has been supported by CEA, CNRS and UVSQ.

Edited by: V. Rath

References

Ambach, W.: Heat balance characteristics and ice ablation, western EGIG-profile, Applied hydrology in the development of northern basins, Danish Society for Arctic Technology, Copenhagen, 1988.

Berger, A.: Long-Term Variations of Daily Insolation and Quaternary Climatic Changes, J. Atmos. Sci., 35, 2362–2367, 1978.

Bonelli, S., Charbit, S., Kageyama, M., Woillez, M.-N., Ramstein, G., Dumas, C., and Quiquet, A.: Investigating the evolution of major Northern Hemisphere ice sheets during the last glacial-interglacial cycle, Clim. Past, 5, 329–345, doi:10.5194/cp-5-329-2009, 2009.

Braithwaite, R. J.: Calculation of degree-days for glacier-climate research, Z. Gletscherk. Glazialgeol., 20, 1–18, 1984.

Braithwaite, R. J.: Positive degree-day factors for ablation on the Greenland ice sheet studies by energy-balance modelling, J. Glaciol., 41, 153–160, 1995.

Braithwaite, R. J. and Zhang, Y.: Sensitivity of mass balance of five Swiss glaciers to temperature changes assessed by tuning a degree-day model, J. Glaciol., 46, 7–14, 2000.

Broccoli, A. J. and Manabe, S.: The influence of continental ice, atmospheric CO_2, and land albedo on the climate of the LGM, Clim. Dynam., 1, 87–99, 1987.

Charbit, S., Dumas, C., Kageyama, M., Roche, D. M., and Ritz, C.: Influence of ablation-related processes in the build-up of simulated Northern Hemisphere ice sheets during the last glacial cycle, The Cryosphere, 7, 681–698, doi:10.5194/tc-7-681-2013, 2013.

Clark, P. U.: Northern Hemisphere Ice-Sheet Influences on Global Climate Change, Science, 286, 1104–1111, 1999.

Clark, P. U., Clague, J., Curry, B., Dreimanis, A., Hicock, S., Miller, G., Berger, G., Eyles, N., Lamothe, M., Miller, B., Mott, R., Oldale, R., Stea, R., Szabo, J., Thorleifson, L., and Vincent, J.-S.: Initiation and development of the Laurentide and Cordilleran Ice Sheets following the last interglaciation, Quaternary Sci. Rev., 12, 79–114, 1993.

Cook, K. H. and Held, I. M.: Stationary Waves of the Ice Age Climate, J, Climate, 1, 807–819, 1988.

Dyke, A. S. and Prest, V. K.: Late Wisconsinan and Holocene History of the Laurentide Ice Sheet, Géographie physique et Quaternaire, 41, 237–263, 1987.

Fausto, R. S., Ahlström, A. P., van As, D., Johnsen, S. J., Langen, P. L., and Steffen, K.: Improving surface boundary conditions with focus on coupling snow densification and meltwater retention in large-scale ice-sheet models of Greenland, J. Glaciol., 55, 869–878, 2009.

Ganopolski, A., Calov, R., and Claussen, M.: Simulation of the last glacial cycle with a coupled climate ice-sheet model of intermediate complexity, Clim. Past, 6, 229–244, doi:10.5194/cp-6-229-2010, 2010.

Hall, N. M. J., Valdes, P. J., and Dong, B.: The Maintenance of the Last Great Ice Sheets: A UGAMP GCM Study, J. Climate, 9, 1004–1019, 1996.

Held, I. M., Ting, M., and Wang, H.: Northern Winter Stationary Waves: Theory and Modeling, J. Climate, 15, 2125–2144, 2002.

Hock, R.: Temperature index melt modelling in mountain areas, J. Hydrol., 282, 104–115, 2003.

Holton, J. R.: An introduction to dynamic meteorology, Academic Press, New York, 1979.

Hutter, K.: Theoretical glaciology: material science of ice and the mechanics of glaciers and ice sheets, Reidel, Terra Scientific Pub. Co., Sold and distributed in the USA and Canada by Kluwer Academic Publishers, 1983.

Kageyama, M. and Valdes, P. J.: Impact of the North American ice-sheet orography on the Last Glacial Maximum eddies and snowfall, Geophys. Res. Lett., 27, 1515, doi:10.1029/1999GL011274, 2000.

Kalnay, E., Kanamitsu, M., Kistler, R., Collins, W., Deaven, D., Gandin, L., Iredell, M., Saha, S., White, G., Woollen, J., Zhu, Y., Leetmaa, A., Reynolds, R., Chelliah, M., Ebisuzaki, W., Higgins, W., Janowiak, J., Mo, K. C., Ropelewski, C., Wang, J., Jenne, R., and Joseph, D.: The NCEP/NCAR 40-Year Reanalysis Project, B. Am. Meteorol. Soc., 77, 437–471, 1996.

Lambeck, K., Purcell, A., Funder, S., Kjaer, K., Larsen, E., and Müller, P.: Constraints on the Late Saalian to early Middle Weichselian ice sheet of Eurasia from field data and rebound modelling, Boreas, 35, 539–575, 2006.

Laîné, A., Kageyama, M., Salas-Mélia, D., Voldoire, A., Rivière, G., Ramstein, G., Planton, S., Tyteca, S., and Peterschmitt, J. Y.: Northern hemisphere storm tracks during the last glacial maximum in the PMIP2 ocean-atmosphere coupled models: energetic study, seasonal cycle, precipitation, Clim. Dynam., 32, 593–614, 2008.

Liakka, J., Nilsson, J., and Löfverström, M.: Interactions between stationary waves and ice sheets: linear versus nonlinear atmospheric response, Clim. Dynam., 38, 1249–1262, 2011.

MacAyeal, D. R.: Large-Scale Ice Flow Over a Viscous Basal Sediment: Theory and Application to Ice Stream B, Antarctica, J. Geophys. Res., 94, 4071–4087, 1989.

Morland, L. W.: Thermomechanical balances of ice sheet flows, Geophys. Astrophys. Fluid Dynam., 29, 237–266, 1984.

Oerlemans, J. and Vugts, H.: A meteorological experiment in the melting zone of the Greenland ice sheet, B. Am. Meteorol. Soc. USA, 74, 355–365, 1993.

Pausata, F. S. R., Li, C., Wettstein, J. J., Kageyama, M., and Nisancioglu, K. H.: The key role of topography in altering North Atlantic atmospheric circulation during the last glacial period, Clim. Past, 7, 1089–1101, doi:10.5194/cp-7-1089-2011, 2011.

Peltier, W.: Global glacial isostasy and the surface of the ice-age Earth: The ICE-5G (VM2) Model and GRACE, Annu. Rev. Earth Planet. Sci., 32, 111–149, 2004.

Petit, J., Jouzel, J., Raynaud, D., Barkov, N., Barnola, J.-M., Basile, I., Bender, M., Chappellaz, J., Davisk, M., Delaygue, G., Delmotte, M., Kotlyakov, V. M., Legrand, M., Lipenkov, V. Y., Lorius, C., Pépin, L., Ritz, C., Saltzman, E., and Stievenard, M.: Climate and atmospheric history of the past 420,000 years from the Vostok ice core, Antarctica, Nature, 399, 429–436, doi:10.1038/20859, 1999.

Petoukhov, V., Ganopolski, A., Brovkin, V., Claussen, M., Eliseev, A., Kubatzki, C., and Rahmstorf, S.: CLIMBER-2: a climate system model of intermediate complexity, Part I: model description and performance for present climate, Clim. Dynam., 16, 1–17, 2000.

Peyaud, V., Ritz, C., and Krinner, G.: Modelling the Early Weichselian Eurasian Ice Sheets: role of ice shelves and influence of ice-dammed lakes, Clim. Past, 3, 375–386, doi:10.5194/cp-3-375-2007, 2007.

Reeh, N.: Parameterization of melt rate and surface temperature on the Greenland ice sheet, Polarforschung, 59, 113–128, 1991.

Ritz, C., Rommelaere, V., and Dumas, C.: Modeling the evolution of Antarctic ice sheet over the last 420,000 years: Implications for altitude changes in the Vostok region, J. Geophys. Res., 106, 31943–31964, 2001.

Rivière, G., Laîné, A., Lapeyre, G., Salas-Mélia, D., and Kageyama, M.: Links between Rossby Wave Breaking and the North Atlantic Oscillation–Arctic Oscillation in Present-Day and Last Glacial Maximum Climate Simulations, J. Climate, 23, 2987–3008, 2010.

Roe, G. H. and Lindzen, R. S.: A one-dimensional model for the interaction between continental-scale ice sheets and atmospheric stationary waves, Clim. Dynam., 17, 479–487, 2001a.

Roe, G. H. and Lindzen, R. S.: The Mutual Interaction between Continental-Scale Ice Sheets and Atmospheric Stationary Waves, J. Climate, 14, 1450–1465, 2001b.

Svendsen, J.: Late Quaternary ice sheet history of northern Eurasia, Quaternary Sci. Rev., 23, 1229–1271, 2004.

Tarasov, L.: Understanding past ice sheet evolution: the challenges in integrating data and glaciological modelling, Quatern. Int., 279–280, 484–485, 2012.

Vallis, G. K.: Atmospheric and Oceanic Fluid Dynamics: Fundamentals and Large-scale Circulation, Cambridge University Press, 2006.

Waelbroeck, C., Labeyrie, L., Michel, E., Duplessy, J., McManus, J., Lambeck, K., Balbon, E., and Labracherie, M.: Sea-level and deep water temperature changes derived from benthic foraminifera isotopic records, Quaternary Sci. Rev., 21, 295–305, 2002.

9

Holocene environmental changes in the highlands of the southern Peruvian Andes (14° S) and their impact on pre-Columbian cultures

K. Schittek[1], M. Forbriger[2], B. Mächtle[3], F. Schäbitz[1], V. Wennrich[4], M. Reindel[5], and B. Eitel[3]

[1] Seminar of Geography and Geographical Education, University of Cologne, Germany
[2] Institute of Geography, University of Cologne, Germany
[3] Geographical Institute, University of Heidelberg, Germany
[4] Institute of Geology and Mineralogy, University of Cologne, Germany
[5] German Archaeological Institute, Commission for Archaeology of Non-European Cultures (KAAK), Bonn, Germany

Correspondence to: K. Schittek (schittek@uni-koeln.de)

Abstract. High-altitude peatlands of the Andes still remain relatively unexploited although they offer an excellent opportunity for well-dated palaeoenvironmental records.

To improve knowledge about climatic and environmental changes in the western Andes of southern Peru, we present a high-resolution record of the Cerro Llamoca peatland for the last 8600 years. The 10.5 m long core consists of peat and intercalated sediment layers and was examined for all kinds of microfossils. We chose homogeneous peat sections for pollen analysis at decadal to centennial resolution. The inorganic geochemistry was analysed in 2 mm resolution (corresponding > 2 years) using an ITRAX X-ray fluorescence core scanner.

We interpret phases of relatively high abundances of Poaceae pollen in our record as an expansion of Andean grasslands during humid phases. Drier conditions are indicated by a significant decrease of Poaceae pollen and higher abundances of Asteraceae pollen. The results are substantiated by changes in arsenic contents and manganese/iron ratios, which turned out to be applicable proxies for in situ palaeoredox conditions.

The mid-Holocene period of 8.6–5.6 ka is characterised by a series of episodic dry spells alternating with spells that are more humid. After a pronounced dry period at 4.6–4.2 ka, conditions generally shifted towards a more humid climate. We stress a humid/relatively stable interval between 1.8 and 1.2 ka, which coincides with the florescence of the Nasca culture in the Andean foothills. An abrupt turn to a sustained dry period occurs at 1.2 ka, which is contemporaneous with the demise of the Nasca/Wari society in the Palpa lowlands. Markedly drier conditions prevail until 0.75 ka, providing evidence of the presence of a Medieval Climate Anomaly. Moister but hydrologically highly variable conditions prevailed again after 0.75 ka, which allowed re-expansion of tussock grasses in the highlands, increased discharge into the Andean foreland and resettling of the lowlands during this so-called late Intermediate Period (LIP).

On a supraregional scale, our findings can ideally be linked to and proved by the archaeological chronology of the Nasca–Palpa region as well as other high-resolution marine and terrestrial palaeoenvironmental records. Our findings show that hydrological fluctuations, triggered by the changing intensity of the monsoonal tropical summer rains emerging from the Amazon Basin in the north-east, have controlled the climate in the study area.

1 Introduction

There is clear evidence that marked, global-scale climatic changes during the Holocene induced significant and complex environmental responses in the central Andes, which repeatedly led to abrupt changes in temperature, precipitation and the periodicity of circulation regimes (Jansen et al., 2007; Bird et al., 2011a). This region is characterised by distinct gradients of several climatic parameters over short distances and hence is particularly sensitive to the effects of environmental change. It hosts a multitude of microenvironments that have varied with climatic changes, resulting in significant responses of vegetation zonation, geomorphodynamics and other variations in biotic and abiotic systems (Grosjean et al., 2001; Garreaud et al. 2003; Grosjean and Veit, 2005).

During the last decade, several studies improved understanding of South American climate, related mechanisms and teleconnections substantially (Baker et al., 2005; Ekdahl et al., 2008; Garreaud et al., 2008; Bird et al., 2011a, b; Vuille et al., 2012). Although considerable efforts have been made to decipher the palaeoenvironmental history of the central Andes, many aspects of the timing, magnitude and origin of past climate changes remain poorly defined (Grosjean et al., 2001; Latorre et al., 2003; Gayo et al., 2012).

Particularly the distorting effects of high-amplitude precipitation changes, such as complex series of humid/dry spells that repeatedly appeared throughout the Holocene, often affect the continuity and resolution of palaeoenvironmental records. Especially in the central Andes, detailed knowledge of the distribution and amplitude of abrupt climatic changes is still sparse and it remains unclear how these climatic oscillations align with the Southern Hemisphere circulation regimes (Baker et al., 2005; Moreno et al., 2007).

Considering the possible relationship between environmental and cultural changes, the emergence, persistence and subsequent collapse of pre-Columbian civilisations offer important insights into human–environment interactions (Binford et al., 1997). The success of pre-Columbian civilisations was closely coupled to areas of geo-ecological favourability, which were directly controlled by distinct regional impacts of large-scale atmospheric circulation mechanisms (Eitel et al., 2005; Mächtle and Eitel, 2013).

A large number of archaeological sites in the northern part of the Río Grande de Nasca drainage were documented by the German Archaeological Institute between 1997 and 2010 (Reindel, 2009; Sossna, 2012; Reindel and Isla, 2013). Based on more than 150 ^{14}C samples, Unkel et al. (2012) presented a numerical chronology for the cultural development of this area that covers the time from the Archaic Period to the late Intermediate Period (LIP) (5.7–0.5 ka/~ 3760 cal BC–AD 1450 cal). This archaeological data source represents a unique prerequisite to facilitate linkages with palaeoenvironmental records obtained from continuous geoarchives in the nearby Andean highlands.

To supplement and specify the current knowledge about climatic and environmental changes in the western Andes of southern Peru, we present a new record from the Cerro Llamoca peatland (CLP) for the last 8.6 ka. The peat and sediment-bearing core was drilled within a Juncaceous cushion peatland in the headwater area of Río Viscas, a tributary to Río Grande, at an altitude of 4200 m a.s.l. The CLP record represents an ideal site for the reconstruction of precipitation changes over long periods that had an influence on desert margin shifts at the Andean footzone. If the record implies distinct dry conditions in the high mountain areas of CLP,

we can assume extreme dryness for the lower parts and a significant reduction of river runoff affecting water availability in the lowlands.

In the Palpa–Nasca region, the rise and fade of pre-Columbian cultures can be an example of the significant impact of environmental changes on cultural development and human settlement strategies (Eitel and Mächtle, 2009; Reindel, 2009).

2 The study area

2.1 Geographical setting, regional climate and vegetation

The investigated CLP is located in the western cordillera of the Peruvian Andes (Fig. 1). The name-giving peak, Cerro Llamoca (14°10′ S, 74°44′ W; 4450 m a.s.l.), is the highest point of the Río Viscas catchment area. As part of the continental divide, water courses on the western flank drain towards the Pacific Ocean.

Geologically, the CLP is situated within an area dominated by Tertiary rocks. The Castrovirreyna formation, formed during the upper Oligocene to early Miocene, consists of andesitic conglomerates intercalated with rhyolitic, dacitic vitric tuffs and thin sandstone layers, followed by andesitic breccias with intermediate andesitic and dacitic tuffs overlain by sandstones and andesitic breccias. Cerro Llamoca itself is a volcanic dyke and part of the early-Pliocene Caudalosa formation. It consists of heavily weathered andesites, andesitic ash tuffs and volcanic conglomerates (Castillo et al., 1993).

Situated in the transition zone between dry and humid Puna (Troll, 1968), an annual rainfall amount based on the data from the Tropical Rainfall Measuring Mission (TRMM) (Bookhagen and Stecker, 2008) of about 200–400 mm a^{-1} is estimated for the Cerro Llamoca area (Schittek et al., 2012).

Precipitation in the study area originates from Atlantic Ocean moist air masses that are transported to the Western Cordillera by upper-level tropical easterly flow. The strength of the easterly flow is controlled by the El Niño–Southern Oscillation (ENSO) system, with increased flow during La Niña episodes (Garreaud et al., 2009). About 90 % of total rainfall is concentrated during the austral summer months between November and March (Garreaud, 2000). This seasonal rainfall variability is connected to the position of the Intertropical Convergence Zone (ITCZ) as well as to the strength of the South American summer monsoon (SASM) (Zhou and Lau, 1998; Maslin and Burns, 2000; Maslin et al., 2011; Vuille et al., 2012). Seasonal water excess in the highlands supports the river oases downstream in the desert, where irrigation agriculture is practised since pre-Columbian times (Mächtle, 2007; Reindel, 2009). Movement of moist air from the Pacific onto the Altiplano is prevented by a strong and persistent temperature inversion maintained by

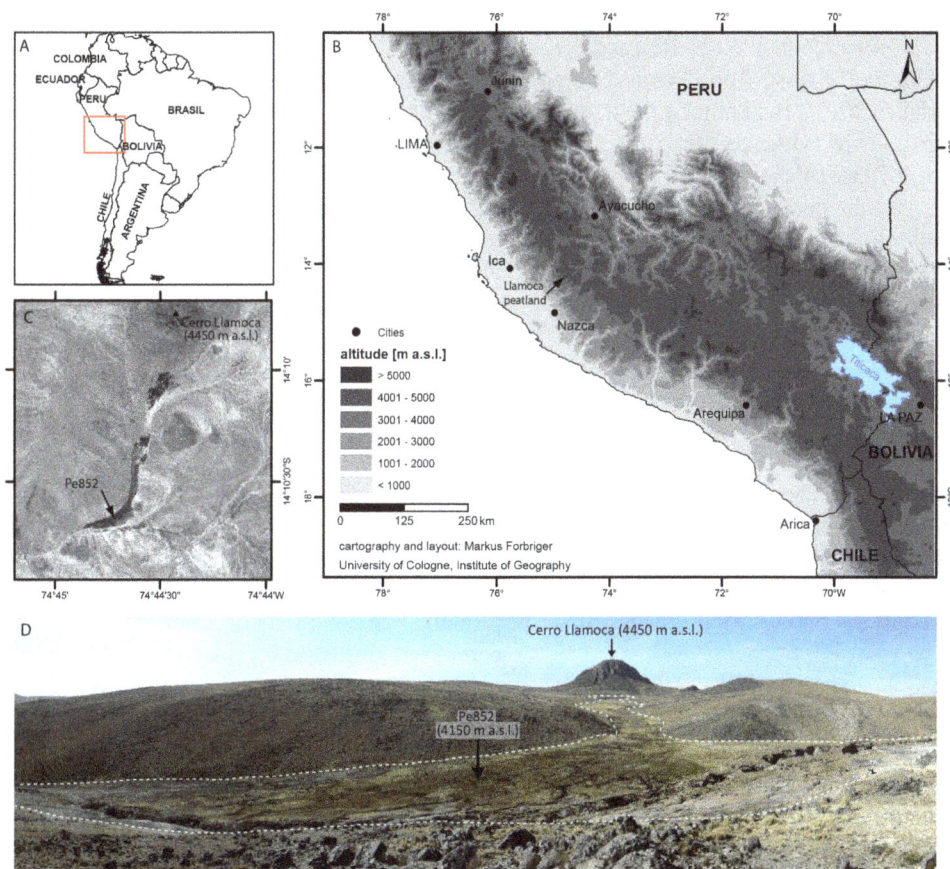

Figure 1. (**a**) Map of Peru and adjacent countries (data source: GLCF World Data). (**b**) The location of Cerro Llamoca peatland (CLP) in the western Andes of southern Peru (data source: DGM-GTOPO 30). (**c**) Aerial photograph of CLP with Cerro Llamoca in the north and location of the coring site of core Pe852 (Servicio Aerofotográfico Nacional – SAN, Lima). (**d**) Panorama of CLP with the name-giving peak and the location of the coring site. Dashed lines indicate the extension of the peat- and sediment-accumulating area. The southern part of the peatland is separated from water supply by a deeply incised gully.

cool waters offshore and large-scale subsidence over the southeastern Pacific (Vuille, 1999).

Several springs in the uppermost headwater zone feed the valley-bottom-type minerotrophic peatland on the southwestern slope of Cerro Llamoca. The peat-accumulating area occupies the upper valley up to its confluence with an incised, tributary stream channel. During heavy rainfall events it carries sediment to the peatland area (Schittek et al., 2012; Höfle et al., 2013). The slopes within the peatland's catchment area, depending on prevailing stable or unstable environmental conditions, represent a source area for allochthonous input to the peat-dominated valley bottom, resulting in a complex intercalation of organic and inorganic sediment layers.

High-altitude cushion peatlands occur along the Andean range with gradually changing floristic composition (Ruthsatz, 2000; Squeo et al., 2006). At CLP, the Juncaceae *Distichia muscoides* and *Oxychloe andina* are the dominant peat-forming cushion plants. They often grow so densely that they can form extensive, stable mats ranging in shape from almost flat to hemispherical. The shoots continue to grow at their tops but die off from the bottom (Rauh, 1988). A more detailed description of the vegetation and present condition of CLP is presented in Schittek et al. (2012).

The natural vegetation of the slopes, mostly dominated by the tussock grasses *Festuca dolichophylla*, *Stipa brachyphylla* and *Stipa ichu*, has been changed significantly by grazing. Today's regional vegetation is dominated by scattered, often grouped stands of xerophilic dwarf shrubs. The overall vegetation cover usually does not exceed 30 %. Especially slopes exposed to the north are scarcely vegetated, which is responsible for an increased erodibility. In areas protected by rocks and where there is less grazing, as well as on slopes exposed to the south-east, tussock grasses still prevail.

2.2 Human settlement history

Pre-Columbian settlement history in the south Peruvian coastal desert dates back to more than 5 ka (Fig. 5). Although there is some evidence of hunting and gathering in the study area since at least the early Holocene, earliest numerically dated human remains in the Palpa–Nasca region reach back to the Archaic Period (5.7–5.0 ka/3760–3060 cal BC) at a site called Pernil Alto, situated in the lower Rio Grande valley (Unkel et al., 2012; Gorbahn, 2013). Here, after a 1600-year hiatus, signs of reoccupation are evidenced during the Initial Period from 3.41 to 2.79 ka/1460 to 840 cal BC. Agriculture is now the primary subsistence strategy. Still, there are no signs of occupation of the higher-elevated Andean area within the Rio Grande catchment. The Initial Period can be considered the basis from which the Paracas culture later emerged (Reindel, 2009).

The early Horizon (2.79–2.21 ka/840–260 cal BC) is subdivided into early, middle and late Paracas. From approximately the late fifth century BC, immigration of people from regions with comparable environmental conditions, experienced with highland farming and camelid herding, led to an intensifying of human activity in the upper slope region of the western cordillera of the Andes called *cabezadas* (Sossna, 2014). Highlights of their cultural activities are petroglyphs and the first geoglyphs on the slopes of the Palpa valley.

The initial Nasca phase (2.21–1.87 ka/260 cal BC–AD 80 cal) was a very dynamic epoch regarding settlement patterns, ceramic technology and textile craft. Between 2.25 and 2.05 ka/300 and 100 cal BC the *cabezadas* were abandoned, whereas downstream along the river oases a strong increase in population occurred during the transition from Paracas to Nasca culture (Reindel, 2009; Sossna, 2014).

Followed by the early Intermediate Period (1.87–1.31 ka/80–AD 640 cal) that is subdivided into early, middle and late Nasca, the early Nasca period (1.87–1.65 ka/80–AD 300 cal) was a time of high cultural development. Settlement density grew, political structures developed and ceramic as well as textile production was intensified and professionalised. Systematic agriculture, partly with irrigation systems, and especially the creation of huge geoglyphs resulted in enormous landscape changes. In contrast to the Initial Period, settlements evolved in the large floodplains at that time, close to the valley border with implications for clear hierarchic features. The cultural evolution climaxed around AD 200, with Cahuachi as a great temple city that was abandoned 100 years later. The highest density of settlements occurred during the middle Nasca (1.65–1.51 ka/300–AD 440 cal). Towards the end of the middle Nasca phase, the settlements tended to shift to the middle valleys and the population underwent a general decline (Reindel, 2009; Sossna, 2014).

This trend continued during the late Nasca phase (1.51–1.31 ka/440–AD 640 cal). The middle Horizon (1.31–1.16 ka/640–AD 790 cal) is documented by only a few findings in the region (Unkel et al., 2012). The lack of archaeological material during the next ca. 400 years indicates a second hiatus in the settlement chronology. During that period the area was influenced by the Wari empire from further east.

From the ninth century AD on, the area was more or less depopulated. At the beginning of LIP (0.77–0.5 ka/1180–AD 1450 cal), human activity rose again markedly along the river oases (Reindel, 2009; Unkel et al., 2012) and at the *cabezadas*. Large groups of immigrants from the highlands recolonised the area (Fehren-Schmitz et al., 2014). The population peaked during the 14th/15th century AD, incident to a massive expansion of agricultural terraces (Sossna, 2014).

3   Methods

The coring field work was carried out in August 2009. For the selection of a suitable coring site within the peatland we applied electrical resistivity tomography (Schittek et al., 2012). Several transects were measured to receive an insight into the peatland's internal structure and depth to bedrock. Multiple cores were drilled at several sites within the whole peatland range by using percussion hammer coring equipment. The retrieved sediment was sealed in liner tubes with a diameter of 5 cm.

This study focuses on the deepest core (Pe852), which reached a depth of 10.5 m. The core was divided into two core halves, photographed and sedimentologically described at the Palaeoecology Laboratory of the Department of Geography and Geographical Education (University of Cologne). One core half was sub-sampled at 5–10 cm intervals (depending on the stratigraphy) from the peat sections for micro- and macrofossil analyses.

The inorganic geochemistry of the other core half was analysed in 2 mm resolution using an ITRAX X-ray fluorescence (XRF) core scanner (Cox Analytical Systems) at the Institute of Geology and Mineralogy (University of Cologne). XRF scanning was performed with a Mo tube at 30 kV and 30 mA, using an exposure time of 20 s per measurement.

For pollen sample preparation we applied an extended protocol. After KOH treatment for deflocculation, the samples were sieved in three additional sections (2 mm, 250 μm, 125 μm). These three size fractions were separated for the study of macrofossils. After spiking with *Lycopodium* markers to allow for concentration calculation, the further pollen preparation followed standard techniques described in Faegri and Iversen (1993). Microfossil samples were mounted in glycerine and pollen was identified under ×400 and ×1000 magnification. A minimum of 300 terrestrial pollen grains was analysed in each sample. Identifications were based on our own reference collection and on published atlases and keys (Heusser, 1971; Markgraf and D'Antoni, 1978; Graf, 1979; Hooghiemstra, 1984; Sandoval et al., 2010; Torres et al., 2012). Pollen and non-pollen palynomorphs data were subjected to numerical zonation using binary

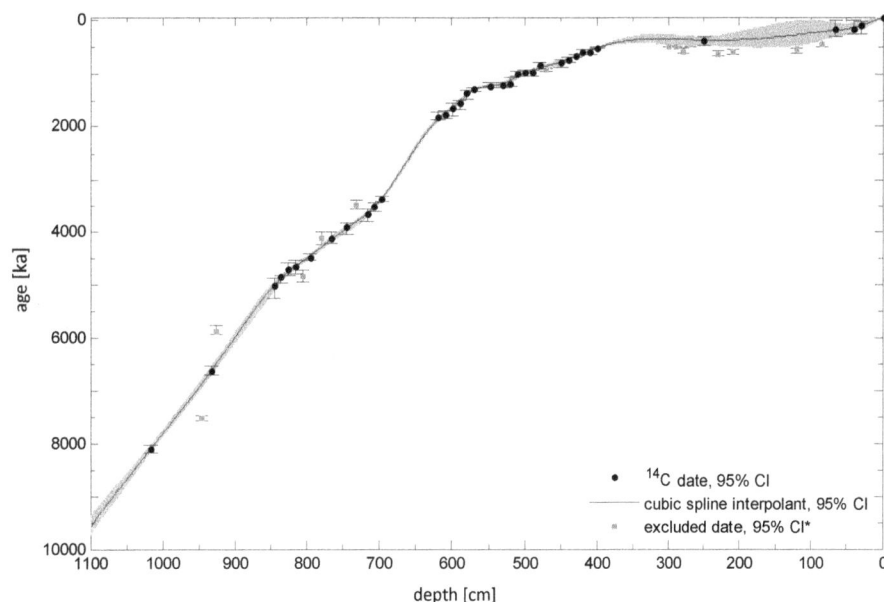

Figure 2. Age-versus-depth model for core Pe852 retrieved from Cerro Llamoca peatland based on 35 ^{14}C dates. The grey band represents the modelled range of dates and the black line the 50th percentile of all runs.

splitting techniques (Hammer et al., 2001), which highlighted five main zones.

Radiocarbon dating was performed from the same samples used for pollen analyses, concerning the 10 cm interval from the peat sections. A total of 50 samples were dated by Dr. Bernd Kromer and Susanne Lindauer (Klaus Tschira Centre for Archaeometry and Heidelberg Academy of Sciences) (Table 1). All radiocarbon dates were calibrated using CALIB 6.0.1 and the IntCal09 data set for Northern Hemisphere calibration (Reimer et al., 2009). Southern Hemisphere calibration is recommended for regions south of the thermal equator (McCormac et al., 2004). As the seasonal shift of the ITCZ brings atmospheric CO_2 from the Northern Hemisphere to the Andes during spring and summer seasons, it is primarily taken up by the vegetation. The age–depth model is based on a Monte Carlo approach to generate confidence intervals that incorporate the probabilistic nature of calibrated radiocarbon dates by using the MCAgeDepth software (Higuera, 2008). The program generates a cubic smoothing spline through all the dates. A total of 800 Monte Carlo simulations were used to generate confidence intervals. The final probability age–depth model is based on the median of all the simulations.

4   Results

4.1   Stratigraphy and chronology

The sedimentary deposits of CLP consist of an interlayered bedding of peat layers and layers of silt, clay and sand in varying compositions and with different contents of plant re-

mains. These are repeatedly interrupted by layers of inorganic debris, which comprise either fine and middle sands or coarse sand and gravel.

The most frequent substrate types are peat and coarse sand. The peat and sediment matrices show variable contents of embedded silt and clay. A rapid change of coarse sediment and layers of silt/clay with variable contents of organic matter characterise the lowermost section (1050–850 cm). The middle section (850–400 cm) shows homogenous peat layers, less frequently interrupted by coarse sediment. The upper 400 cm section contains the highest variability of substrate types and comprises repeated deposition of coarse sediment. This type of peat-debris deposit is typical for high-altitude peatlands in the more arid central and western Altiplano, characterised by an interplay of fan aggradation and peat growth (Schittek et al., 2012).

The age–depth model is based on 50 radiocarbon dates of mostly bulk sediment samples (Fig. 2). Due to redeposition effects, 15 dates were omitted from the model. This is especially the case between 100 and 400 cm, where rapid deposition of allochthonous debris within a short time frame might have eroded and redeposited peat and soil material to the coring site. Ages therefore remain within the same time range.

Nonetheless, the peat sections in particular reveal a continuous chronology, allowing a high-resolution palaeoclimate reconstruction. Sample resolution varies between about 10 and 30 yr cm^{-1} and is highest during periods of peat formation. Pollen analysis allows a palaeoenvironmental reconstruction at decadal to centennial resolution. The inorganic

Table 1. Radiocarbon ages of core Pe852. The calibrated age ranges were calculated using CALIB 6.0.1 and the IntCal09 data set (Reimer et al., 2009). The modelled ages are the result of a probabilistic age–depth model using MCAgeDepth (Higuera, 2008). The range represents the 2σ values, and the median ages are in parentheses.

Lab #	Depth (cm)	Measured ^{14}C	Measured error (±)	2σ calibrated age age (cal yr BP)	MCAgeDepth modelled age (cal yr BP)
MAMS-13291	30.5	96	21	26-(107)-257	28-(108)-188
MAMS-13292	40.5	150	20	8-(177)-276	53-(148)-235
MAMS-11767	65.5	220	28	1-(181)-304	43-(198)-307
MAMS-11768*	84.5	394	30	331-(466)-506	10-(225)-371
MAMS-13293*	120.5	558	23	529-(558)-633	6-(276)-483
MAMS-13294*	147.5	292	31	293-(383)-456	57-(313)-510
MAMS-13295*	209.5	633	26	557-(597)-660	235-(373)-510
MAMS-13296*	229.5	699	31	568-(662)-687	282-(382)-496
MAMS-13297	249.5	334	21	316-(384)-472	311-(383)-489
MAMS-11769*	269.5	392	26	333-(469)-504	295-(377)-477
MAMS-11770*	279.5	574	19	539-(605)-636	286-(372)-473
Hd-29328*	289.5	433	21	471-(502)-518	278-(368)-471
MAMS-10840*	299.5	421	23	346-(496)-514	266-(364)-462
Hd-29296	399.5	428	19	506-(519)-536	506-(520)-539
MAMS-10842	409.5	636	24	559-(595)-660	549-(569)-594
MAMS-10843	419.5	610	25	550-(602)-653	597-(619)-648
MAMS-10844	429.5	729	24	659-(675)-711	662-(675)-708
MAMS-10845	439.5	837	24	694-(742)-789	712-(733)-772
Hd-29297	449.5	893	18	744-(808)-900	750-(783)-842
Hd-29298*	469.5	1016	18	920-(937)-961	813-(854)-917
MAMS-10859	479.5	958	25	799-(854)-926	859-(893)-948
MAMS-10864	489.5	1094	30	940-(1000)-1060	914-(945)-994
MAMS-10857	499.5	1080	25	937-(983)-1053	971-(997)-1038
MAMS-10862	509.5	1115	25	965-(1013)-1068	1030-(1061)-1110
MAMS-10863	519.5	1244	24	1085-(1204)-1262	1100-(1141)-1190
Hd-29340	529.5	1285	19	1181-(1234)-1278	1153-(1203)-1235
MAMS-10861	547.5	1299	24	1182-(1244)-1286	1199-(1246)-1266
Hd-29299	569.5	1398	19	1290-(1304)-1340	1300-(1317)-1347
MAMS-10866	579.5	1499	30	1320-(1378)-1498	1376-(1404)-1446
MAMS-10867	589.5	1673	31	1518-(1577)-1686	1483-(1519)-1570
MAMS-10868	599.5	1764	32	1578-(1669)-1798	1596-(1636)-1687
MAMS-10869	609.5	1845	31	1712-(1779)-1863	1711-(1748)-1800
Hd-29312	619.5	1875	19	1737-(1828)-1872	1816-(1869)-1911
Hd-29313	696.5	3146	22	3335-(3374)-3436	3326-(3358)-3394
MAMS-10905	706.5	3307	33	3458-(3529)-3626	3487-(3515)-3560
MAMS-10906	716.5	3413	34	3579-(3663)-3810	3604-(3647)-3713
MAMS-10907*	730.5	3268	34	3409-(3496)-3574	3757-(3795)-3883
MAMS-10908	744.5	3606	33	3840-(3914)-4051	3879-(3924)-4010
MAMS-10912	766.5	3764	33	4002-(4127)-4233	4085-(4140)-4202
MAMS-10913*	779.5	3764	34	4001-(4127)-4235	4235-(4281)-4338
MAMS-10914	795.5	3997	33	4416-(4476)-4544	4410-(4451)-4510
MAMS-10915*	805.5	4280	34	4739-(4849)-4947	4493-(4546)-4601
Hd-29300	815.5	4123	21	4544-(4650)-4806	4583-(4638)-4704
MAMS-10909	825.5	4171	34	4584-(4713)-4826	4697-(4738)-4808
MAMS-10910	835.5	4299	35	4832-(4859)-4961	4817-(4858)-4943
MAMS-10911	845.5	4429	34	4885-(5017)-5264	4943-(4999)-5129
MAMS-10917*	925.5	5127	28	5761-(5892)-5930	6386-(6483)-6571
MAMS-10958	932.5	5820	29	6527-(6634)-6716	6517-(6619)-6701
MAMS-10959*	946.5	6666	30	7484-(7536)-7584	6768-(6883)-6951
Hd-28899	1016.5	7281	27	8022-(8094)-8167	8010-(8093)-8153

* Ages not used for the age–depth model.

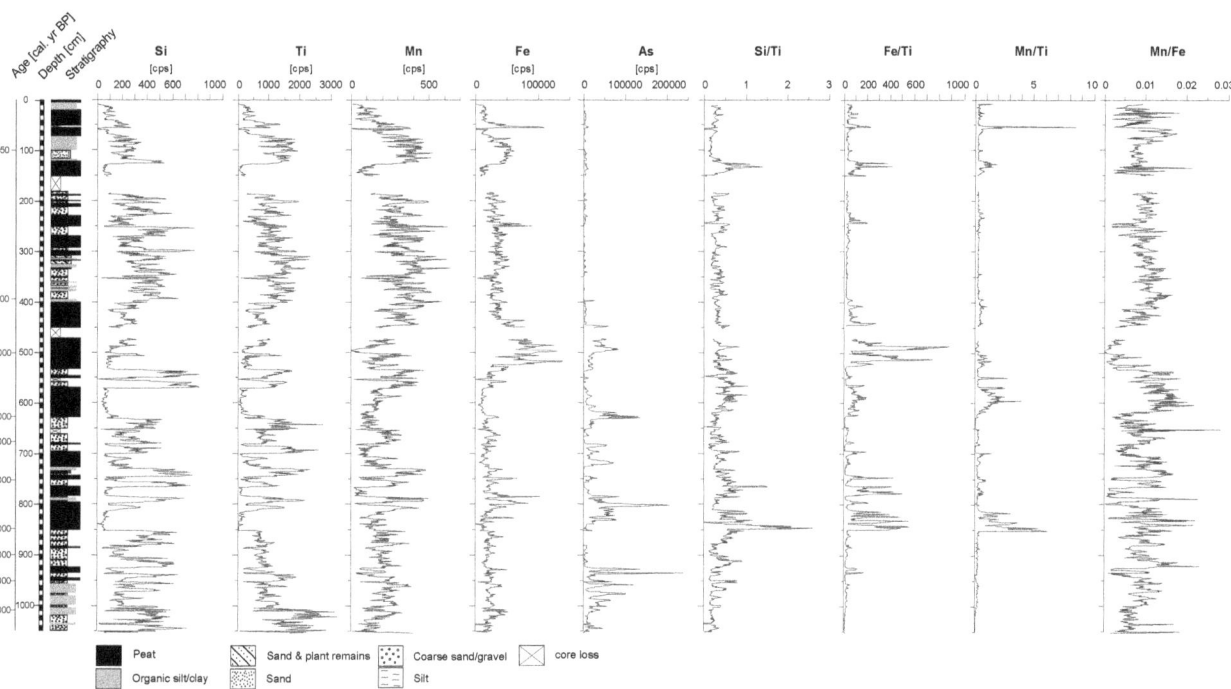

Figure 3. Stratigraphy and selection of elements and elemental ratios measured by the XRF core scanner for core Pe852 of the Cerro Llamoca peatland.

geochemistry (XRF data) was analysed in 2 mm steps, which allows a sub-decadal resolution for the peat sections.

4.2　Geochemical variability of the record

The peatland record is characterised by an interplay of peat accumulation and repeated deposition of inorganic sediments. Several distinct changes can be observed in the XRF signals of the measured elements reflecting the heterogeneous stratigraphy (Fig. 3). Silicon (Si) and titanium (Ti) originate from allochthonous lithogenic material and therefore show highest values in layers dominated by inorganic components. Si is further added by biogenic silica. Cyperaceae and Poaceae are highly abundant components of the peatland vegetation. These Si-accumulating plants deposit significant amounts of amorphous hydrated silica in their tissues as opal phytoliths (Street-Perrot and Barker, 2008). Diatoms represent another source of biosilicification (Servant-Vildary et al., 2001). Hence the Si / Ti ratio is used to discern the biogenic silica amount but might also reflect changes in grain sizes in the clastic sediment. Manganese (Mn) and iron (Fe) are also of lithogenic origin, but contents further depend on environmental factors that control post-depositional processes. By contrast, Ti is considered to be immobile in peat (Muller et al., 2006, 2008). Therefore, the Mn and Fe data were normalised to Ti to better reflect the variations in autochthonous in-peatland dynamics of the record. The Mn / Fe ratio is mainly linked to autochthonous precipitation of iron

oxides and can be used as an indicator of redox conditions (Lopez et al., 2006).

Due to the weathering of volcanic rocks, spring waters in the upper headwater area of CLP (as at many other sites of the area) are enriched with arsenic (As). Recently wetlands, and in particular peatlands, were identified as a trap for As under anoxic redox conditions (Eh) (Langner et al., 2012; Hoffmann et al., 2012). We therefore use As as an indicator for hydrological changes in CLP.

The data show the most marked changes at 1050–930 cm (8.6–6.3 ka) with a high variability of values ranging between 10000 and > 100000 cps (Fig. 3). The following section at 930–840 cm (6.3–4.8 ka) is characterised by very low As values. Significantly higher As values are observed at 840–770 cm (4.8–4.3 ka), peaking at about 810–800 cm (4.5–4.4 ka). Further As peaks, which span shorter periods, are recorded at about 730 cm (3.7 ka), 680 cm (3.0 ka), 620–630 cm (1.9–1.8 ka) and 500–470 cm (1.0–0.9 ka).

Comparable to the As record, Si / Ti ratios are highly variable at 1050–930 cm (8.6–6.3 ka). At about 850–840 cm (5.0–4.9 ka), the Si / Ti ratio reaches its maximum value, corresponding to the Mn / Ti and Fe / Ti ratios. Afterwards, until about 630 cm (2.0 ka), the Si / Ti ratio tends to decrease. Only between 590 and 530 cm (1.9–1.3 ka) does it rise to higher values again, before decreasing towards the present.

Highest Fe / Ti and Mn / Ti ratios are observed during peat-accumulating periods. The Fe / Ti ratio is characterised by a high variability between 850 and 740 cm (5.0–3.9 ka). The

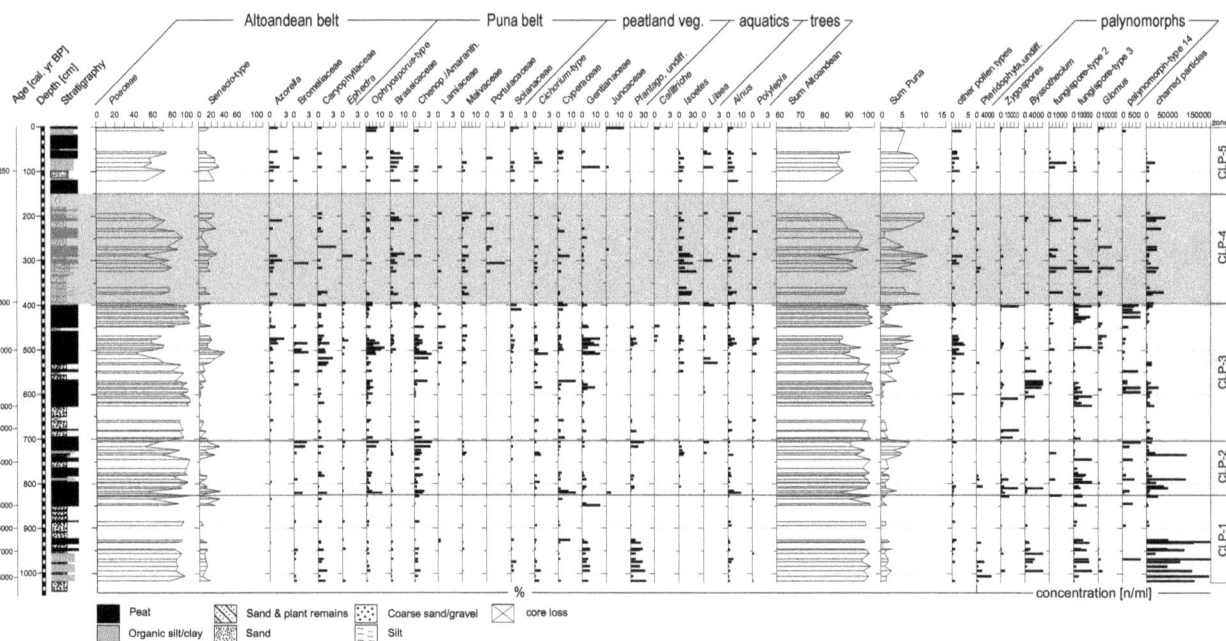

Figure 4. Pollen, palynomorphs and charred particles diagram for the Cerro Llamoca peatland, plotted against depth. Peatland vegetation and aquatic types were excluded from the pollen sum. The shaded zone is mainly composed of redeposited, erosional material.

highest peaks of the record occur at 480 cm (0.9 ka) and 520 cm (1.1 ka). After peaking at 850–840 cm (5.0–4.9 ka), Mn / Ti reaches higher values only between 630 and 570 cm (2.0–1.6 ka). The Mn / Fe ratio is highly variable throughout the record. Periods of low values tend to correspond to periods of higher As concentrations.

4.3 Pollen analysis

The results of the microfossil counts are plotted in Fig. 4. Usually only peat and organic silt/clay layers yielded sufficient pollen for counting. The pollen types are grouped together according to their main regional or local distribution range. As the peatland site is situated in the lower Altoandean altitudinal belt (Ruthsatz, 1977), the overall pollen spectrum is clearly dominated by Poaceae, which make up 40–95 % of the regional pollen assemblage. The other main regional taxa are all typical components of the Altoandean and Puna belts (Reese and Liu, 2005; Kuentz et al., 2007). Apart from *Senecio*-type Asteraceae (5–40 %), only *Ophryosporus*-type Asteraceae, Brassicaceae, Malvaceae and *Alnus* reach percentages > 3 %. Cyperaceae, Gentianaceae and *Plantago* represent local peatland and aquatic vegetation. *Isoetes* spores were included to the pollen counts. All local types are excluded from the pollen sum. Extra-regional pollen types are few throughout the sequence, mainly represented by *Alnus* and *Polylepis*. Other types comprise Ericaceae, Polemoniaceae, Bignoniaceae, Malpighiaceae, *Juglans* and *Podocarpus*, which appeared in very low abundances.

Zone CLP-1 (1050–830 cm; 8.6–4.8 ka) is characterised by a steady presence of Poaceae and *Senecio*-type Asteraceae pollen, both at medium percentage values. Caryophyllaceae are also steadily present with medium percentages. *Plantago* and Gentianaceae pollen are highly abundant. Fern and charred particle concentrations reach the highest values of the whole record within this zone. Due to higher contents of coarse sediment, the upper section of zone CLP-1 mostly lacked countable amounts of pollen.

Zone CLP-2 (830–710 cm; 4.8–3.6 ka) is marked by a high variation of Poaceae and *Senecio*-type Asteraceae pollen percentages. Interestingly, types typical of the Puna belt gain higher abundances towards the upper part of the zone. Cyperaceae pollen peaks at the beginning of zone two.

Zone CLP-3 (710–400 cm; 4.8–0.5 ka) at its initial section is scarce in palaeobotanical evidence due to the characteristics of the sediment. Between 630 and 540 cm (2.0–1.2 ka) high percentages of Poaceae are recorded, dominating the pollen spectrum. At 540–470 cm (1.2–0.8 ka), pollen values of *Senecio*-type Asteraceae and Puna belt types gain higher abundances. Poaceae reach their lowest values of the entire record here. Peatland pollen types show increases of Gentianaceae, Cyperaceae and *Plantago* values. The latter nearly disappears from the record afterwards, whereas *Azorella*, Brassicaceae, Malvaceae and *Isoetes* start to appear more frequently and in higher abundances from this point on. Poaceae are represented in high abundances again at 450–400 cm (0.8–0.5 ka).

Zone CLP-4 (400–150 cm; 0.5–0.3 ka) is mainly composed of re-deposited, erosional material. It therefore remains questionable if this zone can be used for interpretation. Age control reveals that this part of the core was deposited within a short time frame.

Zone CLP-5 (150–0 cm; 0.3 ka to today) represents the youngest section of the CLP record and, at least at its bottom, might be affected by redeposition. The sediment did not always contain sufficient pollen, due to increased decomposition of the peat. Overall, the Altoandean belt types remain at a relatively low level, whereas Puna belt types show high abundances.

5 Discussion

5.1 As, Mn and Fe retention and release under fluctuating water table conditions

Arsenic (As) and its compounds are mobile in the environment (Alonso, 1992; Kumar and Suzuki, 2002; Rothwell et al., 2009; Cumbal et al., 2010). In spring-water samples of the 2009 and 2010 campaigns we measured As contents of 140–270 μg L^{-1} at the head of the peatland; however, the small stream leaving the peatland's main branch further down only contained 4–6 μg L^{-1} (Schittek; unpublished data), which clearly shows that CLP is a sink for As. An analogous remediation of As-bearing waters by peat is reported for a minerotrophic peatland in Switzerland (González et al., 2006).

Langner et al. (2012) report that natural organic matter (NOM) can represent a major sorbent for As in sulfur-rich anoxic environments. They postulate that covalent binding of trivalent As^{3+} to NOM via organic sulfur species is the primary mechanism of As–NOM interactions under sulfate-reducing conditions. Therefore, As mobilisation is suppressed by sorption of As to NOM by the formation of stable inner-sphere complexes.

However, the CLP record shows several significant peaks in As (Figs. 4 and 5). Concerning the last 4000 years, the modelled periods of these events with high As contents strongly correlate with dry events, identified for the central Andes by several authors (Thompson et al., 1995; Rein et al., 2004; Chepstow-Lusty et al., 2009; Bird et al., 2011b). This presumably implies an enhanced As mobility under a climate regime with sustained dry periods, which may be attributed to the fixation of dissolved NOM (Langner et al., 2012) concurrent with a higher humification degree. The increasing sorption capacity to trace elements with increasing decomposition of peat was also found by Klavins et al. (2009) in peatlands in Latvia. The formation of humic acids leads to an increase of functional groups and therefore to an increasing sorption capacity.

Cloy et al. (2009), Rothwell et al. (2009) and Rothwell et al. (2010) report similar As dynamics in Scottish ombrotrophic peatlands. Here, stream water As concentrations are elevated during late summer stormflow periods, when

there has been rewetting of peat after significant water table draw-down. Blodau et al. (2008) demonstrate that rewetting of previously dry minerotrophic peat leads to the rapid release of As and Fe into pore waters coupled with Fe reduction in the peat. Langner et al. (2012) highlight that, under oxic conditions, NOM promotes the release of As from metal (hydr-)oxides, thereby enhancing the mobility of As. Unfortunately, basic information on As biogeochemistry and As dynamics in naturally enriched peat ecosystems is still lacking. The results of this study underline that NOM might play an important role in As dynamics. Further research is needed to identify the exact As retention and release mechanisms. This would not only be of interest to (palaeo-)environmental research, but also be of significance for the protection of ecosystems and water resources.

Changes in Fe and Mn require careful consideration. The behaviour of Fe and Mn strongly depends on pH and water saturation. The portion of Mn^{2+} increases under anoxic conditions and forms soluble complexes with Mn^{2+} humic substances (Graham et al., 2002; Blume et al., 2010). Graham et al. (2002) identified that Mn contents were lowest and in a non-easily reducible form where the extent of humification was greatest. High Fe / Ti ratios indicate an upward movement of Fe^{2+} from the anoxic peat to the upper aerated layers, followed by precipitation as Fe^{3+}-oxide. This process leads to an enrichment of Fe in the zone of water table fluctuations (Damman et al., 1992; Margalef et al., 2013). As the peatland environment is naturally highly enriched with Fe, it strongly precipitates under oxic conditions and thus lowers the Mn / Fe ratio. Low values, indicating prevailing water table fluctuations and a more frequent occurrence of oxic conditions, correlate with As peaks. At about 1.8–1.2 ka, an outstanding period of high Mn / Fe ratios prevails, which indicates a period of steady saturation of the peat deposits at this site.

5.2 Mid-Holocene and late Holocene palaeoenvironmental changes

Selected proxies from the CLP record were plotted on the temporal scale and compared with published records from the Cariaco Basin (Haug et al., 2001) and Huascarán ice core (Thompson et al., 1995) (Fig. 5). The dominant driver of long-term climatic variations in the tropical Andes during the Holocene is the ITCZ (Haug et al., 2001; Ledru et al., 2009; Bird et al., 2011a, b; Vuille et al., 2012). Similarities in the δ^{18}O isotopic signatures from speleothems (van Breukelen et al., 2008; Cruz et al., 2009; Reuter et al., 2009), lake records (Ekdahl et al., 2008; Bird et al., 2011a, b; Placzek et al., 2011) and glacier ice cores (Thompson et al., 1998) in the South American tropics and subtropics indicate that water had a common main origin. The methodological advances in the application of multi-proxy approaches and the increasing number of palaeoclimatic studies in the tropical/subtropical Andes underline the hypothesis of Haug et al. (2001) that

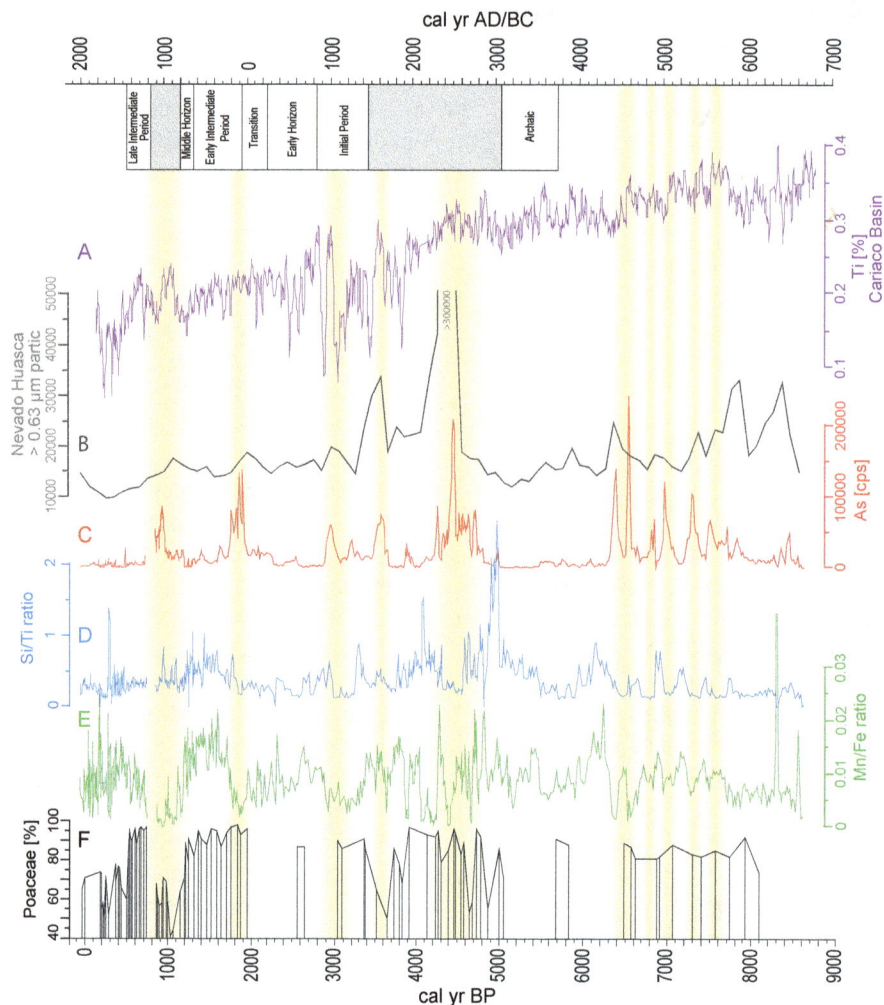

Figure 5. The archaeological chronology of the pre-Columbian cultures in the Palpa valleys (Unkel et al., 2012) compared with the bulk Ti content of Cariaco Basin sediments (**a**; Haug et al., 2001) and dust particle concentrations of Huascarán ice core (**b**; Thompson et al., 1995). (**c**), (**d**), (**e**) In situ geochemical parameters and the Poaceae pollen record (**f**) of the Cerro Llamoca peatland.

changes in precipitation relate to shifts in the mean latitude of the ITCZ. A more southerly position of the ITCZ triggers moisture flux into the tropical lowlands, which enhances convective activity in the Amazon Basin.

Data on mid-Holocene palaeoclimate in the central Andes remain discontinuous and still only provide snapshots of information. Moreno et al. (2007) identified the interval between 8.6 and 6.4 ka as the driest episode of the Chungará lake record in the northern Chilean Altiplano. They point out that dry conditions were not constant but rather characterised by a series of short and rapid dry spells. This finding coincides very well with the CLP record for nearly the same period. Here dry spells, indicated by marked As peaks, alternate with humid spells, indicated by a higher degree of anoxic conditions (Mn / Fe ratio) and higher amounts of biogenic silica (Si / Ti ratio). Grass pollen percentages remain at medium values.

That the generally dry conditions were repeatedly interrupted by short-lived, abrupt moisture changes was also found in other central Andean lake and sediment archives (Grosjean et al., 2001). Nonetheless, records of mid-Holocene climate conditions are not synchronous in the central Andes (Betancourt et al., 2000; Holmgren et al., 2001; Abbott et al., 2003; Latorre et al., 2003; Kuentz et al., 2011). Discrepancies in the exact timing of climatic changes and the interpretation of their causes are common as proxy records are obtained from different archives and geographically heterogeneous localities. A major constraint of most palaeoclimate records of the central Andes is that they do not show a significant variability on multi-centennial to millennial scales (Lamy et al., 2001), which is needed to compare them with other continent-scale, high-resolution records.

Based on oxygen isotope ratios, Bird et al. (2011a) suggest weak SASM precipitation at Lake Pumaqucha from 7.0 to 5.0 ka, which corresponds to a low stand of Lake Titicaca from 7.5 to 5.0 ka inferred from seismic profiling and sediment $\delta^{13}C$ (Seltzer et al., 1998; Rowe et al., 2003). Following the concept of Haug et al. (2001), the ITCZ had remained at a relatively stable northern position throughout the middle Holocene. Thus, monsoon intensity might have been predominantly weak. However, minor intensifications in the southward migration of the ITCZ might have temporarily increased moisture availability at CLP during the middle Holocene, as visible by the episodically higher levels of Si / Ti and Mn / Fe ratios (Fig. 5).

The CLP record does not offer clear palaeoenvironmental evidence for the period of 6.4 to 5.1 ka due to the dominance of coarse sediment in the record and a lack of pollen. Higher Mn / Fe and Si / Ti ratios suggest moister conditions at around 6.3–6.0 ka and again starting from 5.5 ka. At 5.0–4.9 ka a significant transition to wetter conditions is evidenced in the CLP record by a pronounced Si / Ti peak. This abrupt climate change has been recognised in several records from the tropical Andes (Abbott et al., 2003; Thompson et al., 2006; Ekdahl et al., 2008; Buffen et al., 2009). The onset of this cool and wet period led to an expansion of the Quelccaya ice cap (Thompson et al., 2006) and an increase of the Lake Titicaca lake level (Baker et al., 2001). A climatic transition ~ 5000 years ago is also notable, e.g. in eastern equatorial Africa and the eastern Mediterranean region (Thompson et al., 2006). The colder conditions promoted an increased peat growth at CLP but remained highly variable and unstable as evidenced by the high fluctuations of the pollen and Mn / Fe ratio in the sediment record. The humid period between 5.4 and 4.9 ka probably culminated in the formation of a palaeosoil within a loess sequence in the desert margin area of southern Peru. This indicates stable conditions with weathering processes and a dense vegetation cover in an area now characterised by extremely arid conditions (Mächtle and Eitel, 2013).

The cool and wet period is followed by a marked dry period at about 4.6–4.2 ka, as indicated by the extremely high As contents in the CLP record. Peaking at 4.5–4.4 ka, the As record coincides with a peak of insoluble dust concentrations evidenced in the Huascarán ice core (Thompson et al., 1995). The further As peaks in the CLP record strongly correlate to dry events identified at the lake site of Marcacocha (Chepstow-Lusty et al., 2003, 2009), which started to accumulate lake sediments after 4.2 ka. Based on inorganic contents and Cyperaceae pollen concentrations, drier episodes, coinciding with the CLP record, occurred around 3.6–3.5, 3.1–2.9, 2.0–1.8 and 1.2–0.8 ka. The highly variable Mn / Fe and Si / Ti ratios prior to 4.5 ka suggest unstable climatic conditions until 1.8 ka. The pollen record in this section is rather fragmentary due to the dominance of coarse sediments in the retrieved cores.

After about 2.0 ka, Mn / Fe ratios declined at CLP and remained low until about 1.8 ka (Fig. 6). Elevated As contents point to a pronounced dry period. Vinther et al. (2009) recorded higher temperatures in Greenland at exactly the same time span. The timing and extent of this dry period

can be well correlated to the Roman Warm Period (RWP) (Zolitschka et al., 2003; Ljungqvist, 2010). Warmer and drier conditions in South America during that period have been found by Jenny et al. (2002), based on geochemical, sedimentology and diatom assemblage data derived from sediment cores extracted from Laguna Aculeo (central Chile). Similar observations have also been made by Chepstow-Lusty et al. (2003), who evidenced the RWP as a period of 100 to 200 years of relative warmth and dryness in comparison to the periods before and after.

The occurrence of a sustained cold period in South America after about 1.8 ka is evidenced by concomitant glacier expansions in the Peruvian (Wright, 1984; Seltzer and Hastorf, 1990; Thompson et al., 1995) and Bolivian Andes (Abbott et al., 1997). Chepstow-Lusty et al. (2003) noted a suppression of agriculture at Lake Marcacocha, which is suggested to be a direct reflection of a period of colder climate conditions leading to significantly reduced human population in that area. Poaceae pollen percentages and Mn / Fe ratios remain at elevated and stable levels in the CLP record from 1.8 to 1.2 ka.

A harsh return to drier conditions at CLP at around 1.2–1.15 ka can be inferred from a sudden reduction in Poaceae pollen percentages and Mn / Fe ratios, which must have severely affected the peatland water regime and the vegetation cover of the surrounding high-Andean grasslands. Grass percentages dropped down to the lowest value of the record at 1.05 ka. The period of extreme drought lasts until about 0.75–0.7 ka, when grass pollen become highly abundant again. Rein et al. (2004), who presented a high-resolution marine record from the Peruvian shelf west of Lima, also discussed this sustained dry period contemporary to the Medieval Climate Anomaly (MCA) (Fig. 6). Based on lithic concentrations, they identified this period as characterised by a lack of strong flooding, because of reduced river runoff, from 1.15 to 0.7 ka. Bird et al. (2011a) suggested a considerable weakening of the SASM during 1.05 and 0.85 ka and linked this event with the Northern Hemisphere MCA and a northward position of the Atlantic ITCZ.

Starting shortly after 0.75 ka, grass pollen abundance at CLP was back to levels > 90 % and remained being highly abundant until about 0.5 ka. Mn / Fe ratios indicate variable redox conditions in the peatland.

After 0.5 ka the proxy signals of CLP underwent strong and repeated shifts. These changes are likely to be linked to the destabilisation of slopes within the CLP water catchment area (Schittek et al., 2012). More than 3 m of debris were deposited upon the peatland sediments between 0.5 and 0.25 ka. The cooling of the Little Ice Age (LIA) might have altered the resilience of the peatland and its water catchment area to erosion and triggered the fluvial input of alluvial sediment by very strong episodic rainfall events and by a reduction of vegetation cover on the slopes due to aridity and/or increased pasturing. Debris flows usually occurred during periods of slow vegetation growth on the slopes of the water catchment

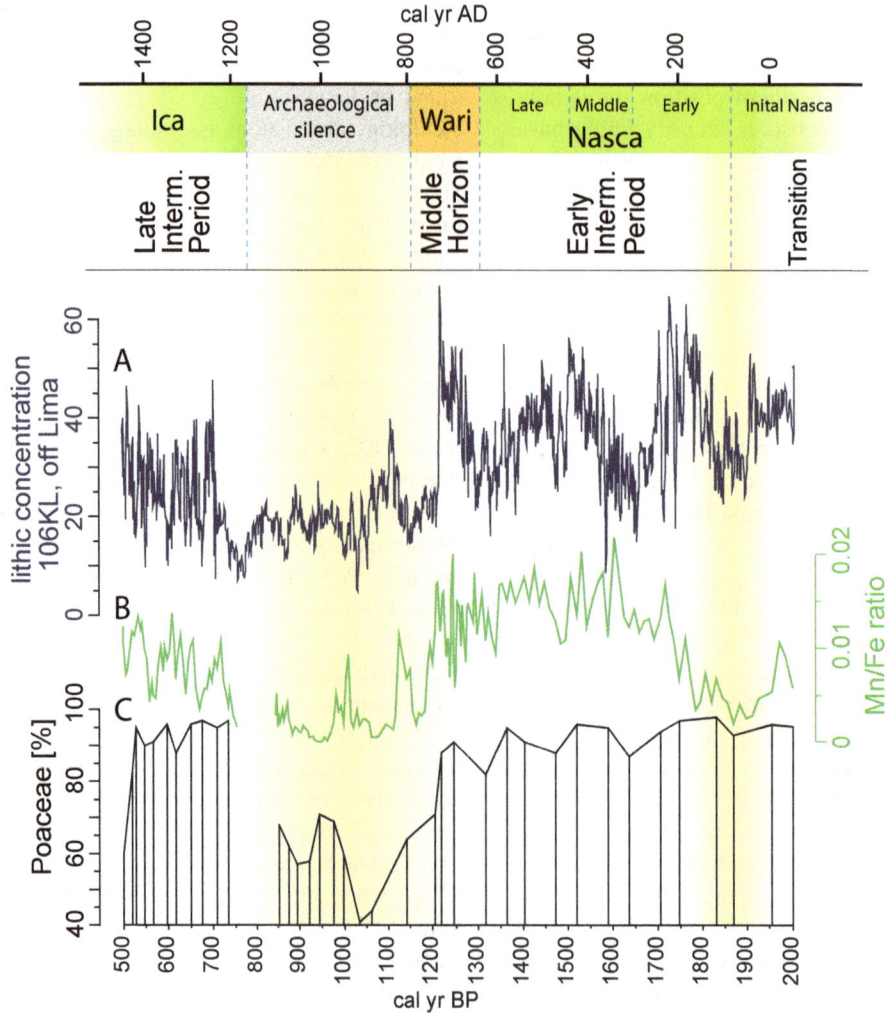

Figure 6. The archaeological chronology of the last 2000 years in the Palpa valleys (Unkel et al., 2012) in comparison to lithic concentration of a marine core from the Peruvian shelf west of Lima (**a**; Rein et al., 2004), Mn / Fe ratios (**b**) and Poaceae pollen percentages (**c**) from the Cerro Llamoca peatland.

area because of climatic changes and/or soil degradation by overgrazing (Schittek et al., 2012).

Bird et al. (2011b) noted a pronounced decrease in Pumaqucha δ^{18}O between 0.55 and 0.13 ka, which was likely in response to a southward displacement of the Atlantic ITCZ, associated with cooler temperatures and significantly increased precipitation. However, Morales et al. (2012) pointed out that the LIA was not a persistent period of wet/cool conditions. Moreover, several severe droughts occurred during that period.

The CLP sequence represents an exemplary record of long-term trajectories between periods of landscape stability and transitional phases of landscape destabilisation. Periods of relative landscape stability under a more humid and balanced climate regime with less pronounced droughts would promote soil accumulation and the establishment of a dense grassland vegetation cover on the surrounding moun-

tain slopes, which significantly slows down overland water runoff.

The abundant presence of grass pollen reflects very well the predominance of grasses in the high-Andean vegetation belt. The density of the grass cover diminishes during drier periods and better-adapted high-mountainous vegetation components like Asteraceae (mostly *Senecio*-type), Brassicaceae, Caryophyllaceae and Chenopodiaceae/Amaranthaceae become more evident in the pollen spectrum. Gentianaceae and *Plantago* typically spread in oxidised sections of Andean peatlands where water table fluctuations prevail.

The dynamics of the SASM appears to be a conceivable driver for moisture fluctuations in the investigated area. Vuille et al. (2012) pointed out that the intensity of the SASM, and thus the amount of Andean rainfall, is sensitive to the position of the ITCZ, which depends on sea surface temperatures in the North Atlantic and the eastern tropical Pacific. Episodes of water table fluctuations at CLP, as reflected by low Mn/Fe ratios and high As contents, and, in some cases, a reduction of grass pollen abundances, tend to correlate with northward positions of the ITCZ and hence a reduced SASM intensity (Fig. 5). A reduced convection in the Amazonian lowlands might shorten the rainy season at CLP and result in an enforced seasonality with a concentration of rainfall in summer and a prolonged dry phase during the rest of the year. These conditions trigger erosion and, consequently, the deposition of debris upon the peatland after heavy episodic rainfall events. Although moisture transport is closely connected to ITCZ dynamics, SASM intensity is more or less determined by further modes of climatic variability like ENSO, the Pacific Decadal Oscillation and latitudinal shifts of the southern westerlies. The role of Pacific modes of variability as a control over precipitation in the tropical/subtropical Andes is still controversially discussed (Mann et al., 2009).

5.3 The impact of climate change on pre-Columbian cultures

Constructing comparable chronologies between past climate and cultural changes based on palaeoenvironmental and archaeological records should be exercised with utmost discernment. Only when both records are sufficiently continuous and reliable might they offer the opportunity for correlation (Seltzer and Hastorf, 1990).

The numerical archaeological chronology of the Palpa–Nasca valleys presented by Unkel et al. (2012) is one of the best-dated chronologies available in South America and serves as the backbone of our integrated approach in the climatically sensitive western margin of the Andes. The CLP palaeoenvironmental record evidences repeated climatic fluctuations, concurrent with changes in cultural florescence of local people. This is especially the case between the early and late Intermediate periods (Fig. 6) when the resolution of the CLP record is highest and most reliable.

When tracking climate and settlement patterns it is important to remember that the success of early cultures in areas close to desert margins mainly depended on the availability of water. Although civilisations were able to expand natural limits by inventing and adapting techniques of water harvesting and agriculture, they were not able to master long-lasting periods of severe drought. A typical settlement pattern in the Palpa–Nasca valleys is that the main settlement and agricultural production area repeatedly shifted between the Andean foothills/lower valleys and the highland section (cabezadas) (Reindel, 2009; Sossna, 2012, 2014).

At present, human activity and environmental changes cannot be linked during the early phase (∼ 5.25 ka), due to the low population of non-sedentary pre-Columbian people and very little archaeological evidence. The earliest well-dated occupation period in Pernil Alto (∼ 5.7–5.0 ka) coincides with a transition to wetter conditions, starting at around 5.5 ka. River runoff from the high Andes may have allowed for hunters and gatherers to temporarily settle in pit houses close to the floodplains and revisit that site over a long period of time (Reindel, 2009; Unkel et al., 2012; Gorbahn, 2013). At 4.6 ka, the onset of repeated dry periods could have been the trigger for a temporary abandonment of the settlement.

During the Initial Period (∼ 3.4–2.8 ka), a reoccupation of Pernil Alto is evidenced by the construction of buildings with depots, hearths and workshops out of adobe, which displays permanent sedentary lifestyle apparently based on agriculture (Reindel, 2009).

The CLP record suggests unstable conditions during the transition of the Initial Period and the early Horizon, manifested by a contemporaneous, high heterogeneity of the stratigraphic record, which limits the informative value of the pollen sequence. Monsoon intensity might have been temporarily strong, and conditions at the desert margin were anything but stable. Population density was low and settlements clustered in the upper Palpa valley (Sossna, 2014). This significantly changed in the middle Paracas phase, when settlement activity boomed and a large number of settlements were established in the formerly almost uninhabited cabezadas. This coincides with the onset of a period of less marked ITCZ shifts at around 2.5 ka. At CLP, stratigraphy and variable Mn/Fe ratios suggest unstable conditions, but climate was more humid than during the Initial Period, as evidenced by overall higher Mn/Fe ratio values after 2.5 ka (Fig. 5). The end of the late Paracas phase and the earlier Initial Nasca phase (around 2.25–2.05 ka) is characterised by the abandonment of many settlements in the cabezadas and a dramatic increase of population density at the foothills. This trend continued during the Initial Nasca phase (2.21–1.87 ka).

Due to the gradual character of cultural transitions and a limited number of ^{14}C dates, the emergence of the Initial Nasca phase and its transition into the early Nasca phase is still inaccurately dated. It therefore remains uncertain if the short dry episode, evidenced at about 2.0–1.75 ka at CLP, had any effect on the evolution of the Nasca culture. At least it very clearly marks the transition between the two phases.

General precipitation and increased river runoff can be assumed for 1.75–1.2 ka (see Sect. 5.2; Rein et al., 2004), which led to a concentration of settlements along the river oases (Eitel and Mächtle, 2009; Sossna, 2014). Widespread agricultural terraces fed a large population. These terraces were mainly irrigated by local rainfall reaching beyond today's desert margin.

The short decrease in Poaceae pollen percentages, as well as a decrease in lithic concentration (Rein et al., 2004), can be correlated with a short decrease in lowland population. During this dry episode, river runoff was reduced, which led to a re-occupation of the cabezadas (Sossna, 2014). In the lowlands, the most important centres were abandoned, and the focus of the settlements shifted to the middle and upper valleys. The processes again coincide with oscillations registered in the CLP pollen record and the marine core at the same time by Rein et al. (2014) (Fig. 6).

During the late Nasca phase, new centres of power were once again re-established in the lowlands. But population levels stayed relatively low and further decreased towards the end of the phase, which again might have been triggered by reduced precipitation as evidenced by reduced Mn / Fe ratios and pollen percentages at CLP around 1.3 ka (Rein et al., 2004).

This significant dry period, which was fully developed around 1.05 ka, again corresponds to the archaeological silence when the foothills may have been depopulated almost completely. This archaeological silence without any findings in the study area is recorded in the CLP record by a massive decrease of Poaceae pollen and is also corroborated within the results of Rein et al. (2004). A rapid shift to more humid conditions during the late Intermediate Period (Ica) triggered a massive migration from the highlands to lower valleys and the river oases (Fehren-Schmitz et al., 2014) where agricultural terraces expanded to the slopes (Sossna, 2014), which were also fed by water from local rainfall in an area where hyper-arid conditions now prevail (Mächtle, 2007). Still many people remained in the highlands and the upper valleys.

A significant low stand of the Mn / Fe ratio around 0.66 ka, indicating a temporary decrease in precipitation, correlates with the construction of water harvesting systems in the Palpa lowlands (Mächtle et al., 2013). The unstable conditions during the LIP might favour the establishment of manifold storages, even above the household level (Reindel, 2009; Sossna, 2014).

6 Conclusions

This investigation supports the assumptions made by Bird et al. (2011a, b) and Vuille et al. (2012), who suggest that a more southerly position of the ITCZ triggers moisture flux into the tropical Amazonian lowlands, which leads to an intensification of the SASM and thus a stronger easterly moisture transport towards the western range of the Andes. Since the mid-Holocene, increased moisture flux has repeatedly reached the headwaters of some rivers that drain to the Pacific and bring water to the lowland river oases. Here, with a strong dependence on the moisture derived from across the Andes, pre-Columbian cultures developed or declined.

The sediment deposits of CLP in the high-Andean headwater of Río Viscas represent a high-resolution archive for the reconstruction of the palaeoenvironmental history in the western Peruvian Andes and their adjacent lowlands. The heterogeneity of the deposits reflects the sensitivity of high-Andean ecosystems towards environmental changes. Especially in the subarid western Andes of southern Peru, climatic changes have a strong influence on the surface geomorphic

features, which led to repeated fan aggradation upon the peat-accumulating area.

The content of As and Mn / Fe ratios turned out to be valuable new proxies for in situ palaeoredox conditions. Verified by pollen analysis, the archaeological chronology for the cultural development in the valleys of Palpa and several independent, continent-scale proxy archives, the CLP record evidences prominent mid- and late Holocene climate oscillations.

The mid-Holocene period of 8.6–5.6 ka was identified as being characterised by highly variable moisture conditions with a series of episodic dry spells alternating with spells that were more humid. After a pronounced cool and humid spell at 5.0–4.9 ka, conditions generally remained unstable, being frequently interrupted by pronounced dry periods that enhanced erosional processes. Periods of cultural bloom in the Palpa–Nasca lowlands coincide with stable, humid periods at 1.8–1.2 ka and 0.75–0.5 ka at CLP. Our findings therefore show that past fluctuations in SASM intensity had a significant influence on the cultures within the Río Grande de Nasca drainage.

Acknowledgements. The authors give their sincere thanks to Jonathan Hense (Palynological Laboratory, University of Cologne) for the preparation of the pollen slides and to Armine Shahnazarian (Institute of Geology and Mineralogy, University of Cologne) for support and help with the interpretation of the XRF results. We also thank Christian Mikutta (Institute of Biogeochemistry and Pollutant Dynamics, ETH Zürich) for helpful explanations concerning the cycling of arsenic in peat. We further acknowledge the great support of the "Comunidad Campesina de Laramate" during fieldwork in Peru. Funding by the German Federal Ministry for Education and Research within the project "Andean transect – Climate Sensitivity of pre-Columbian Man–Environment Systems" (BMBF-01UA0804B) is greatly appreciated.

The manuscript has greatly benefitted from comments by Martha Bell, Rolf Kilian, Marcelo Morales (reviewers) and Finn Viehberg (communicating editor).

Edited by: F. Viehberg

References

Abbott, M. B., Seltzer, G. O., Kelts, K. R., and Southon, J.: Holocene paleohydrology of the tropical Andes from lake records, Quaternary Res., 47, 70–80, 1997.

Abbott, M. B., Wolfe, B. B., Wolfe, A.P., Seltzer, G. O., Aravena, R., Mark, B. G., Polissar, P. J., Rodbell, D. T., Rowe, H. D., and Vuille, M.: Holocene paleohydrology and glacial history of the central Andes using multiproxy lake sediment studies, Palaeogeogr. Palaeocl., 194, 123–138, 2003.

Alonso, H.: Arsenic enrichment in superficial waters. II. Región Northern Chile, in: Proceedings of the International Seminar on Arsenic in the Environment and Its Incidence on Health, Santiago: Universidad de Chile, 101–108, 1992.

Baker, P. A., Seltzer, G. O., Fritz, S. C., Dunbar, R. B., Grove, M. J., Tapia, P. M., Cross, S. L., Rowe, H. D., and Broda, J. P.: The history of South American tropical precipitation for the past 25 000 years, Science, 291, 640–643, 2001.

Baker, P. A., Fritz, S. C., Garland, J., and Ekdahl, E.: Holocene hydrologic variation at Lake Titicaca, Bolivia/Peru, and its relationship to North Atlantic climate variation, J. Quaternary Sci., 20, 655–662, 2005.

Betancourt, J. L., Latorre, C., Rech, J. A., Quade, J., and Rylander, K. A.: A 22,000-year record of monsoonal precipitation from Northern Chile's Atacama Desert, Science, 289, 1542–1546, 2000.

Binford, M. W., Kolata, A. L., Brenner, M., Janusek, J. W., Seddon, M. T., Abbott, M., and Curtis, J. H.: Climate variation and the rise and fall of an Andean civilization, Quaternary Res., 47, 235–248, 1997.

Bird, B. W., Abbott, M. B., Vuille, M., Rodbell, D. T., Stansell, N. D., and Rosenmeier, M. F.: A 2,300-year-long annually resolved record of the South American summer monsoon from the Peruvian Andes, Proc. Natl. Acad. Sci. USA, 21, 8583–8588, 2011a.

Bird, B. W., Abbott, M. B., Rodbell, D. T., and Vuille, M.: Holocene tropical South American hydroclimate revealed from a decadally resolved lake sediment $\delta 18O$ record, Earth Planet. Sci. Lett., 310, 192–202, 2011b.

Blodau, C., Fulda, B., Bauer, M., and Knorr, K.-H.: Arsenic speciation and turnover in intact organic soils during experimental drought and rewetting, Geochim. Cosmochim. Ac., 72, 3991–4007, 2008.

Blume, H.-P., Brümmer, G. W., Horn, R., Kandeler, E., Kögel-Knabner, I., Kretzschmar, R., Stahr, K., and Wilke, B.-M.: Scheffer/Schachtschabel: Lehrbuch der Bodenkunde, Springer, Heidelberg, 570 pp., 2010.

Bookhagen, B. and Strecker, M. R.: Orographic barriers, high-resolution TRMM rainfall, and relief variations along the eastern Andes, Geophys. Res. Lett., 35, L06403, doi:10.1029/2007GL032011, 2008.

Buffen, A. M., Thompson, L. G., Mosley-Thompson, E., and Huh, K. I.: Recently exposed vegetation reveals Holocene changes in the extent of the Quelccaya Ice Cap, Peru, Quaternary Res., 72, 157–163, 2009.

Castillo, J., Barreda, A., and Vella, C.: Geología de los cuadrángulos de Laramate y Santa Ana, hojas 29-n; 29-ñ, Ingemmet Boletin No. 45, Seria A. Editoral Allamanda, Lima, p. 66 , 1993.

Chepstow-Lusty, A., Frogley, M. R., Bauer, B. S., Bush, M. B., and Herreras, A. T.: A late Holocene record of arid events from the Cuzco region, Peru, J. Quaternary Sci., 18, 491–502, 2003.

Chepstow-Lusty, A. J., Frogley, M. R., Bauer, B. S., Leng, M. J., Boessenkool, K. P., Carcaillet, C., Ali, A. A., and Gioda, A.: Putting the rise of the Inca Empire within a climatic and land management context, Clim. Past, 5, 375–388, doi:10.5194/cp-5-375-2009, 2009.

Cloy, J. M., Farmer, J. G., Graham, M. C., and MacKenzie, A. B.: Retention of As and Sb in Ombrotrophic Peat Bogs: Records of As, Sb, and Pb Deposition at Four Scottish Sites, Environ. Sci. Technol., 43, 1756–1762, 2009.

Cruz, F. W., Karmann, I., Viana, O., Burns, S. J., Ferrari, J. A., Vuille, M., Sial, A. N., and Moreira, M. Z.: Stable isotope study of cave percolation waters in subtropical Brazil: Implications for paleoclimate inferences from speleothems, Chem. Geol., 220, 245–262, 2005.

Cumbal, L., Vallejo, P., Rodriguez, B., and Lopez, D.: Arsenic in geothermal sources at the north-central Andean region of Ecuador: concentrations and mechanisms of mobility, Environ. Earth Sci., 61, 299–310, 2010.

Damman, A. W. H., Tolonen, K., and Sallantaus, T.: Element retention and removal in ombrotrophic peat of Hadetkeidas, a boreal Finnish peat bog, Suo, 43, 137–145, 1992.

Eitel, B. and Mächtle, B.: Man and Environment in the Eastern Atacama Desert (Southern Peru): Holocene Climate Changes and Their Impact on Pre-Columbian Cultures, in: New Technologies for Archaeology, Natural Science in Archaeology, edited by: Reindel, M. and Wagner, G. A., Springer-Verlag Berlin Heidelberg, 17–23, 2009.

Eitel, B., Hecht, S., Mächtle, B., Schukraft, G., Kadereit, A., Wagner, G. A., Kromer, B., Unkel, I., and Reindel, M.: Geoarchaeological evidence from desert loess in the nazca-palpa region, southern Peru: Palaeoenvironmental changes and their impact on pre-Columbian cultures, Archaeometry, 47, 137–158, 2005.

Ekdahl, E. J., Fritz, S. C., Baker. P.A., Rigsby, C. A., and Coley, K.: Holocene multidecadal- to millennial-scale hydrologic variability on the South America Altiplano, Holocene, 18, 867–876, 2008.

Faegri, K. and Iversen, J.: Bestimmungsschlüssel für die nordwesteuropäische Pollenflora, Gustav Fischer Verlag, Jena, 1993.

Fehren-Schmitz, L., Haak, W., Mächtle, B., Masch, F., Llamas, B., Tomasto Cagigao, E., Sossna, V., Schittek, K., Isla Cuadrado, J., Eitel, B., and Reindel, M.: Climate change underlies global demographic, genetic, and cultural transitions in pre-Columbian southern Peru, Proc. Natl. Acad. Sci. USA, 111, 9443–9448, 2014.

Garreaud, R. D.: Intraseasonal variability of moisture and rainfall over the South American Altiplano, Mon. Weather Rev., 128, 3337–3346, 2000.

Garreaud, R. D., Vuille M., and Clement A. C.: The climate of the Altiplano: observed current conditions and mechanisms of past changes, Palaeogeogr. Palaeocl., 194, 5–22, 2003.

Garreaud, R., Barichivich, J., Christie, D. A., and Maldonado, A.: Interannual variability of the coastal fog at Fray Jorge relict forests in semiarid Chile, J. Geophys. Res.-Biogeosciences, 113, G04011, doi:10.1029/2008JG000709, 2008.

Garreaud, R. D., Vuille, M., Compagnucci, R., and Marengo, J.: Present-day South American climate, Palaeogeogr. Palaeocl., 281, 180–195, 2009.

Gayo, E. M., Latorre, C., Jordan, T. E., Nester, P. L., Estay, S. A., Ojeda, K. F., and Santoro, C. M.: Late Quaternary hydrological and ecological changes in the hyperarid core of the northern Atacama Desert (similar to 21° S), Earth-Sci. Rev., 113, 120–140, 2012.

González, Z. I., Krachler, M., Cheburkin, A. K., and Shotyk, W.: Spatial distribution of natural enrichments of arsenic, selenium, and uranium in a mineratrophic peatland, Gola di Lago, Canton Ticino, Switzerland, Environ. Sci. Technol., 40, 6568–6574, 2006.

Gorbahn, H.: The Middle Archaic Site of Pernil Alto, Southern Peru: The beginnings of horticulture and sedentariness in mid-Holocene conditions, Diálogo Andino, 41, 61–82, 2013.

Graf, K.: Untersuchungen zur rezenten Pollen- und Sporenflora in der nördlichen Zentalkordillere Boliviens und Versuch einer Auswertung von Profilen aus postglazialen Torfmooren, Habil., Univ. Zürich, 1979.

Graham, M. C., Gavin, K. G., Farmer, J. G., Kirika, A., and Britton, A.: Processes controlling the retention and release of manganese in the organic-rich catchment of Loch Bradan, SW Scotland, Appl. Geochem., 17, 1061–1067, 2002.

Grosjean, M. and Veit, H.: Water Resources in the Arid Mountains of the Atacama Desert (Northern Chile): Past Climate Changes and Modern Conflicts, in: Global Change and Mountain Regions – An Overview of Current Knowledge, Adv. Glob. Change Res., edited by: Huber, U. M., Bugmann, H. K. M., and Reasoner, M. A. Springer, Heidelberg, 23, 93–104, 2005.

Grosjean, M., van Leeuwen, J. F. N., van der Knaap, W. O., Geyh, M. A., Ammann, B., Tanner, W., Messerli, B., Núñez, L. A., Valero-Garcés, B. L., and Veit, H.: A 22 000 ^{14}C year BP sediment and pollen record of climate change from Laguna Miscanti (23° S), northern Chile, Glob. Planet Change, 28, 35–51, 2001.

Hammer, Ø., Harper, D. A. T., and Paul D. R.: Past: Paleontological Statistics Software Package for Education and Data Analysis. Palaeontologia Electronica, 4, 9 pp., 2001.

Haug, G. H., Hughen, K. A., Sigman, D. M., Peterson, L. C., and Rohl, U.: Southward migration of the intertropical convergence zone through the Holocene, Science, 293, 1304–1308, 2001.

Heusser, C. J.: Pollen and spores of Chile, University of Arizona Press, 167 pp., Tucson, 1971.

Higuera, P. E.: MCAgeDepth 0.1: Probabilistic age-depth models for continuous sediment records, available at: https://code.google.com/p/mcagedepth/downloads/detail?name=MCAgeDepth_UsersGuide.pdf, 2008.

Höfle, B., Griesbaum, L., and Forbriger, M.: GIS-Based Detection of Gullies in Terrestrial LiDAR Data of the Cerro Llamoca Peatland (Peru), Remote Sens., 5, 5851–5870, 2013.

Hoffmann, M., Mikutta, C., and Kretzschmar, R.: Bisulfide reaction with natural organic matter enhances arsenite sorption: Insights from X-ray absorption spectroscopy, Environ. Sci. Technol., 46, 11788–11797, 2012.

Holmgren, C. A., Betancourt, J. L., Rylander, K. A., Roque, J., Tovar, O., Zeballos, H., Linares, E., and Quade, J.: Holocene vegetation history from fossil rodent middens near Arequipa, Peru, Quaternary Res., 56, 242–251, 2001.

Hooghiemstra, H.: Vegetational and climatic history of the high plain of Bogotá, Colombia: a continuous record of the last 3.5 million years, Dissertationes Botanicae, 79, J. Cramer, Vaduz, 138 pp. 1984.

Jansen, E., Overpeck, J., Briffa, K. R., Duplessy, J. C., Joos, F., Masson-Delmotte, V., Olago, D., Otto-Bliesner, D., Peltier, W. R., Rahmstorf, S., Ramesh, R., Raynaud, R., Rind, D., Solomina, O., Villalba, R. and Zhang, D.: Palaeoclimate, in: Solomon, S., Qin, D., Manning, M., Chen, Z., Marquis, M., Averyt, K. B., Tignor, M., and Miller, T. L.: Climate Change 2007: The Physical Science Basis. Contribution of Working Group I to the Forth Assessment Report of the Intergovernmental Panel on Climate Change. Cambridge University Press, Cambridge and New York, 433–497, 2007.

Jenny, B., Valero-Garces, B. L., Urrutia, R., Kelts, K., Veit, H., Appleby, P. G., and Geyh, M.: Moisture changes and fluctuations of the Westerlies in Mediterranean Central Chile during the last

2000 years: The Laguna Aculeo record (33°50′ S), Quaternary Int., 87, 3–18, 2002.

Klavins, M., Silamikele, I., Nikodemus, O., Kalnina, L., Kuske, E., Rodinov, V., and Purmalis, O.: Peat properties, major and trace element accumulation in bog peat in Latvia, Baltica, 22, 37–49, 2009.

Kuentz, A., Galán de Mera, A., Ledru, M., and Thouret, J.: Phytogeographical data and modern pollen rain of the puna belt in southern Peru (Nevado Coropuna, Western Cordillera), J. Biogeogr., 34, 1762–1776, 2007.

Kuentz, A., Ledru, M.-P., and Thouret, J.-C.: Environmental changes in the highlands of the western Andean Cordillera, southern Peru, during the Holocene, Holocene, 22, 1215–1226, 2011.

Kumar, B. and Suzuki, K. T.: Arsenic round the world: a review, Talanta, 58, 201–235, 2002.

Lamy, F., Hebbeln, D., Röhl, U., and Wefer, G.: Holocene rainfall variability in southern Chile: A marine record of latitudinal shifts of the Southern Westerlies, Earth Planet. Sci. Lett., 185, 369–382, 2001.

Langner, P., Mikutta, C., and Kretzschmar, R.: Arsenic sequestration by organic sulphur in peat, Nat. Geosci., 5, 66–73, 2012.

Latorre, C., Betancourt, J. L., Rylander, K. A., Quade, J., and Matthei, O.: A vegetation history from the arid prepuna of northern Chile (22–23° S) over the last 13 500 years, Palaeogeogr. Palaeocl., 194, 223–246, 2003.

Ledru, M.-P., Mouruiart, P., and Riccomini, C.: Related changes in biodiversity, insolation and climate in the Atlantic rainforest since the last interglacial, Palaeogeogr. Palaeocl., 271, 140–152, 2009.

Ljungqvist, F. C.: A New Reconstruction Of Temperature Variability In The Extra-Tropical Northern Hemisphere During The Last Two Millennia, Geogr. Ann., 92, 339–351, 2010.

Lopez, P., Navarro, E., Marce, R., Ordoñez, Caputo, L., and Armengol, J.: Elemental ratios in sediments as indicators of ecological processes in Spanish reservoirs, Limnetica, 25, 499–512, 2006.

Mächtle, B.: Geomorphologisch-bodenkundliche Untersuchungen zur Rekonstruktion der holozänen Umweltgeschichte in der nördlichen Atacama im Raum Palpa/Südperu. Ph.D., Heidelberger Geographische Arbeiten, Heidelberg, 123, 227 pp., 2007.

Mächtle, B. and Eitel, B.: Fragile landscapes, fragile civilizations – How climate determined societies in the pre-Columbian south Peruvian Andes, Catena, 103, 62–73, 2013.

Mächtle, B., Ross, K., and Eitel, B.: The Khadin water harvesting system of Peru – an ancient example for future adaption to climatic change, in: Hydrogeology of Arid Environments, edited by: Rausch, R., Schüth, C., and Himmelsbach, T., Proceedings, 76–80, 2012.

Mann, M. E., Zhang, Z., Rutherford, S., Bradley, R., Hughes, Malcolm K., Shindell, D., Ammann, C., Faluvegi, G., and Ni, F.: Global Signatures and Dynamical Origins of the Little Ice Age and Medieval Climate Anomaly, Science, 326, 1256–1260, 2009.

Margalef, O., Canellas-Bolta, N., Pla-Rabes, S., Giralt, S., Pueyo, J. J., Joosten, H., Rull, V., Buchaca, T., Hernandez, A., Valero-Garces, B. L., Moreno, A., and Saez, A.: A 70 000 year multiproxy record of climatic and environmental change from Rano Aroi peatland (Easter Island), Glob. Planet Change, 108, 72–84, 2013.

Markgraf, V. and D'Antoni, H. L.: Pollen flora of Argentina. University of Arizona Press, 208 pp., 1978.

McCormac, F. G., Hogg, A. G., Blackwell, P. G., Buck, C. E., Higham, T. F. G., and Reimer, P. J.: SHCal04 Southern Hemisphere calibration, 0–11.0 cal kyr BP, Radiocarbon, 46, 1087–1092, 2004.

Maslin, M. A. and Burns, S. J.: Reconstruction of the Amazon Basin effective moisture availability over the past 14 000 years, Science, 290, 2285–2287, 2002.

Maslin, M. A., Ettwein, V. E., Wilson, K. E., Guilderson, T. P., Burns, S. J., and Leng, M. J.: Dynamic boundary-monsoon intensity hypothesis: evidence from the deglacial Amazon River discharge record, Quaternary Sci. Rev., 30, 3823–3833, 2011.

Morales, M.S., Christie, D. A., Villalba, R., Argollo, J., Pacajes, J., Silva, J. Z., Alvarez, C. A., Llancabure, J. C., and Soliz Gamboa, C. C.: Precipitation changes in the South American Altiplano since 1300 AD reconstructed by tree-rings, Clim. Past, 8, 653–666, doi:10.5194/cp-8-653-2012, 2012.

Moreno, A., Giralt, S., Valero-Garcés, B., Sáez, A., Bao, R., Prego, R., Pueyo, J. J., González-Sampériz, P., and Taberner, C.: A 14 kyr record of the tropical Andes: The Lago Chungará sequence (18° S, northern Chilean Altiplano), Quat. Int., 161, 4–21, 2007.

Muller, J., Wüst, R. A. J., Weiss, D., and Hu, Y.: Geochemical and stratigraphic evidence of environmental change at Lynch's Crater, Queensland, Australia, Glob. Planet Change 53, 269–277, 2006.

Muller, J., Kylander, M., Wüst, R. A. J., Weiss, D., Martinez Cortizas, A., LeGrande, A. N., Jennerjahn, T., Behling, H., Anderson, W. T., and Jacobson, G.: Possible evidence for wet Heinrich phases in tropical NE Australia: the Lynch's Crater deposit, Quaternary Sci. Rev., 27, 468–475, 2008.

Placzek, C. J., Quade, J., and Patchett, P. J.: Isotopic tracers of paleohydrologic change in large lakes of the Bolivian Altiplano, Quat. Res., 75, 231–244, 2011.

Rauh, W: Tropische Hochgebirgspflanzen: Wuchs- und Lebensformen, Springer-Verlag, 206 pp., 1988.

Reese, C. A. and Liu, K.: A modern pollen rain study from the central Andes region of South America, J. Biogeogr., 32, 709–718, 2005.

Reimer, P. J., Baillie, M. G. L., Bard, E., Bayliss, A. Beck, J. W., Blackwell, P. G., Ramsey, C. B., Buck, C. E., Burr, G. S., Edwards, R. L., Friedrich, M., Grootes, P. M., Guilderson, T. P., Hajdas, I., Heaton, T. J., Hogg, A. G., Hughen, K. A., Kaiser, K. F., Kromer, B., Mccormac, F. G., Manning, S. W., Reimer, R. W., Richards, D. A., Southon, J. R., Talamo, S., Turney, C. S. M., Van Der Plicht, J. M., and Weyhenmeyer, C. E.: IntCal09 and Marine09 radiocarbon age calibration curves, 0–50 000 years cal BP, Radiocarbon, 51, 1111–1150, 2009.

Rein, B., Luckge, A., and Sirocko, F.: A major Holocene ENSO anomaly during the Medieval period, Geophys. Res. Lett, 31, L17211, doi:10.1029/2004GL020161, 2004.

Reindel, M.: Life at the Edge of the Desert – Archaeological Reconstruction of the Settlement History in the Valley of Palpa, Peru, in: New Technologies for Archaeology, Natural Science in Archaeology, Reindel, M., and Wagner, G. A., 25, Springer-Verlag Berlin Heidelberg, 439–461, 2009.

Reindel, M. and Isla, J.: Cambio climático y patrones de asentamiento en la vertiente occidental de los Andes del sur del Perú, Diálogo Andino, 41, 83–99, 2013.

Reuter, J., Stott, L., Khider, D., Sinha, A., Cheng, H., and Edwards, R. l.: A new perspective on the hydroclimate variability in northern South America during the Little Ice Age, Geophys. Res. Lett., 36, L21706, doi:10.1111/j.1365-2699.2005.01183.x, 2009.

Rothwell, J. J., Dise, N. B., Taylor, K.G., Allott, T. E. H., Scholefield, P., Davies, H., and Neal, C.: A spatial and seasonal assessment of river water chemistry across North West England, Sci. Total Environ., 408, 841–855, 2009.

Rothwell, J. J., Taylor, K. G., Chenery, S. R. N., Evans, M. G., and Allott, T. E. H.: Storage and Behaviour of As, Sb, Pb, and Cu in Ombotrophic Peat Bogs under Constrasting Water Table Conditions, Environ. Sci. Technol., 44, 8497–8502, 2010.

Rowe, H. D., Guilderson, T. P., Dunbar, R. B., Southon, J. R., Seltzer, G. O., Mucciarone, D. A., Fritz, S. C., and Baker P. A.: Late Quaternary lake-level changes constrained by radiocarbon and stable isotope studies on sediment cores from Lake Titicaca, South America, Glob. Planet Change, 38, 273–290, 2003.

Ruthsatz, B.: Pflanzengesellschaften und ihre Lebensbedingungen in den Andinen Halbwüsten Nordwest-Argentiniens. Dissertationes Botanicae, 39, J. Cramer, Vaduz, 168 pp., 1977.

Ruthsatz, B.: Die Hartpolstermoore der Hochanden und ihre Artenvielfalt, Berichte der Reinhold-Tüxen-Gesellschaft, 12, 351–371, 2000.

Sandoval, A. P., Marconi, L., and Ortuno, T.: Flora Polínica de Bofedales y Áreas Aldanas del Tuni Condoriri, Herbario Nacional de Bolivia, Weinberg S.R.L., no. 4-2-2908-10, La Paz, 2010.

Schittek, K., Forbriger, M., Schäbitz, F., and Eitel, B.: Cushion Peatlands – Fragile Water Resources in the High Andes of Southern Peru, in: Water – Contributions to Sustainable Supply and Use, Landscape and Sustainable Development, Workinggroup Landscape and Sustainable Development, edited by: Weingartner, H., Blumenstein, O., and Vavelidis, M., Salzburg, Austria, 63–84, 2012.

Seltzer, G. O. and Hastorf, C. A.: Climatic Change and its Effect on Prehispanic Agriculture in the Central Peruvian Andes, J. Field Archaeol., 17, 397–414, 1990.

Seltzer, G. O., Baker, P., Cross, S., Dunbar, R., and Fritz, S.: High-resolution seismic reflection profiles from Lake Titicaca, Peru-Bolivia: Evidence for Holocene aridity in the tropical Andes, Geology, 26, 167–170, 1998.

Servant-Vildary, S., Servant, M., and Jimenez, O.: Holocene hydrological and climatic changes in the southern Bolivian Altiplano according to diatom assemblages in paleowetlands, Develop. Hydrobiol., 162, 267–277, 2001.

Squeo, F. A., Warner, B. G., Aravena, R., and Espinoza, D.: Bofedales: High altitude peatlands of the central Andes, Rev. Chil. Hist. Nat., 79, 245–255, 2006.

Sossna, V.: Los patrones de asentamiento del Periodo Intermedio Temprano en Palpa, costa sur del Perú, Zeitschrift für Archäologie Außereuropäischer Kulturen, 4, 207–280, 2012.

Sossna, V.: Impacts of Climate Variability on Pre-Hispanic Settlement Behavior in South Peru – The Northern Río Grande de Nasca Drainage between 1500 BCE and 1532 CE. Ph.D., Christian-Albrechts-Universität zu Kiel, 293 pp., 2014.

Street-Perrott, F. and Barker, P.: Biogenic silica: a neglected component of the coupled global continental biogeochemical cycles of carbon and silicon. Earth Surf. Process. Landforms, 33, 1436–1457, 2008.

Thompson, L. G., Mosley-Thompson, E., Davis, M. E., Lin, P.-N., Henderson, K. A., Cole-Dai, J., Bolzan, J. F., and Liu, K.-B.: Late Glacial Stage and Holocene Tropical Ice Core Records from Huascarán, Peru, Science, 269, 46–50, 1995.

Thompson, L. G., Davis, M. E., Mosley-Thompson, E., Sowers, T. A., Henderson, K. A., Zagorodnov, V. S., Lin, P. N., Mikhalenko, V. N., Campen, R. K., Bolzen, J. F., Cole-Dai, J., and Francou, B.: A 25 000-year tropical climate history from Bolivian ice cores, Science, 282, 1858–1864, 1998.

Thompson, L. G., Mosley-Thompson, E., Brecher, H., Davis, M., León, B., Les, D., Lin, P.-N., Mashiotta, T., and Mountain, K.: Abrupt tropical climate change: Past and present, Proc. Natl. Acad. Sci. USA, 103, 10536–10543, 2006.

Torres, G. R., Lupo, L. C., Sánchez, A. C., and Schittek, K.: Aportes a la flora polínica de turberas altoandinas, Provincia de Jujuy, noroeste argentino, Gayana Bot., 69, 30–36, 2012.

Troll, C.: The cordilleras of the tropical Andes. Aspects of climatic, phytogeographical and agrarian ecology, in: Geo-Ecology of the mountainous regions of the tropical Americas, Troll, C., Colloquium Geographicum, 9, Bonn, 1968.

Unkel, I., Reindel, M., Gorbahn, H., Isla Cuadrado, J., Kromer, B., and Sossna, V.: A comprehensive numerical chronology for the pre-Columbian cultures of the Palpa valleys, south coast of Peru, J. Archaeol. Sci., 39, 2294–2303, 2012.

van Breukelen, M. R., Vonhof, H. B., Hellstrom, J. C., Wester, W. C. G., and Kroon, D.: Fossil dripwater in stalagmites reveals Holocene temperature and rainfall variation in Amazonia, Earth Planet Sc. Lett., 275, 54–60, 2008.

Vinther, B. M., Burchardt, S. L., Clausen, H. B., Dahl-Jensen, D., Johnsen, S. J., Fisher, D. A., Koerner, R. M., Raynaud, D., Lipenkov, V., Andersen, K. K., Blunier, T., Rasmussen, S. O., Steffensen, J. P., and Svensson, A. M.: Holocene thinning of the Greenland ice sheet, Nature, 461, 385–388, 2009.

Vuille, M.: Atmospheric circulation over the Bolivian Altiplano during dry and wet periods and extreme phases of the Southern Oscillation, Int. J. Climatol., 19, 1579–1600, 1999.

Vuille, M., Burns, S. J., Taylor B. L., Cruz, F. W., Bird, B. W., Abbott, M. B, Kanner, L. C., Cheng, H., and Novello, V. F.: A review of the South American Monsoon history as recorded in stable isotopic proxies over the past two millennia, Clim. Past, 8, 1309–1321, doi:10.5194/cp-8-1309-2012, 2012.

Wright Jr., H. E.: Late Glacial and late Holocene moraines in the Cerros Cuchpanga, central Peru, Quat. Res, 21, 275–285, 1984.

Zhou, J. and Lau, K. M.: Does a Monsoon Climate Exist over South America?, J. Climate, 11, 1020–1040, 1998.

Zolitschka, B., Behre, K. E., and Schneider, J.: Human and climatic impact on the environment as derived from colluvial, fluvial and lacustrine archives – examples from the Bronze Age to the Migration Period, Germany, Quat. Sci. Rev., 22, 81–100, 2003.

10

East Asian Monsoon controls on the inter-annual variability in precipitation isotope ratio in Japan

N. Kurita[1], Y. Fujiyoshi[2], T. Nakayama[3], Y. Matsumi[3], and H. Kitagawa[1]

[1]Graduate School of Environmental Studies, Nagoya University, Furo-cho, Nagoya, 464-8601, Japan
[2]Institute of Low Temperature Science, Hokkaido University, Kita-19 Nishi-8, Sapporo, 060-0819, Japan
[3]Solar-Terrestrial Environment Laboratory, Nagoya University, Furo-cho, Nagoya, 464-8601, Japan

Correspondence to: N. Kurita (kurita.naoyuki@e.mbox.nagoya-u.ac.jp)

Abstract. To elucidate the mechanism for how the East Asian Monsoon (EAM) variability have influenced the isotope proxy records in Japan, we explore the primary driver of variations of precipitation isotopes at multiple temporal scales (event, seasonal and inter-annual scales). Using a new 1-year record of the isotopic composition of event-based precipitation and continuous near-surface water vapor at Nagoya in central Japan, we identify the key atmospheric processes controlling the storm-to-storm isotopic variations through an analysis of air mass sources and rainout history during the transport of moisture to the site, and then apply the identified processes to explain the inter-annual isotopic variability related to the EAM variability in the historical 17-year long Tokyo station record in the Global Network of Isotopes in Precipitation (GNIP).

In the summer, southerly flows transport moisture with higher isotopic values from subtropical marine regions and bring warm rainfall enriched with heavy isotopes. The weak monsoon summer corresponds to enriched isotopic values in precipitation, reflecting higher contribution of warm rainfall to the total summer precipitation. In the strong monsoon summer, the sustaining Baiu rainband along the southern coast of Japan prevents moisture transport across Japan, so that the contribution of warm rainfall is reduced. In the winter, storm tracks are the dominant driver of storm-to-storm isotopic variation and relatively low isotopic values occur when a cold frontal rainband associated with extratropical cyclones passes off to the south of the Japan coast. The weak monsoon winter is characterized by lower isotopes in precipitation, due to the distribution of the cyclone tracks away from the southern coast of Japan. In contrast, the northward shift of the cyclone tracks and stronger development of cyclones during the strong monsoon winters decrease the contribution of cold frontal precipitation, resulting in higher isotopic values in winter precipitation. Therefore, year-to-year isotopic variability in summer and winter Japanese precipitation correlates significantly with changes in the East Asian summer and winter monsoon intensity ($R = -0.47$ for summer, $R = 0.42$ for winter), and thus we conclude that the isotope proxy records in Japan should reflect past changes in the East Asian Monsoon. Since our study identifies the climate drivers controlling isotopic variations in summer and winter precipitation, we highlight the retrieval of a record with seasonal resolution from paleoarchives as an important priority.

1 Introduction

The East Asian Monsoon (EAM) is a dominant climatic phenomenon over the East Asia region. The thermal contrast between the Asian continent and the Pacific Ocean causes a seasonal reversal of the monsoon winds and alternation of cold winter and warm summer (e.g., Webster et al., 1998; Ding and Chan, 2005; Trenberth et al., 2006). The EAM is the main factor that determines climate and its variability over East Asia, covering eastern China, South Korea, and Japan. Monsoon variability has extensive influence on the agriculture, human lives, and economics of East Asian countries (e.g., Huang et al., 2007; Tao et al., 2008). Thus, knowledge of the future variability of the EAM system that accompanies global change is of great concern for sustainable development in this region. An understanding of its past variability

a) JJA

b) DJF

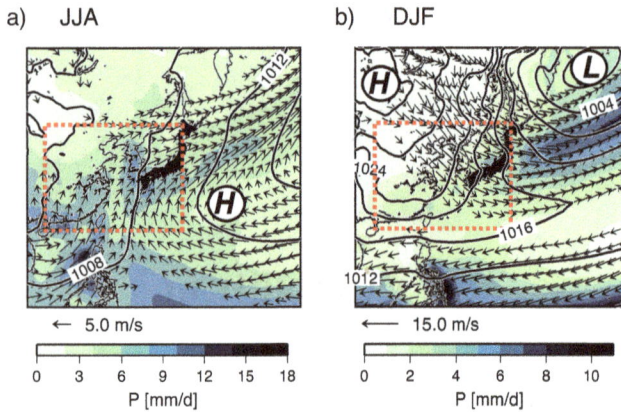

Figure 1. Climatological fields of precipitation (shading), sea level pressure (solid contours, contour interval 4 hPa) and surface wind (vectors: m s^{-1}) for East Asia during (**a**) summer (JJA) and (**b**) winter (DJF). The symbols H and L are centers of high and low surface air pressure, respectively. The dotted box represents the East Asian monsoon region defined by Ding and Chan (2005).

and relationship to climate change is of pivotal importance for assessing the impact of future climate change.

Over the past few decades, many studies, using various natural archives such as loess-paleosol (e.g., An et al., 1991; Porter, 2001), lake sediments (e.g., Nakagawa et al., 2006; Yancheva et al., 2007), marine sediments (Kubota et al., 2014), tree ring cellulose (Liu et al., 2004, 2008), and stalagmites (e.g., Wang et al., 2001; Yuan et al., 2004; Hu et al., 2008; Zhang et al., 2008; Cheng et al., 2009), have documented the evolution and variability of the EAM. Among these paleoarchives, stalagmites have been commonly used to reconstruct the EAM because they offer a precise chronology and a high temporal resolution record. Generally, the EAM is divided into a warm and wet summer monsoon (EASM) and a cold and dry winter monsoon (EAWM) (Fig. 1). The stalagmite oxygen isotope records are regarded as a proxy for the amount of the EASM rainfall or the EASM intensity (e.g., Wang et al., 2005; Cai et al., 2010; Liu et al., 2014). The strong EASM periods correspond to a depleted oxygen isotope value in stalagmite, reflecting increased EASM rainfall with relatively lower oxygen isotope values. In the case of the weak EASM periods, decreasing EASM rainfall leads to richer isotopic values in stalagmite. However, these paleomonsoon data are not obtained from the whole EAM region, but are concentrated in China. As shown in Fig. 1, the EAM region receives moisture from both the Pacific Ocean and the South China Sea (SCS) during the summer rainy season. The moisture for summer precipitation over China is transported by southwesterlies from the SCS. Consequently, the oxygen isotope variations in Chinese stalagmites may be influenced to large extent by the variability of southwesterlies from the SCS (Yang et al., 2014). To understand the past EASM variability in the whole region,

the monsoon activity over the Pacific should be considered. Over the Pacific Ocean, the variability of the southerly monsoonal flow is associated with the development of the Bonin High, which is the western part of the North Pacific Subtropical High. Southerly winds along the western edge of the Bonin High transport moisture for precipitation to northeast Asia, including Japan and Korea (Fig. 1). Anomalous intensification/weakening of the Bonin High induces change in precipitation during the summer monsoon (rainy) season in the northeastern countries (Ha and Lee, 2007; Kosaka et al., 2011). Due to the precipitation change in response to monsoon activity, we anticipate that reconstruction of the past monsoon variability in this region from stalagmite or tree ring cellulose, which preserve a record of isotopic variation in precipitation, is possible. However, a major difference of the isotopic composition in between Chinese and Japanese precipitation lies in seasonal contrasts. For example, at the Hulu Cave site in China, 80 % of the precipitation falls during the summer monsoon season (June to September), and the oxygen isotope values in summer precipitation is 10 ‰ lower than those in winter precipitation (Wang et al., 2001). In contrast, in Tokyo, Japan, the oxygen isotopic value in summer rainy season is not lower than that in the dry winter season and the seasonal difference is less than 3 ‰ (Araguás-Araguás et al., 1998). Thus, to reconstruct the past monsoon variability in the northeastern East Asian monsoon countries from stalagmites or tree ring cellulose, we need to clarify the relationship between the isotopic content in precipitation and the monsoon circulation, and to provide a comprehensive physical explanation for how the East Asian Monsoon variability influences the isotopic composition of precipitation.

1.1 Atmospheric circulation controls on the isotopic composition of mid-latitude precipitation

Northeast Asia is positioned on the boundary between the area where the seasonal variation in the precipitation isotope ratio is controlled by temperature (high latitude regions) and the area controlled by precipitation amount (lower latitude regions; Bowen, 2008). For mid-latitude regions, the number of studies using event or daily-based isotope data are rapidly increasing in an effort to identify the climate drivers controlling isotopic variability on a seasonal to an annual time scale. For example, in the United States, interevent isotopic variations in response to changes in storm track have been reported (Lawrence et al., 1982; Friedman, 2002; Burnett et al., 2004). Recently, Berkelhammer et al. (2012) showed that changes in vapor sources are manifest in the isotopic composition of precipitation that falls along the western coast of the US. The most isotopically enriched values occur over California when storms supply subtropical marine air with relatively enriched isotopic values from the Pacific. In contrast, storms bringing moisture from the northern Gulf of Alaska are related to the most depleted events. The same feature has been reported for Irish (Dublin) precipitation in northwestern Europe (Baldini et al., 2010). The results of observation in southern Australia revealed the close relationship of isotopes in precipitation to the prevailing synoptic scale weather pattern (Treble et al., 2005; Barras and Simmonds, 2008, 2009; Crawford et al., 2013). These results demonstrate that isotopic variability in individual local precipitation is closely related to the synoptic scale meteorological conditions. This interpretation is consistent with the results from analyses of controls on inter-annual variability of isotopes in mid-latitude precipitation using historical monthly-based isotope records archived in the GNIP, coordinated by the International Atomic Energy Agency (IAEA) in cooperation with the World Meteorological Organization (WMO; e.g., IAEA-WMO, 2013). Recent studies showed that inter-annual isotopic variability in European winter precipitation is related to the North Atlantic Oscillation (NAO) index (Baldini et al., 2008; Casado et al., 2013) and hemisphere-wide teleconnections associated with the Arctic Oscillation (AO; Field, 2010). Birks and Edwards (2009) found a strong Pacific–North American (PNA) control on isotopic composition of Canadian winter precipitation. These new findings demonstrate that the variability in the isotopic composition of seasonal precipitation occurs in response to changes in the pattern of meridional atmospheric circulation. Therefore, coupled with the findings from the event-based record of precipitation isotopes, we can say that isotopic variation in mid-latitude precipitation is not directly controlled by temperature and precipitation amounts, but that it mostly depends on meridional atmospheric circulation. The EAM is characterized by the seasonal reversal of the meridional moisture flow, so that the isotopic variation in northeast Asian precipitation would reflect the EAM activity. Proving the last statement in order to provide a seamless explanation of isotopic variability in association with monsoon circulation at multiple temporal scales (inter-annual, seasonal, and event scales) will require clearly showing that changes in short-term isotopic variability affect inter-annual isotopic variations.

1.2 Outline

The central purpose of this study is to improve our knowledge of how the EAM variability influences the isotopic composition of northeast Asian precipitation. In this study, using new 1-year long event-based isotope values in precipitation and long-term records of monthly isotopes in precipitation in central Japan, we sought to unveil the mechanism that exerts monsoon circulation controls on isotopic composition in precipitation at multiple time scales from event to inter-annual scales. A historical 17-year record of oxygen and hydrogen isotopic composition of precipitation at Tokyo station is archived in the GNIP data set. The 17-year-long Tokyo station GNIP record provides enough information for a discussion of the physical structure of the inter-annual isotopic

variability in response to changes in monsoon circulation. First, we identified key atmospheric processes controlling the storm-to-storm isotopic variations through an analysis of back trajectories and classification of precipitation systems. Then we applied the identified processes to explain inter-annual variability and examine how changes in the EAM activity affect the isotopic composition in Japanese precipitation.

2 Data and methods

2.1 Climate summary at observation site

The isotope observation was conducted at Nagoya (35.15° N 136.97° E, 50 m a.s.l.) located on the Pacific side of central Japan (Fig. 2a). The mean annual rainfall at Nagoya is about 1500 mm, falling almost entirely as rain. The precipitation in summer is much greater than that in winter. GNIP Tokyo station is also located on the Pacific side. The climate in Japan has clear seasonal differences due to the influence of the seasonal reversal of monsoon wind. The summer monsoon season is characterized by two active rainfall periods separated by a break phase (see Ding and Chan, 2005; Ninomiya and Murakami, 1987, for a detailed review). The first rainy period is the northward-migrating rainband known as Baiu in Japan. This Baiu rainband forms on the boundary between the maritime tropical air mass and both continental and maritime polar air masses, and is maintained by moisture supplied by southerly monsoon winds. The Baiu rainband appears in mid-May in the southernmost regions of Japan, and then migrates slowly northward across Japan from early June to mid-July. In mid-July, the Baiu rainband rapidly jumps to northern China and Korea, ending the first rainy period. At the end of August, the Baiu, which had moved northward in early summer, retreats southward to begin the second rainy period that runs from the end of August to early October (called *Akisame* in Japan). At the Pacific side in central Japan, the second rainy period is not so obvious compared with the first period. In addition to the Baiu and *Akisame* rainfall, typhoons occasionally bring large amounts of rainfall to Japan in the summer season. The number of typhoons approaching the Japanese archipelago is much higher in boreal and late summer than in the first rainy period.

In winter, the low-level winds reverse primarily from southerlies to northwesterlies (Fig. 1b). The Siberian anticyclone (called the Siberian High) sits over the eastern Siberia with a strong Aleutian low pressure to its east, pushing away the accumulated cold from Siberia with northwesterlies along the eastern frank of the Siberian High (e.g., Chang and Hitchman, 1982). The subsequent cold and dry airflow into the relatively warmer ocean triggers enhanced evaporation and convective activity over the Japan Sea (e.g., Manabe, 1957; Ninomiya, 1968). Therefore, winters are snowy all along the Japan Sea coast. In contrast, the weather is sunny

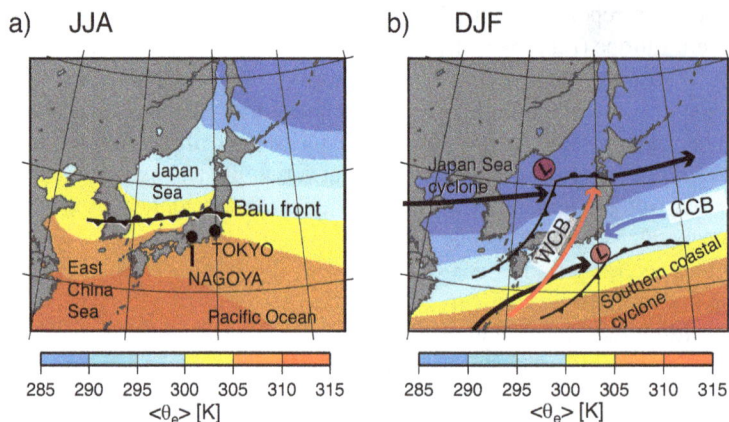

Figure 2. Seasonal mean field of vertically averaged (850–925 hPa) equivalent potential temperatures $\langle\theta_e\rangle$ (shading over the oceans: K) during **(a)** summer (JJA) and **(b)** winter (DJF). The black circle represents the observation site locations (Nagoya and Tokyo). The solid black lines represent typical cyclone tracks (southern coastal cyclone and Japan Sea cyclone) over the Japan archipelago during winter. The red and blue arrows represent the warm conveyor belt (WCB), and the cold conveyor belt (CCB), respectively.

and bright along the Pacific side because most of the snow falls out as the air mass traverses the mountain range that runs centrally along Japan. On the Pacific side, winter precipitation is usually brought by extratropical cyclones advecting off the southern coast of Japan (called southern coastal cyclones) and associated fronts embedded in the cyclone passing through the Japan Sea (called Japan Sea cyclone; Chen et al., 1991; Kusaka and Kitahara, 2009; Adachi and Kimura, 2007; see Fig. 2b).

2.2 Isotope observation

A 1-year long event-based precipitation sampling and continuous measurements of water isotopes were conducted at the campus of Nagoya University from June 2013 to June 2014. Throughout this paper, the isotopic concentrations are expressed in δ notation: δD or $\delta^{18}O = (\frac{R_{sample}}{R_{V\text{-}SMOW}} - 1) \times 1000$, where R is the isotopic ratio (HDO/H_2O or $H_2^{18}O/H_2^{16}O$). V-SMOW is Vienna Standard Mean Ocean Water.

Precipitation samples were collected in a capped high-density polyethylene (HDPE) bottle with a plastic funnel when precipitation was not observed in the previous few hours or at 8 a.m. local time (LT) if precipitation had ended at midnight. To prevent post-evaporation from the collected samples, stable isotope analysis was performed as soon as possible after the collection using cavity ring-down spectroscopy isotopic water analysis (model L1102-i; Picarro Inc., Sunnyvale, CA, USA) with a CTC Analytics autosampler (model HTC-PAL; Leap Technologies, Carrboro, NC, USA). Measurement precision was better than $\pm 0.2‰$ for $\delta^{18}O$ and $\pm 2.0‰$ for δD. Analytical uncertainty in d-excess for our measurements was better than 2.6‰.

Water vapor isotopic composition was measured using both a WVIA (model DLT-100) manufactured by Los Gatos Research Inc. (LGR Inc., Mountain View, CA, USA) and

a conventional cold trap method, which was used in Kurita (2013). Outdoor air (15 m above the ground) was drawn to the WVIA and the cold trap system with a 5 m length of Teflon tubing via the external pump at a flow rate of $1.5\,L\,min^{-1}$. As for laser-based measurement, δD and $\delta^{18}O$ in the ambient air were recorded by the WVIA at 1 Hz. On the other hand, water vapor samples were collected in the trap at 8 a.m. LT every day and were analyzed by the same method as for precipitation samples. Data calibration for laser-based measurement followed the procedure developed by Kurita et al. (2012), without correction for time-dependent isotope drift. For δD measurements, the instrumental drift was less than the analytical error related to the cold trap method, although the instrumental drift in $\delta^{18}O$ values cannot be deemed negligible compared with the results obtained from the cold trap approach. Here, therefore, we focus on δD data. The H_2O concentration-δD response was evaluated by humidity bias defined as the δD difference of water vapor between WVIA data and cold trap samples at each H_2O concentration. We calculated a second order polynomial fitting curve using 1-year long humidity bias data, and then applied it for calibrating the H_2O-concentration dependence. To evaluate the validity of WVIA-measured δD values, the corrected WVIA data were compared to the results obtained from the cold trap approach. We calculated the time-averaged WVIA-measured δD values during vapor trapping and then compared them with the cold trap values for the same sampling periods. The day-to-day variation of the δD from cold trap samples matched the WVIA values as expected, although the short-term variation (within a day) could not be resolved by the measurement using the trapping approach. The mean value of the deviations of 298 samples was $0.4 \pm 2.8‰$. This value is worse than the analytical error associated with the cold trap method ($\pm 2.0‰$), but is acceptably smaller than the natural variability. The optimum average time of the WVIA was examined using the Allan variance method (Werle et al., 1993) by Sturm and Knohl (2010) who concluded that the highest precision was obtained from a 10 to 15 min average. We therefore used 10 min-average data for the analysis.

The long-term records of the oxygen isotopic composition of precipitation ($\delta^{18}O$) at the Tokyo station (35.7° N 139.8° E, 4 m a.s.l.) archived in the GNIP database were used to examine the inter-annual seasonal isotopic variation. The data consists of monthly precipitation samples and are available from 1962 to 1979. The variation in the oxygen isotopic content of precipitation mirrors that of the hydrogen isotopic content.

2.3 Meteorological data

Surface meteorological data were obtained from the local meteorological observatory located nearest each isotope monitoring station (Nagoya University and GNIP Tokyo station). Hourly meteorological variables such as barometric pressure, relative humidity, air temperature, wind speed, wind direction, incoming solar radiation, and precipitation amounts are available from the Japan Meteorological Agency (JMA; http://www.data.jma.go.jp/gmd/risk/obsdl). As for large-scale precipitation field data, radar precipitation data calibrated with rain gauge observations (Radar-AMeDAS precipitation data) provided by JMA was used in this study. Radar-AMeDAS is 10 min data that entirely covers all the Japanese islands and the adjacent oceans. The horizontal resolution of this data is 5 km. Data is available from 1988. The Japanese 55-year reanalysis project (JRA-55) data set (Ebita et al., 2011) were used to examine synoptic scale weather condition. The JRA-55 data are on a horizontal 1.25° × 1.25° grid with 37 vertical layers from 1000 to 1 hPa.

2.4 Back trajectory analysis

To estimate cumulative precipitation during the transport along the air mass pathway, backward air mass trajectories were calculated for each precipitation event using the Hybrid Single-Particle Lagrangian Integrated Trajectory (HYSPLIT) Model (Version 4.0) provided by the National Oceanographic and Atmospheric Administration Air Resources Laboratory (NOAA ARL; Draxler and Rolph, 2003). Wind fields were provided by NCEP (National Center for Environmental Prediction). Global data assimilation system (GDAS), which is a 1.0° × 1.0° horizontal grid, were used as forcing data. Trajectories were calculated from nine different points around the observation site (0.5° × 0.5°) at hourly intervals while rainfall was recorded at the site. Each 30 min position along the 24 h back trajectories was archived. This 24 h duration was selected as it is sufficient to record the final transport of air masses from oceanic source regions to the site. The cumulative rainfall (P_{cumul}) was calculated as the sum of pre-

cipitation along the trajectories before arriving at the site. In order to obtain the time average value during each event, the calculated cumulative precipitation was weighted by the observed hourly precipitation amount at the measurement site:

$$P_{cumul} = \frac{\sum_{t=0}^{m} R_t (\sum_{i=0}^{n} P_i / n)}{\sum_{t=0}^{m} R_t}, \qquad (1)$$

where the subscript m indicates total rainfall hours and n represents the number of trajectories at each hour, P_i is the cumulative rainfall for each trajectory, and R_t is the recorded rainfall amount at the site. In this study, cumulative rainfall corresponds to total precipitation, while air masses travel through the precipitation system that provides rainfall at the observation site. The Japanese precipitation mostly originates from the surrounding oceans (Yoshimura et al., 2004). The spatial distribution of isotopic composition of marine vapor is homogeneous in these areas (Kurita, 2013). Based on these findings, we assume that the isotopic content in source vapor that feeds the precipitation system is less sensitive to the past rainout history over several days. As precipitation intensity exhibits remarkable spatiotemporal variability over a precipitating area, high spatial and temporal resolution precipitation data was used to obtain precise cumulative rainfall. Here, we used 30 min averaged Radar-AMeDAS data, which was re-gridded to a 0.5° resolution to estimate precipitation amount at each trajectory position.

2.5 Precipitation classification

In the EASM season, a quasi-stationary rainband appears at the boundary between warm air and relatively cold air and is characterized by a large gradient of equivalent potential temperature θ_e (Ninomiya, 1984; see Fig. 2a). At the center of this rainband, θ_e in the lower atmosphere is around 335 K and θ_e gradually decreases (increases) toward the north (south; Tomita et al., 2011; Kanada et al., 2012). In this study, we calculated the vertically averaged (925–850 hPa) equivalent potential temperature $\langle \theta_e \rangle$ around the observation site, and then the summer precipitation was divided into warm and cold events depending on the $\langle \theta_e \rangle$ of the air mass-induced precipitation. The warm rainfall event produced by (sub)tropical maritime air mass was defined as (1) those that $\langle \theta_e \rangle$ exceed 335 K around the observation site, or (2) the southerlies from the subtropics regions ($\langle \theta_e \rangle > 335$ K) that transport moisture to feed rain-bearing systems over the observation site. On the other hand, cold rainfall events were defined as residual cases, or those that did not satisfy both conditions.

In the EAWM season, the precipitation events were classified into three types of cyclonic precipitation and into others (O-type). The surface cyclones were identified using the 6-hourly sea level pressure (SLP) field in the JRA-55 data set, and the cyclonic precipitation events were classified depending on the route for cyclones as follows: Japan Sea cyclone (J-type); southern coastal cyclone (S-type); and inter-

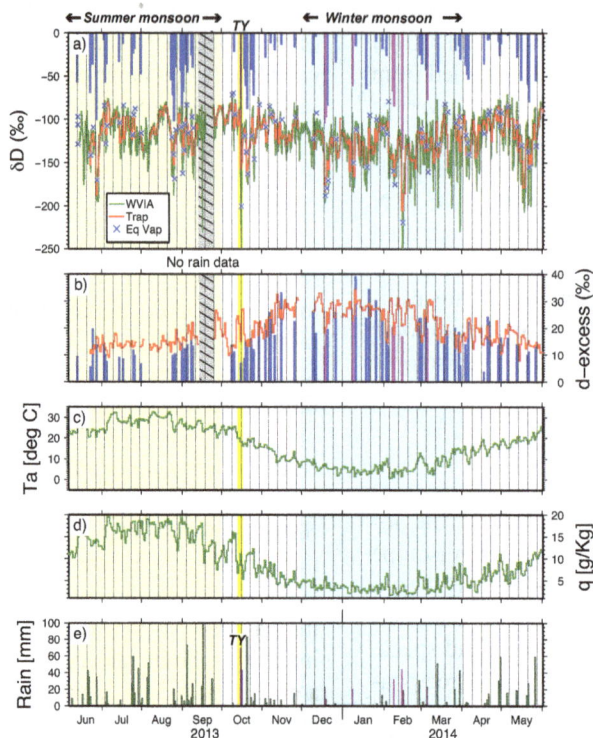

Figure 3. Time series of isotopic values and surface meteorological variables at Nagoya in Japan for the period of June 2013 to June 2014 (summer: shaded in light yellow; winter: shaded in sky blue). (**a**) δD in precipitation (bar) and surface vapor measured by the laser instrument (green line) and by the conventional cold trap method (red line). The black cross represents the calculated δD of vapor in isotopic equilibrium with precipitation at surface air temperature. The pink-colored bars represent rainfall from Nangan cyclones. (**b**) The d-excess in precipitation and surface vapor measured by the cold trap method. (**c**) Air temperature, (**d**) mixing ratio and (**e**) precipitation observed at the nearest meteorological station. The "TY" at the top of the figure corresponds to the passage of a typhoon at the observation site.

mediate cyclone type (I-type). The Japan Sea cyclone type corresponds to the rainfall from the frontal system related to the Japan Sea cyclone. The southern coastal cyclone type is a cold rainfall or snowfall event that occurs at the north of the surface warm front moving eastward along the southern coast of Japan. Intermediate type cyclones travel over the main island of Japan, with the center of the cyclones passing near the study site. Although most precipitation events are driven by extratropical cyclone connected with fronts, occasionally small-scale convective clouds traveled from the coast of the Sea of Japan, bringing weak snowfall or rainfall to the site (O-type).

3   Results

3.1   Seasonal cycle

One-year records of δD in both water vapor and precipitation are shown in Fig. 3, together with temporal variation in meteorological variables. Continuous δD variation in surface vapor shows that the seasonal cycle is weak, and that sub-monthly or intra-seasonal variation is dominant. This contrasts with meteorological variables (surface air tempera-

ture, water vapor concentration, and precipitation amounts), which exhibit distinct seasonality with the maximum value in summer and the minimum in winter. Additionally, clear seasonality of the d-excess obtained from cold trap samples, with higher values in winter and lower values in summer, is seen in Fig. 3b. This reflects the change in d-excess over oceanic source regions by the reversal of the monsoon wind direction. In winter, northerly winds push cold and dry continental air over Japan and the adjacent ocean. Evaporation occurs under conditions with a large humidity deficit and a strong temperature contrast between the surface air temperature and the sea surface temperature. As noted previously by numerous studies (e.g., Gat et al., 2003; Uemura et al., 2008), this leads to relatively high d-excess in marine vapor. In contrast, in summer, because evaporation occurs under humid condition, d-excess in a maritime air surrounding Japan is much lower than that in winter (Kurita, 2013). Weak seasonality of δD in surface vapor means that isotopic changes over the oceanic regions have only a minor impact on the δD values in surface vapor. Meanwhile, intraseasonal δD variation is associated with precipitation events. When raindrops fall through the atmosphere, diffusive exchange between raindrops and the surrounding vapor and evaporation of the rain-

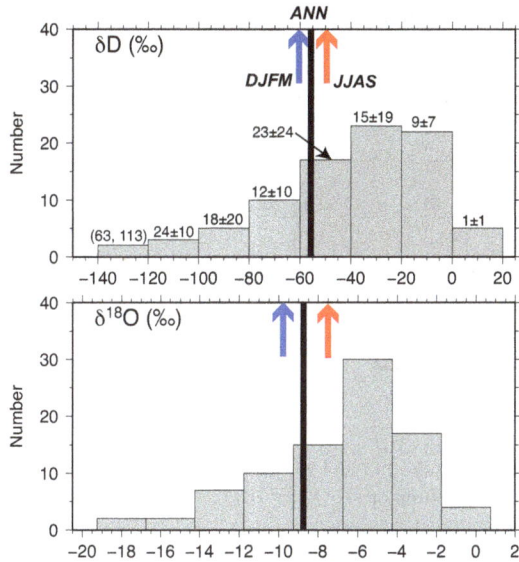

Figure 4. Histograms of (upper panel) δD and (lower panel) δ^{18}O values in event-based precipitation. The black line represents the annual mean value weighted by precipitation amount. Blue and red arrows correspond to the winter (DJFM) and summer (JJAS) mean value. The numbers on each bar in the top figure represent average precipitation amount at each bin.

drops takes place, and the δD in vapor tends to be close to the isotopic equilibrium with precipitation (Fig. 3a). The temporal isotopic variability of surface vapor during a year therefore mostly reflects the variation in precipitation. Our time series data highlights the control by short-term isotopic variability on seasonal or more long-term mean isotopic values. This time scale matches well with the synoptic disturbance activity. Hereafter, we focus most of the analysis on storm-to-storm isotopic variability in precipitation.

The distribution of δD in individual precipitation events (daily precipitation amount) is shown in Fig. 4a. Interestingly, the precipitation-weighted annual mean δD value ($-56\,‰$) does not correspond to the dominant bin of δD value, and is significantly lower than the arithmetic mean ($-40\,‰$). The winter (DJFM) and summer (JJAS) mean values are also lower than the arithmetic mean annual precipitation. The same features can be seen in δ^{18}O (Fig. 4b). These findings suggest that the precipitation events with lower δD values make a larger contribution to the seasonal and annual mean values, and the variability of annual or seasonal mean δD value is attributable to the change in relative contribution of storm events. However, for all the precipitation events ($n = 78$), the negative correlation between δD in precipitation and local precipitation amount (the amount effect) is weak ($R^2 = 0.167$). The storm-to-storm isotopic variation may be controlled by not only the local rainout, but also by upstream atmospheric conditions such as moisture sources and rainout history during the transport.

3.2 Summer precipitation

During the summer rainy season, sub-synoptic scale rain fronts form between a warm air mass from subtropical maritime regions and a cold air mass from both continental and maritime polar regions. In the first rainy season (Baiu), rainfall occurs in association with the north-south displacement of this front. Notably different equivalent potential temperatures (θ_e) characterize the cold and warm air masses (Fig. 2a). θ_e at the observation site varies considerably depending on the north-south displacement of this front. In addition, as shown in Fig. 5, the temporal variation of δD in surface vapor is associated with changes in vertical average of θ_e ($\langle\theta_e\rangle$). The highest δD peaks correspond to the southerly flows supplying warm maritime air to the site, while abrupt δD decreases appear in association with a southward shift of cold air mass.

Rainfall events during the second rainy season (*Akisame*) are usually derived from the low-pressure systems connected with fronts. Low-level southerly winds east of the cyclone transport warm air with high $\langle\theta_e\rangle$, while relatively cold air with low $\langle\theta_e\rangle$ is supplied by northerly winds. In Fig. 5, the short-term increases and decreases in $\langle\theta_e\rangle$ correspond to the passage of the frontal systems and troughs, and similar to those occurring during the first rainy season, δD in surface vapor relates well to the $\langle\theta_e\rangle$ variations. These features are identical with the results obtained from the observation in 2010 (Kurita et al., 2013). In addition, this temporal isotopic variation mirrors that of precipitation. Relatively high δD in precipitation matches the warm rainfall events, and lower δD is observed when cold air inflow occurred. These are classified as cold rainfall events. We conclude from this that the δD difference between warm- and cold-type rainfall reflects changes in air mass sources. On 24 August (labeled 1 in Fig. 5), δD incoming precipitation was relatively low, although $\langle\theta_e\rangle$ was similar to that in subtropical air (> 335 K). The weather map showed that southerly winds from the subtropics did not arrive at the observation site and the frontal system was displaced to the south of the observation site. The rainfall event therefore corresponds to cold rain and not warm rain.

We investigate the influence of rainout history from the upstream region. The relationship between the δD in individual precipitation events and cumulative rainfall along the trajectories (P_{cumul}) over 9 h is displayed in Fig. 6a. Although δD of warm rainfall events are plotted above those from cold events, the gradual decreasing trend in δD with P_{cumul} is obvious. The correlation coefficient gradually improves with increasing integration time and then reaches the highest value ($R^2 = 0.654$) at 9 h (see inset in Fig. 6a). The sample labeled "2" in Fig. 6a (same as labeled "2" in Fig. 5) was ignored from the regression analysis because this sample was obtained from the downstream-propagating fully developed convective line, in which the influence of vapor recycling associated with mesoscale convective systems may signifi-

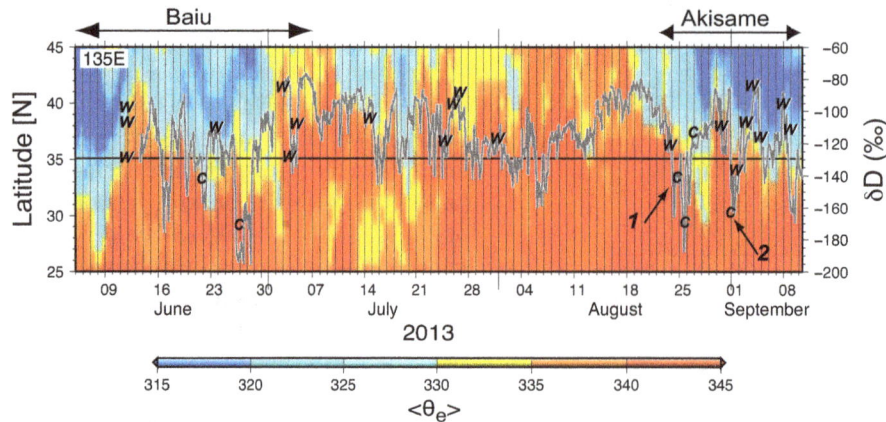

Figure 5. Time series of the value of δD in surface vapor observed at Nagoya University (gray line) and vertically averaged (850–925 hPa) equivalent potential temperatures $\langle\theta_e\rangle$ (K) between 25° N and 45° N along 135° E during summer (JJAS) in 2013. The labeled W's (C's) represent the value of δD in surface vapor in equilibrium with warm (cold) type precipitation at the same site.

cantly reduce rainfall δD (Kurita, 2013). The lowest δD was observed with the maximum P_{cumul} when an intense typhoon passed southeast of the site in the middle of October (labeled TY in Fig. 6a). An intense precipitation area extended from east and north of the center of the typhoon along the cyclonic flow with the site being located in the downstream area of the precipitation (the northwestern side of the typhoon). Also, since northeasterly winds at the northern side of the typhoon transport low $\langle\theta_e\rangle$ air from the north, the contribution of this air to the typhoon precipitation may act to further decrease δD. For summer precipitation, including October, good correlation between δD in precipitation and P_{cumul} indicates that cumulative precipitation while air mass passed through the rain-bearing systems is a major driver of the storm-to-storm variability in δD.

3.3 Winter precipitation

In winter, precipitation occurs in association with the passage of eastward-moving extratropical cyclones connected with fronts (see Fig. 2b). We examine the isotopic variability arising from changes in moisture sources. According to the conveyor belt concept (Carlson, 1980), a winter cyclone consists of three major air streams: (a) the warm conveyor belt (WCB); (b) the cold conveyor belt (CCB); and (c) the dry intrusion. The WCB is a stream of relatively warm moist air and originates over the warm waters of Pacific. This air flows northward toward the center of the cyclone and supplies moisture to a warm frontal rainband. The CCB originates to the northeast of the cyclone, and easterly or northeasterly winds north of the cyclone transport this air westward, and feeds a cold frontal rainband. Therefore, the precipitation associated with the cold front mainly originates from the mid- or high-latitude regions.

In the case of the Japan Sea cyclones, the low pressure centers move across the middle of Japan Sea. Figure 7 shows

that warm air with high $\langle\theta_e\rangle$ is injected into the high-latitude region across the observation site when this type of precipitation appeared. The fact that these warm air injections correspond to the highest peaks of δD in surface vapor indicates that the WCB supplies moisture northward with enriched δD. Two events of intermediate type (I-type) also show similar features to the Japan Sea type (J-type). This indicates that the WCB reaches near the center of the cyclone. In contrast, the δD depletions occur in association with the passage of the southern coastal cyclones. The major route for the southern coastal cyclone is off the southern coast of Japan from the East China Sea. Unlike the Japan Sea cyclone, a cold rainband fed by the CCB is primarily responsible for precipitation along the Pacific coast of east Japan (Takano, 2002). The CCB transports moisture originating from mid- and high-latitude regions, and as shown by Kurita (2013), the isotopic content of marine surface vapor decreases toward the high latitude regions. Therefore, precipitation in association with the southern coastal cyclone (S-type) is characterized by relatively lower δD values than the others. We further examine the influence of rainout history along the trajectories as an additional source of the δD variability in winter. Excepting S-type events, a robust trend of lower δD with increasing P_{cumul} is evident in Fig. 6b. However, δD in S-type precipitation are distributed without a clear trend in relation to P_{cumul} and are plotted below the regression line obtained from both J-type and I-type samples. Two S-type events observed in March (plotted near the regression line) were considered to be I-type rather than S-type because warm air with high $\langle\theta_e\rangle$ was transported beyond the center of the cyclone and reached near the observation site (Fig. 7). These results indicate that the progressive depletion of δD in vapor-produced rainfall along the pathway from the moisture source region is not a major contributor to large δD depletion in S-type precipitation. Distinctive changes in the isotopic composition of moisture that

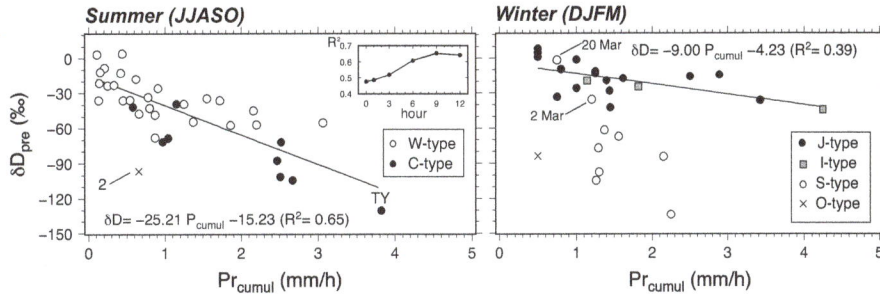

Figure 6. Relationships between δD values in individual precipitation events and cumulative precipitation amount (P_{cumul}) over 9 h back trajectories of the air mass launched from the observation site in **(a)** summer (from June to October) and **(b)** winter (December to March). Symbols indicate each different precipitation type (see detail in the text). Inset: variation in the correlation coefficient (R^2) for δD-P_{cumul} relationship with variation in the period of accumulation. The highest R^2 values were observed with 9 h of cumulative precipitation in summer.

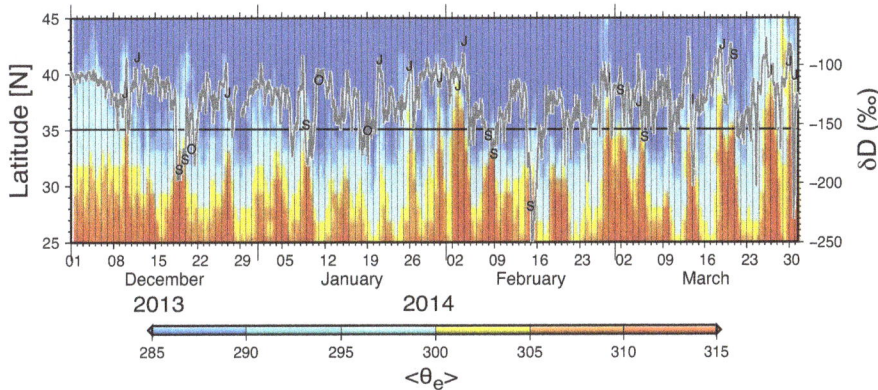

Figure 7. Same as Fig. 5, but represents the time series for the winter (DJFM) season. The labeled characters (J, I, S, O) represent the type of precipitation event as follows: J, Japan Sea-type; I, Intermediate-type; S, southern coastal-type; and O, others-type.

feeds precipitation systems is likely a more important factor influencing the large isotopic variability in winter.

4 Inter-annual isotopic variation

Here, using the historical 17-year record of isotopes in precipitation from the GNIP Tokyo station, we examine whether the identified key drivers can account for the inter-annual isotopic variations. In this study we use oxygen isotopes $\delta^{18}O$ as a substitute for δD, because the oxygen isotope record of precipitation in Tokyo has been recorded much longer than for δD. The $\delta^{18}O$ variation of precipitation mirrors that of hydrogen isotopic content.

In summer, δD ($\delta^{18}O$) in individual rainfall events varies widely from event to event, ranging from close to 0‰ (0‰ for $\delta^{18}O$) to less than −100‰ (−13‰), and the wide range of isotopic variability is mainly attributed to the cumulative rainfall along the trajectories (Fig. 6a). To investigate whether this effect is sufficient to explain the inter-annual isotopic variability, the relationship between isotopic content in precipitation and regional average precipitation amounts is examined. Because the observation period for

when monthly precipitation was collected in the GNIP Tokyo station (1962–1979) occurred prior to the availability of the Radar-AMeDAS data, we cannot calculate cumulative rainfall amounts in the same way as mentioned above. The insert in Fig. 6a shows gradual improvement of the correlation coefficient with increasing integration time of P_{cumul}. However P_{cumul} along the 9 h trajectories exhibits a good correlation with the values at 0 h back trajectories ($R^2 = 0.653$, p value < 0.01), corresponding to area averaged precipitation (0.5° × 0.5°). This means that we can use area-averaged precipitation amounts as a substitute for cumulative rainfall. The area-averaged rainfall (1.0° × 1.0°) is calculated using Global Precipitation Climatology Center monthly precipitation (GPCC) data set version 6 (Schnider et al., 2011) and then compared with $\delta^{18}O$ in summer precipitation. The result shows that inter-annual variability in monthly rainfall $\delta^{18}O$ is completely independent from the precipitation amount. In addition, $\delta^{18}O$ in the summer-averaged precipitation shows no significant isotopic depletion with precipitation amount ($R^2 = 0.076$). Different from the storm-to-storm isotopic variability, rainout history seems not to have a significant influence on the inter-annual variability in the summer rainfall $\delta^{18}O$. Next, we explore the influence of moisture sources on the inter-annual isotopic variation. In summer, southerly flows transport moisture with relatively higher $\delta^{18}O$, with the warm rainfall type being relatively enriched in heavy isotopes compared with the other rainfall events (Fig. 5). An increase in the contribution of warm rainfall to the total summer precipitation must therefore lead to richer isotopic values in summer average precipitation. Here, we divide daily precipitation into warm and cold events using the $\langle \theta_e \rangle$ field, and then calculate the cumulative warm rainfall amounts for the summer. To examine how changes in contribution of warm rainfall are manifested in the isotopic values in summer precipitation, we define the warm rain ratio (R_{WR}) as the ratio of cumulative warm rainfall to the total summer precipitation. During the period of 1962 to 1979, the R_{WR} changes widely year by year, ranging from 0.35 to 0.85 and the inter-annual variation in $\delta^{18}O$ in summer precipitation matches well with R_{WR} (Fig. 8a). The correlation between $\delta^{18}O$ and R_{WR} for each individual month is as follows: June ($R^2 = 0.630$); July ($R^2 = 0.719$); August ($R^2 = 0.413$); and September ($R^2 = 0.719$). This indicates that year-to-year variations in R_{WR} particularly influence the inter-annual variability in rainfall $\delta^{18}O$ during the first rainy season (June–July). With more than half of summer precipitation occurring during the first rainy season, a high correlation coefficient ($R^2 = 0.627$, p value < 0.01) is observed for the precipitation-weighted summer average (Fig. 9a). From this we can conclude that variations in the southerly moisture flux are the primary driver of inter-annual variations in the summer precipitation $\delta^{18}O$.

Winter precipitation is usually related to extratropical cyclones, and inter-event isotopic variability is linked to their storm tracks. Southerly storms classified as southern coastal-type induce cold precipitation with the most depleted isotope values, as northerly winds bring cold and humid air with lower isotopic values to the observation site. Clearly, the greater the contribution by southern coastal-type precipitation to the total winter precipitation, the more depleted the isotopic content in winter average precipitation will be. Using the SLP field, we identify the date when southern coastal cyclones passed through the site, and then calculate cumulative daily precipitation produced by them during winter season. In this calculation, well-developed cyclones moving off the southern coast of Japan are classified as I-types, because the strengthening southerly flow, corresponding to the WCB, injects warm air further north. As expected, year-to-year variation in the precipitation-weighted $\delta^{18}O$ in winter precipitation are negatively correlated with the relative contribution of southern coastal-type precipitation to total winter precipitation ($R_{S\text{-type}}$) in Fig. 8d. Here, the winter of 1961 denotes December 1961, January 1962, and February 1962. The positive (negative) peaks of $R_{S\text{-type}}$ (1965, 1969, 1973 for positive; 1966, 1970, 1976 for negative) match the relatively depleted (enriched) $\delta^{18}O$ in winter precipitation. Moreover, an abrupt increase of $\delta^{18}O$ after 1975 occurs concurrently with a clear

decreasing trend of $R_{S\text{-type}}$. Consequently, a statistically robust negative correlation ($R^2 = 0.483$, p value < 0.01) is observed during the 1962 to1979 period (Fig. 9b). We can conclude that frequency and intensity of the southern coastal cyclones have a significant impact on both individual precipitation and the winter averaged precipitation $\delta^{18}O$. Nakamura et al. (2012) have reported that the tracks of southern coastal cyclones varied in association with the meridional shift of the warm ocean current (called Kuroshio) flowing in the south of Japan. The straight path of Kuroshio along the south coast of Japan changed to a meander path state after the year 1975. The abrupt decrease of $R_{S\text{-type}}$ after 1975 may be related with this transition of the Kuroshio path.

5 EAM controls on inter-annual isotopic variability in precipitation

The historical 17-year record of isotopes in precipitation demonstrates that isotopic variability on an inter-annual time scale is primarily driven by changes in the recurrence of synoptic-scale variations. Here, we examine whether these synoptic-scale changes are forced by the EAM variability. During the summer monsoon season (JJA), a strong or weak EASM year is determined by the activity of the Bonin High (e.g., Ha and Lee, 2007). Anomalous intensification (weakening) of the Bonin High enhances (reduces) the EASM and thus leads to an increase (decrease) in summer precipitation over northeast Asia. Following Ha and Lee (2007), we define the Bonin High index (BHI) by the 500 hPa geopotential height anomaly over the western Pacific area (140–145° E, 25–30° N). Three strong EASM years (1963, 1969, and 1979) and three weak EASM years (1967, 1972, and 1974) are identified using the BHI time series during 1962 to 1978. Interestingly, these strong and weak EASM years correspond well with the positive and negative R_{WR} peaks (Fig. 8b). The year-to-year variability of the normalized BHI exhibits a significant negative correlation with R_{WR} variation from 1962 to 1978 ($R = -0.66$, p value < 0.05) except for the year of 1966. The lowest R_{WR} peak in 1966 was influenced by the passage of an intense typhoon on the southeast side in June. More than 200 mm of rainfall was induced by the strong northeasterly wind from the northwestern side of the typhoon. A negative BHI- R_{WR} relationship indicates that the enhanced southerly monsoon flow decreases the relative ratio of precipitation fed by subtropical moisture. To explain this relationship, composite maps of atmospheric circulation and anomalies of the $\langle \theta_e \rangle$ field for strong and weak monsoon summers are shown in Fig. 10. The strong monsoon summers are characterized by low-level southwesterlies along the northwestern rim of the Bonin High. The enhanced moisture supply to the Baiu rainband in turn augments precipitation around Japan. Anomalous diabatic heating associated with precipitation induces a positive $\langle \theta_e \rangle$ anomaly along the southern coast of Japan (Fig. 10). This positive anomaly

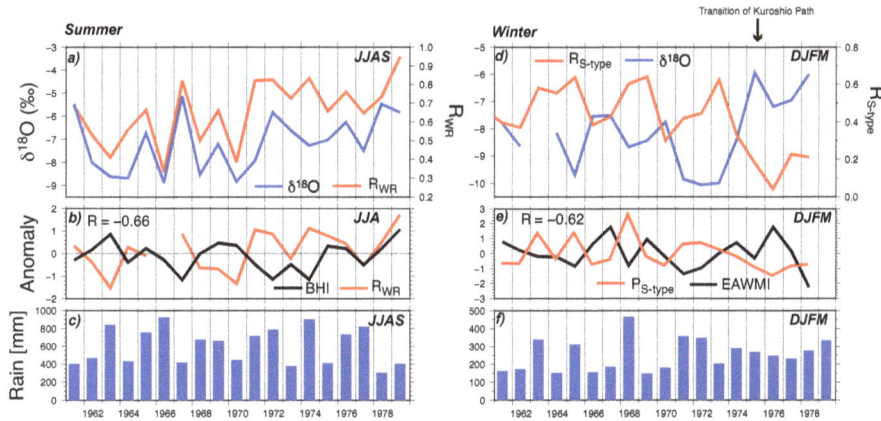

Figure 8. Time series of seasonal average data for the period 1961 to 1979. (**a**) Precipitation-weighted $\delta^{18}O$ in summer (June–September) precipitation at Tokyo station and the ratio of the warm rainfall to the total summer precipitation R_{WR}. (**b**) The R_{WR} anomaly during boreal summer (June–August) and the Bonin High index (BHI) normalized by standard deviation. (**c**) The total summer precipitation amount at a meteorological station located near the GNIP Tokyo station. (**d**) Same as (**a**), but for winter (December–March) precipitation and for the ratio of the southern coastal cyclone to total winter precipitation $R_{S\text{-type}}$. (**e**) Same as (**b**), but for $R_{S\text{-type}}$ anomaly during winter and for East Asian Winter Monsoon index (EAWMI). (**f**) Same as (**c**), but for winter precipitation.

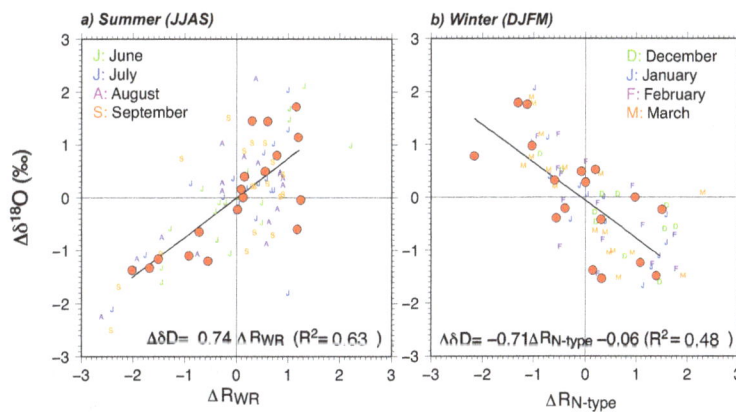

Figure 9. Anomalies of $\delta^{18}O$ in precipitation at GNIP Tokyo station vs. (**a**) the ratio of warm rainfall to total summer precipitation (R_{WR}) and (**b**) the ratio of the southern coastal precipitation ratio to the total winter precipitation ($R_{S\text{-type}}$). Anomalies for monthly and seasonal averaged values were calculated by subtracting the long-term mean for that month and season and then dividing by the standard deviation. The long-term means were calculated from all available records (1962–1979) at GNIP Tokyo. Red circles represent the precipitation-weighted values in summer (from June to September) and in winter (from December to March).

indicates that the Baiu rainband was frequently located to the south of the observation site, and thus the contribution of the warm-type rainfall would be less than usual. By contrast, the weak EASM years are characterized by southerly winds. Anomalous southerlies over the Pacific may induce large-scale positive $\langle \theta_e \rangle$ anomaly area over the Pacific and increases the fraction of warm-type rainfall.

In winter, the EAWM intensity is controlled by the pressure gradients between the Aleutian Low and the Siberian High. A steeper pressure gradient results in a more vigorous EAWM over northeast Asia. In this study, the strength of the EAWM is evaluated by the EAWM index (EAWMI), which reflects the 300 hPa meridional wind shear associated with the jet stream (Jhun and Lee, 2004). Figure 8e exhibits

the time series of the normalized EAWMI and the S-type precipitation anomalies during the winter monsoon season (DJFM) from 1961 to 1978. The strong winter monsoon years (1967, 1969, 1976) and weak winter monsoon years (1968, 1971, 1972) match the negative and positive anomalies of S-type precipitation. The EAWMI variations are negatively correlated with changes in the S-type precipitation ($R = -0.62$, p value < 0.05). This suggests that the precipitation brought by the southern coastal cyclones is smaller in the strong EAWM years than in the weak EAWM years. The composite maps of $\langle \theta_e \rangle$ anomalies for strong and weak EAWM shows a clear contrast between the strong and weak monsoon winter years (Fig. 10). In the strong EAWM years, remarkable negative $\langle \theta_e \rangle$ anomalies caused by cold air out-

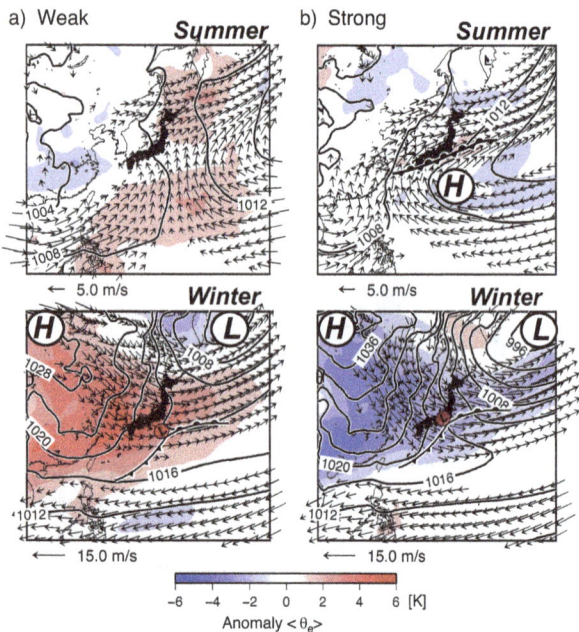

Figure 10. Composite maps of equivalent potential temperatures $\langle \theta_e \rangle$ anomaly (shading: K) sea level pressure (solid contours, contour interval 4 hPa) and surface wind (vectors: $m\,s^{-1}$) for: weak monsoon summers (top left panel); strong monsoon summers (top right panel); weak monsoon winters (bottom left panel); and strong monsoon winters (bottom right panel).

breaks cover the entire northeastern Asia region and the adjacent ocean. The weak EAWM years exhibit exactly opposite behavior. In this region, the heat and moisture flux from the warm ocean is enhanced during the strong EAWM, resulting in favorable conditions for the development of southern coastal cyclones (Yoshiike and Kawamura, 2009). These authors reported that the number of well-developed cyclones is larger in the strong EAWM than in the weak EAWM. As noted in the previous section, the well-developed cyclones are classified as I-types because the strengthening southerly flow transports warm air further north. Therefore, the development of southern coastal cyclones results in decreased S-type precipitation to the site. In addition, it is noteworthy that the storm tracks tend to be concentrated in the vicinity of the south coast of Japan during the strong EAWM years. This indicates that the classification of the extratropical cyclones moving eastward along the Japanese coast tend to be of the I-type. This is consistent with snowfall or cold precipitation events that occur only if the tracks of southern coastal cyclones are distributed away from the south coast of Japan (Nakamura et al., 2012). From these indications, intensified EAWM considerably influences the development and the tracks of the southern coastal cyclones, resulting in decreased S-type precipitation. In contrast, in the weak EAWM years, the distribution of the cyclone tracks away from the southern coast of Japan provides suitable condition for S-type precipitation.

The goal of our study is to unveil the relationship between $\delta^{18}O$ variability in precipitation and the EAM. We have shown that summer and winter monsoon circulation influence to a large extent the key processes controlling isotopic variability in Japanese precipitation. Therefore, year-to-year variation in the precipitation-weighted $\delta^{18}O$ in summer precipitation correlates negatively with the BHI between 1962 and 1978 ($R = -0.47$, p value < 0.05). The winter precipitation $\delta^{18}O$ is positively correlated with the EAWMI ($R = 0.42$, p value < 0.1). These significant correlations are attributed to changes in meridional moisture transport associated with monsoon activity due to the distinctive difference in $\delta^{18}O$ between low- and high-latitude moisture.

6   Conclusions

This study elucidates the mechanism by which the EAM variability influences the isotopic composition of Japanese precipitation. During the summer rainy season, the sub-synoptic scale rainband (Baiu precipitation) forms on the boundary between the warm air mass with relatively enriched isotopic values and the cold air mass with lower isotopic values. The storm-to-storm isotopic variation is linked to a north-south displacement of this rainband; the higher (lower) isotopic values correspond to the rainband moving north (south) of Japan. On an inter-annual time scale, the Baiu precipitation varies in association with EASM variability. The year-to-year isotopic variability in Japanese precipitation is mostly associated to changes in monsoon circulation. The strong EASM years are characterized by the development of stronger low-level southwesterlies. The enhanced supply of moisture from southwesterlies creates favorable conditions for sustaining the Baiu precipitation along the southern coast of Japan through the rainy season. Consequently, the isotopic composition of summer precipitation is then lower than that of a normal year. The prevailing southwesterly winds in the strong EASM change to southerly winds in the weak EASM. Southerly flows then transport moisture with relatively higher isotopic values from subtropical marine regions, and bring warm precipitation relatively enriched with heavy isotopes compared with the others. As a result, we observe a positive correlation between $\delta^{18}O$ variability in summer precipitation and the EASM.

In the winter monsoon season, low isotopic values occur when a cold frontal rainband associated with extratropical cyclones (southern coastal cyclones) passes off to the south of the Japan coast. Easterly or northeasterly winds north of the cyclone transport relatively cold air from the mid- or high-latitude regions to the Pacific side of Japan, and feed the cold frontal rainband. Therefore, the precipitation related to the southern coastal cyclone is characterized by relatively lower isotopic values than those from another type of cyclone. It follows that the occurrence of southern coastal cyclones is the most likely contributor to changes in the mean isotopic composition of precipitation in the winter. The EAWM variability influences the activity and tracks of the southern coastal cyclones on an inter-annual time scale. The northward shift of the cyclone tracks and stronger development of cyclones during the strong EAWM years are responsible for decreasing the contribution of cold frontal rainfall fed by northerly winds and for increasing isotopic values in winter precipitation. In contrast, in the weak EAWM years, the distribution of cyclone tracks away from the southern coast of Japan provide suitable conditions for cold precipitation, resulting in lower isotopic values in winter precipitation. These results indicate that inter-annual isotopic variability in summer and winter Japanese precipitation is tightly related to the meridional monsoon circulation, which is due to the distinctive difference in isotopic composition between low- and high-latitude moisture. Therefore, the isotopic composition of summer and winter precipitation are significantly correlated with the EASM and EAWM index during the period of 1962 to 1978. Our data suggest that processes driving the intra-seasonal variability, evidenced thanks to our high resolution event-based sampling, also explain inter-annual variations, in each season. We emphasize however one exception, which is the sharp increase in winter precipitation isotopic composition recorded in 1975. In this case, we attribute this signal to the impact of a transition of the Kuroshio's path.

This improved understanding of processes controlling the isotopic composition of precipitation highlights the importance of large scale drivers. As a result, records of past precipitation isotopic composition that could be obtained from the wealth of natural archives in central Japan (e.g., stalagmites, tree ring cellulose, leaf wax in lake sediments) should have a strong potential to expand the documentation of past changes in summer and winter EAM. Because our study elucidates the climate drivers controlling isotopic variations in summer and winter precipitation, it is important to retrieve a record with seasonal time resolution from paleoarchives.

Acknowledgements. This study has been funded by a grant for environmental research projects from the Sumitomo Foundation (Japan). Most special thanks to our editor, Valérie Masson-Delmotte, for her valuable suggestions to improve this manuscript as well as for enthusiastic support throughout the editorial process. We also thank two anonymous reviewers for their constructive comments. The authors gratefully acknowledge the NOAA Air Resources Laboratory (ARL) for the provision of the HYSPLIT transport and dispersion model and the relevant input files for generation of back trajectories. The Radar-AMeDAS data was acquired from the Research Institute for Sustainable Humanosphere (RISH), Kyoto University.

Edited by: V. Masson-Delmotte

References

Adachi, S. and Kimura, F.: a 36-year climatology of surface cyclogenesis in east Asia using high-resolution reanalysis data, SOLA, 3, 113–116, 2007.

An, Z., Kukla, G., Porter, S., and Xiao, J.: Magnetic susceptibility evidence of monsoon variation on the Loess Plateau of central China during the last 130,000 years, Quaternary Res., 36, 29–36, 1991.

Araguás-Araguás, L., Froehlich, K., and Rozanski, K.: Stable isotope composition of precipitation over southeast Asia, J. Geophys. Res., 103, 28721–28752, doi:10.1029/98JD02582, 1998.

Baldini, L. M., McDermott, F., Foley, A. M., and Baldini, J. U. L.: Spatial variability in the European winter precipitation δ^{18}O-NAO relationship: Implications for reconstructing NAO-mode climate variability in the Holocene, Geophys. Res. Lett., 35, L04709, doi:10.1029/2007GL032027, 2008.

Baldini, L. M., McDermott, F., Baldini, J. U. L., Fischer, M. J., and Mröllhoff, M.: An investigating of the controls on Irish precipitation δ^{18}O values on monthly and event timescales, Clim. Dyn., 35, 977–993, 2010.

Barras, V. and Simmonds, I.: Observation and modeling of stable water isotopes as diagnostics of rainfall dynamics over southeastern Australia, J. Geophys. Res., 114, D23308, doi:10.1029/2009JD012132, 2009.

Barras, V. J. I. and Simmonds, I.: Synoptic controls upon δ^{18}O in southern Tasmanian precipitation, Geophys. Res. Lett., 35, L02707, doi:10.1029/2007GL031835, 2008.

Berkelhammer, M., Stott, L., Yoshimura, K., Johnson, K., and Shinha, A.: Synoptic and mesoscale controls on the isotopic composition of precipitation in the western United States, Clim. Dyn., 38, 433–454, 2012.

Birks, S. and Edwards, T.: Atmospheric circulation controls on precipitation isotope climate relations in western Canada, Tellus, 61B, 566–576, 2009.

Bowen, G.: Spatial analysis of the intra-annual variation of precipitation isotope ratios and its climatological corollaries, J. Geophys. Res., 113, D05113, doi:10.1029/2007JD009295, 2008.

Burnett, A. W., Mullins, H. T., and Patterson, W. P.: Relationship between atmospheric circulation and winter precipitation δ^{18}O in central New York State, Geophys. Res. Lett., 31, L22209, doi:10.1029/2004GL021089, 2004.

Cai, Y., Tan, L., Cheng, H., An, Z., Edwards, R., Kelly, M., Kong, X., and Wang, X.: The variation of summer monsoon precipitation in central China since the last deglaciation, Earth Planet. Sci. Lett., 291, 21–31, 2010.

Carlson, T.: Airflow through midlatitude cyclones and the comma cloud pattern, Mon. Wea. Rev., 108, 1498–1509, 1980.

Casado, M., Ortega, P., Masson-Delmotte, V., Risi, C., Swingedouw, D., Daux, V., Genty, D., Maignan, F., Solomina, O., Vinther, B., Viovy, N., and Yiou, P.: Impact of precipitation intermittency on NAO-temperature signals in proxy records, Clim. Past, 9, 871–886, doi:10.5194/cp-9-871-2013, 2013.

Chang, J. H. and Hitchman, M. H.: On the role of successive downstream development in East Asian polar air outbreaks, Mon. Wea. Rev., 110, 1224–1237, 1982.

Chen, S.-J., Kuo, Y.-H., Zhang, P.-Z., and Bai, Q.-F.: Synoptic climatology of cyclogenesis over east Asia, 1958–1987, Mon. Wea. Rev., 119, 1407–1418, 1991.

Cheng, H., Edwards, R., Broecker, W., Denton, G., Kong, X., Wang, Y., Zhang, R., and Wang, X.: Ice Age Terminations, Science, 326, 248–252, 2009.

Crawford, J., Hughes, C. E., and Parkes, S. D.: Is the isotopic composition of event based precipitation driven by moisture source or synoptic scale weather in the Sydney Basin, Australia?, J. Hydrol., 507, 213–226, 2013.

Ding, Y. and Chan, J.: The east Asian summer monsoon: and overview, Meteor. Atmos. Phys., 89, 117–142, 2005.

Draxler, R. and Rolph, G.: Hybrid Singlee-Particle Lagrangian Integrated Trajectory (HYSPLIT), Model released from NOAA ARL READY NOAA Air Resources Laboratory, Silver Spring, MD., http://ready.arl.noaa.gov/HYSPLIT.php (last access: 30 April 2014), 2003.

Ebita, A., Kobayashi, S., Ota, Y., Moriya, M., Kumabe, M. R., Onogi, K., Harada, Y., Yasui, S., Miyaoka, K., Takahashi, K., Kamahori, H., Kobayashi, C., Endo, H., Soma, M., Oikawa, Y., and Ishimizu, T.: The Japanese 55-year Reanalysis JRA-55:An Interim Report, SOLA, 7, 149–152, doi:10.2151/sola.2011-038, 2011.

Field, R. D.: Observed and modeled controls on precipitation $\delta^{18}O$ over Europe: From local temperature to the Northern Annular Mode, J. Geophys. Res., 115, D12101, doi:10.1029/2009JD013370, 2010.

Friedman, I.: Stable isotpe composition of waters in the Great Basin, United States 1, Air-mass trajectories, J. Geophys. Res., 107, 1–14, doi:10.1029/2001JD000565, 2002.

Gat, J. R., Klein, B., Kushnir, Y., Roether, W., Wernli, H., Yam, R., and Shemesh, A.: Isotope composition of air moisture over the Mediterranean Sea: and index of the air-sea interaction pattern, Tellus B, 55, 953–965, 2003.

Ha, K.-J. and Lee, S.-S.: On the interannual variability of the Bonin high associated with the east Asian summer monsoon rain, Clim. Dyn., 28, 67–83, 2007.

Hu, C., Henderson, G., Huang, J., Xie, S., Sun, Y., and Johnson, K.: Quantification of Holocene Asian monsoon rainfall from spatially separated cave records, Earth. Planet. Sci. Lett., 266, 221–232, 2008.

Huang, R., Chen, J., and Huang, G.: Characteristics and variations of the East Asian Monsoon system and its impacts on climate disasters in China, Adv. Atmos. Sci., 24, 993–1023, 2007.

IAEA-WMO: Global Network of Isotopes in Precipitation, The GNIP database, http://www-naweb.iaea.org/napc/ih/IHS_resources_gnip.html (last access: 31 March 2013), 2013.

Jhun, J.-G. and Lee, E.-J.: a new east Asian winter monsoon index and associated characteristics of the winter monsoon, J. Clim., 17, 711–726, 2004.

Kanada, S., Nakano, M., and Kato, T.: Projections of future changes in precipitation and the vertical structure of the frontal zone during the Baiu season in the vicinity of Japan using a 5-km-mesh regional climate model, J. Meteor. Soc. Japan, 90A, 65–86, 2012.

Kosaka, Y., Xie, S.-P., and Nakamura, H.: Dynamics of interannual variability in summer precipitation over east Asia, J. Clim., 24, 5435–5453, 2011.

Kubota, Y., Tada, R., and Kimoto, K.: Quantitative reconstruction of East Asian summer monsoon precipitation during the Holocene based on oxygen isotope mass-balance calculation in the East China Sea, Clim. Past Discuss., 10, 1447–1492, doi:10.5194/cpd-10-1447-2014, 2014.

Kurita, N.: Water isotopic variability in response to mesoscale convective system over the tropical ocean, J. Geophys. Res., 118, 10376–10390, doi:10.1002/jgrd.50754, 2013.

Kurita, N., Newman, B. D., Araguas-Araguas, L. J., and Aggarwal, P.: Evaluation of continuous water vapor δD and $\delta^{18}O$ measurements by off-axis integrated cavity output spectroscopy, Atmos. Meas. Tech., 5, 2069–2080, doi:10.5194/amt-5-2069-2012, 2012.

Kurita, N., Fujiyoshi, Y., Wada, R., Nakayama, T., Matsumi, Y., Hiyama, T., and Muramoto, K.: Isotopic variations associated with north-south displacement of the Baiu Front, SOLA, 9, 187–190, doi:10.2151/sola.2013-042, 2013.

Kusaka, H. and Kitahara, H.: Synoptic-scale climatology of cold frontal precipitation system during the passage over central Japan, SOLA, 5, 61–64, 2009.

Lawrence, J. R., Gedzelman, S. D., White, J. W. C., Smiley, D., and Lazov, P.: Storm trajectories in eastern US D/H isotopic composition of precipitation, Nature, 296, 638–640, 1982.

Liu, W., Feng, X., Liu, Y., Zhang, Q., and An, Z.: $\delta^{18}O$ values of tree rings as a proxy of monsoon precipitation in arid northwest China, Chem. Geol., 206, 73–80, 2004.

Liu, Y., Cai, Q., Liu, W., Yang, Y., Sun, J., and Song, H.: Monsoon precipitation variation recorded by tree-ring $\delta^{18}O$ in arid northwest China since AD 1878, Chem. Geol., 252, 56–61, 2008.

Liu, Z., Wen, X., Brady, E., Otto-Bliesner, B., Yu, G., Lu, H., Cheng, H., Wang, Y., Zheng, W., Ding, Y., Edwards, R., Cheng, J., Liu, W., and Yang, H.: Chinese cave records and the east Asia summer monsoon, Quat. Sci. Rev., 83, 115–128, 2014.

Manabe, S.: On the modification of airmass over the Japan Sea when the outburst of cold air predominates, J. Meteor. Soc. Japan, 35, 311–326, 1957.

Nakagawa, T., Tarasov, P., Kitagawa, H., Yasuda, Y., and Gotanda, K.: Seasonal specific responses of the east Asian monsoon to deglacial climate changes, Geology, 34, 521–524, 2006.

Nakamura, H., Nishina, A., and Minobe, S.: Response of storm track to bimodal Kuroshio path states south of Japan, J. Clim., 25, 7772–7779, 2012.

Ninomiya, K.: Heat and water budget over the Japan Sea and the Japan island in winter season, J. Meteor. Soc. Japan, 46, 343–372, 1968.

Ninomiya, K.: Characteristics of Baiu front as a predominant subtropical front in the summer northen hemisphere, J. Meteor. Soc. Japan, 62, 880–894, 1984.

Ninomiya, K. and Murakami, T.: The early summer rainy season (Baiu) over Japan, Oxford University Press, 1987.

Porter, S.: Chinese loess record of monsoon climate during the last glacial-interglacial cycle, Earth Sci. Rev., 54, 115–128, 2001.

Schnider, U., Becker, A., Finger, P., Meyer-Christoffer, A., Rudolf, B., and Ziese, M.: GPCC Full Data Reanalysis Version 6.0 at 0.5°: Monthly Land-Surface Precipitation from Rain-Gauges built on GTS-based and Historic Data, doi:10.5676/DWD_GPCC/FD_M_V6_050, 2011.

Sturm, P. and Knohl, A.: Water vapor δD and $\delta^{18}O$ measurements using off–axis integrated cavity output spectroscopy, Atmos. Meas. Tech., 3, 67–77, doi:10.5194/amt-3-67-2010, 2010.

Takano, I.: Analysis of an intense winter extratropical cyclone that advanced along the south coast of Japan, J. Meteor. Soc. Japan., 80, 669–695, 2002.

Tao, F., M. Yokozawa, J. L., and Zhang, Z.: Chimate-crop yield relationships at provincial scales in China and the impacts of recent climate trends, Clim. Res., 38, 83–94, 2008.

Tomita, T., Yamaura, T., and Hashimoto, T.: Interannual variability of the Baiu season near Japan evaluated from the equivalent potential temperature, J. Meteor. Soc. Japan, 89, 517–537, 2011.

Treble, P. C., Budd, W. F., Hope, P. K., and Rustomji, P. K.: Synoptic-scale climate patterns associated with rainfall $\delta^{18}O$ in southern Australia, J. Hydrol., 302, 270–282, 2005.

Trenberth, K., Hurrell, J., and Stepaniak, D.: The Asian monsoon: Global perspectives, in: The Asian Monsoon, edited by Wang, B., chap. 2, 67–87, Springer, 2006.

Uemura, R., Matsui, Y., Yoshimura, K., Motoyama, H., and Yoshida, N.: Evidence of deuterium excess om water vapor as a indicator of ocean surface condition, J. Geophys. Res., 113, D19114, doi:10.1029/2008JD010209, 2008.

Wang, Y., Cheng, H., Edwards, R., He, Y., Kong, X., An, Z., Wu, J., Kelly, M. J., Dykoski, C., and Li, X.: The Holocene Asian monsoon:Links to solar changes and north Atlantic climate, Science, 308, 854–857, 2005.

Wang, Y. J., Cheng, H., Edwards, R. L., An, Z. S., Wu, J. Y., Shen, C.-C., and Dorale, J. A.: a high resolution absolute-dated late Pleistocene monsoon record from Hulu cave, China, Science, 294, 2346–2348, 2001.

Webster, P., Magaña, V., Palmer, T., Shukla, J., Tomas, R., Yanai, M., and Yasunari, T.: Monsoons:Processes, predictability, and the prospects for prediction, J. Geophys. Res., 103, 14451–14510, 1998.

Werle, P., Mücke, R., and Slemr, F.: The limits of signal averaging in atmospheric trace-gas monitoring by tunale diode-laser absorption spectroscopy (TDLAS), Appl. Phys. B, 57, 131–139, 1993.

Yancheva, G., Nowaczyk, N., Mingram, J., Dulski, P., Negendank, J., Liu, J., abd L.C. Peterson, D. S., and Haug, G.: Influence of the intertropical convergence zone on the east Asian monsoon, Nature, 445, 74–77, 2007.

Yang, X., Liu, J., Liang, F., Yuan, D., Yang, Y., Lu, Y., and Chen, F.: Holocene stalagmite $\delta^{18}O$ records in the East Asian monsoon region and their correlation with those in the Indian monsoon region, The Holocene, 24, 1657–1664, 2014.

Yoshiike, S. and Kawamura, R.: Influence of wintertime large-scale circulation on the explosively developing cyclones over the western North Pacific and their downstream effects, J. Geophys. Res., 114, D13110, doi:10.1029/2009JD011820, 2009.

Yoshimura, K., Oki, T., Ohte, N., and Kanae, S.: Colored moisture analysis estimates of variations in 1998 Asian monsoon water sources, J. Meteorol. Soc. Japan, 82, 1315–1329, 2004.

Yuan, D., Cheng, H., Edwards, R., Dykoski, C. A., Kelly, M., Zhang, M., Qing, J., Lin, Y., Wang, Y., Wu, J., Dorale, J., An, Z., and Cai, Y.: Timing, Duration, and Transitions of the Last Interglacial Asian Monsoon, Science, 304, 575–578, 2004.

Zhang, P., Cheng, H., Edwards, R., Chen, F., Wang, Y., Yang, X., Liu, J., Tan, M., Wang, X., Liu, J., An, C., Dai, Z., Zhou, J., Zhang, D., Jia, J., Jin, K., and Johnson, K.: a Test of Climate, Sun, and Culture Relationships from an 1810-Year Chinese Cave Record, Science, 322, 940–942, 2008.

11

Late Pliocene lakes and soils: a global data set for the analysis of climate feedbacks in a warmer world

M. J. Pound[1], J. Tindall[2], S. J. Pickering[2], A. M. Haywood[2], H. J. Dowsett[3], and U. Salzmann[1]

[1]Department of Geography, Faculty of Engineering and Environment, Northumbria University, Ellison Building, Newcastle upon Tyne, UK
[2]School of Earth and Environment, University of Leeds, Woodhouse Lane, Leeds, UK
[3]Eastern Geology and Paleoclimate Science Center, US Geological Survey, Reston, Virginia 20192, USA

Correspondence to: M. J. Pound (matthew.pound@northumbria.ac.uk)

Abstract. The global distribution of late Pliocene soils and lakes has been reconstructed using a synthesis of geological data. These reconstructions are then used as boundary conditions for the Hadley Centre General Circulation Model (HadCM3) and the BIOME4 mechanistic vegetation model. By combining our novel soil and lake reconstructions with a fully coupled climate model we are able to explore the feedbacks of soils and lakes on the climate of the late Pliocene. Our experiments reveal regionally confined changes of local climate and vegetation in response to the new boundary conditions. The addition of late Pliocene soils has the largest influence on surface air temperatures, with notable increases in Australia, the southern part of northern Africa and in Asia. The inclusion of late Pliocene lakes increases precipitation in central Africa and at the locations of lakes in the Northern Hemisphere. When combined, the feedbacks on climate from late Pliocene lakes and soils improve the data to model fit in western North America and the southern part of northern Africa.

1 Introduction

1.1 Background

The late Pliocene (Piacenzian: 3.6–2.6 Ma) is the most recent geological time period of considerable global warmth, before the onset of the glacial–interglacial cycles of the Pleistocene (Dowsett et al., 1992, 1994; Haywood et al., 2011b; Salzmann et al., 2011). As it is, geologically speaking, relatively recent, it represents a recognisable world (in terms of its geography, orography and bathymetry) in which aspects of the Earth's climate can be explored through proxies and modelling studies to better understand the feedbacks, processes and impacts of sustained global warmth (Dowsett et al., 1996, 2012; Salzmann et al., 2009; Lunt et al., 2010, 2012). The focus on the late Pliocene palaeoclimates has been driven by the PRISM (Pliocene Research Interpretations and Synoptic Mapping) project of the US Geological Survey. PRISM3 (the third iteration of the PRISM project, which includes a three-dimensional ocean reconstruction) provides palaeoenvironmental reconstructions of the Pliocene world, from geological data, in a form suitable for use in climate modelling studies (Dowsett et al., 2010). With the availability of boundary conditions (aspects of the world required to initialise climate modelling experiments) from PRISM, it has been possible to undertake meaningful climate modelling studies to explore Pliocene climates (e.g. Chandler et al., 1994; Sloan et al., 1996; Haywood et al., 2009; Lunt et al., 2012). Building on single model studies of the Pliocene, PlioMIP (Pliocene Model Intercomparison Project) has brought together ten different climate modelling groups to simulate identical experiments and investigate not only the climate of the Pliocene but also inter-model variability and uncertainty (Haywood et al., 2010, 2011a; Dowsett et al., 2012, 2013; Salzmann et al., 2013). PlioMIP uses the palaeoenvironmental reconstructions from PRISM3, which includes palaeogeography, orography, bathymetry, vegetation, ice sheet configuration and oceanic temperatures. However, there is currently no information on late Pliocene global soils or lakes. In PlioMIP experiments 1 and 2, global soils were specified in a manner consistent with the vegetation, or they were kept as modern (Haywood et al., 2010; Contoux et al., 2012). Lakes were specified as absent and not included in any of the PlioMIP experiments. In this paper we present global data sets and palaeoenvironmental reconstructions of global late Pliocene soils and lakes. These are intended to be incorporated into the PRISM4 Pliocene global reconstruction and future PlioMIP experimental design.

1.2 The importance of soils and lakes in palaeoclimate studies

Albedo-related soil and vegetation feedbacks are key uncertainties in the Earth System and climate models differ considerably in estimating their strength (e.g. Haywood and Valdes, 2006; Knorr and Schnitzler, 2006). For the terrestrial realm, large inland water bodies and wetlands have also been shown to significantly affect surface temperatures and energy balance in past and present climate systems (e.g. Sloan, 1994; Delire et al., 2002; Sepulchre et al., 2009; Burrough et al., 2009; Krinner et al., 2012). Studies of the African Humid Period in the Holocene have found that lakes and wetlands contribute to the "greening of the Sahara" by increasing regional precipitation (Krinner et al., 2012). A similar increase in regional precipitation was found for the late Pleistocene Lake Makgadikgadi in the middle Kalahari (Burrough et al., 2009). In deeper time palaeoclimate studies, Sloan (1994) found that, when simulating the early Eocene of North America, the addition of a lake had as much impact on the climate of the continental interior as the 1680 ppmv CO_2 in the model's atmosphere. The addition of a modest lake deflected the winter freezing line north and improved the data-to-model fit for winter temperatures (Sloan, 1994).

Soil albedo has also been shown to have a large impact on regional precipitation (Knorr and Schnitzler, 2006). A series of climate model experiments on the mid- to late Holocene showed that soil albedo in the Sahara had a larger effect on regional precipitation than orbital forcing and sea surface temperatures (Knorr and Schnitzler, 2006). Other experiments have shown that wetter and darker soils in the mid-Holocene Sahara would have facilitated the northward movement of the African monsoon, creating a positive feedback (Levis et al., 2004).

Current palaeoclimate modelling studies of the late Pliocene often struggle to generate sufficient precipitation, particularly in the semi-arid and arid tropical and subtropical regions, to match proxy data (Salzmann et al., 2008; Haywood et al., 2009; Pope et al., 2011). As some studies have shown that lakes and soils have had significant regional impacts on mid-Holocene precipitation (e.g. Knorr and Schnitzler, 2006; Krinner et al., 2012), it stands to reason that similar effects could be seen in the late Pliocene. Conversely, recent work focussing on the Megalake Chad region during the late Pliocene did not show significant increases in precip-

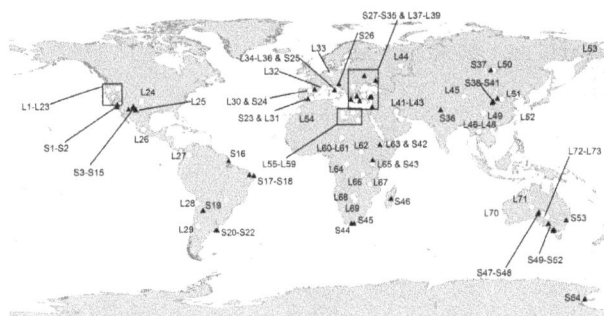

Fig. 1. The location of the soils and lake data used in this study to reconstruct global late Pliocene land surface features. Soil data have the prefix S (Supplement Table S1), whilst lake data have the prefix L (Supplement Table S2).

itation (Contoux et al., 2013). However, Holocene palaeoclimate studies benefit from comprehensive published data sets of soils and lakes (e.g. Hoelzmann et al., 1998), and up to present no such data have been available for late Pliocene climate model studies (Haywood et al., 2010). In this paper we present the first global data sets of late Pliocene soil and lake distributions, and these data sets have been transformed into climate model boundary conditions suitable for exploring the feedbacks of soils and lakes in a warmer world. The data sets provide previously missing boundary conditions for late Pliocene palaeoclimate modelling studies. We present the initial results of the first late Pliocene palaeoclimate model studies using the new realistic soil and lake boundary conditions. Finally, we compare BIOME4 (Kaplan, 2001) output, from our new simulations, to the global vegetation database of Salzmann et al. (2008, 2013) to qualitatively evaluate data–model similarity.

2 Methods

2.1 Construction of the lake and soil database

Late Pliocene lake and soil data have been collected and synthesised into an internally consistent format using a Microsoft Access–ArcGIS database that is based on the vegetation database TEVIS (Salzmann et al., 2008). The soil and lake data have been compiled from published literature: soil data (Fig. 1; Supplement Table S1) are based upon paleosol occurrences (e.g. Mack et al., 2006), whereas evidence for lakes (Fig. 1; Supplement Table S2) comes from sedimentology (e.g. Müller et al., 2001), dynamic elevation models and topographic studies (e.g. Drake et al., 2008), fauna (e.g. Otero et al., 2009) or a combination of these (e.g. Adam et al., 1990). Both the soil and lake data are recorded with a latitude–longitude (for lakes this represents the centre), a maximum and minimum age in millions of years (Ma) and the method used to date the deposit. The documented soil data also include a soil type, which is based upon the orders

Table 1. The colour (for albedo) and texture translations for the soil orders used in the modelling of late Pliocene soils.

Soil	Colour	Texture	Albedo
Gelisol	Intermediate	Medium	0.17
Histosol	Dark	Fine	0.11
Spodosol	Intermediate	Medium/Coarse	0.17
Oxisol	Intermediate	Fine/Medium	0.17
Vertisol	Dark	Fine	0.11
Aridisol	Light	Coarse	0.35
Ultisol	Intermediate	Fine/Medium	0.17
Mollisol	Dark	Medium	0.35
Alfisol	Intermediate	Medium	0.17

of the US Department of Agriculture soil taxonomy scheme (Soil Survey Staff, 1999). The lake data also record an estimated surface area extent, the shape of the lake and for lakes with a surface area greater than 1500 km^2 the latitude–longitude of its northern-, eastern-, southern- and western-most points. In addition to this, any reported information on water chemistry, details of inflows and outflows or whether the lake was ephemeral have also been recorded. Full details of the databasing methodology and the data sets are available in the Supplement.

2.2 Preparing the data for inclusion in a climate model

From the geological data recorded in the late Pliocene lakes and soil database we have produced three maps to allow the inclusion of lakes and soils in palaeoclimate modelling experiments. The three maps are a global soil map, which is accompanied by a table providing preferred soil characteristics (Fig. 2a; Table 1); a dry-lakes scenario (Fig. 2b); and a wet-lakes scenario (Fig. 2c). All maps use a grid cell size of 2.5° latitude × 3.75° longitude; this equates to a spatial resolution of 278 km × 417 km at the Equator.

To develop a global late Pliocene soil map (Fig. 2a) from the 54 palaeosol occurrences recorded in the database (Supplement Table S1), we combined the soil data with the Piacenzian biome reconstruction presented in Salzmann et al. (2008). This allowed a soil type to be assigned to each model grid cell even if late Pliocene palaeosol data had not been reported from that region. This technique of combining a realistic biome reconstruction with palaeosol data uses the knowledge that at a global scale the distribution of each soil order mirrors certain vegetation biomes (Soil Survey Staff, 1999). When the palaeosol data were combined with the vegetation reconstruction, there were no mismatches between a palaeosol occurrence and a biome that we did not expect to be associated with that soil order.

Both the late Pliocene dry-lake scenario and wet-lake scenario are based upon the estimated surface area of the palaeolakes, translated into a percentage of a model grid cell

(Fig. 2b, c). A wet and a dry scenario have been generated to compensate for the uncertainty in the dating of many of these features, which often cover several orbital cycles. By producing a dry- and a wet-scenario map it is possible for climate modelling experiments to explore the impacts of late Pliocene lakes in a warm-wet climate period or a cold-dry climate; this follows the vegetation work of Salzmann et al. (2013). Late Pliocene lake surface areas have been either taken from the published literature or calculated from published estimates of lake extent. These were then translated into percentages of grid cells by calculating how much of a grid cell would be occupied by each lake. Where megalakes occupied more than one grid cell, the geographic distribution of the lake was based upon the published shape and the distalmost latitude–longitude points of the reconstructed lake.

2.3 Uncertainties in reconstructing soils and lakes from geological data

This study is the first to present realistic late Pliocene soil and lake maps derived from the synthesis of geological data (Fig. 2). Despite these maps being the current state of the art it is important to discuss the uncertainty involved in them. For the global soils reconstruction the greatest uncertainty comes from the limited geographic distribution of data (Fig. 1) and the reliance on the global biome reconstruction (Salzmann et al., 2008) to fill in the gaps. However, there were no soil data points coinciding with a vegetation type they would not normally be associated with, so we can have confidence in our methodology to generate a global map. One discrepancy between the late Pliocene reconstruction and the modern global soil order map is that the late Pliocene does not have any regions of Inceptisols or Entisols (Fig. 2a). Today these two soil orders make up about 30 % of the land surface and represent undeveloped or moderate pedogenic development (Soil Survey Staff, 1999). Producing a global soil order map for the late Pliocene requires the incorporation of many distinct palaeosol occurrences, and most of these are preserved as a pedogenically developed soil order, rather than preserved as an undeveloped or moderately developed soil (e.g. Gürel and Kadir, 2006). Where inceptisol palaeosols are preserved they are commonly associated with a fully developed soil order (e.g. Mack et al., 2006). Further to their limited geological preservation, Inceptisols and Entisols are not commonly associated with particular vegetation types, being a product of limited soil development rather than ambient climate and biome (Soil Survey Staff, 1999). This made them near impossible to plot on a late Pliocene soil map with the available data and methodology. Inceptisols and Entisols were therefore omitted from the reconstruction, but it should be noted that they were not absent during the Pliocene (Sangode and Bloemendal, 2004; Mack et al., 2006). It is difficult to assess the likely impacts on albedo and texture that the addition of Inceptisols and Entisols may have had during the late Pliocene. As these soil orders have limited pedogenic

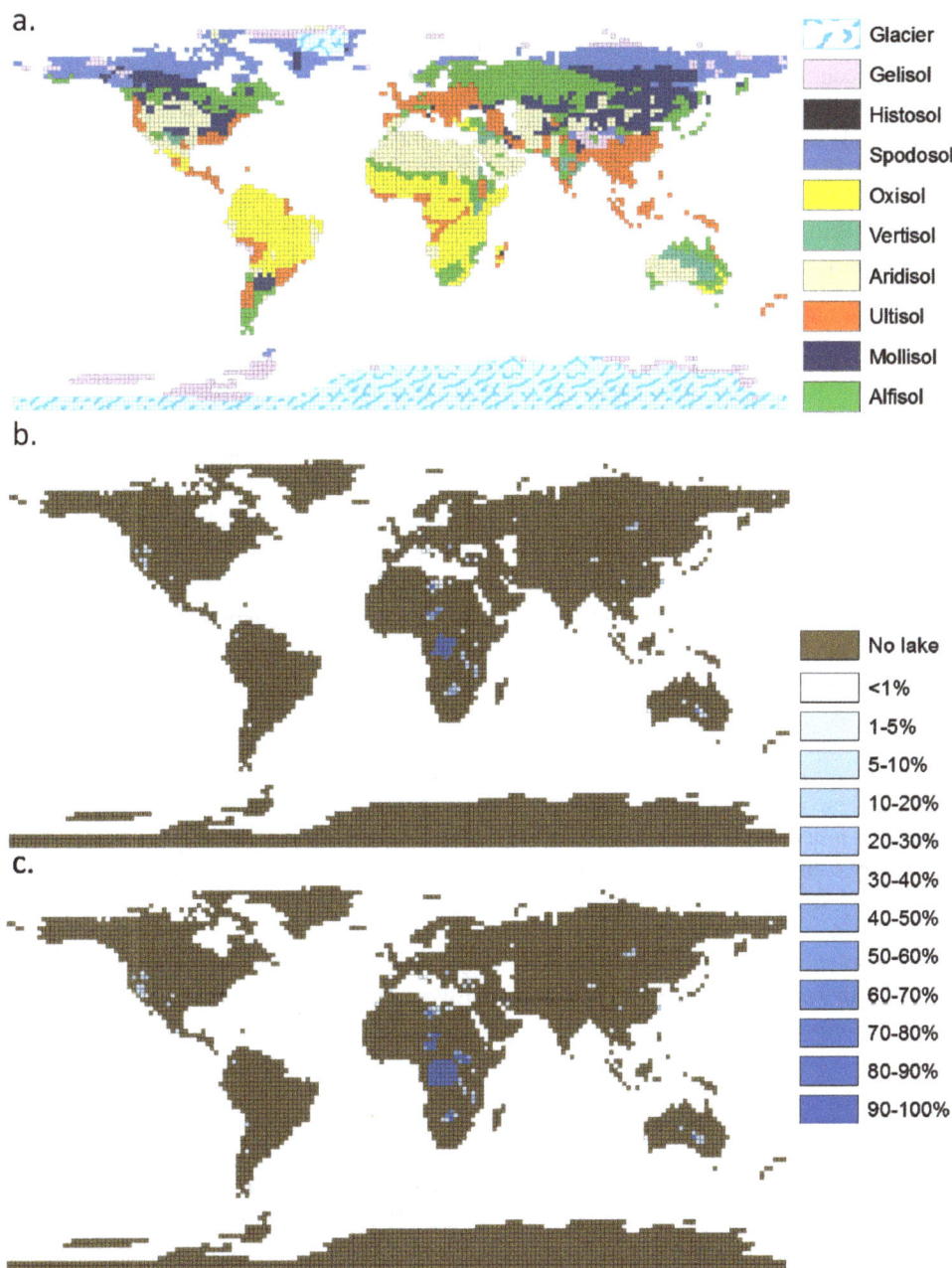

Fig. 2. The reconstructed distribution of (**A**) late Pliocene soils, (**B**) late Pliocene dry-lakes scenario and (**C**) late Pliocene wet-lakes scenario.

development, they are more intimately tied to their parent material than other soil orders. This could mean that Inceptisols and Entisols could have had any combination of albedo and texture depending on late Pliocene surface geology.

The late Pliocene biome reconstruction from Salzmann et al. (2008) is a hybrid map combining 240 palaeobotanical data sites with a "best fit to data" BIOME4 output (forced by HadAM3-predicted late Pliocene climate), which were merged using expert knowledge. Although this means that limited regions of the biome reconstruction do rely more on model predictions than real data, the overall product is pri-

marily based on an exhaustive database of late Pliocene plant fossil localities. Developing a global late Pliocene soil map with only 54 paleosol localities required either extensive interpolation (with all the possible errors that may have come with that) or the use of another data set (the hybrid biome reconstruction) and the knowledge that most soil orders (with the exception of Inceptisols and Entisols) are related to particular vegetation types.

The reconstructed late Pliocene lakes represent a synthesis of the published geological data. What is most obvious is the vast areas with no percentage of the grid cell covered by surface water (Fig. 2). This is not meant to mean that these regions were without any surface water, but that there are no published records of lake sediments with the lake extent estimated. However, it is highly likely that during wetter climates in the past many presently arid regions were covered with river systems, as has been shown for the Sahara during the Eemian (MIS 5e) and Holocene (e.g. Hoelzmann et al. 1998; Coulthard et al., 2013). With continued research into late Pliocene lake faunas, floras and sediments these regions will contain more evidence for surface water. Of the lakes presented in the reconstructions of this study the one with the greatest uncertainty is the substantial Megalake Zaire (Fig. 2). This megalake has long been speculated due to all the major tributaries of the Congo River being orientated to the centre of the basin (Summerfield, 1991; Goudie, 2005) and the presence of a submarine canyon rather than a delta (Cahen, 1954; Peters and O'Brien, 2001; Goudie, 2005). However, it has not been ground-truthed with recent geological data and this should be an imperative (Peters and O'Brien, 2001). For further discussion on the uncertainties surrounding the reconstructions of late Pliocene soils and lakes please see Supplement 1.

2.4 Modelling

The potential effects of the new lake and soil databases on the Pliocene climate were tested using modelling simulations with the UK Hadley Centre General Circulation Model (GCM), HadCM3. This is a coupled atmosphere–ocean GCM described by Gordon et al. (2000) and Pope et al. (2000), with horizontal resolution of $3.75° \times 2.5°$ in the atmosphere and $1.25° \times 1.25°$ in the ocean. The atmospheric component has 19 levels in the vertical and 30 min time steps, while the oceanic component has 20 levels in the vertical and 1 h time steps.

To investigate the impacts of realistic soil and lake distributions on the late Pliocene climate we analyse the results of 5 simulations: a control simulation of 850 yr (PRISM3 control), a simulation with late Pliocene lake levels from the wet-lakes scenario (PRISM3 + wet-lakes scenario) (Fig. 2c), a simulation with late Pliocene lake levels from the dry-lakes scenario (PRISM3 + dry-lakes scenario), a simulation with late Pliocene soils (PRISM3 + soils) and a simulation with soils and the wet-lakes scenario (PRISM3 + soils + wet-lakes scenario). The PRISM3 + wet-lakes scenario, PRISM3 + dry-lakes scenario, PRISM3 + soils, and PRISM3 + soils + wet-lakes scenario simulations were all started 500 yr into the PRISM3 control simulation and were run for a further 350 yr. This is sufficient to spin up all atmosphere and vegetation parameters of interest (Hughes et al., 2006). The control experiment comprises boundary conditions from PRISM3; these include a near-modern orography (except for areas of the Andes) and a reduced Greenland Ice Sheet (for full details please see Dowsett et al., 2010; Haywood et al., 2010, 2011). Although the preferred boundary conditions for PRISM3 are

to remove the West Antarctic Ice Sheet, here we utilise the alternative PRISM3 boundary conditions, which remove all ice from the West Antarctic Ice Sheet and reduce the topography to sea level (Haywood et al., 2011). The control experiment has a modern orbit and CO_2 levels of 405 ppmv. The initial vegetation patterns for the control run were prescribed from PRISM3; however the version of HadCM3 used here comprises the MOSES2.1 land surface scheme and the TRIFFID dynamic vegetation model (Cox et al., 1999; Cox, 2001) such that vegetation dynamically changes with the climate, and the relative proportions of different vegetation types adjust throughout a long simulation. TRIFFID was run in equilibrium mode for the first 50 yr of the control run; after this TRIFFID was run in dynamic mode throughout. For the control simulation, soil parameters were set to be the same as modern and there were assumed to be no lakes. The lake and soil experiments were initialised from a state 500 yr into the PRISM3 control run; hence their initial vegetation patterns were predicted by HadCM3 + MOSES2/TRIFFID for the PRISM3 climate. Vegetation continued to respond dynamically in all experiments.

The PRISM3 + wet-lakes scenario simulation was identical to the control except that the high-level lakes were included in a very simple way. In the MOSES2.1/TRIFFID version of HadCM3, each grid box is assigned a fractional coverage of 9 different surface types (broadleaf trees; needleleaf trees; shrubs; C_4 grasses; C_3 grasses; ice; urban – not used in the late Pliocene; bare soil; or water); lakes are included in a grid box by increasing the water surface type, while the fractions of all other surface types are reduced as appropriate. It is noted that, although trees, grasses and shrubs can dynamically change throughout the model simulation, the lake fraction of the grid box will remain constant and will be neither increased by precipitation nor decreased by evaporation (Essery et al., 2001). This means that the lakes in the simulation are static and do not depend on precipitation/evaporation/runoff patterns. The prescribed albedo of the lakes is 0.06, whilst roughness length is 0.0003 m.

The PRISM3 + soils simulation required changes to several HadCM3 boundary conditions. The simplest of these is soil albedo, which was determined from the colour of the soil type shown in Table 1. Following Jones (2008) light soils were prescribed an albedo of 0.35, medium soils an albedo of 0.17 and dark soils an albedo of 0.11. These albedo values are based on the assumption that medium and dark soils have average wetness, while light soils are dry. It is noteworthy that, although there are differences in soil types between the late Pliocene and the modern, the simple way that these soil types are incorporated into HadCM3 means that their potential for changing the climate is limited. Table 1 shows that, of the 9 soil types to be included in the PRISM4 database, five are intermediate colour, three are dark and one is light. This means that, even though a soil type may change between the Pliocene and the modern, it is only if the soil type changes to one of a different colour that the climate can be

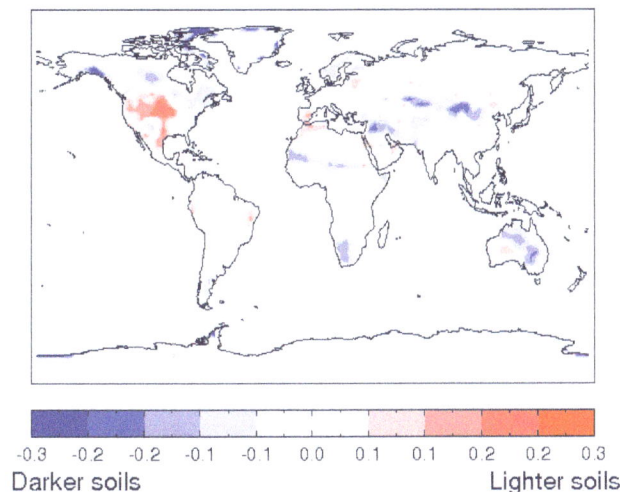

Fig. 3. The changes in soil albedo between the PRISM3 control experiment and the new late Pliocene soils reconstruction. Differences in other soil components can be found in Supplement 2.

altered via albedo feedbacks (Fig. 3). Other soil parameters used in HadCM3 (Clapp–Hornberger exponent, saturated hydraulic conductivity, saturated soil water suction, volumetric soil moisture concentrations at critical, saturation and wilting points, dry soil volumetric heat capacity and dry soil thermal conductivity) are prescribed values depending on soil texture as suggested by Cox et al. (1999). Again, it is noteworthy that, even though a soil type may have changed between the modern and the late Pliocene, it is where a soil changes to one of a different texture that will have the potential to impact the climate.

Although HadCM3 dynamically predicts vegetation patterns, this is limited to only 5 types of vegetation, and these are difficult to compare with data sets of late Pliocene vegetation such as PRISM3 (Salzmann et al., 2008). To facilitate a better comparison with palaeobotanical proxy data the climate output from HadCM3 were used to drive the offline vegetation model BIOME4. The BIOME4 model (Kaplan, 2001) is a mechanistic global vegetation model which predicts the distribution of 28 global biomes based on the monthly means of temperature, precipitation, cloudiness and absolute minimum temperature. The model includes 12 plant functional types (PFTs) from cushion forb to tropical evergreen tree (Prentice et al., 1992). It is the bioclimatic tolerances of these that determine which is dominant in a grid cell and, from this, which biome is predicted. The BIOME4 model was run in the anomaly mode with a CO_2 of 405 ppmv. The model was driven from the average annual climate data obtained from the last 100 yr of each HadCM3 experiment, to assess which PFTs were feasible in each grid box and to allocate an appropriate biome at each location.

3   Results

In this section we will first describe the geographic distribution of late Pliocene soils and lakes that have been reconstructed from geological data and then the results of including these reconstructions in a series of GCM simulations using the PRISM3 boundary conditions.

3.1   Late Pliocene soils

During the late Pliocene there were significant differences in the global distribution of soils (Fig. 2a). Overall, the distribution of soils reflects the warmer and wetter world seen in the vegetation reconstruction. Gelisols, associated with tundra type vegetation are restricted to very high latitude areas of North America, Greenland and Eurasia, as well as coastal regions of Antarctica (Fig. 2a). The more northern distribution of boreal and temperate forests is accompanied by extensive high-latitude Spodosols and Alfisols at higher than modern latitudes (Fig. 2a). There is evidence supporting Alfisols at 54° N from around Lake Baikal, where grey forest soils are preserved (Mats et al., 2004). The extensive grassland and savannas in the continental interiors of North America and Eurasia are translated into extensive Mollisols (Fig. 2a). South of the Alfisol, in North America, there were Ultisols along the west coast and in the southeast of the continent (Fig. 2a). The centre of North America contains a large region of Aridisols, and a mixture of Alfisols (Abbott, 1981) and Vertisols (Mack et al., 2006) along the southern margin. In Europe there is evidence for Alfisols (Icole, 1970; Günster and Skowronek, 2001), Histosols (Basilici, 1997; Bechtel et al., 2003) and extensive Ultisols (Gerasimenko, 1993). At the eastern end of the Mediterranean there was a region of mixed Alfisols (Quade et al., 1994), Oxisols (Kelepertsis, 2002), Ultisols (Paepe et al., 2004) and Vertisols (Graef et al., 1997) (Fig. 2a). The Indian subcontinent contained a mixture of Ultisols and Vertisols during the late Pliocene (Fig. 2a). There is also evidence for Alfisols close to the Himalayas (Sangode and Bloemendal, 2004). In southeast Asia the biome reconstruction translates into extensive Ultisols across this region (Fig. 2a).

In South America there is a large region of Oxisols as evidenced from palaeosols (Mabesoone and Lobo, 1980) and the biome reconstruction. In southern South America Ultisols, Molisols and Alfisols dominated during the Piacenzian (Fig. 2a). The soils reconstruction for Africa relies heavily on the biome reconstruction, except for direct evidence of Vertisols in east Africa (Wynn, 2000; Campisano and Feibel, 2008), Oxisols in southern Africa (Helgren and Butzer, 1977) and Histosols in Madagascar (Lenoble, 1949). Combining these palaeosol occurrences with the biome reconstruction, the distribution of soils in Africa is shown to be dominated by Aridisols in northern Africa and Oxisols in central and southern Africa (Fig. 2a). Palaeosol evidence in Australia shows the presence of Aridisols in the middle of the continent (Hou et al., 2008) and Oxisols in the southeast (Firman, 1994; Hughes et al., 1999). The biome reconstruction of Salzmann et al. (2008) suggests the presence of Alfisols in the south-west and north of the continent, Vertisols in the east and Aridisols in the west and south (Fig. 2a).

3.2 Late Pliocene lakes

The global distribution of late Pliocene lakes is dominated by megalakes in Africa and Australia (Fig. 2b, c). In Africa the largest megalake, in both the wet and dry scenario, is Lake Zaire (Beadle, 1974; Peters and O'Brien, 2001; Goudie, 2005). This large water body is reconstructed in the wet scenario as occupying the majority of the modern river drainage basin (Fig. 2c), whereas it is reconstructed smaller in the dry scenario (Fig. 2b). To the east of Lake Zaire was Lake Sudd, which was large during wet phases of the Pliocene (Fig. 2c). However, it is reported to have been a very shallow lake (Salama, 1987) and is therefore considerably reduced in the dry scenario (Fig. 2b). In southern Africa there is evidence for surface water in the region of the modern Okavango Delta and the Makgadikgadi Pan (Ringrose et al., 2002, 2005), both of which have a reduced surface area in the dry scenario (Fig. 2b, c). In east Africa there is evidence for Lake Malawi (Dixey, 1927) and Lake Tanganyika (Cohen et al., 1997), both of these occupy multiple grid cells of the reconstruction (Fig. 2b, c). There were also smaller (sub-grid-cell) lakes associated with parts of the Rift Valley (Deino et al., 2006). Northern Africa was dominated by Lake Chad, which was considerably bigger than in modern times during the late Pliocene (Schuster et al., 2009; Otero et al., 2010). The sedimentology of the Chad Basin shows that there were considerable shifts from lake to sub-aerial environments during the late Neogene (Schuster et al., 2009), and we have represented this by using the reported sediment sections to define a wet-scenario (Fig. 2c) and dry-scenario (Fig. 2b) late Pliocene egalake Chad. Further north there was Lake Fazzan and several smaller lakes in Libya, which were associated with an extensive river system (Drake et al., 2008). The lakes formed in topographic lows as volcanic eruptions restricted and blocked the flow of the river system (Drake et al., 2008).

In Australia there was a series of large lakes in the centre of the continent (Fig. 2b, c). The largest of these was Megalake Eyre (Simon-Coinçon et al., 1996; Alley, 1998; Martin, 2006). To the east of Lake Eyre was Lake Tarka-rooloo, another large water body (Callen, 1977), whilst to the northwest was Lake Amadeus, which may have fed Megalake Eyre (Chen et al., 1993).

There is limited evidence for late Pliocene lakes in South America and those reported are modest in size and associated with the Andes Mountains (Fig. 1; Supplement Table S2). The Quillagua–Llamara Basin, Chile, records an ephemeral late Pliocene lake with evaporates present (Sáez et al., 1999). In Argentina a small saline, though permanent, lagoon is preserved at Llancanelo (Violante et al., 2010) and a small

lake is reported from Bogota in Colombia (Wijninga and Kuhry, 1993).

In North America there is a swarm of small to modest-sized lakes associated with the valleys of the Rocky Mountains (Fig. 2b, c). The largest of these was Glenn's Ferry in Idaho, which has been reconstructed from sediments and the distribution of fossil fishes (Smith, 1981; Thompson, 1992). There are many small lakes across Eurasia, but only Lake Baikal and Lake Suerkuli covered multiple grid cells (Fig. 2b, c). Lake Baikal is reconstructed as having a similar size to the modern lake; however there was a change in sedimentation related to tectonic activity at 3.15 Ma (Müller et al., 2001). Lake Suerkuli, located on the northern Tibetan Plateau, had an estimated surface area of 4800 km^2, but was destroyed by activity of the middle Altyn Fault (Chang et al., 2012).

3.3 Impact of soils and lakes on simulating late Pliocene climate and vegetation

Using the boundary conditions described in the previous section a series of modelling experiments was undertaken. The GCM simulations were designed to explore the impacts on late Pliocene climate and vegetation that using realistic soil and lake boundary conditions can have. In this section we will present the differences in temperature and precipitation of the three experiments – PRISM3 + soils, PRISM3 + wet lakes and PRISM3 + soils + wet lakes – from the standard PRISM3 control. All results use average values from the final 100 yr of each simulation and only those results deemed to be significant at the 0.1 % confidence level following a Student's t test are presented. This means that in a statistical sense there is only a 0.1 % chance that the results shown are due to intrinsic model variability; any differences in climate that are less significant than this are not presented. However, it is noted that, while the results shown are very likely to be due to the soil and lake boundary conditions imposed, climate model output does generally fulfil all the criteria for accurate significance testing (e.g. variables normally distributed, and independent). This means that some features may occur in the results that cannot be fully attributed to lakes and soils. Nonetheless by only presenting results that are significant at the 0.1 % confidence level, noise can be removed from the results and the effects of lakes and soils on the climate can be more clearly seen. We also show how the inclusion of soils and lakes has effected biome distributions in Pliocene simulations. The PRISM3 + dry-lakes experiment showed results comparable to, though not as pronounced as, the PRISM3 + wet-lakes experiment. To avoid repetition we have not presented the results of the PRISM3 + dry-lakes experiment below, but have included it in Supplement 3.

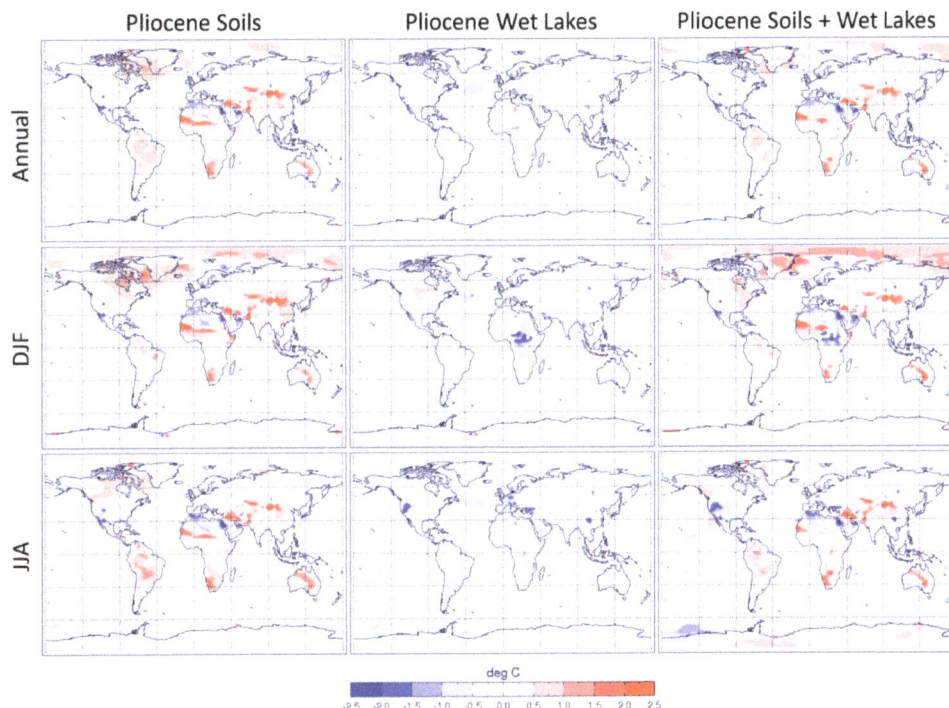

Fig. 4. Mean annual and seasonal surface air temperature for the soil and lake experiments. Anomalies relative to the standard PRISM3 control run (with modern lake and soil distribution). Everything plotted is significant at the 0.1 % confidence level.

3.3.1 PRISM3 + soils

The mean annual surface air temperature (SAT) in the PRISM3 + soils experiment shows a 1 °C cooling across northern Africa, a 1–2 °C cooling across southwest North America, a warming of 0.5–1 °C across northern South America, up to 2.5 °C warming in the southern part of northern Africa and in southern Africa, a warming of between 1 °C and 3 °C across east Asia, a 1 °C warming in central Australia and small changes around the Middle East (Fig. 4). With the exception of South America, all of these SAT changes can be attributed to soil albedo changes which determine the proportion of incoming shortwave radiation that can be absorbed (Fig. 3). During December-January-February (DJF) the differences in SAT with the PRISM3 control run are generally the same as in the annual ones, but they vary in magnitude (Fig. 4). This is also true during June-July-August (JJA), where we see an additional 0.5–1 °C warming in central northern North America, relating to a decrease in soil albedo (Figs. 3, 4), which is not visible in the annual mean.

In the experiment with late Pliocene soils there is a small increase in mean annual precipitation (MAP) in the southern part of northern Africa and in eastern Africa of around 10 mm month^{-1} and a reduction in the MAP across the Amazon region of up to 30 mm month^{-1} (Fig. 5). This reduction in MAP across the Amazon region is also associated with a reduction in evaporation and a reduction in soil moisture. This change in MAP across South America is robust and oc-

curs throughout the simulation. The only boundary condition to have occurred in the Amazon region between the soil experiment and the control run is the Clapp–Hornberger exponent (Supplement 2). This parameter can affect the availability of soil moisture for evaporation and in this experiment appears to be inhibiting moisture recycling over the Amazon. It is noted that the Amazon region has high precipitation with substantial internal model variability, and is a particularly sensitive region in HadCM3 (Good et al., 2013); however the changes seen here are substantial and the possibility of the changes occurring due to the influence of changes to soil parameters at remote locations cannot be discounted. It will be interesting to see whether the results over South America are replicated with other GCMs that make use of the new soil database.

Seasonal changes of precipitation attributable to soils shown in Fig. 5 include a 10 mm month^{-1} increase in rainfall over central Africa during DJF and an increase of around 8 mm month^{-1} in a narrow band in the southern region of the Sahara during JJA (Fig. 5). These annual and seasonal climatological changes translate into very modest biome changes (Fig. 6). The main differences in this experiment from the PRISM3 control run is an increase of the tropical xerophytic shrubland biome in northern Africa and southern Africa (Fig. 6). In northern Africa this replaces desert and relates to the increased JJA rainfall (Fig. 5). In southern Africa tropical xerophytic shrubland replaces warm-temperate forest biome and temperate sclerophyll woodland biome (Fig. 6). These

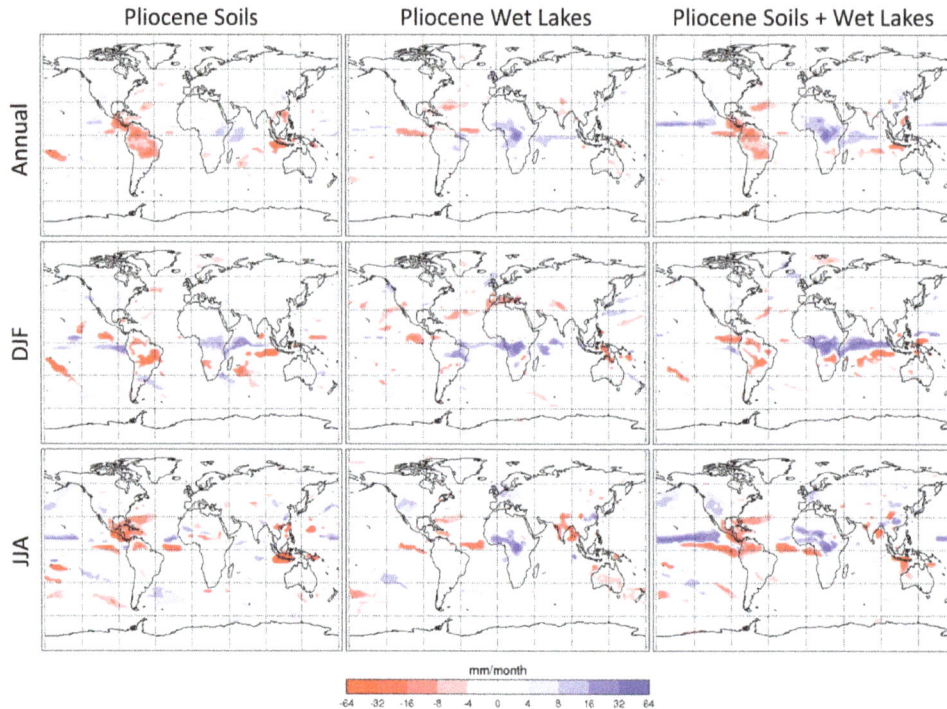

Fig. 5. Mean annual and seasonal precipitation for the soil and lake experiments. Anomalies relative to the standard PRISM3 control run (with modern lake and soil distribution). Everything plotted is significant at the 0.1 % confidence level.

changes are probably the result of increased winter temperatures and no change in MAP, leading to higher evaporation and limiting the development of lusher biome types. BIOME4 also predicts an expansion of desert in coastal Brazil and Australia, based upon the climate simulated by HadCM3 (Fig. 6).

3.3.2 PRISM3 + wet-lakes scenario

Figure 4 suggests that the inclusion of Pliocene lakes leads to a cooling in mean annual SAT of up to 2 °C that is confined to the immediate vicinity of the lakes; however there are seasonal variations. For example, the decreases in SAT (up to 2.5 °C) around the large lakes in northern and central Africa are most pronounced in DJF (Fig. 4). The exception to this is near Lake Fazzan (northern Africa), where a year-round warming is observed (Fig. 4). There are also regional decreases of up to 2.5 °C in SAT associated with lakes in the mid-latitudes of the Northern Hemisphere during JJA. The largest differences in SAT in the mid-latitudes occur in the warmest season. This implies that the cooling effects of using energy to evaporate lake water are larger than the warming that would be caused by the lake surface having a lower albedo than the vegetation it replaces. The large seasonal changes in central Africa do not conform to the above hypothesis as they are in a uniformly warm regional climate. For this region the decrease in DJF SAT is related to the average relative humidity. In the PRISM3 control run the average relative humidity during DJF was ca. 20 %, whereas JJA had an average relative humidity of ca. 80 %. Therefore it appears it was only possible to evaporate more lake water during DJF as the higher average relative humidity in JJA meant the air was already saturated with water vapour. For Lake Fazzan the change in surface albedo from bare soil to a lake (bare soil = 0.35, whereas surface water = 0.06) appears to be dominating over the cooling influence of evaporation. It is not understood why the late Pliocene lakes of Australia had no impact on SAT.

Changes in MAP between PRISM3 + wet lakes and the PRISM3 control run appear to be relatively widespread; however we only consider those changes proximal to lakes to be true signals (Fig. 5). All MAP anomalies in areas away from the reconstructed late Pliocene lakes should be considered as the result of model variability. This means that the only significant change in MAP associated with our late Pliocene lakes reconstruction is an up to 53 mm month^{-1} increase associated with Megalake Zaire (Fig. 5). This we attribute to regional recycling of evaporated lake water. Seasonal changes associated with late Pliocene lakes in the Northern Hemisphere are mainly constrained to JJA, when there is enough energy to evaporate lake water and increase local precipitation (Fig. 5).

The biome predictions indicate an expansion of the temperate conifer forest biome and open conifer woodland biome in western North America; this replaces the temperate xerophytic shrubland biome (Fig. 6). Cool mixed forest

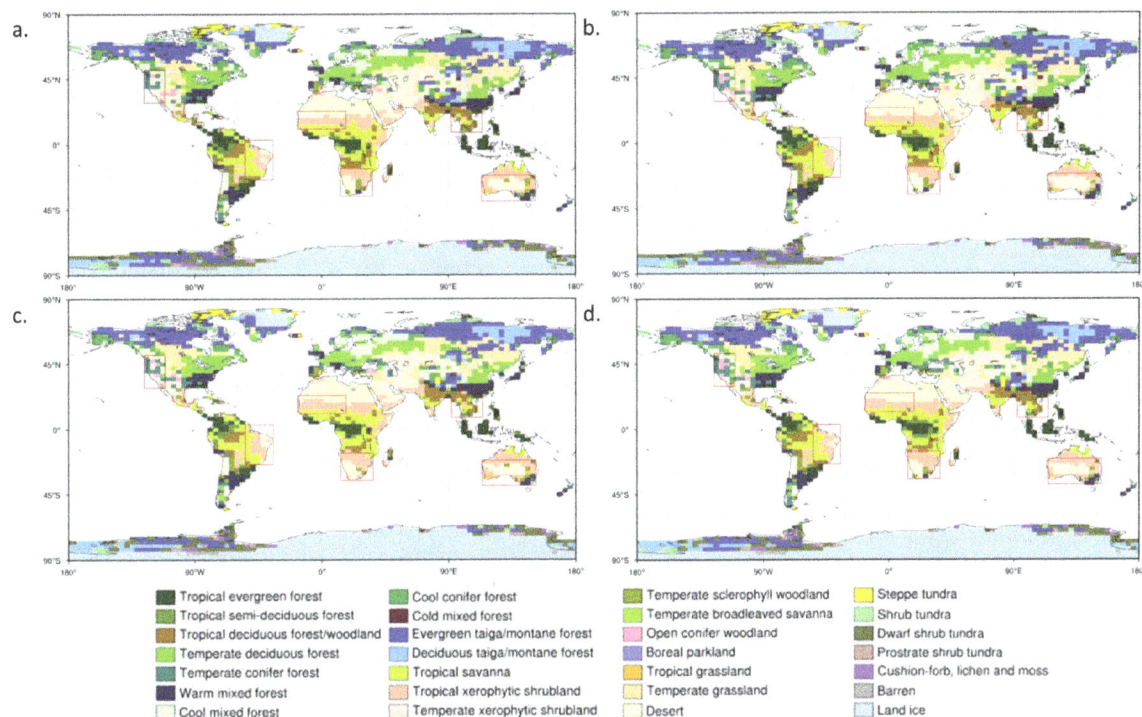

Fig. 6. The predicted biomes for the late Pliocene soil and lake experiments with red boxes to highlight key regions discussed in the text. (a) PRISM3 control, (b) PRISM3 + wet lakes, (c) PRISM3 + soils and (d) PRISM3 + soils + wet lakes.

is replaced with temperate deciduous forest in the region of Europe–Russia. In Africa, there is a small change in biomes along the Sahel–Sahara transition and an expansion of tropical forest biomes at the expense of savanna around Megalake Zaire (Fig. 6).

3.3.3 PRISM3 + soils + wet-lakes scenario

Since the results from the PRISM3 + soils experiment (Sect. 3.3.1) and the PRISM3 + wet-lakes experiment (Sect. 3.3.2) generally occur in disparate regions, the effects of adding both soils and lakes to the boundary conditions of the model is essentially a linear combination of adding the soils and lakes separately. This can be seen in temperature (Fig. 4) and precipitation (Fig. 5) and is the case for all seasons.

As expected the biome reconstruction from this experiment also shows a combination of the biome plots of PRISM3 + soils and PRISM3 + wet-lakes scenario experiments (Fig. 6). There is an increase in temperate conifer forest and woodland in western North America, which is associated with the lakes. Coastal Brazil has a significant increase of the tropical xerophytic shrubland biome, seen in the PRISM3 + soils experiment and the slight increase in desert seen in both experiments (Fig. 6). The Sahara is slightly reduced and has changed into tropical xerophytic shrubland biome (Fig. 6). An expansion of the xerophytic shrubland biome in southern Africa is at the expense of other biome

types, except desert (Fig. 6). The tropical xerophytic shrubland biome in southeast Asia is replaced by the savannah biome and the Australian desert has a small reduction, which was also seen in the PRISM3 + wet-lakes scenario (Fig. 6).

4 Discussion

The global distribution of late Pliocene soils and lakes (Fig. 2) is significantly different from the present day (Soil Survey Staff, 1999; Lehner and Döll, 2004). Whereas soils reflect the global distribution of late Pliocene biomes (Salzmann et al., 2008), the distribution and size of late Pliocene lakes (Fig. 2) contribute to a land-surface covering dramatically different from the present day (Lehner and Döll, 2004). The increase in the number and size of lakes is a response to the generally wetter global climate of the late Pliocene (e.g. Salzmann, 2008) and in places different tectonic conditions (e.g. Chang et al., 2012). However, application of these boundary conditions into HadCM3, alongside the other PRISM3 boundary conditions, produced subtle results (Figs. 4, 5, 6). The changes in climate and vegetation attributable to Pliocene soils and lakes seen in this study are generally less than seen in similar studies of the mid-Holocene. Palaeoclimate studies on the mid-Holocene have shown that it is possible to double regional precipitation with large lakes (Krinner et al., 2012) and move the African monsoon northwards by modifying soil albedo in the Sahara (Levis et al., 2004).

Although the results of our study are not as obvious as those from the Holocene, they do offer a glimpse that some of the processes, previously identified in the Holocene, may operate in the Pliocene. Although the addition of late Pliocene lakes did not double regional precipitation, as was simulated for the Holocene (Krinner et al., 2012), our experiments do show a 50 % increase around Lake Chad and notable summer increases in western North America. We also show a similar austral summer increase in precipitation around Megalake Makgadikgadi to that which Burrough et al. (2009) produced in their late Quaternary experiments. These similarities suggest that similar forcings are operating in the Pliocene and the Holocene, with regard to large lakes. However, in a study on Megalake Chad during the late Pliocene Contoux et al. (2013) found that the presence of a large lake did not sufficiently influence late Pliocene climate to impact on regional vegetation.

The parameterisation of Alfisol along the southern margin of the late Pliocene Sahara led to a summer increase in precipitation (Figs. 2a, 4), which led to a change from the desert biome type to a tropical xerophytic shrubland (Fig. 6). The Alfisol is a darker soil type than the Aridisol (Table 1), and this result is comparable, although more subtle, to those presented by Levis et al. (2004). When the soil and wet-lake boundary conditions were combined, this precipitation signal is increased, but the vegetation response remains the same (Figs. 5, 6). A northwards shift of the Sahel–Sahara boundary is consistent with the few vegetation data available for this region (Leroy and Dupont, 1994; Salzmann et al., 2008).

The climatological changes in western North America (Figs. 4, 5) have changed the regional vegetation from drier open biomes to a region dominated by the temperate conifer forest biome and the conifer parkland biome, which is consistent with reconstructions from palaeobotanical data (Thompson, 1991; Fleming, 1994; Salzmann et al., 2008). The main driving force for these vegetation changes appears to be increased evaporation of the lakes, reducing summer temperature and increasing summer precipitation. When soils and lakes are combined there is a further increase in less seasonal biome types (Fig. 6). This suggests that, although the differences between the control experiment and the PRISM3 + soils experiment are localised, the combination of realistic late Pliocene soils and lakes has positive feedbacks that facilitate the expansion of less seasonal biome types. Despite the limited improvements in the modelled climates and biome distribution of the late Pliocene, there are positive changes between the experiments containing the soils and lake data and the control run. We therefore encourage the use of the late Pliocene soil and lake boundary conditions in future climate modelling studies, including the future PlioMIP experiments (Haywood et al., 2010, 2011a).

5 Conclusions

Through a synthesis of geological data we have reconstructed the global distribution of late Pliocene soils and lakes. From these reconstructions we have conducted a suite of climate modelling experiments to test the impacts of realistic soils and lakes on the climate of the late Pliocene. The inclusion of soils and lakes does not significantly modify global climate, but does have important regional impacts. Some of these regions have previously been simulated as too dry (e.g. western North America), when compared to palaeobotanical data. We see improvements in the seasonal amounts of precipitation in the southern part of northern Africa and in western North America, which results in the model-predicted biomes comparing more favourably with vegetation proxy data. We strongly encourage the use of these newly developed boundary conditions in future late Pliocene climate research, and the boundary conditions will be made available on the PRISM4 website (http://geology.er. usgs.gov/eespteam/prism/index.html). These new boundary conditions improve regional data–model comparisons, and their feedbacks in a warmer world should be explored further in future palaeoclimate modelling studies.

Acknowledgements. M. J. Pound, U. Salzmann and A. M. Haywood acknowledge funding received from the Natural Environment Research Council (NERC Grant NE/I016287/1). J. Tindall, S. J. Pickering and A. M. Haywood acknowledge that the research leading to these results has received funding from the European Research Council under the European Union's Seventh Framework Programme (FP7/2007-2013)/ERC grant agreement no. 278636. H. J. Dowsett acknowledges support from the US Geological Survey Land Use and Climate Change R&D Program and the Powell Center for Analysis and Synthesis. We wish to thank C. Contoux, P. Hoelzman and E. Stone for their reviews which greatly improved the manuscript.

Edited by: V. Brovkin

References

Abbott, P. L.: Cenozoic paleosols San Diego area, California, CATENA, 8, 223–237, 1981.

Adam, D. P., Bradbury, J. P., Rieck, H. J., and Sarna-Wojcicki, A. M.: Environmental changes in the Tule Lake Basin, Siskiyou and Modoc Counties, California, from 3 to 2 Million years before present, USGS Bulletin, 1933, 1–13, 1990.

Alley, N. F.: Cainozoic stratigraphy, palaeoenvironments and geological evolution of the Lake Eyre Basin, Palaeogeogr. Palaeocli. Palaeoecol., 144, 239–263, 1998.

Basilici, G.: Sedimentary facies in an extensional and deep-lacustrine depositional system: the Pliocene Tiberino Basin, Central Italy, Sediment. Geol., 109, 73–94, 1997.

Beadle, L. C.: The inland waters of tropical Africa: an introduction to tropical limnology, Longman, London, 365 pp., 1974.

Bechtel, A., Sachsenhofer, R. F., Markic, M., Gratzer, R., Lücke, A., and Püttmann, W.: Paleoenvironmental implications from biomarker and stable isotope investigations on the Pliocene Velenje lignite seam (Slovenia), Org. Geochem., 34, 1277–1298, 2003.

Burrough, S. L., Thomas, D. S. G., and Singarayer, J. S.: Late Quaternary hydrological dynamics in the Middle Kalahari: Forcing and feedbacks, Earth-Sci. Rev., 96, 313–326, 2009.

Cahen, L.: Geologie du Congo Belge. Liège: Vaillant-Carmanne, 577 pp., 1954.

Callen, R. A.: Late Cainozoic environments of part of northeastern South Australia, J. Geol. Soc. Aust., 24, 151–169, 1977.

Campisano, C. J. and Feibel, C. S.: Depositional environments and stratigraphic summary of the Pliocene Hadar Formation at Hadar, Afar Depression, Ethiopia, The Geol. Soc. Am. Special Paper, 446, 179–201, 2008.

Chandler, M., Rind., D., and Thompson, R.: Joint investigations of the middle Pliocene climate II: GISS GCM Northern Hemisphere results, Global Planet. Change, 9, 197–219, 1994.

Chang, H., Ao, H., An, Z., Fang, X., Song, Y., and Qiang, X.: Magnetostratigraphy of the Suerkuli Basin indicates Pliocene (3.2 Ma) activity of the middle Altyn Tagh Fault, northern Tibetan Plateau, J. Asian Earth Sci., 44, 169–175, 2012.

Chen, X. Y., Bowler, J. M., and Magee, J. W.: Late Cenozoic stratigraphy and hydrologic history of Lake Amadeus, a central Australian playa, Aust. J. Earth Sci., 40, 1–14, 1993.

Cohen, A. S., Lezzar, K. E., Tiercelin, J. J., and Soreghan, M.: New palaeogeographic and lake-level reconstructions of Lake Tanganyika: implications for tectonic, climatic and biological evolution in a rift lake, Basin Res., 9, 107–132, 1997.

Contoux, C., Ramstein, G., and Jost, A.: Modelling the mid-Pliocene Warm Period climate with the IPSL coupled model and its atmospheric component LMDZ5A, Geosci. Model Dev., 5, 903–917, doi:10.5194/gmd-5-903-2012, 2012.

Contoux, C., Jost, A., Ramstein, G., Sepulchre, P., Krinner, G., and Schuster, M.: Megalake Chad impact on climate and vegetation during the late Pliocene and the mid-Holocene, Clim. Past, 9, 1417–1430, doi:10.5194/cp-9-1417-2013, 2013.

Coulthard, T. J., Ramirez, J. A., Barton, N., Rogerson, M., and Brücher, T.: Were Rivers Flowing across the Sahara During the Last Interglacial? Implications for Human Migration through Africa, PLoS ONE 8, e74834, doi:10.1371/journal.pone.0074834, 2013.

Cox, P.: Description of the "TRIFFID" Dynamic Global Vegetation Model. Hadley Centre technical note 24, 1–16, 2001.

Cox, P. M., Betts, R., Bunton, C. B., Essery, R. L. H., Rowntree, P. R., and Smith, J.: The impact of new land surface physics on the GCM simulation of climate and climate sensitivity, Clim. Dynam., 15, 183–203, 1999.

Deino, A. L., Kingston, J. D., Glen, J. M., Edgar, R. K., and Hill, A.: Precessional forcing of lacustrine sedimentation in the late Cenozoic Chemeron Basin, Central Kenya Rift, and calibration of the Gauss/Matuyama boundary, Earth Planet. Sci. Lett., 247, 41–60, 2006.

Delire, C., Levis, S., Bonan, G., Foley, J. A., Coe, M., and Vavrus, S.: Comparison of the climate simulated by the CCM3 coupled to two different land-surface models, Clim. Dynam., 19, 657–669, 2002.

Dixey, F.: The Tertiary and Post-Tertiary Lacustrine Sediments of the Nyasan Rift-Valley, Q. J. Geol. Soc., 83, 432–442, 1927.

Dowsett, H. J., Cronin, T. M., Poore, R. Z., Thompson, R. S., Whatley, R. C., and Wood, A. M.: Micropaleontological evidence for Increased meridional heat transport in the North Atlantic Ocean during the Pliocene, Science, 258, 1133–1135, 1992.

Dowsett, H. J., Thompson, R., Barron, J., Cronin, T., Fleming, F., Ishman, S., Poore, R., Willard, D., and Holtz Jr., T.: Joint investigations of the Middle Pliocene climate I: PRISM paleoenvironmental reconstructions, Global Planet. Change, 9, 169–195, 1994.

Dowsett, H. J., Barron, J., and Poore, R.: Middle Pliocene sea surface temperatures: a global reconstruction, Mar. Micropaleontol., 27, 13–26, 1996.

Dowsett, H. J., Robinson, M., Haywood, A., Salzmann, U., Hill, D., Sohl, L., Chandler, M., Williams, M., Foley, K., and Stoll, D.: The PRISM3D paleoenvironmental reconstruction, Stratigraphy, 7, 123–139, 2010.

Dowsett, H. J., Robinson, M. M., Haywood, A. M., Hill, D. J., Dolan, A. M., Stoll, D. K., Chan, W.-L., Abe-Ouchi, A., Chandler, M. A., Rosenbloom, N. A., Otto-Bliesner, B. L., Bragg, F. J., Lunt, D. J., Foley, K. M., and Riesselman, C. R.: Assessing confidence in Pliocene sea surface temperatures to evaluate predictive models, Nat. Clim. Change, 2, 365–371, 2012.

Dowsett, H., Foley, K., Stoll, D., Chandler, M., Sohl, L., Bentsen, M., Otto-Bliesner, B., Bragg, F., Chan, W.-L., Contoux, C., Dolan, A., Haywood, A., Jonas, J., Jost, A., Kamae, Y., Lohmann, G., Lunt, D., Nisancioglu, K., Abe-Ouchi, A., Ramstein, G., Riesselman, C., Robinson, M., Rosenbloom, N., Salzmann, U., Stepanek, C., Strother, S., Ueda, H., Yan, Q., and Zhang, Z.: Sea surface temperature of the mid-Piacenzian ocean: A data-model comparison, Sci. Reports, 3, 1–8, doi:10.1038/srep02013, 2013.

Drake, N. A., El-Hawat, A. S., Turner, P., Armitage, S. J., Salem, M. J., White, K. H., and McLaren, S.: Palaeohydrology of the Fazzan Basin and surrounding regions: The last 7 million years, Palaeogeogr. Palaeocli. Palaeoecol., 263, 131–145, 2008.

Essery, R. L. H., Best, M., and Cox, P.: MOSES 2.2 technical documentation, Hadley Centre technical note, 30, 1–30, 2001.

Firman, J. B.: Paleosols in laterite and silcrete profiles Evidence from the South East Margin of the Australian Precambrian Shield, Earth-Sci. Rev., 36, 149–179, 1994.

Fleming, R. F.: Palynological records from Pliocene sediments in the California region: Centerville Beach, DSDP Site 32, and the Anza-Borrego desert. Pliocene Terrestrial Environments and Data/Model Comparisons, edited by: Thompson, R. S., 11–15, USGS Open-File Report, 1994.

Gerasimenko, N.: Vegetation development cycles of the Ukrainian forest-steppe zone in the Middle – Late Pliocene, in: Paleofloristic and paleoclimatic changes during Cretaceous and Tertiary, edited by: Planderova, E., Konzalova, M., Kvacek, Z., Sitar, V., Snopkova, P., and Suballyova, D., Geologicky Ustav Dionyza Stura, Bratislava, 199–204, 1993.

Good, P., Jones, C., Lowe, J., Betts, R., and Gedney, N.: Comparing tropical forest projections from two generations of Hadley

Centre Earth System Models, HadGEM2-ES and HadCM3LC, J. Climate, 26, 495–511, 2013.

Gordon, C., Cooper, C., Senior, C. A., Banks, H., Gregory, J. M., Johns, T. C., Mitchell, J. F. B., and Wood, R. A.: The simulation of SST, sea ice extents and ocean heat transports in a version of the Hadley Centre coupled model without flux adjustments, Clim. Dynam., 16, 147–168, 2000.

Goudie, A. S.: The drainage of Africa since the Cretaceous, Geomorphology, 67, 437–456, 2005.

Graef, F., Singer, A., Stahr, K., and Jahn, R.: Genesis and diagenesis of paleosols from Pliocene volcanics on the Golan Heights, CATENA, 30, 149–167, 1997.

Günster, N. and Skowronek, A.: Sediment–soil sequences in the Granada Basin as evidence for long- and short-term climatic changes during the Pliocene and Quaternary in the Western Mediterranean, Quaternary Int., 78, 17–32, 2001.

Gürel, A. and Kadir, S.: Geology, mineralogy and origin of clay minerals of the Pliocene fluvial-lacustrine deposits in the Cappadocian Volcanic Province, central Anatolia, Turkey, Clays Clay Miner., 54, 555–570, 2006.

Haywood, A. M. and Valdes, P. J.: Vegetation cover in a warmer world simulated using a dynamic global vegetation model for the Mid-Pliocene, Palaeogeogr. Palaeocli. Palaeoecol., 237, 412–427, 2006.

Haywood, A. M., Chandler, M. A., Valdes, P. J., Salzmann, U., Lunt, D. J., and Dowsett, H. J.: Comparison of mid-Pliocene climate predictions produced by the HadAM3 and GCMAM3 General Circulation Models, Global Planet. Change, 66, 208–224, 2009.

Haywood, A. M., Dowsett, H. J., Otto-Bliesner, B., Chandler, M. A., Dolan, A. M., Hill, D. J., Lunt, D. J., Robinson, M. M., Rosenbloom, N., Salzmann, U., and Sohl, L. E.: Pliocene Model Intercomparison Project (PlioMIP): experimental design and boundary conditions (Experiment 1), Geosci. Model Dev., 3, 227–242, doi:10.5194/gmd-3-227-2010, 2010.

Haywood, A. M., Dowsett, H. J., Robinson, M. M., Stoll, D. K., Dolan, A. M., Lunt, D. J., Otto-Bliesner, B., and Chandler, M. A.: Pliocene Model Intercomparison Project (PlioMIP): experimental design and boundary conditions (Experiment 2), Geosci. Model Dev., 4, 571–577, doi:10.5194/gmd-4-571-2011, 2011a.

Haywood, A. M., Ridgwell, A., Lunt, D. J., Hill, D. J., Pound, M. J., Dowsett, H. J., Dolan, A. M., Francis, J. E., and Williams, M.: Are there pre-Quaternary geological analogues for a future greenhouse warming? Philosophical Transactions of the Royal Society A: Mathematical, Phys. Eng. Sci., 369, 933–956, 2011b.

Helgren, D. M. and Butzer, K. W.: Paleosols of the southern Cape Coast, South Africa: Implications for laterite definition, genesis, and age, Geogr. Rev., 67, 430–445, 1977.

Hoelzmann, P., Jolly, D., Harrison, S. P., Laarif, F., Bonnefille, R., and Pachur, H. J.: Mid-Holocene land-surface conditions in northern Africa and the Arabian Peninsula: A data set for the analysis of biogeophysical feedbacks in the climate system, Global Biogeochem. Cy., 12, 35–51, 1998.

Hou, B., Frakes, L. A., Sandiford, M., Worrall, L., Keeling, J., and Alley, N. F.: Cenozoic Eucla Basin and associated palaeovalleys, southern Australia – Climatic and tectonic influences on landscape evolution, sedimentation and heavy mineral accumulation, Sedim. Geol., 203, 112–130, 2008.

Hughes, M. J., Carey, S. P., and Kotsonis, A.: Lateritic weathering and secondary gold in the Victorian Gold Province, Regolith, 98, 155–172, 1999.

Hughes, J. K., Valdes, P. J., and Betts, R.: Dynamics of a global-scale vegetation model, Ecol. Modell., 198, 452–462, 2006.

Icole, M.: Une nouvelle methode pour la paleopedologie du Pliocene et du Villafranchien des Pyrenees centrales, Bulletin de l'Association française pour l'etude du Quaternaire, 1970, 135–143, 1970.

Jones, C. P.: Ancillary file data sources, Unified Model Documentation, 70, 1–38, 2008.

Kaplan, J.: Geophysical applications of vegetation modeling, Lund University, Lund, p. 128, 2001.

Kelepertsis, A.: Mineralogy and geochemistry of the pliocene iron-rich laterite in the Vatera area, Lesvos Island, Greece and its genesis, Chinese J. Geochem., 21, 193–205, 2002.

Knorr, W. and Schnitzler, K.-G.: Enhanced albedo feedback in North Africa from possible combined vegetation and soil-formation processes, Clim. Dynam., 26, 55–63, 2006.

Krinner, G., Lézine, A.-M., Braconnot, P. Sepulchre, P. Ramstein, G. Grenier, C., and Gouttevin, I.: A reassessment of lake and wetland feedbacks on the North African Holocene climate, Geophys. Res. Lett., 39, L07701, doi:10.1029/2012GL050992, 2012.

Lehner, B. and Döll, P.: Development and validation of a global database of lakes, reservoirs and wetlands, J. Hydrol., 296, 1–22, 2004.

Lenoble, A.: Les depots lacustres Pliocenes-Pleistocenes de L'Ankaratra (Madagascar): Etude geologique, Ann. Geolog. Serv. Mines, 18, 9–82, 1949.

Leroy, S. and Dupont, L.: Development of vegetation and continental aridity in northwestern Africa during the Late Pliocene: the pollen record of ODP site 658, Palaeogeogr. Palaeocli. Palaeoecol., 109, 295–316, 1994.

Levis, S., Bonan, G. B., and Bonfils, C.: Soil feedback drives the mid-Holocene North African monsoon northward in fully coupled CCSM2 simulations with a dynamic vegetation model, Clim. Dynam., 23, 791–802, 2004.

Lunt, D. J., Haywood, A. M., Schmidt, G. A., Salzmann, U., Valdes, P. J., and Dowsett, H.J.: Earth system sensitivity inferred from Pliocene modelling and data, Nat. Geosci., 3, 60–64, 2010.

Lunt, D. J., Haywood, A. M., Schmidt, G. A., Salzmann, U., Valdes, P. J., Dowsett, H. J., and Loptson, C. A.: On the causes of mid-Pliocene warmth and polar amplification, Earth Planet. Sci. Lett., 321–322, 128–138, 2012.

Mabesoone, J. M. and Lobo, H. R. C.: Paleosols as stratigraphic indicators for the cenozoic history of northeastern Brazil, CATENA, 7, 67–78, 1980.

Mack, G. H., Seager, W. R., Leeder, M. R., Perez-Arlucea, M., and Salyards, S. L.: Pliocene and Quaternary history of the Rio Grande, the axial river of the southern Rio Grande rift, New Mexico, USA, Earth-Sci. Rev., 79, 141–162, 2006.

Martin, H. A.: Cenozoic climatic change and the development of the arid vegetation in Australia, J. Arid Environ., 66, 533–563, 2006.

Mats, V. D., Lomonosova, T. K., Vorobyova, G. A., and Granina, L. Z.: Upper Cretaceous–Cenozoic clay minerals of the Baikal region (eastern Siberia), Appl. Clay Sci., 24, 327–336, 2004.

Müller, J., Oberhänsli, H., Melles, M., Schwab, M., Rachold, V., and Hubberten, H. W.: Late Pliocene sedimentation in Lake Baikal:

implications for climatic and tectonic change in SE Siberia, Palaeogeogr. Palaeocli. Palaeoecol., 174, 305–326, 2001.

Otero, O., Pinton, A., Mackaye, H. T., Likius, A., Vignaud, P., and Brunet, M.: Fishes and palaeogeography of the African drainage basins: Relationships between Chad and neighbouring basins throughout the Mio-Pliocene, Palaeogeogr. Palaeocli. Palaeoecol., 274, 134–139, 2009.

Otero, O., Pinton, A., Mackaye, H. T., Likius, A., Vignaud, P., and Brunet, M.: The early/late Pliocene ichthyofauna from Koro-Toro, Eastern Djurab, Chad. Geobios, 43, 241–251, 2010.

Paepe, R., Mariolakos, I., Van Overloop, E., Nassopoulou, S., Hus, J., Hatziotou, M., Markopoulos, T., Manutsoglu, E., Livaditis, G., and Sabot, V.: Quaternary soil-geological stratigraphy in Greece, B. Geol. Soc. Greece, 36, 856–863, 2004.

Peters, C. R. and O'Brien, E. M.: Palaeo-lake Congo: Implications for Africa's Late Cenozoic climate – some unanswered questions, Palaeoecol. Africa, 27, 11–18, 2001.

Pope, V. D., Gallani, M. L., Rowntree, P. R., and Stratton, R. A.: The impact of new physical parametrizations in the Hadley Centre climate model: HadAM3, Clim. Dynam., 16, 123–146, 2000.

Pope, J. O., Collins, M., Haywood, A. M., Dowsett, H. J., Hunter, S. J., Lunt, D. J., Pickering, S. J., and Pound, M. J.: Quantifying Uncertainty in Model Predictions for the Pliocene (Plio-QUMP): Initial results, Palaeogeogr. Palaeocli. Palaeoecol., 309, 128–140, 2011.

Quade, J., Solounias, N., and Cerling, T. E.: Stable isotopic evidence from paleosol carbonates and fossil teeth in Greece for forest or woodlands over the past 11 Ma, Palaeogeogr. Palaeocli. Palaeoecol., 108, 41–53, 1994.

Ringrose, S., Kampunzu, A. B., Vink, W., Matheson, W., and Downe, W. S.: Origin and palaeo-environments of calcareous sediments in the Moshaweng dry valley, southeast Botswana, Earth Surf. Process. Landf., 27, 591–611, 2002.

Ringrose, S., Huntsman-Mapila, P., Basira Kampunzu, A., Downey, W., Coetzee, S., Vink, B., Matheson, W., and Vanderpost, C.: Sedimentological and geochemical evidence for palaeoenvironmental change in the Makgadikgadi subbasin, in relation to the MOZ rift depression, Botswana, Palaeogeogr. Palaeocli. Palaeoecol., 217, 265–287, 2005.

Sáez, A., Cabrera, L., Jensen, A., and Chong, G.: Late Neogene lacustrine record and palaeogeography in the Quillagua-Llamara basin, Central Andean fore-arc (northern Chile), Palaeogeogr. Palaeocli. Palaeoecol., 151, 5–37, 1999.

Salama, R. B.: The evolution of the River Nile. The buried saline rift lakes in Sudan – I. Bahr El Arab Rift, the Sudd buried saline lake, J. Afr. Earth Sci., 6, 899–913, 1987.

Salzmann, U., Haywood, A. M., Lunt, D. J., Valdes, P. J., and Hill, D. J.: A new global biome reconstruction and data-model comparison for the Middle Pliocene, Global Ecol. Biogeogr., 17, 432–447, 2008.

Salzmann, U., Haywood, A. M., and Lunt, D. J.: The past is a guide to the future? Comparing Middle Pliocene vegetation with predicted biome distributions for the twenty-first century, Philos. Trans. Royal Soc. A, 367, 189–204, 2009.

Salzmann, U., Williams, M., Haywood, A. M., Johnson, A. L. A., Kender, S., and Zalasiewicz, J.: Climate and environment of a Pliocene warm world, Palaeogeogr. Palaeocli. Palaeoecol., 309, 1–8, 2011.

Salzmann, U., Dolan, A. M., Haywood, A. M., Chan, W.-L., Voss, J., Hill, D. J., Abe-Ouchi, A., Otto-Bliesner, B., Bragg, F. J., Chandler, M. A., Contoux, C., Dowsett, H. J., Jost, A., Kamae, Y., Lohmann, G., Lunt, D. J., Pickering, S. J., Pound, M. J., Ramstein, G., Rosenbloom, N. A., Sohl, L., Stepanek, C., Ueda, H., and Zhang, Z.: Challenges in quantifying Pliocene terrestrial warming revealed by data-model discord, Nat. Clim. Change, doi:10.1038/nclimate2008, 2013.

Sangode, S. J. and Bloemendal, J.: Pedogenic transformation of magnetic minerals in Pliocene–Pleistocene palaeosols of the Siwalik Group, NW Himalaya, India, Palaeogeogr. Palaeocli. Palaeoecol., 212, 95–118, 2004.

Schuster, M., Duringer, P., Ghienne, J.-F., Roquin, C., Sepulchre, P., Moussa, A., Lebatard, A.-E., Mackaye, H. T., Likius, A., Vignaud, P., and Brunet, M.: Chad Basin: Paleoenvironments of the Sahara since the Late Miocene, Compt. Rendus Geosci., 341, 603–611, 2009.

Sepulchre, P., Ramstein, G., and Schuster, M.: Modelling the impact of tectonics, surface conditions and sea surface temperatures on Saharan and sub-Saharan climate evolution, Compt. Rendus Geosci., 341, 612–620, 2009.

Simon-Coinçon, R., Milnes, A. R., Thiry, M., and Wright, M. J.: Evolution of landscapes in northern South Australia in relation to the distribution and formation of silcretes, J. Geolog. Soc., 153, 467–480, 1996.

Smith, G. R.: Late Cenozoic Freshwater Fishes of North America, Ann. Rev. Ecol. System., 12, 163–193, 1981.

Soil Survey Staff: Soil taxonomy: A basic system of soil classification for making and interpreting soil surveys United States Department of Agriculture Agriculture Handbook 436, 871 pp., 1999.

Sloan, L. C.: Equable climates during the early Eocene: Significance of regional paleogeography for North American climate, Geology, 22, 881–884, 1994.

Sloan, L. C., Crowley, T. J., and Pollard, D.: Modeling of Middle Pliocene climate with the NCAR Genesis general circulation model, Mar. Micropaleontol., 27, 51–61, 1996.

Summerfield, M. A.: Global Geomorphology, Longman, Harlow, 1991.

Thompson, R. S.: Pliocene environments and climates in the western United States, Quaternary Sci. Rev., 10, 115–132, 1991.

Thompson, R. S.: Palynological data from a 989-FT (301-M) core of Pliocene and Early Pleistocene sediments from Bruneau, Idaho, USGS Open File Report 92–713, 1–28, 1992.

Violante, R., Osella, A., Vega, M. d. l., Rovere, E., and Osterrieth, M.: Paleoenvironmental reconstruction in the western lacustrine plain of Llancanelo Lake, Mendoza, Argentina, J. South Am. Earth Sci., 29, 650–664, 2010.

Wijninga, V. M. and Kuhry, P.: Late Pliocene paleoecology of the Guasca Valley (Cordillera Oriental, Colombia). Review of Palaeobotany and Palynology 78, 69-127, 1993.

Wynn, J. G.: Paleosols, stable carbon isotopes, and paleoenvironment interpretation of Kanapoi, Northern Kenya, J. Human Evol., 39, 411–432, 2000.

12

Variability of summer humidity during the past 800 years on the eastern Tibetan Plateau inferred from $\delta^{18}O$ of tree-ring cellulose

J. Wernicke, J. Grießinger, P. Hochreuther, and A. Bräuning

Institute of Geography, Friedrich-Alexander-University Erlangen-Nuremberg, Erlangen, Germany

Correspondence to: J. Wernicke (jakob.wernicke@fau.de)

Abstract. We present an 800-year $\delta^{18}O$ chronology from the eastern part of the Tibetan Plateau (TP). The chronology dates back to AD 1193 and was sampled in AD 1996 from living *Juniperus tibetica* trees. This first long-term tree-ring-based $\delta^{18}O$ chronology for eastern Tibet provides a reliable archive for hydroclimatic reconstructions. Highly significant correlations were obtained with hydroclimatic variables (relative humidity, vapour pressure, and precipitation) during the summer season. We applied a linear transfer model to reconstruct summer season relative humidity variations over the past 800 years. More moist conditions prevailed during the termination of the Medieval Warm Period while a systematic shift during the Little Ice Age is not detectable. A distinct trend towards more dry conditions since the 1870s is apparent. The moisture decline weakened around the 1950s but still shows a negative trend. The mid-19th century humidity decrease is in good accordance with several multiproxy hydroclimate reconstructions for south Tibet. However, the pronounced summer relative humidity decline is stronger on the central and eastern TP. Furthermore, the relative humidity at our study site is significantly linked to the relative humidity at large parts of the TP. Therefore, we deduce that the reconstructed relative humidity is mostly controlled by local and mesoscale climatic drivers, although significant connections to the higher troposphere of west-central Asia were observed.

1 Introduction

The variation in strength, timing, and duration of the Asian summer monsoon (ASM) system affects the life and economy of many millions of people living in south and east Asia (Immerzeel et al., 2010; Zhang et al., 2008). In remote areas, such as the Tibetan Plateau (TP), reliable climate records are short and scattered. Nevertheless, a recent weakening trend of the ASM precipitation amount was reported in several studies (Bollasina et al., 2011; Sano et al., 2011; Zhou et al., 2008b). The decline in air humidity was explained by a reduction in the thermal gradient between the surface temperatures of the Indian Ocean and the TP due to global warming (Sun et al., 2010). Different locations and climate archives reveal contemporaneous strengthened monsoonal precipitation (Anderson et al., 2002; Kumar et al., 1999; Zhang et al., 2008). This discrepancy may be explained by the high variability of the monsoon circulation itself and also by the limited number of available palaeoclimate studies and resulting climate modelling uncertainties. Thus, for a better understanding of the circulation system as a whole, but also for the verification of climate change scenarios, a keen demand for reliable climate reconstructions exists for the TP. With increasing numbers of palaeoclimatic records, forecast and climate projection precision increases and can be helpful for facilitating targeted decision-making regarding water and resource management.

The northward movement of the Intertropical Convergence Zone (ITCZ) in the Northern Hemisphere in boreal summer is amplified over the Asian continent by the thermal contrast between the Indian Ocean and the TP (Webster et al., 1998). Convective rainfalls during the summer monsoon season between June and September are strongly altered by the complex topography of the Himalayas and western Chinese mountain systems (e.g. Böhner, 2006; Maussion et al., 2014; Thomas and Herzfeld, 2004). Extreme climatic

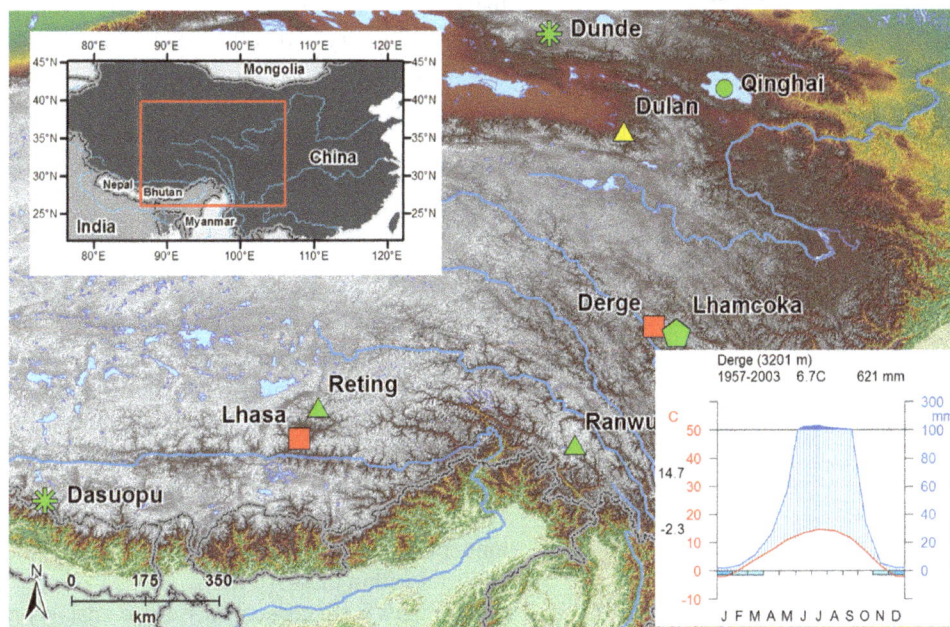

Figure 1. Location of the study site Lhamcoka (green pentagon) and other proxy archives mentioned in the text. Green triangles: tree-ring $\delta^{18}O$ chronologies; yellow triangle: tree-ring width chronology; green asterisk: ice cores; green circle: lake sediments. Red rectangles indicate climate stations.

events that may have devastating effects and the long-term trends of ASM intensity are therefore in the focus of numerous climate reconstruction efforts (e.g. Cook et al., 2010; Xu et al., 2006b; Yang et al., 2003). Most of these studies use tree-ring width as a proxy for palaeoclimate reconstructions. Nonetheless, several studies demonstrated that $\delta^{18}O$ of wood cellulose is a strong indicator of hydroclimatic conditions (McCarroll and Loader, 2004; Roden et al., 2000; Saurer et al., 1997; Sternberg, 2009). Even if tree stands might have been influenced by external disturbances (e.g. competition, insect attacks, or geomorphological processes) they still reflect variations of the local hydroclimate accurately (Sano et al., 2013). Recently published tree-ring $\delta^{18}O$ chronologies from the TP show a common strong response to regional moisture changes. Grießinger et al., 2011 successfully reconstructed August precipitation over the past 800 years. They demonstrated reduced precipitation during the Medieval Warm Period (MWP), stronger rainfalls during the Little Ice Age (LIA), decreasing precipitation rates since the 1810s, and slightly wetter conditions since the 1990s. In addition, shorter $\delta^{18}O$ chronologies from the central Himalayas showed consistent negative correlations to summer precipitation (Sano et al., 2010, 2011, 2013). The detected recent reduction of monsoonal precipitation has been interpreted as a reaction to increased sea surface temperatures (SSTs) over the tropical Pacific and Indian Ocean (Zhou et al., 2008a). Strong responses to regional cloud cover changes were found for tree-ring $\delta^{18}O$ chronologies from the south-eastern TP (Liu et al., 2013; Liu et al., 2014; Shi et al., 2012). The local moisture reduction starting in the middle of the 19th century

is less pronounced than for south-west Tibet, and is associated with complex El Niño–Southern Oscillation (ENSO) teleconnections (Liu et al., 2012). Existing tree-ring $\delta^{18}O$ chronologies on the north-eastern part of the TP respond to local precipitation and relative humidity (Wang et al., 2013; Liu et al., 2008). Except for a relatively short summer moisture-sensitive time series (An et al., 2014), no long-term $\delta^{18}O$ chronologies or reliable reconstructions have been conducted for the eastern TP so far. It still remains unclear to what extent the MWP, LIA, and the modern humidity decrease are reflected in tree-ring $\delta^{18}O$ on the eastern TP, where the influence of the ASM, the Indian Summer monsoon and the westerlies overlap.

We present a new, well-replicated 800-year $\delta^{18}O$ chronology, representing a unique archive for studying the past hydroclimate in eastern Tibet. We applied response and transfer functions and obtained a reliable reconstruction of summer relative humidity (July + August). We compared the long-term trend of our chronology to other moisture-sensitive proxy archives from several sites over the TP and discuss climatic control mechanisms on the relative humidity.

2   Material and methods

2.1   Study site – Lhamcoka

Lhamcoka is located on the eastern TP (see Fig. 1 green pentagon). During a field campaign in 1996, 16 living *Juniperus tibetica* trees were cored twice in order to enhance the

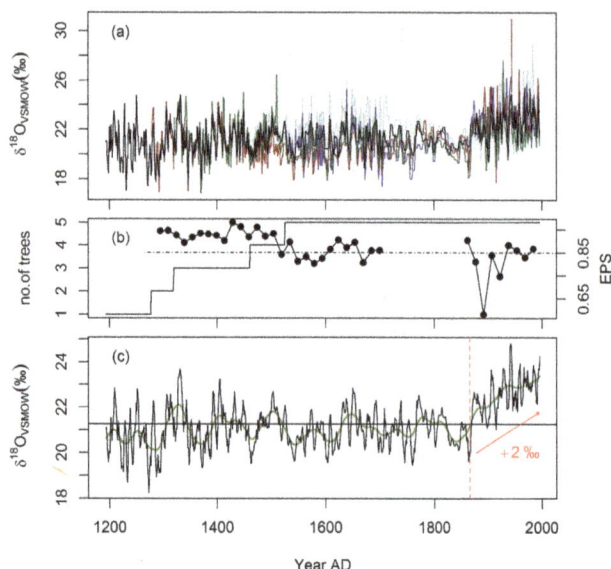

Figure 2. Lhamcoka tree-ring $\delta^{18}O$ isotope chronology. (**a**) Individual $\delta^{18}O$ time series of five individuals. The coarse resolution between 1707 and 1864 results from shifted block pooling. (**b**) Running EPS (calculated for 25-year intervals, lagged by 10 years) and number of trees used for the reconstruction (solid line). Dashed line represents the theoretical EPS threshold of 0.85. (**c**) Tree-ring $\delta^{18}O$ chronology spanning the period AD 1193–1996. Green solid line represents a 50-year smoothing spline. The red dashed line marks the turning point towards heavier isotope ratios after $\sim$ 1870.

chance of detecting missing rings. The samples were collected from a steep, south-east exposed slope at an elevation of 4350 m a.s.l. (31°49′ N, 99°06′ E). The oldest tree is 801 years old, resulting in an overall chronology time span of AD 1193–1996. The average single core length is 633 years, with single segment lengths of 801 yr, 697 yr, 668 yr, 528 yr, and 469 yr. The chronology is not biased by an age trend as it was supposed for different high-altitude mountain ecosystems (Esper et al., 2010; Treydte et al., 2006). We applied a spline-based trend analysis and revealed non-systematic trends during the first 100 years after germination (graph not shown here). Therefore, a "juvenile" effect is not likely to affect our chronology, justifying the retention of the oldest parts of each single core. Juniper forms the upper timberline in the region due to its cold temperature tolerance (Bräuning, 2001). The species' annual tree-ring growth is limited by temperature and spring precipitation (February–April) (see Lhamcoka E site description in Bräuning, 2006). Therefore, the early wood formation is negatively affected by spring conditions, leading to growth reduction of the annual growth rings. Due to the steep slope angle of more than 30° and well-drained substrate properties at the study site, ground water influence can be excluded. Therefore, we assume the trees $\delta^{18}O$ source water properties are mainly controlled by the oxygen isotope configuration of summer precipitation, although it is known that snow-derived meltwater input affects

the source water properties of trees (Treydte et al., 2006). According to dry and cold winter monsoon conditions (see climate diagram in Fig. 1), a high and persistent snow cover at our study site is not likely. Hence, 13 % of potential solid precipitation falling between October and April will probably not strongly influence the source water properties at our study site.

Lhamcoka is influenced by the Indian summer monsoon system with typical maxima of temperature and precipitation during the summer months (see climate diagram in Fig. 1). The nearby climate station Derge (3201 m a.s.l., 50 km from the sampling site) records 78 % (541 mm) of annual precipitation between June and September, which is in accordance with common monsoonal climate properties (Böhner, 2006). The Derge climate record (data provided by the China Meteorological Administration) revealed increasing temperatures of about 0.6 °C during the period 1956–1996, whereas the amount and interannual variability of precipitation remained constant within these 41 years.

Five trees were chosen for isotope analysis to adequately capture inter-tree variability of $\delta^{18}O$ (Leavitt, 2010). The trees were selected for the (i) old age of the cores, to maximize the length of the derived reconstruction, (ii) avoidance of growth asymmetries due to slope processes, (iii) sufficient amounts of material (samples with wider rings were favoured), and (iv) high inter-correlation among the tree-ring width series of the respective cores.

2.2 Sample preparation

We used the tree-ring width master chronology of Bräuning, 2006 in order to date each annual ring precisely. The dated tree-rings were cut with a razor blade under a microscope. $\delta^{18}O$ values were measured from each tree individually in annual resolution. During periods of the chronology with extremely narrow rings, we used shifted block pooling to obtain sufficient material (Böttger and Friedrich, 2009). Pooling was applied between the years 1707 and 1864 (see chronology parts with missing expressed population signal (EPS) in Fig. 2). To obtain pure α-cellulose, we followed the chemical treatment presented in Wieloch et al., 2011. The α-cellulose was homogenized with an ultrasonic unit and the freeze-dried material was loaded into silver capsules (Laumer et al., 2009). The ratio of $^{18}O/^{16}O$ was determined in a continuous flow mass spectrometer (Delta V Advantage; Thermo Fisher Scientific Inc.). The standard deviation for the repeated measurement of an internal standard was better than 0.25‰.

2.3 Statistical analyses

We used standard dendrochronology techniques of chronology building, model building, and verification for reliable climate reconstruction (Cook and Kairiukstis, 1990). All analysis were conducted with the open source statistical software

R (http://cran.r-project.org/). The stable isotope chronology was calculated within the "dplR" package developed by Bunn, 2008 and the dendroclimatological correlation and response analyses were conducted by the "bootRes" package (Zang and Biondi, 2012). The pooling method we executed required a running mean calculation. Thus, the presented chronology has a quasi-annual resolution, smoothed with a 5-year running mean filter. To evaluate the isotope chronology reliability, the EPS (introduced by Wigley et al., 1984) and the *Gleichläufigkeit* (GLK) were computed. The EPS expresses the variance fraction of a chronology in comparison with a theoretically infinite tree population, whereas the GLK specifies the proportion of agreements/disagreements of interannual growth tendencies among the trees of the study site. The EPS is interrupted within our $\delta^{18}O$ chronology at parts where we applied shifted block pooling.

3 Results

3.1 Chronology characteristics

The Lhamcoka $\delta^{18}O$ chronology is defined by a mean of 21.27‰ and global minima/maxima of 18.24‰/24.83‰. The values are similar to results from nearby studies (An et al., 2014; Liu et al., 2012; Liu et al., 2013). Moreover, the trees within the chronology are characterized by a common signal that is expressed by an EPS of 0.88 and a highly significant GLK of 0.57 ($p<0.01$). Thus, our selected trees are likely to be affected by a common force, a prerequisite for compiling a reliable mean $\delta^{18}O$ chronology. The chronology can be sub-divided into two parts (see Fig. 2). The younger section (AD 1868–1996) shows a pronounced trend of about 2‰ towards heavy isotope ratios. Within this segment, the year with the heaviest ratio was detected in AD 1943 (24.8‰). Before the late 1870s, the isotope $\delta^{18}O$ values oscillate around the chronology mean. A phase of considerable low $\delta^{18}O$ values is obvious from AD 1200 to 1300. Within this section, the lightest isotope ratio was detected in AD 1272 (18.2‰). The signal strength (EPS) occasionally drops below the commonly accepted threshold of 0.85 during several periods. One reason might be the imprecise cutting of very narrow rings (ring width <0.2 mm). A mix of several rings produces a signal that cannot be related with certainty to a specific year, a well-known problem when using very old trees (Berkelhammer and Stott, 2012; Xu et al., 2013). Nevertheless, we have confidence in the Lhamcoka chronology due to an EPS above the threshold during the period AD 1300–1700.

3.2 Climatic response of tree-ring stable oxygen isotopes

We conducted linear correlation analyses between the $\delta^{18}O$ values and monthly climate data as well as calculated seasonal means of climate elements. The available climate

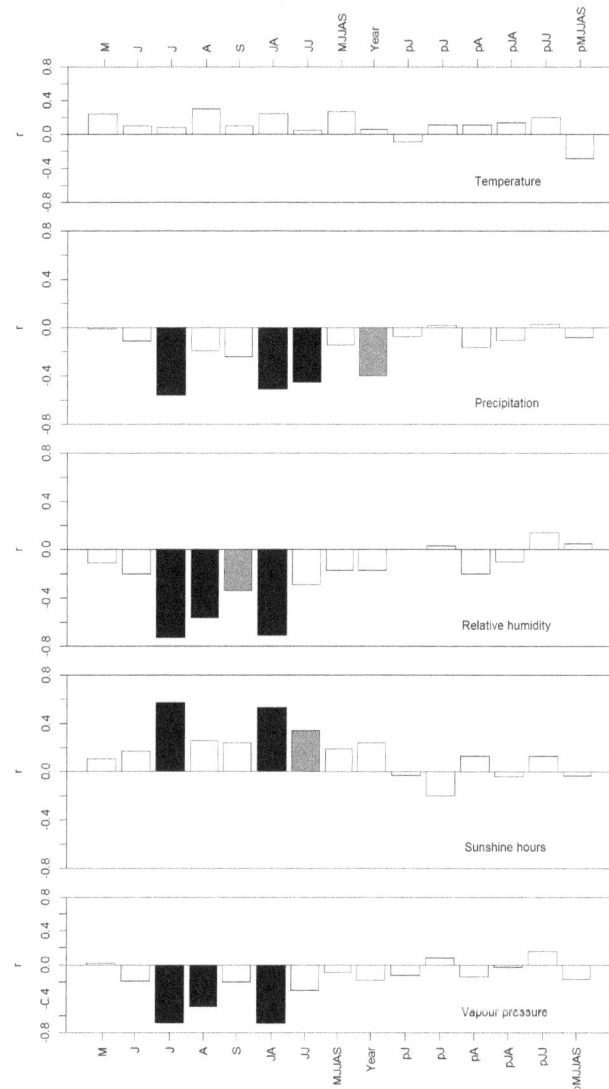

Figure 3. Response of tree-ring $\delta^{18}O$ to monthly/seasonal temperature, precipitation, relative humidity, sunshine hours, and vapour pressure over the period AD 1956–1996. Gray and black bars indicate correlations significant at $p<0.05$ and $p<0.01$, respectively; p indicates months/seasons of the previous year.

record of station Derge covers 41 years (AD 1956–1996) and correlations were calculated for temperature (mean), precipitation, relative humidity, sunshine hours (duration of global radiation > 120 W m^{-2}), and vapour pressure (see Fig. 3).

Summer moisture conditions explain most of the variance of the $\delta^{18}O$ chronology during the calibration period (AD 1956–1996). The stable oxygen isotopes are highly significantly ($p<0.01$) correlated with precipitation, relative humidity, sunshine hours, and vapour pressure during July and August. Highest (negative) correlations were obtained with relative humidity during July ($r = -0.73$) and July/August ($r = -0.71$). Thus, if relative humidity is high, transpiration is lowered and the depletion of light 16O due to leaf

Table 1. Verification statistics according to the linear transfer model of $\delta^{18}O$ and relative humidity within the calibration period AD 1956–1996.

Sign test (ST)	0.73 ($p<0.1$)
Product-moment correlation (PMC)	0.67 ($p<0.01$)
Product means test (PMT)	3.3 ($p<0.01$)
Reduction of error (RE)	0.45
Coefficient of efficiency (CE)	0.45

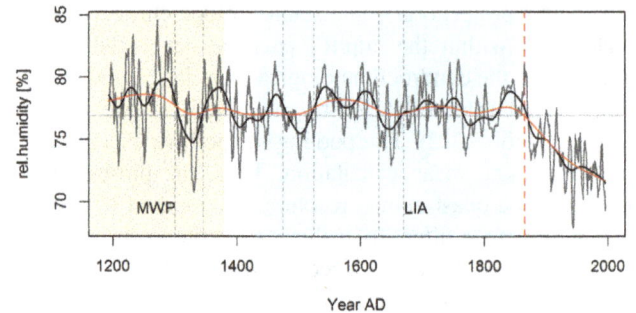

Figure 4. Summer (July + August) relative humidity reconstruction AD 1193–1996 for the eastern TP. Solid black and red lines represent 50-year and 150-year smoothing splines, respectively. Red dashed line emphasizes the turning point towards drier conditions ($\sim$ 1870s). The horizontal gray line illustrates mean relative summer humidity (RH = 72.4 %). Vertical dashed lines mark relatively dry periods. The Medieval Warm Period (MWP) and Little Ice Age (LIA) are emphasized in yellow and blue.

water fractionation is reduced. Additionally, weak and non-significant relationships were found with the mean temperature in all months/seasons. Thus, concepts of integrated temperature–moisture indexes, e.g. the vapour pressure difference (VPD: Kahmen et al., 2011), are unlikely to explain more of the variance in our data. However, we calculated the VPD as the difference between water vapour saturation pressure (E) and vapour pressure (e) and correlated the VPD time series against the $\delta^{18}O$ during the calibration period. From this we obtained significant but slightly weaker relationships with VPD ($r = 0.68$, $p<0.01$) since relative humidity and VPD are both influenced by temperature. Moreover, sunshine hours are positively related to the $\delta^{18}O$ variation. This association of high sunshine hours, less cloudiness, decreased relative humidity, and thus increased $\delta^{18}O$ values was corroborated by findings for south-east Tibet (Shi et al., 2012). Very weak correlations were found with climate conditions during the previous year. Therefore, plant physiological carryover effects as well as stagnating soil water can be regarded as insignificant factors for tree-ring $\delta^{18}O$ variations. The explained variance of the linear regression model between annual stable oxygen isotope values and relative humidity is 53 %. Hence, the $\delta^{18}O$ value mainly depends on relative humidity, which is in accordance with the findings of Roden and Ehleringer, 2000. Although highest correlations were obtained with single months (July: $r = -0.73$ ($p<0.01$)), the reconstruction was established for the summer season (mean relative humidity of July and August). In terms of using the wood cellulose of a single year, the humidity reconstruction of the major growing season is more robust than for single months.

3.3 Reconstruction of relative humidity

We employed a linear model for the reconstruction of relative humidity over the past 800 years. The linear relationship was achieved for the $\delta^{18}O$ values and instrumental records of relative humidity at climate station Derge between AD 1956 and 1996. The model was validated according to the standard methods presented in Cook and Kairiukstis, 1990, and Cook et al., 1994. We applied the leave-one-out validation procedure due to the short time period of available climate data. The model statistics are summarized in Table 1.

The validation tests indicated that (1) the number of agreements between the reconstructed climate series and the meteorological record is determined according to the sign orientation significantly different from a pure chance driven binomial test (ST); (2) the cross-correlation between the reconstruction and the measurement is highly significant (PMC, PMT), and (3) the reconstruction is reliable due to a positive RE and CE, indicating the reconstruction is better than the calibration period mean (Cook et al., 1994). Thus, our linear model is suitable for climate reconstruction purposes. The model related to the reconstruction of summer relative humidity is described as $RH_{JA} = -2.3 \cdot \delta^{18}O + 125.3$ (RH_{JA} expressed in percentage). A negative relationship between tree-ring stable oxygen isotopes and relative humidity was documented properly in several studies around the globe and among different species (Anderson et al., 1998; Burk and Stuiver, 1981; Ramesh et al., 1986; Tsuji et al., 2006). However, due to varying environmental settings (e.g. climate, soil) and different biological leaf properties (Kahmen et al., 2009), the slopes of the regression function differ significantly among study sites and species. Hence, $\delta^{18}O$ inferred model parameters from a neighbouring summer relative humidity reconstruction (June–August) using Abies trees differ from our regression model (An et al., 2014).

Our reconstruction reveals several phases of high and low summer humidity (see Fig. 4). Negative deviations from the mean value (72.4 %; sd = 4.9 %) occurred during AD 1300–1345, AD 1475–1525, AD 1630–1670, and AD 1866–1996 (periods are emphasized with dashed vertical lines in Fig. 4). The most pronounced relative humidity depression started in the late 1870s (dashed red line in Figure 4) and lasts until approximately the 1950s. The period is characterized by the driest summer in AD 1943 (RH = 68.4 %). The remarkable moisture reduction since the end of the LIA has been validated for the southern and south-eastern part of the TP (Liu

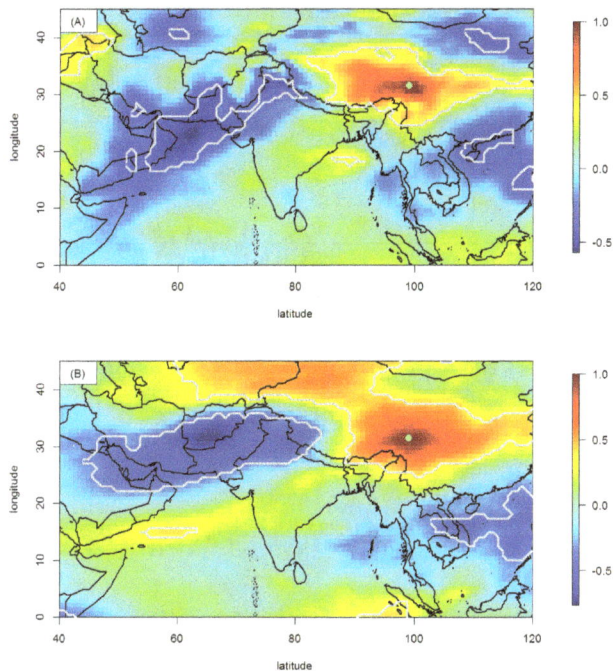

Figure 5. Spatial correlation of July–August relative humidity (ERA interim data, AD 1979–2013) at the (**a**) 500 hPa and (**b**) 300 hPa pressure level. Colour code represents the Pearson correlation coefficient. White lines delineate the 95 % significance level. Proxy location is shown by the light green dot.

et al., 2014; Xu et al., 2012; Zhao and Moore, 2006). After approximately the 1950s, a clear trend towards even drier conditions is attenuated (trend slope − 0.01, $p = 0.63$). This finding is in accordance with results from the central and south-eastern TP (Grießinger et al., 2011; Liu et al., 2013; Shi et al., 2012) and might be caused by uneven warming trends of the northern and equatorial Indian Ocean sea surface temperatures (Chung and Ramanathan, 2006). More humid periods were detected during AD 1193–1300, AD 1345–1390, AD 1455–1475, and AD 1740–1750, with the highest relative humidity in AD 1272 (RH = 83.5 %). Thus, the MWP is characterized by the highest humidity values within the past 800 years. Similar conditions were observed for inner Asia and the northern TP (Pederson et al., 2014; Yang et al., 2013) but were not corroborated for the central TP (Grießinger et al., 2011). The moderate oscillation of our humidity reconstruction during the LIA contrasts results of increasing and decreasing moisture trends at different parts of the TP (Grießinger et al., 2011; Shao et al., 2005; Yao et al., 2008). We identified extreme interannual humidity variations by calculating the third standard deviation of the first differences. Years with humidity variations above 10 % were detected in AD 1960/1961, AD 1946/1947, AD 1941/1942, AD 1706/1707, AD 1253/1254, AD 1238/1239, AD 1233/1234, AD 1230/1231, and AD 1225/1226.

4   Discussion

Lhamcoka is located at the assumed boundary zone of air masses from the Indian Ocean, the South and North Pacific, and Central Asia (Araguás-Araguás et al., 1998). Thus, our study site is likely influenced by the monsoon circulation (Indian and Southeast Asian monsoon) as well as by the westerlies (Morrill et al., 2003). In particular, the long-term spatiotemporal modulation of the monsoon circulation systems has been intensively studied (e.g. Kumar et al., 1999; Wang et al., 2012; Webster et al., 1998) and may significantly control the moisture availability at our study site. The precondition for the formation of the monsoon is the land–sea surface temperature gradient between the Asian land mass and the surrounding oceans (Kumar et al., 1999). However, the monsoon circulation system shows variations at interannual and intraseasonal timescales (Webster et al., 1998). In particular, the ENSO impact on the monsoon circulation has been studied extensively (e.g. Cherchi and Navarra, 2013; Kumar et al., 2006; Park and Chiang, 2010). We tested the influence of ENSO on our humidity reconstruction and achieved no significant relationships, implying an ENSO decoupled climate variability at our proxy site (see interactive discussion of this paper Wernicke et al., 2014). On an intraseasonal timescale, the Madden–Julian Oscillation (MJO) modulates the monsoonal precipitation (Madden and Julian, 1994), where the 30- to 90-day zonal propagation of cloud clusters causes breaks in and strengthening of the monsoonal precipitation (Zhang, 2005). More recently, the monsoon circulation system has been affected by greenhouse gas and aerosol emissions (Hu et al., 2000; Lau et al., 2006). Both induce a positive anomaly of monsoonal precipitation due to the strengthening of the thermal gradient in the upper troposphere.

However, in this study, we primarily focus on the controls of relative humidity at our study site, rather than targeting large-scale atmospheric circulation influences immediately. Therefore, we conducted correlation analysis of the July–August relative humidity at the grid cell of our study site with the July–August relative humidity in the area of 0–45° N, 40–120° E (ERA Interim data: http://apps.ecmwf. int/datasets/data). Beforehand, we examined the accordance of our summer relative reconstruction and the ERA interim data (mean relative humidity July–August). The significant relationship ($r = 0.77$, $p < 0.01$) suggests that the ERA interim data are likely to represent our relative humidity reconstruction. As shown in Fig. 5a, significant correlations at the 500 hPa pressure level are found with almost the entire TP. This suggests a regional signal, reflecting the strong connection of moisture variability at our study site with moisture variability over the whole TP. However, significant negative relationships were found with the south-west and south-east Asian regions. These correlations are even more evident on the 300 hPa level (Fig. 5b) and show a remarkable spatial pattern. Interestingly, the negative correlation in south-west

Asia contains the region where Ding and Wang, 2005 defined an index for the westerly wave activity (west central Asia: 35–40° N, 60–70° E). The significance of this finding is corroborated by strong correlations of the mean summer relative humidity in 200 hPa of the west central Asian region and our proxy record ($r = -0.58$, $p < 0.05$). Several studies highlight the general influence of the ASM as the major driver for Tibetan moisture variability (Araguás-Araguás et al., 1998; Hren et al., 2009; Tian et al., 2007). However, the results of Ding and Wang, 2005, Saeed et al., 2011, Mölg et al., 2014, and our findings indicate that the mid-latitude westerlies influence should be taken into consideration in future studies.

For an analysis of the regional representativeness of our data set, we compared the Lhamcoka $\delta^{18}O$ chronology with six moisture-sensitive proxies from the TP (see Fig. 6 and locations in Fig. 1), including normalized tree-ring (TR) $\delta^{18}O$ records (Ranwu TR: Liu et al., 2013; Reting TR: Grießinger et al., 2011), tree-ring width data (Dulan TR: Sheppard et al., 2004), accumulation records (Dasuopu and Dunde ice cores: Thompson et al., 2000) and lake sediments (Qinghai sediment: Xu et al., 2006a). We found significant positive correlations between our time series and the Ranwu ($r = 0.55$, $p < 0.01$), Reting ($r = 0.23$, $p < 0.01$), Dunde ($r = 0.16$, $p < 0.01$) and Qinghai ($r = 0.22$, $p < 0.1$) data sets. Only the tree-ring width series of Dulan is negatively correlated with the $\delta^{18}O$ values of Lhamcoka ($r = -0.16$, $p < 0.01$). The snow accumulation rate of Dasuopu ice core has no relationship with our $\delta^{18}O$ chronology ($r = -0.04$, $p = 0.3$). In the case of weak correlations ($|r| < 0.2$) and due to the degrees of freedom (DF >100), significance levels alone might be misleading and indicate only a statistical and not a causal relationship. However, strong relationships between the tree-ring $\delta^{18}O$ chronologies of Lhamcoka and Ranwu, and partially Reting, are reasonable since moisture reconstructions from these sites rely on the same proxy ($\delta^{18}O$ of tree-ring cellulose) and the trees grown under similar climate conditions. Relationships with the more northern sites (Dunde, Dulan, Qinghai) are difficult to verify according to a clearly detectable westerly influence at these sites. We adapted the colour scheme of Fig. 4 and highlighted the MWP (yellow polygon), LIA (blue polygon), and the remarkable humidity decline since the late 1870s (dashed red line) in Fig. 7. The MWP is characterized by more humid conditions on the eastern TP (Lhamcoka), a drier phase on the central plateau (Reting) and moderate humidity conditions on the northern plateau (Dulan). During the LIA, a remarkable moisture increase occurred at the central and southern plateau (Reting, Dasuopu). Although humidity was high according to these archives, the ASM was weak during that time (Anderson et al., 2002; Gupta et al., 2003).

Thus, the findings for Reting and Dasuopu revealed moisture conditions during cold phases and even drier circumstances during warm periods which might be contrary to findings of Meehl, 1994 and Zhang and Qiu, 2007. The sudden moisture decrease since the late 1870s affects the eastern

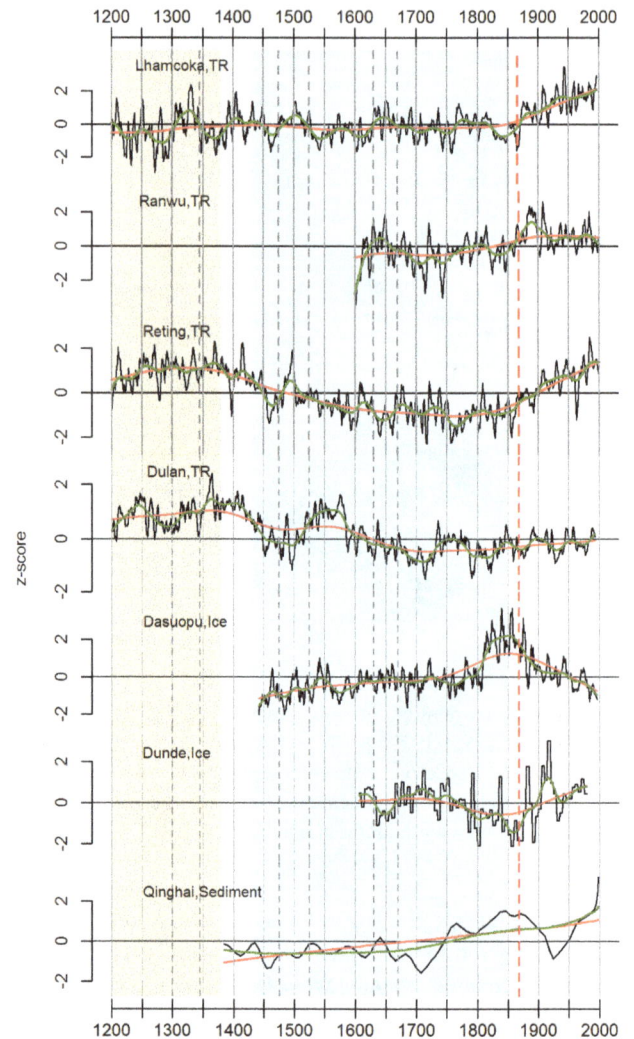

Figure 6. Multiproxy comparison of tree-ring data (TR), ice core, and lake sediment data. TR: Lhamcoka, this study; Ranwu, Liu et al., 2013; Reting, Grießinger et al., 2011; Dulan, Sheppard et al., 2004. Ice: Dasuopu and Dunde, Thompson et al., 2000. Sediment: Qinghai, Xu et al., 2006a. Locations of the several proxies are shown in Fig. 1. Z-scores were derived from raw proxy data and not from reconstructions. High positive Z-scores indicating dry conditions for TR and sediment records, whereas high Z-scores of ice accumulations represent humid conditions.

(Lhamcoka), southern (Dasuopu), and central (Reting) parts of the TP. Reasons for the sudden moisture decline were discussed in detail by Xu et al., 2012. They address the effect of the moisture decrease on the reduction in the thermal gradient induced by uneven land-ocean temperature rise caused by aerosol and greenhouse gas loads. In fact, under rising northern hemispheric air temperatures (Shi et al., 2013), the air moisture load over sea is increased and the meridional moisture transport is contemporaneously hampered due to the black aerosol induced solar dimming effect (Sun et al., 2010). In addition, Zhao and Moore, 2006 attributed the moisture

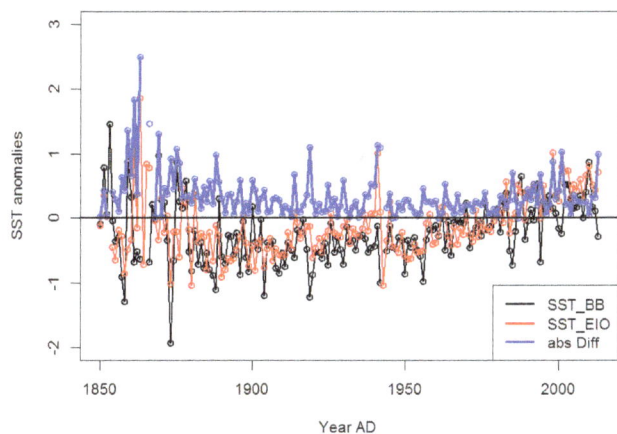

Figure 7. Sea surface temperature anomalies in different regions of the Indian Ocean: Bay of Bengal–North Indian ocean (SST BB: equatorial and northern Indian Ocean (2.5°N–2.5° S, 52.5–112.5° E; 22.5−27.5° N, 52.5–112.5° E) and equatorial Indian Ocean (SST EIO: 2.5°N–2.5° S, 52.5–112.5° E) (Rayner et al., 2006). Difference between the two time series is marked with a blue line.

5   Conclusions

We demonstrated that our 800-year $\delta^{18}O$ chronology is suitable for reliable reconstruction of summer relative humidity. Long-term air humidity variations revealed more humid conditions during the termination of the MWP, relatively stable humidity during the LIA, and a sudden decrease in summer humidity beginning in the 1870s. After approximately the 1950s, the trend towards heavier oxygen isotope ratios has been mitigated due to the restrengthening of the ISM. These findings are in accordance with other reconstructions of moisture conditions for the central and eastern TP. Spatial correlations indicate a significant relationship of summer relative humidity at our study site and major parts of the TP. Additionally, a negative correlation within the higher atmosphere over the west central Asia region imply a westerly influence. Furthermore, the thermal contrast between the equatorial and northern Indian Ocean, which is assumed to control moisture supply during the ISM, is slightly stable over time. Thus, to comprehensively indicate reasons for the distinct approximately 1870s moisture decline, more detailed climate dynamic studies and highly resolved spatiotemporal hydroclimate reconstructions are needed.

Acknowledgements. The authors thank the German Federal Ministry of Education and Research (BMBF) for financial support. We also thank Roswitha Höfner-Stich for her efficient and precise work with the mass spectrometer. We would additionally like to thank Thomas Mölg for his inspiring and helpful suggestions.

Edited by: D. Fleitmann

decline to the weakening of the easterly trade wind system along the equatorial Pacific since the middle of 19th century. Moreover, decreasing varve thicknesses imply a weakening Asian summer monsoon over the past 160 years (Chu et al., 2011). The aforementioned analysis revealed a link to warm phases of ENSO and an anomalous regional Hadley circulation. However, their explanation approach remains incomplete due to dynamic issues associated with rising temperatures and a weakening South Asian summer monsoon. Therefore, a terminal explanation is not given yet and should be discussed in future studies.

In comparison to tree-ring sites located further south (e.g. Liu et al., 2013; Sano et al., 2013; Shi et al., 2012), the distinct humidity decline is more pronounced on the central and eastern TP. From that observation, Sano et al., 2013, concluded that there is a weakening of the monsoon over the last 100–200 years due to uneven SST variation (equatorial vs. northern Indian Ocean regions). To test this hypothesis, we calculated the averaged SST anomalies of the equatorial and northern Indian Ocean (2.5°N–2.5° S, 52.5–112.5° E; 22.5−27.5° N, 52.5–112.5° E). As shown in Fig. 7, a slight SST increase in both regions beginning in approximately the 1950s is obvious. Besides, the gradient constantly decreases, but has restrengthened since approximately the 1970s. This finding contrasts with a generally weakening monsoon circulation over the past 100–200 years deduced from a thermal gradient reduction. Therefore, the various moisture variations of the southern and central/eastern TP during the last 100–200 years might show influences of varying local air mass characteristics.

References

An, W., Liu, X., Leavitt, S., Xu, G., Zeng, X., Wang, W., Qin, D., and Ren, J.: Relative humidity history on the Batang–Litang Plateau of western China since 1755 reconstructed from tree-ring $\delta^{18}O$ and δD, Clim. Dynam., 42, 2639–2654, doi:10.1007/s00382-013-1937-z, 2014.

Anderson, D., Overpeck, J., and Gupta, A.: Increase in the Asian Southwest Monsoon During the Past Four Centuries, Science, 297, 596–599, doi:10.1126/science.1072881, 2002.

Anderson, W., Bernasconi, S., and McKenzie, J.: Oxygen and carbon isotopic record of climatic variability in tree ring cellulose (*Picea abies*)' An example from central Switzerland (1913–1995), J. Geophys. Res., 103, 31625–31636, 1998.

Araguás-Araguás, L., Fröhlich, K., and Rozanski, K.: Stable isotope composition of precipitation over southeast Asia, J. Geophys. Res. Lett., 103, 721–728, 1998.

Berkelhammer, M. and Stott, L.: Secular temperature trends for the southern Rocky Mountains over the last five centuries, Geophys. Res. Lett., 39, 1–6, doi:10.1029/2012GL052447,, 2012.

Bollasina, M., Ming, Y., and Ramaswamy, V.: Anthropogenic aerosols and the weakening of the South Asian Summer

Monsoon, Science, 334, 502–505, doi:10.1126/science.1204994, 2011.

Böhner, J.: General climatic controls and topoclimatic variations in Central and High Asia, Boreas, 35, 279–295, 2006.

Böttger, T. and Friedrich, M.: A new serial pooling method of shifted tree ring blocks to construct millennia long tree ring isotope chronologies with annual resolution, Isot. Environ. Healt. S., 45, 68–80, 2009.

Bräuning, A.: Climate history of the Tibetan Plateau during the last 1000 years derived from a network of Juniper chronologies, Dendrochronologia, 19, 127–137, 2001.

Bräuning, A.: Tree-ring evidence of "Little Ice Age" glacier advances in southern Tibet, The Holocene, 16, 369–380, 2006.

Bunn, A.: A dendrochronology program library in R (dplR), Dendrochronologia, 26, 115–124, doi:10.1016/j.dendro.2008.01.002, 2008.

Burk, R. and Stuiver, M.: Oxygen isotope ratios in trees reflect mean annual temperature and humidity, Science, 27, 1417–1419, 1981.

Cherchi, A. and Navarra, A.: Influence of ENSO and of the Indian Ocean Dipole on the Indian summer monsoon variability, Clim. Dynam., 41, 81–103, doi:0.1007/s00382-012-1602-y, 2013.

Chu, G., Sun, Q., Yang, K., Li, A., Yu, X., Xu, T., Yan, F., Qang, X., Xie, M., Lin, Y., and Liu, Q.: Evidence for decreasing South Asian summer monsoon in the past 160 years from varved sediment in Lake Xinluhai, Tibetan Plateau, J. Geophys. Res., 116, doi:10.1029/2010JD014454, 2011.

Chung, C. and Ramanathan, V.: Weakening of north Indian SST gradients and the Monsoon rainfall in India and the Sahel, J. Climate, 19, 2036–2045, 2006.

Cook, E. and Kairiukstis, L.: Methods of dendrochronology, Kluwer Academic Publishers, Dordrecht, Boston, London, 1990.

Cook, E., Briffa, K., and Jones, P.: Spatial regression methods in dendroclimatology: A review and comparison of two techniques, Int. J. Clim., 14, 379–402, 1994.

Cook, E., Anchukaitis, K., Buckley, B., D'Arrigo, R., Jacoby, G., and Wright, W.: Asian Monsoon Failure and Megadrought During the Last Millennium, Science, 328, 486–489, doi:10.1126/science.1185188, 2010.

Ding, Q. and Wang, B.: Circumglobal teleconnection in the northern hemisphere summer, J. Climate, 18, 3483–3505, 2005.

Esper, J., Frank, D., Battipaglia, G., Büntgen, U., Holert, C., Treydte, K., Siegwolf, R., and Saurer, M.: Low-frequency noise in $\delta^{13}C$ and $\delta^{18}O$ tree ring data: A case study of Pinus uncinata in the Spanish Pyrenees, Glob. Biogeochem. Cy., 24, 1–11, doi:10.1029/2010GB003772, 2010.

Grießinger, J., Bräuning, A., Helle, G., Thomas, A., and Schleser, G.: Late Holocene Asian summer monsoon variability reflected by $\delta^{18}O$ in tree-rings from Tibetan junipers, Geophys. Res. Lett., 38, 1–5, doi:10.1029/2010GL045988, 2011.

Gupta, A., Anderson, D., and Overpeck, J.: Abrupt changes in the Asian southwest monsoon during the Holocene and their links to the North Atlantic Ocean, Nature, 421, 354–357, 2003.

Hren, M., Bookhagen, B., Blisniuk, P., Booth, A., and Chamberlain, C.: $\delta^{18}O$ and δD of streamwaters across the Himalaya and Tibetan Plateau: Implications for moisture sources and paleoelevation reconstructions, Earth Planet. Sci. Lett., 288, 20–32, doi:10.1016/j.epsl.2009.08.041, 2009.

Hu, Z.-Z., Latif, M., Roeckner, E., and Bengtsson, L.: Intensified Asian summer monsoon and its variability in a coupled model

forced by increasing greenhouse gas concentrations, Geophys. Res. Lett., 27, 2681–2684, 2000.

Immerzeel, W., van Beek, L., and Bierkens, M.: Climate Change will affect the Asia Water Towers, Science, 328, 1382–1385, doi:10.1126/science.1183188, 2010.

Kahmen, A., Simonin, K., Tu, K., Goldsmith, G., and Dawson, T.: The influence of species and growing conditions on the 18-O enrichment of leaf water and its impact on effective path length, New Phytol., 184, 619–630, doi:10.1111/j.1469-8137.2009.03008.x, 2009.

Kahmen, A., Sachse, D., Arndt, S., Tu, K., Farrington, H., Vitousek, P., and Dawson, T.: Cellulose $\delta^{18}O$ is an index of leaf-to-air vapor pressure difference (VPD) in tropical plants, P. Natl. Acad. Sci., 108, 1981–1986, www.pnas.org/cgi/doi/10.1073/pnas.1018906108, 2011.

Kumar, K., Rajagopalan, B., and Cane, M.: On the weakening relationship Between the Indian Monsoon and ENSO, Science, 284, 2156–2159, doi:10.1126/science.284.5423.2156, 1999.

Kumar, K., Rajagopalan, B., Hoerling, M., Bates, G., and Cane, M.: Unraveling the mystery of Indian Monsoon failure during El Nino, Science, 314, 115–119, doi:10.1126/science.1131152, 2006.

Lau, K., Kim, M., and Kim, K.: Asian summer monsoon anomalies induced by aerosol direct forcing: the role of the Tibetan Plateau, Clim. Dynam., 26, 855–864, doi:10.1007/s00382-006-0114-z, 2006.

Laumer, W., Andreau, L., Helle, G., Schleser, G., Wieloch, T., and Wissel, H.: A novel approach for the homogenization of cellulose to use micro-amounts for stable isotope analyses, Rapid Commun. Mass Spectrom., 23, 1934–1940, doi:10.1002/rcm.4105, 2009.

Leavitt, S.: Tree-ring C-H-O isotope variability and sampling, Sci. Tot. Environ., 408, 5244–5253, 2010.

Liu, X., An, W., Treydte, K., Shao, X., Leavitt, S., Hou, S., Chen, T., Sun, W., and Qin, D.: Tree-ring $\delta^{18}O$ in southwestern China linked to variations in regional cloud cover and tropical sea surface temperature, Chem. Geol., 291, 104–115, doi:10.1016/j.chemgeo.2011.10.001, 2012.

Liu, X., Zeng, X., Leavitt, S., Wang, W., An, W., Xu, G., Sun, W., Wang, Y., Qin, D., and Ren, J.: A 400-year tree-ring $\delta^{18}O$ chronology for the southeastern Tibetan Plateau: Implications for inferring variations of the regional hydroclimate, Glob. Planet. Change, 104, 23–33, doi:10.1016/j.gloplacha.2013.02.005, 2013.

Liu, X., Xu, G., Grießinger, J., An, W., Wang, W., Zeng, X., Wu, G., and Qin, D.: A shift in cloud cover over the southeastern Tibetan Plateau since 1600: evidence from regional tree-ring $\delta^{18}O$ and its linkages to tropical oceans, Quaternary Sci. Rev., 88, 55–68, doi:10.1016/j.quascirev.2014.01.009, 2014.

Liu, Y., Cai, Q., Liu, W., Yang, Y., Sun, J., Song, H., and Li, X.: Monsoon precipitation variation recorded by tree-ring $\delta^{18}O$ in arid Northwest China since AD 1878, Chem. Geol., 252, 56–61, doi:10.1016/j.chemgeo.2008.01.024, 2008.

Madden, R. and Julian, P.: Observations of the 40-50 day Tropical Oscillation- A review, Month. Weather Rev., 122, 814–837, 1994.

Maussion, F., Scherer, D., Mölg, T., Collier, E., Curio, J., and Finkelnburg, R.: Precipitation seasonality and variability over the

Tibetan Plateau as resolved by the High Asia Reanalysis, J. Climate, 27, 1910–1927, doi:10.1175/JCLI-D-13-00282.1, 2014.

McCarroll, D. and Loader, N.: Stable isotopes in tree rings, Quaternary Science Review, 23, 771–801, 2004.

Meehl, G.: Influence of the land surface in the Asian Summer Monsoon: External conditions versus internal feedbacks, J. Climate, 7, 1033–1049, 1994.

Mölg, T., Maussion, F., and Scherer, D.: Mid-latitude westerlies as a driver of glacier variability in monsoonal High Asia, Nat. Clim. Change, 4, 68–73, doi:10.1038/NCLIMATE2055, 2014.

Morrill, C., Overpeck, J., and Cole, J.: A synthesis of abrupt changes in the Asian summer monsoon since the last deglaciation, The Holocene, 13, 465–476, 2003.

Park, H.-S. and Chiang, J.: The delayed effect of major El Nino Events on Indian Monsoon Rainfall, J. Climate, 23, 932–946, doi:10.1175/2009JCLI2916.1, 2010.

Pederson, N., Hessl, A., Baatarbileg, N., Anchukaitis, K., and Di Cosmo, N.: Pluvials, droughts, the Mongol Empire, and modern Mongolia, Proceedings of the National Academy of Sciences of the United States of America (PNAS), 111, 4375–4379, doi:10.1073/pnas.1318677111/-/DCSupplemental, 2014.

Ramesh, R., Bhattacharya, S., and Gopalan, K.: Climatic correlations in the stable isotope records of silver fir (Abies pindrow) trees from Kashmir, India, Earth Planet. Sci. Lett., 79, 66–74, 1986.

Rayner, N., Brohan, P., Parker, D., Folland, C., Kennedy, J., Vanicek, M., Ansell, T., and Tett, S.: Improved analyses of changes and uncertainties in sea surface temperature measured in situ since the mid-nineteenth century: the HadSST2 data set, J. Climate, 19, 446–469, 2006.

Roden, J. and Ehleringer, J.: Hydrogen and oxygen isotope ratios of tree ring cellulose for field-grown riparian trees, Oecologia, 123, 481–489, 2000.

Roden, J., Lin, G., and Ehleringer, J.: A mechanistic model for interpretation of hydrogen and oxygen isotope ratios in tree-ring cellulose, Geochim. Cosmochim. Ac., 64, 21–35, 2000.

Saeed, S., Müller, W., Hagemann, S., and Jacob, D.: Circumglobal wave train and the summer monsoon over northwestern India and Pakistan: the explicit role of the surface heat low, Clim. Dynam., 37, 1045–1060, doi:10.1007/s00382-010-0888-x, 2011.

Sano, M., Sheshshayee, M., Managave, S., Ramesh, R., Sukumar, R., and Sweda, T.: Climatic potential of δ18O of Abies spectabilis from the Nepal Himalaya, Dendrochronologia, 28, 93–98, doi:10.1016/j.dendro.2009.05.005, 2010.

Sano, M., Ramesh, R., Sheshshayee, M., and Sukumar, R.: Increasing aridity over the past 223 years in the Nepal Himalaya inferred from a tree-ring δ18O chronology, The Holocene, 1, 1–9, doi:10.1177/0959683611430338, 2011.

Sano, M., Tshering, P., Komori, J., Fujita, K., Xu, C., and Nakatsuka, T.: May–September precipitation in the Bhutan Himalaya since 1743 as reconstructed from tree ring cellulose δ18O, J. Geophys. Res. Atmos., 118, 8399–8410, doi:10.1002/jgrd.50664, 2013.

Saurer, M., Aellen, K., and Siegwolf, R.: Correlating δ13C and δ18O in cellulose of trees, Plant Cell Enviro., 20, 1543–1550, 1997.

Shao, X., Huang, L., Liu, H., Liang, E., Fang, X., and Wang, L.: Reconstruction of precipitation variation from tree rings in recent 1000 years in Delingha, Qinghai, Science in China Ser. D Earth Sciences, 48, 939–949, doi:10.1360/03yd0146, 2005.

Sheppard, P., Tarasov, P., Graumlich, L., Heussner, K.-U., Wagner, M., Österle, H., and Thompson, L.: Annual precipitation since 515 BC reconstructed from living and fossil juniper growth of northeastern Qinghai Province, China, Clim. Dynam., 23, 869–881, 2004.

Shi, C., Daux, V., Zhang, Q.-B., Risi, C., Hou, S.-G., Stievenard, M., Pierre, M., Li, Z., and Masson-Delmotte, V.: Reconstruction of southeast Tibetan Plateau summer climate using tree ring δ18O: moisture variability over the past two centuries, Clim. Past, 8, 205–213, doi:10.5194/cp-8-205-2012, 2012.

Shi, F., Yang, B., Mairesse, A., von Gunten, L., Li, J., Bräuning, A., Yang, F., and Xiao, X.: Northern Hemisphere temperature reconstruction during the last millennium using multiple annual proxies, Clim. Res., 56, 231–244, doi:10.3354/cr01156, 2013.

Sternberg, L.: Oxygen stable isotope ratios of tree-ring cellulose: the next phase of understanding, New Phytol., 181, 553–562, 2009.

Sun, Y., Ding, Y., and Dai, A.: Changing links between South Asian summer monsoon circulation and tropospheric land-sea thermal contrasts under a warming scenario, Geophys. Res. Lett., 37, 1–5, 2010.

Thomas, A. and Herzfeld, U.: REGEOTOP: New climatic data fields for east asia based on localized relief information and geostatistical methods, Int. J. Climatol., 24, 1283–1306, doi:10.1002/joc.1058, 2004.

Thompson, L., Yao, T., Mosley-Thompson, E., Davis, M., Henderson, K., and Lin, P.-N.: A high- resolution millennial record of the South Asian monsoon from Himalayan ice cores, Science, 289, 1916–1919, doi:10.1126/science.289.5486.1916, 2000.

Tian, L., Yao, T., MacClune, K., White, J., Schilla, A., Vaughn, B., Vachon, R., and Ichiyanagi: Stable isotopic variations in west China: A consideration of moisture sources, J. Geophys. Res., 112, 1–12, doi:10.1029/2006JD007718, 2007.

Treydte, K., Schleser, G., Helle, G., Frank, D., Winiger, M. Haug, G., and Esper, J.: The twentieth century was the wettest period in northern Pakistan over the past millennium, Nature, 440, 1179–1182, doi:10.1038/nature04743, 2006.

Tsuji, H., Nakatsuka, T., and Takagi, K.: δ18O of tree-ring cellulose in two species (spruce and oak) as proxies of precipitation amount and relative humidity in northern Japan, Chem. Geol., 231, 67–76, doi:10.1016/j.chemgeo.2005.12.011, 2006.

Wang, W., Liu, X., Xu, G., Shao, X., Qin, D., Sun, W., An, W., and Zeng, X.: Moisture variations over the past millennium characterized by Qaidam Basin tree-ring δ18O, Chinese Sci. Bull., 58, 3956–3961, doi:10.1007/s11434-013-5913-0, 2013.

Wang, Y., Jian, Z., and Zhao, P.: Extratropical modulation on Asian summer monsoon at precessional bands, Geophys. Res. Lett., 39, 1–6, doi:10.1029/2012GL052553, 2012.

Webster, P., Magana, V., Palmer, T., Shukla, J., Tomas, R., Yanai, M., and Yasunari, T.: Monsoons: Processes, predictability, and the prospects for prediction, J. Geophys. Res., 103, 14451–14510, doi:10.1029/97JC02719, 1998.

Wernicke, J., Grießinger, J., Hochreuther, P., and Bräuning, A.: Variability of summer humidity during the past 800 years on the eastern Tibetan Plateau inferred from δ18O of tree-ring cellulose, Clim. Past Discuss., 10, 3327–3356, doi:10.5194/cpd-10-3327-2014, 2014.

Wieloch, T., Helle, G., Heinrich, I., Voigt, M., and Schyma, P.: A novel device for batch-wise isolation of α-cellulose from small-amount wholewood samples, Dendrochronologia, 29, 115–117, doi:10.1016/j.dendro.2010.08.008, 2011.

Wigley, T., Briffa, K., and Jones, P.: On the Average Value of Correalted Time Series, with Application in Dendroclimatology and Hydrometeorology, J. Clim. Appl. Meteorol., 23, 201–213, 1984.

Xu, C., Sano, M., and Nakatsuka, T.: A 400-year record of hydroclimate variability and local ENSO history in northern Southeast Asia inferred from tree-ring $\delta^{18}O$, Palaeogeogr. Palaeocl., 286, 588–598, doi:10.1016/j.palaeo.2013.06.025, 2013.

Xu, H., Ai, L., Tan, L., and An, Z.: Stable isotopes in bulk carbonates and organic matter in recent sediments of Lake Qinghai and their climatic implications, Chem. Geol., 235, 262–275, doi:10.1016/j.chemgeo.2006.07.005, 2006a.

Xu, H., Hong, Y., Lin, Q., Zhu, Y., Hong, B., and Jiang, H.: Temperature responses to quasi-100-yr solar variability during the past 6000 years based on $\delta^{18}O$ of peat cellulose in Hongyuan, eastern Qinghai-Tibet plateau, China, Palaeogr. Palaeocl., 230, 155–164, doi:10.1016/j.palaeo.2005.07.012, 2006b.

Xu, H., Hong, Y., and Hong, B.: Decreasing Asian summer monsoon intensity after 1860 AD in the global warming epoch, Clim. Dynam., 39, 2079–2088, doi:10.1007/s00382-012-1378-0, 2012.

Yang, B., Bräuning, A., and Yafeng, S.: Late Holocene temperature fluctuations on the Tibetan Plateau, Quat. Sci. Rev., 22, 2335–2344, doi:10.1016/S0277-3791(03)00132-X, 2003.

Yang, B., Qin, C., Wang, J., He, M., Melvin, T., Osborn, T., and Briffa, K.: A 3.500-year tree-ring record of annual precipitation on the northeastern Tibetan Plateau, P. Natl. Acad. Sci. USA, 111, 2903–2908, doi:10.1073/pnas.1319238111, 2013.

Yao, T., Duan, K., Xu, B., Wang, N., Guo, X., and Yang, X.: Precipitation record since AD 1600 from ice cores on the central Tibetan Plateau, Clim. Past, 4, 175–180, doi:10.5194/cp-4-175-2008, 2008.

Zang, C. and Biondi, F.: Dendroclimatic calibration in R: The bootRes package for response and correlation function analysis, Dendrochronologia, 31, 68–74, doi:10.1016/j.dendro.2012.08.001, 2012.

Zhang, C.: Madden-Julian oscillation, Reviews of Geophysics, pp. 1–36, 2005.

Zhang, P., Cheng, H., Edwards, R., Chen, F., Wang, Y., Yang, X., Liu, J., Tan, M., Wang, X., Liu, J., An, C., Dai, Z., Zhou, J., Zhang, D., Jia, J., Jin, L., and Johnson, K.: A test of climate, sun, and culture relationships from an 1810-year chinese cave record, Science, 322, 940–942, 2008.

Zhang, Q.-B. and Qiu, H.: A millennium-long tree-ring chronology of Sabina przewalskii on northeastern Qinghai-Tibetan Plateau, Dendrochronologia, 24, 91–95, doi:10.1016/j.dendro.2006.10.009, 2007.

Zhao, H. and Moore, G.: Reduction in Himalayan snow accumulation and weakening of the trade winds over the Pacific since the 1840s, Geophys. Res. Lett., 33, 1–5, doi:10.1029/2006GL027339, 2006.

Zhou, T., Yu, R., Li, H., and Wang, B.: Ocean forcing to changes in global Monsoon precipitation over the recent half-century, J. Climate, 21, 3833–3852, doi:10.1175/2008JCLI2067.1, 2008a.

Zhou, T., Zhang, L., and Li, H.: Changes in global land monsoon area and total rainfall accumulation over the last half century, Geophys. Res. Lett., 35, 1–6, doi:10.1029/2008GL034881, 2008b.

Permissions

The contributors of this book come from diverse backgrounds, making this book a truly international effort. This book will bring forth new frontiers with its revolutionizing research information and detailed analysis of the nascent developments around the world.

We would like to thank all the contributing authors for lending their expertise to make the book truly unique. They have played a crucial role in the development of this book. Without their invaluable contributions this book wouldn't have been possible. They have made vital efforts to compile up to date information on the varied aspects of this subject to make this book a valuable addition to the collection of many professionals and students.

This book was conceptualized with the vision of imparting up-to-date information and advanced data in this field. To ensure the same, a matchless editorial board was set up. Every individual on the board went through rigorous rounds of assessment to prove their worth. After which they invested a large part of their time researching and compiling the most relevant data for our readers.

The editorial board has been involved in producing this book since its inception. They have spent rigorous hours researching and exploring the diverse topics which have resulted in the successful publishing of this book. They have passed on their knowledge of decades through this book. To expedite this challenging task, the publisher supported the team at every step. A small team of assistant editors was also appointed to further simplify the editing procedure and attain best results for the readers.

Apart from the editorial board, the designing team has also invested a significant amount of their time in understanding the subject and creating the most relevant covers. They scrutinized every image to scout for the most suitable representation of the subject and create an appropriate cover for the book.

The publishing team has been an ardent support to the editorial, designing and production team. Their endless efforts to recruit the best for this project, has resulted in the accomplishment of this book. They are a veteran in the field of academics and their pool of knowledge is as vast as their experience in printing. Their expertise and guidance has proved useful at every step. Their uncompromising quality standards have made this book an exceptional effort. Their encouragement from time to time has been an inspiration for everyone.

The publisher and the editorial board hope that this book will prove to be a valuable piece of knowledge for researchers, students, practitioners and scholars across the globe.

List of Contributors

L. Menviel
Climate Change Research Centre, University of New South Wales, Sydney, Australia
ARC Centre of Excellence in Climate System Science, Australia

A. Timmermann
International Pacific Research Center, University of Hawaii, Honolulu, USA

T. Friedrich
International Pacific Research Center, University of Hawaii, Honolulu, USA

M. H. England
Climate Change Research Centre, University of New South Wales, Sydney, Australia
ARC Centre of Excellence in Climate System Science, Australia

M. M. Telesiński
GEOMAR Helmholtz Centre for Ocean Research Kiel, Wischhofstrasse 1–3, 24148 Kiel, Germany

R. F. Spielhagen
GEOMAR Helmholtz Centre for Ocean Research Kiel, Wischhofstrasse 1–3, 24148 Kiel, Germany
Academy of Sciences, Humanities, and Literature, 53151 Mainz, Germany

H. A. Bauch
GEOMAR Helmholtz Centre for Ocean Research Kiel, Wischhofstrasse 1–3, 24148 Kiel, Germany
Academy of Sciences, Humanities, and Literature, 53151 Mainz, Germany

J. T. Andrews
INSTAAR and Department of Geological Sciences, University of Colorado, Boulder, CO 80309, USA

A. E. Jennings
INSTAAR and Department of Geological Sciences, University of Colorado, Boulder, CO 80309, USA

E. Dietze
Section 5.2 Climate Dynamics and Landscape Evolution, GFZ German Research Centre for Geosciences, Potsdam, Germany

F. Maussion
Chair of Climatology, Technische Universität Berlin, Berlin, Germany

M. Ahlborn
Physical Geography, Institute of Geography, Friedrich-Schiller-Universität Jena, Jena, Germany

B. Diekmann
Alfred Wegener Institute for Polar and Marine Research, Research Unit Potsdam, Potsdam, Germany

K. Hartmann
Institute of Geographical Sciences, EDCA, Freie Universität Berlin, Berlin, Germany

K. Henkel
Physical Geography, Institute of Geography, Friedrich-Schiller-Universität Jena, Jena, Germany

T. Kasper
Physical Geography, Institute of Geography, Friedrich-Schiller-Universität Jena, Jena, Germany

G. Lockot
Institute of Geographical Sciences, EDCA, Freie Universität Berlin, Berlin, Germany

S. Opitz
Institute for Earth and Environmental Sciences, Universität Potsdam, Potsdam, Germany

T. Haberzettl
Physical Geography, Institute of Geography, Friedrich-Schiller-Universität Jena, Jena, Germany

R. P. M. Topper
Department of Earth Sciences, Utrecht University, Budapestlaan 4 3584CD Utrecht, the Netherlands now at: MARUM – Center for Marine Environmental Sciences and Department of Geosciences, University of Bremen, P.O. Box 330440, 28334, Germany

P. Th. Meijer
Department of Earth Sciences, Utrecht University, Budapestlaan 4 3584CD Utrecht, the Netherlands now at: MARUM – Center for Marine Environmental Sciences and Department of Geosciences, University of Bremen, P.O. Box 330440, 28334, Germany

A. Schmittner
College of Earth, Ocean, and Atmospheric Sciences, Oregon State University, Corvallis, Oregon, USA

D. C. Lund
Department of Marine Sciences, University of Connecticut, USA

P. A. Araya-Melo
Université catholique de Louvain, Earth and Life Institute, Georges Lemaître Centre for Earth and Climate Research, 1348 Louvain-la-Neuve, Belgium

M. Crucifix
Université catholique de Louvain, Earth and Life Institute, Georges Lemaître Centre for Earth and Climate Research, 1348 Louvain-la-Neuve, Belgium

N. Bounceur
Université catholique de Louvain, Earth and Life Institute, Georges Lemaître Centre for Earth and Climate Research, 1348 Louvain-la-Neuve, Belgium

P. Beghin
Laboratoire des Sciences du Climat et de l'Environnement, CEA-CNRS – UMR8212, Gif-sur-Yvette, France

S. Charbit
Laboratoire des Sciences du Climat et de l'Environnement, CEA-CNRS – UMR8212, Gif-sur-Yvette, France

C. Dumas
Laboratoire des Sciences du Climat et de l'Environnement, CEA-CNRS – UMR8212, Gif-sur-Yvette, France

M. Kageyama
Laboratoire des Sciences du Climat et de l'Environnement, CEA-CNRS – UMR8212, Gif-sur-Yvette, France

D. M. Roche
Laboratoire des Sciences du Climat et de l'Environnement, CEA-CNRS – UMR8212, Gif-sur-Yvette, France
Earth and Climate Cluster, Faculty of Earth and Life Sciences, Vrije Universiteit Amsterdam, Amsterdam, the Netherlands

C. Ritz
Laboratoire de Glaciologie et de Géophysique de l'Environnement, CNRS, Saint Martin d'Hérès, France

K. Schittek
Seminar of Geography and Geographical Education, University of Cologne, Germany

M. Forbriger
Institute of Geography, University of Cologne, Germany

B. Mächtle
Geographical Institute, University of Heidelberg, Germany

F. Schäbitz
Seminar of Geography and Geographical Education, University of Cologne, Germany

V. Wennrich
Institute of Geology and Mineralogy, University of Cologne, Germany

M. Reindel
German Archaeological Institute, Commission for Archaeology of Non-European Cultures (KAAK), Bonn, Germany

B. Eitel
Geographical Institute, University of Heidelberg, Germany

N. Kurita
Graduate School of Environmental Studies, Nagoya University, Furo-cho, Nagoya, 464-8601, Japan

Y. Fujiyoshi
Institute of Low Temperature Science, Hokkaido University, Kita-19 Nishi-8, Sapporo, 060-0819, Japan

T. Nakayama
Solar-Terrestrial Environment Laboratory, Nagoya University, Furo-cho, Nagoya, 464-8601, Japan

Y. Matsumi
Solar-Terrestrial Environment Laboratory, Nagoya University, Furo-cho, Nagoya, 464-8601, Japan

H. Kitagawa
Graduate School of Environmental Studies, Nagoya University, Furo-cho, Nagoya, 464-8601, Japan

M. J. Pound
Department of Geography, Faculty of Engineering and Environment, Northumbria University, Ellison Building, Newcastle upon Tyne, UK

J. Tindall
School of Earth and Environment, University of Leeds, Woodhouse Lane, Leeds, UK

S. J. Pickering
School of Earth and Environment, University of Leeds, Woodhouse Lane, Leeds, UK

A. M. Haywood
School of Earth and Environment, University of Leeds, Woodhouse Lane, Leeds, UK

H. J. Dowsett
Eastern Geology and Paleoclimate Science Center, US Geological Survey, Reston, Virginia 20192, USA

U. Salzmann
Department of Geography, Faculty of Engineering and
Environment, Northumbria University, Ellison Building,
Newcastle upon Tyne, UK

J. Wernicke
Institute of Geography, Friedrich-Alexander-University
Erlangen-Nuremberg, Erlangen, Germany

J. Grießinger
Institute of Geography, Friedrich-Alexander-University
Erlangen-Nuremberg, Erlangen, Germany

P. Hochreuther
Institute of Geography, Friedrich-Alexander-University
Erlangen-Nuremberg, Erlangen, Germany

A. Bräuning
Institute of Geography, Friedrich-Alexander-University
Erlangen-Nuremberg, Erlangen, Germany